Organic and Hybrid Photonic Crystals

Davide Comoretto
Editor

Organic and Hybrid Photonic Crystals

Springer

Editor
Davide Comoretto
Department of Chemistry
and Industrial Chemistry
University of Genova
Genova, Italy

ISBN 978-3-319-16579-0 ISBN 978-3-319-16580-6 (eBook)
DOI 10.1007/978-3-319-16580-6

Library of Congress Control Number: 2015944977

Springer Cham Heidelberg New York Dordrecht London

Printed on acid-free paper

Springer International Publishing AG Switzerland is part of Springer Science+Business Media (www.springer.com)

To my Mother and my Father

Preface

The research field of *Photonic Crystals*, i.e., composite structures where materials possessing different refractive index are assembled into a highly ordered dielectric lattice with submicrometric periodicity, was founded in 1987 by the seminal papers by E. Yablonovitch and S. John published in the same volume of Physical Review Letters just 3 weeks one after the other. They provided the tools to rationalize the dielectric lattices optics within a new formalism that not only is able to extend their theoretical description but also becomes a source of inspiration for novel systems, structures, and applications.

As it was already observed in other research fields, the development of novel *Photonic Crystals* structures was based on the use of top-down approaches to impart the dielectric structure into inorganic insulators and semiconductors. Such techniques enable the fabrication of photonic structures possessing extraordinary precision and finely tailored properties for selected technological applications.

Several books that have been so far published in the field of *Photonic Crystals* are usually tuned to a specialist readership mainly composed of Physicists and Engineers. Even though colloidal chemists and block copolymer scientists provided important contributions to the field, a cultural and communication gap still exists between fundamental Physics and Chemistry, as well as other disciplines of potential interest to the *Photonic Crystals* field. For instance, novel organic and hybrid materials that are revolutionizing the field of electronics and sensing can hardly be nanostructured in the form of *Photonic Crystals* with top-down techniques. Furthermore, biomedical applications could greatly benefit from the developments of the field. In this respect, and in particular when organic and hybrid materials are used, the use of the bottom-up approach as well as the exploitation of the chemistry of the self-assembling process, widely exploited in *Nature*, provides an important step forward to the field. This book, ***Organic and Hybrid Photonic Crystals***, was conceived as a bridge between different communities in order to establish a common set of fundamental concepts and a language to be shared between Physicists, Chemists, Biologists, Engineers, and Material Scientists. This bridge has two main by-products: (1) it is a common playground where fundamental concepts from different know-hows can be shared among all researchers, the language being adapted to allow a mutual understanding; (2) it shares important fundamental concepts thus becoming a possible educational tool for Ph.D. students or postdocs approaching this research field.

We are aware that the field of *Photonic Crystals* is still dominated by inorganic materials. However, the "flexibility" of both organic materials (in terms of mechanical properties and chemical tunability) and colloidal composites makes ***Organic and Hybrid Photonic Crystals*** promising for several applications and devices, avoiding the typical drawbacks of the inorganic world. For instance, the tunability provided by self-supporting structures, their flexibility, their controlled porosity, and their chemical functionalization are unprecedented for inorganic insulators or semiconductors. Moreover, the smart mixing of organic materials to selected inorganic ones (noble metals, semiconductors, quantum dots) allows the development of hybrid structures with enhanced properties and functionalities.

Last, but not least, the presence of photonic structures in *Nature* has been considered as a source of inspiration for novel devices. In particular, it has been highlighted in this book for biosensing applications and for the effects of phase separation in blends, which plays a major role in self-assembling processes.

The research field of ***Organic and Hybrid Photonic Crystals*** is intrinsically multidisciplinary, and it requires skills of fundamental optics, radiation-matter interaction, chemical properties of polymers and colloids, as well as photonic devices and biology. The contributors of the book have provided to colleagues from different disciplines the fundamental concepts used to develop ***Organic and Hybrid Photonic Crystals*** as well as main results reported in literature and future perspectives. The understanding of the different strategies pursued by chemists, physicists, engineers, material scientists, and biologists is then the tool to foster the development of the field and of its technological by-products.

To the best of our knowledge, none of the books on photonic crystals published to date are easily readable by chemists, thus excluding a large part of such scientists from the field. Only reviews in Chemistry journals have been so far reported. In the same way, physicists are excluded from the know-how of the chemical tools to develop organic photonic crystal structures. From an engineering point of view, the interest is limited to the beautiful inorganic structures, while organics may provide structures unconceivable in the inorganic world. In addition, the biological approach should be a source of inspiration for novel applications. **The ambitious goal of this book is to become a melting pot of multidisciplinary knowledge in the field of Organic and Hybrid Photonic Nanostructures**.

The book is organized into three main parts:

I. Introduction to Organic and Hybrid Photonic Crystal Properties
II. Growth and Properties
III. Chemistry and Physics of Photonic Crystals at Work

The three chapters in Part I provide the essential theoretical tools for understanding *Photonic Crystal* concepts thus allowing an easy reading of the following

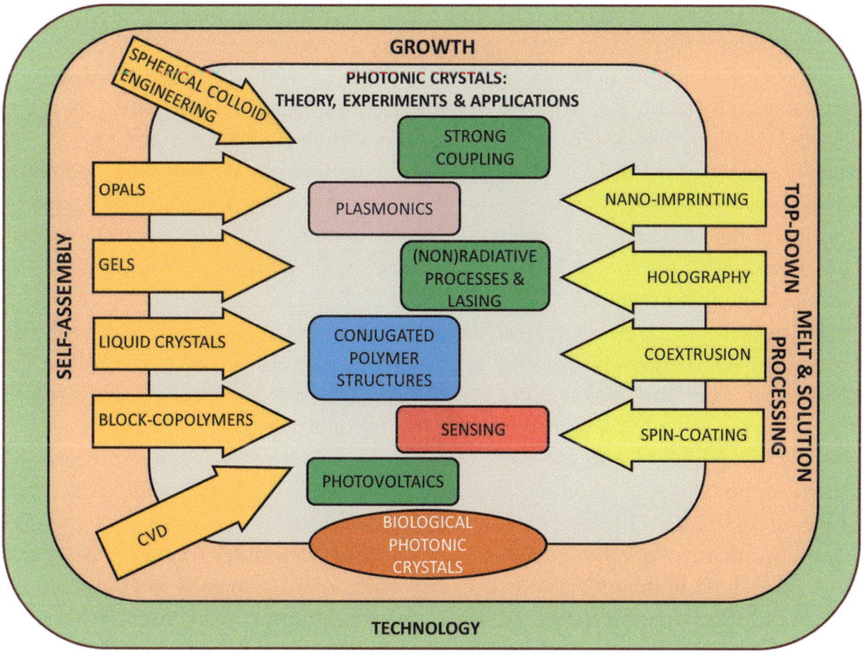

Fig. 1 Scheme of ***Organic and Hybrid Photonic Crystals***

chapters. Moreover, they introduce in a very elegant way the potential of organic materials in the field of photonics, from polymer to carbonaceous systems. Finally, photonic crystal structures in plants and animals are reviewed, being considered as hybrid nanostructured material systems, with inorganic and organic components and appropriate refractive index contrast. General principles of structure formation, the influence of disorder, and the achievable biological functions are elucidated through different examples. In the scheme of the book reported in Fig. 1, Part I constitutes the deep core.

Part II is composed of seven chapters and is devoted to the description of methods used to fabricate ***Organic and Hybrid Photonic Crystals***, i.e., self-assembling, top-down, and melt or solution processing. Among self-assembling processes to obtain ***Organic and Hybrid Photonic Crystals***, different approaches and materials are described, such as the engineering of spherical colloids in order to prepare opal-like structures, the use of sol–gel responsive systems, the supramolecular structure of liquid crystals or block copolymers, as well as Chemical Vapour Deposition (CVD). Top-down methods suitable for organic and hybrid materials are very different from those—time and energy consuming—used for inorganic materials. Indeed, nano-imprinting and holography are widely used, since they are very powerful, simple, and relatively cheap. When polymer materials are solution- or melt-processable, spin-coating and coextrusion are widely employed. The first is efficient to engineer structures and materials over relatively wide areas, while the latter allows preparing very large area multilayered structures.

Part III is composed of 11 chapters and deals with different physical and technological issues. The role of light–matter interaction is described in these chapters with emphasis on fundamental issues such as strong coupling, energy transfer, and modification to the radiative recombination as well as to photonic applications, e.g., laser systems, photovoltaic heterostructures, and different kind of sensors (both chemical and biological). Moreover, specific features and opportunities provided by the smart mixing (engineering) of photonic crystal systems to plasmonic nanostructures are described.

We did not plan a technological section in the book, since different devices are permeating all chapters where they are described as proof-of-concept for different applications. For this reason, in each chapter, specific applications are reported discussing the role of materials and structures in improving device performances. A more detailed insight into the technological issues concerning the synthesis and specific applications of photonic nanostructures can be achieved taking advantage of the large number of references cited for each chapter.

Dimensionality of the ***Organic and Hybrid Photonic Crystals*** structures deserves a final comment. The processing techniques described in the book are mainly suitable to the preparation of one- and two-dimensional systems. Among 1D Photonic Crystals, the Distributed Bragg Reflector is easily achieved by spin-coating, coextrusion, CVD, as well as by self-assembling of liquid crystals and block copolymers. A variant of the DBR structure is the microcavity, where a periodicity defect is sandwiched between two multilayers. Nano-imprinting allows easily preparing 1D molded Distributed FeedBack structures. This technique should be extended to 2D and 3D photonic structures. However, the widely used 3D structure is the opal one, which is obtained by the self-assembling of monodisperse nanospheres and the inverse-opal one, as obtained upon infiltration by using opal templates. Face centered cubic and compact hexagonal structures can be easily obtained. Several examples are discussed in depth within the book. An additional opportunity for different 3D photonic crystal structures is provided by holographic lithography, where suitable patterns are obtained by laser beam interference into the photosensitive materials. 2D self-assembled structures obtained by both block copolymers and microsphere arrays are also briefly described.

In order to help the reader to approach different problems/materials/structures from different perspectives, an extended keyword list is provided allowing an easy cross-reference to different chapters. Additional more specific cross-references can be also found within each chapter.

Genova, Italy
12 January 2015

Davide Comoretto

Acknowledgments

This work has been developed within the framework of the Italian Ministry of University and Scientific and Technological Research project PRIN 2010XLLNM3. Such project was not only a source of funding but also a very stimulating playground to set up new ideas and concepts on organic and hybrid photonic crystals. Moreover, this project has received funding from the European Union's Horizon 2020 research and innovation programme under the Marie Sklodowska-Curie grant agreement No 643238.

A special thanks to all colleagues who helped me in chapter revision and provided me suggestions and hints to improve the book quality. Discussions with chapter authors during all phases of book preparation have been very fruitful.

Thanks to Ania, Associate Editor at Springer, who took care of all procedures during the one-year long book editing period.

I am also indebted to many friends and colleagues, who in recent years stimulated my interest in this field with discussions, suggestions, works, and criticisms; among them, I am particularly pleased to remember those in Pavia where my curiosity in *Photonic Crystals* was born.

In spite of my personal interest, the work we did in the field of ***Organic and Hybrid Photonic Crystals*** systems could not be developed without the generous, tireless, and bracing contribution of all master students succeeded one another in the last years in the research group. To all of them, I want to convey my grateful thank as well as my best wishes for a bright future.

Contents

Contributors

Lucio Claudio Andreani Dipartimento di Fisica, Università di Pavia, Pavia, Italy

Giuseppe Barillaro Dipartimento di Ingegneria dell'Informazione, Università di Pisa, Pisa, Italy

Rafael Betancur ICFO-Institut de Ciencies Fotoniques, Mediterranean Technology Park, Castelldefels, Barcelona, Spain

Luca Boarino Nanofacility Piemonte, Istituto Nazionale di Ricerca Metrologica, Torino, Italy

Robert Brückner Center for Advancing Electronics Dresden Technische Universität Dresden, Dresden, Germany

Institut für Angewandte Photophysik, Dresden, Germany

Franco Cacialli Department of Physics and Astronomy and London Centre for Nanotechnology, University College London, London, UK

Dario Cavallo Department of Chemistry and Industrial Chemistry, University of Genoa, Genoa, Italy

Koen Clays Department of Chemistry, Laboratory of Molecular Photonics and Electronics, KU Leuven University, Heverlee, Flemish Brabant, Belgium

Anna Maria Coclite Institute of Solid State Physics, Graz University of Technology, Graz, Austria

David M. Coles Department of Materials, University of Oxford, Oxford, UK

Silvia Colodrero ICFO-Institut de Ciencies Fotoniques, Mediterranean Technology Park, Castelldefels, Barcelona, Spain

Davide Comoretto Department of Chemistry and Industrial Chemistry, University of Genova, Genova, Italy

Luigino Criante Center for Nano Science and Technology@PoliMi, Istituto Italiano di Tecnologia, Milan, Italy

Natascia De Leo Nanofacility Piemonte, Istituto Nazionale di Ricerca Metrologica, Torino, Italy

Francesco Di Stasio Nanochemistry Department, Istituto Italiano di Tecnologia, Genoa, Italy

Pieter-Jan Demeyer Department of Chemistry, Laboratory of Molecular Photonics and Electronics, KU Leuven University, Heverlee, Flemish Brabant, Belgium

Emiliano Descrovi Department of Applied Science and Technology, Politecnico di Torino, Torino, Italy

Xavier Elias ICFO-Institut de Ciencies Fotoniques, Mediterranean Technology Park, Castelldefels, Barcelona, Spain

J. Patrick A. Fairclough Department of Mechanical Engineering, The University of Sheffield, Sheffield, UK

Lucia Fornasari Physics Department, University of Pavia, Pavia, Italy

Ana Frangolho Plasmore s.r.l., Ranco (Varese), Italy

Thomas Fuhrmann-Lieker Macromolecular Chemistry and Molecular Materials, Institute of Chemistry and Center for Interdisciplinary Nanostructure Science and Technology, University of Kassel, Kassel, Germany

Seiichi Furumi Faculty of Science, Department of Applied Chemistry, Tokyo University of Science (TUS), Tokyo, Japan

Applied Photonic Materials Group, National Institute for Materials Science (NIMS), Tsukuba, Ibaraki, Japan

Juan F. Galisteo-López Instituto de Ciencia de Materiales de Madrid (ICMM-CSIC), Madrid, Spain

Instituto de Ciencia de Materiales de Sevilla (ICMS-CSIC), Sevilla, Spain

Serena Gazzo Dipartimento di Chimica e Chimica Industriale, Università di Genova, Genoa, Italy

Luz K. Gil Instituto de Ciencia de Materiales de Madrid (ICMM-CSIC), Madrid, Spain

Fabrizio Giorgis Department of Applied Science and Technology, Politecnico di Torino, Torino, Italy

Silvia Giudicatti Physics Department, University of Pavia, Pavia, Italy

Institute for Integrative Nanosciences, IFW Dresden, Dresden, Germany

Han Goossens Department of Chemical Engineering and Chemistry, Eindhoven University of Technology, Eindhoven, The Netherlands

Michael Grimann Macromolecular Chemistry and Molecular Materials, Institute of Chemistry and Center for Interdisciplinary Nanostructure Science and Technology, University of Kassel, Kassel, Germany

Marta Ibisate Instituto de Ciencia de Materiales de Madrid (ICMM-CSIC), Madrid, Spain

Toshimitsu Kanai Faculty of Engineering, Yokohama National University, Yokohama, Kanagawa, Japan

Robert J. Knarr III Dipartimento di Chimica e Chimica Industriale, Università di Genova, Genoa, Italy

Cefe López Instituto de Ciencia de Materiales de Madrid (ICMM-CSIC), Madrid, Spain

Michele Laus Dipartimento di Scienze e Innovazione Tecnologica (DISIT), Università del Piemonte Orientale “A. Avogadro”, Alessandria, Italy

Karl Leo Institut für Angewandte Photophysik, Technische Universität Dresden, Dresden, Germany

Quan Li Liquid Crystal Institute and Chemical Physics Interdisciplinary Program, Kent State University, Kent, OH, USA

David G. Lidzey Department of Physics and Astronomy, University of Sheffield, Sheffield, UK

Marco Liscidini Dipartimento di Fisica, Università di Pavia, Pavia, Italy

Paola Lova Energy Research Institute at NTU (Eri@n) and Interdisciplinary Graduate School, Nanyang Technological University, Singapore, Singapore

Vadim G. Lyssenko Institut für Angewandte Photophysik, Technische Universität Dresden, Dresden, Germany

Giovanni Manfredi Dipartimento di Chimica e Chimica Industriale, Università di Genova, Genoa, Italy

Paola Mantilla-Pérez ICFO-Institut de Ciencies Fotoniques, Mediterranean Technology Park, Castelldefels Barcelona, Spain

Franco Marabelli Physics Department, University of Pavia, Pavia, Italy

Marina Mariano ICFO-Institut de Ciencies Fotoniques, Mediterranean Technology Park, Castelldefels Barcelona, Spain

Alberto Martínez-Otero ICFO-Institut de Ciencies Fotoniques, Mediterranean Technology Park, Castelldefels Barcelona, Spain

Jordi Martorell ICFO-Institut de Ciencies Fotoniques, Mediterranean Technology Park, Castelldefels, Barcelona, Spain

Departament de Fisica i Enginyeria Nuclear, Universitat Politècnica de Catalunya, Terrassa, Spain

Han E.H. Meijer Department of Mechanical Engineering, Eindhoven University of Technology, Eindhoven, The Netherlands

Shabbir Mian Department of Physics, McDaniel College, Westminster, MD, USA

Andrew J. Parnell Department of Physics and Astronomy, The University of Sheffield, Sheffield, UK

Paola Pellacani Plasmore s.r.l., Ranco (Varese), Italy

Department of Applied Physics, Universidad Autònoma de Madrid, Cantoblanco, Madrid, Spain

Paola Rivolo Department of Applied Science and Technology, Politecnico di Torino, Torino, Italy

Valentina Robbiano Department of Physics and Astronomy and London Centre for Nanotechnology, University College London, London, UK

Sergei G. Romanov Institute of Optics, Information and Photonics, University of Erlangen-Nuremberg, Erlangen, Germany

Ioffe Physical Technical Institute, St. Petersburg, Russia

Pablo Romero-Gómez ICFO-Institut de Ciencies Fotoniques, Mediterranean Technology Park, Castelldefels, Barcelona, Spain

Francesco Scotognella Dipartimento di Fisica, Politecnico di Milano, Milan, Italy

Center for Nano Science and Technology@PoliMi, Istituto Italiano di Tecnologia, Milan, Italy

Cesare Soci Division of Physics and Applied Physics, Centre for Disruptive Photonic Technologies (CDPT) and School of Physical and Mathematical Sciences, Nanyang Technological University, Singapore, Singapore

Katia Sparnacci Dipartimento di Scienze e Innovazione Tecnologica (DISIT), Università del Piemonte Orientale "A. Avogadro", Alessandria, Italy

Salvatore Surdo Dipartimento di Ingegneria dell'Informazione, Università di Pisa, Pisa, Italy

Andrea Valsesia Plasmore s.r.l., Ranco (Varese), Italy

Institute for Health and Consumer Protection, European Commission, Joint Research Center, Ispra, Italy

Simone Varo Dipartimento di Fisica, Politecnico di Milano, Milan, Italy

Center for Nano Science and Technology@PoliMi, Istituto Italiano di Tecnologia, Milan, Italy

Ling Wang Liquid Crystal Institute and Chemical Physics Interdisciplinary Program, Kent State University, Kent, OH, USA

Kam Sing Wong Department of Physics, Hong Kong University of Science and Technology, Hong Kong, P.R. China

Lijun Wu Guangdong Provincial Key Laboratory of Nanophotonic Functional Materials and Devices, School for Information and Optoelectronic Science and Engineering, South China Normal University, Guangzhou, P.R. China

Yi Xu Department of Electronic Engineering, College of Information Science and Technology, Jinan University, Guangzhou, P.R. China

Part I
Introduction to Organic and Hybrid Photonic Crystal Properties

Photonic Crystals: An Introductory Survey

Marco Liscidini and Lucio Claudio Andreani

Abstract Photonic crystals, namely materials whose dielectric function is periodic in real space, are very suited to control light propagation and emission, as well as many linear and nonlinear processes involving radiation–matter interaction. This chapter gives an introduction to a few basic theoretical concepts related to photonic crystals of various dimensionalities. After presenting the general properties of Maxwell equations in periodic media, leading to the concepts of photonic bands and photonic band gaps (PBGs), we discuss the main features of photonic crystals in one, two, and three dimensions and in waveguide-embedded photonic crystals. We shall also treat defects and microcavities in photonic crystals, which lead to localized defect modes within a PBG.

Keywords DBR • Distributed bragg reflectors • 1D photonic crystals • 2D photonic crystals • 3D photonic crystals • Microcavity • Theory • Lasing • Light emission • Functional materials @ photonic crystals • Purcell effect

1 Introduction

Photonic crystals are materials whose dielectric function is periodic in one, two, or three spatial dimensions (1D, 2D, 3D). The concept of photonic crystal (PhC) is a very general one and it applies also to absorbing media, like metals. In this chapter we shall focus on transparent materials, that are characterized by a real dielectric function (i.e., no absorption) which is non-dispersive or only weakly dispersive. The propagation properties of photons in such periodic media are analogous to those of electrons in crystalline solids: in particular, allowed and forbidden frequency regions are found. Because of translational invariance, the Bloch–Floquet theorem holds and the allowed states of the electromagnetic field are organized into *photonic bands* with a characteristic frequency-wavevector dispersion. Those photonic bands are analogous to electron bands in periodic solids. A frequency window for which light cannot propagate is named a *photonic gap* and periodic dielectric media are often called *photonic band-gap materials*.

M. Liscidini • L.C. Andreani (✉)
Dipartimento di Fisica, Università di Pavia, 27100 Pavia, Italy
e-mail: marco.liscidini@unipv.it; lucio.andreani@unipv.it

D. Comoretto (ed.), *Organic and Hybrid Photonic Crystals*,
DOI 10.1007/978-3-319-16580-6_1

The concept of a 1D periodic dielectric medium is a familiar one in optics, where it is known as a distributed Bragg reflector (DBR, or dielectric mirror) with well-defined stop bands [1]. However, a stop band (or photonic gap) in 1D forbids propagation of light only in a limited angular cone around the direction of periodicity. The birth of the field of photonic crystals is considered to be in 1987, with two seminal papers by E. Yablonovitch and S. John dealing with the control of spontaneous emission and with light localization in three dimensions [2, 3]. Indeed, the presence of a full PBG in 3D leads to the suppression of vacuum fluctuations of the electromagnetic field and therefore to the inhibition of spontaneous emission of a light source at frequencies within the band gap. Moreover, the presence of disorder may lead to weak localization of light at frequencies close to a band edge. After more than 25 years of intense research, photonic crystals are still of great interest for basic physical properties related to the control of light propagation and emission, radiation–matter interaction, and quantum electrodynamic effects, nonlinear properties, as well as for applications to optoelectronic and photonic devices like lasers and LEDs, optical fibers, filters, integrated optical interconnects, and various other photonic devices. Photonic crystals allow to tailor and control the propagation of light in many ways—somewhat like semiconductors allow to control the electron current—and for this reason they can be viewed as "semiconductors of light" [4].

In this chapter we start with an introductory discussion of the photonic band structure taking the example of 1D photonic crystals. We then present the general properties of the photonic band structure. Then we illustrate the main features of 2D and 3D photonic crystals, as well as photonic crystals embedded in planar waveguides (also known as photonic crystal slabs). Line and point defect states in photonic crystals, which form linear waveguides and nanocavities for light propagation and confinement, are also considered. Finally, we briefly review a few methods for measuring the photonic band dispersion, and present concluding remarks.

2 Basic Principles

The treatment of electromagnetic propagation in photonic crystals starts with Maxwell equations for the electromagnetic field in matter [5]. Considering the case with no free charges or currents, these equations have the form (Gaussian units)

$$\nabla \cdot \mathbf{D} = 0 \tag{1}$$

$$\nabla \times \mathbf{E} = -\frac{1}{c}\frac{\partial \mathbf{B}}{\partial t} \tag{2}$$

$$\nabla \cdot \mathbf{B} = 0 \tag{3}$$

$$\nabla \times \mathbf{H} = \frac{1}{c}\frac{\partial \mathbf{D}}{\partial t}, \tag{4}$$

where $\mathbf{E}$ is the electric field, $\mathbf{D}$ is the displacement field, and $\mathbf{B}, \mathbf{H}$ are the magnetic induction and intensity of magnetic field, respectively. Throughout this chapter we assume unit magnetic permeability, so that $\mathbf{B} \equiv \mathbf{H}$. The constitutive relation between displacement and electric fields is taken to be

$$\mathbf{D}(\mathbf{r}) = \varepsilon(\mathbf{r})\mathbf{E}(\mathbf{r}), \tag{5}$$

where $\varepsilon(\mathbf{r})$ is the spatially dependent dielectric constant. The above equations can be cast into a second-order equation for the harmonic components of the electric field,

$$\nabla \times \nabla \times \mathbf{E} = \frac{\omega^2}{c^2}\varepsilon(\mathbf{r})\mathbf{E}(\mathbf{r}), \tag{6}$$

or an analogous equation for the magnetic field,

$$\nabla \times \left[\frac{1}{\varepsilon(\mathbf{r})}\nabla \times \mathbf{H}\right] = \frac{\omega^2}{c^2}\mathbf{H}(\mathbf{r}). \tag{7}$$

The first equation should be combined with the divergence equation $\nabla \cdot (\varepsilon(\mathbf{r})\mathbf{E}(\mathbf{r})) = 0$, while the second one requires $\nabla \cdot \mathbf{H}(\mathbf{r}) = 0$. The equation for the magnetic field is usually taken as the starting point for photonic band structure computations, for two reasons. First, it has the form of an eigenvalue problem $\Theta\mathbf{H}(\mathbf{r}) = \frac{\omega^2}{c^2}\mathbf{H}(\mathbf{r})$, with Θ being a Hermitian operator. Second, the divergence equation for the magnetic field does not contain the dielectric function and it is easier to implement. Working with a Hermitian eigenvalue problem is convenient from a computational point of view and has strong analogies with the matrix formulation of quantum mechanics.

The basic electromagnetic equations have the very useful property of scale invariance. If $\mathbf{H}(\mathbf{r})$ is a solution at frequency ω corresponding to the dielectric constant $\varepsilon(\mathbf{r})$, then the scaled system with dielectric constant $\varepsilon'(\mathbf{r}) = \varepsilon(\mathbf{r}/s)$ has a solution $\mathbf{H}(\mathbf{r}) = \mathbf{H}(\mathbf{r}/s)$ with frequency $\omega' = s\omega$. Reducing the length scale of the system by a factor s leads to eigenfrequencies that are multiplied by the same factor. Thus there is no fundamental length scale for the photonic problem (unlike for electrons in solids, for which the Bohr radius is the natural unit of length).

A periodic dielectric medium (in 3D, say) is invariant under translations by vectors $\mathbf{R} = n_1\mathbf{a}_1 + n_2\mathbf{a}_2 + n_3\mathbf{a}_3 (n_1, n_2, n_3 \in Z)$, where $\mathbf{a}_1, \mathbf{a}_2, \mathbf{a}_3$ are primitive vectors and the set of all **R**s forms a Bravais lattice. The dielectric constant satisfies the relation

$$\varepsilon(\mathbf{r}+\mathbf{R}) = \varepsilon(\mathbf{r}) \tag{8}$$

and the magnetic field has the form implied by Bloch–Floquet theorem:

$$\mathbf{H}_{n\mathbf{k}}(\mathbf{r}+\mathbf{R}) = e^{i\mathbf{k}\cdot\mathbf{r}}\mathbf{H}_{n\mathbf{k}}(\mathbf{r}), \tag{9}$$

where **k** is the Bloch vector (which may be restricted to the first Brillouin zone, or Wigner–Seitz cell of the reciprocal lattice) and n is a discrete band index. The frequency eigenvalues have the form $\omega = \omega_n(\mathbf{k})$ and the set of all frequencies for **k** spanning the first Brillouin zone is called a *photonic band.* In addition to translational invariance leading to Bloch vector conservation, discrete rotational symmetries of the lattice may also be used to analyze the photonic bands, again in close similarity to the electronic problem. The density of states (DOS) is defined as

$$\rho(\omega) = \frac{1}{(2\pi)^3}\sum_n \int_{\mathrm{BZ}} \delta(\omega - \omega_n(\mathbf{k}))\, d\mathbf{k} \tag{10}$$

and is strongly modified as compared to the photonic DOS in vacuum or in a homogeneous medium, which increases with the square of the frequency.

Among the methods for calculating the band structures and the optical properties we may mention the plane-wave expansion, the finite-difference time domain (FDTD) method, Fourier-modal methods, and many others: for these aspects we refer the reader to one of the many available books on the subject [5–10]. Generally speaking, plane-wave expansion is often the preferred method for band-structure calculations. Frequency-domain approaches like Fourier-modal methods are very suited for calculating the optical properties, but the band structure can be obtained as well using proper formulations. The FDTD method is a general purpose approach based on discretizing Maxwell equations on a temporal and spatial grid, it can be used for band structure and optical properties when proper boundary conditions are applied and also for simulating time-dependent phenomena. It is often rather computationally intensive and it is especially powerful for complex geometries and device simulations, where the dielectric structure is best defined in real space.

A very important concept for treating the interaction between a localized emitter and the electromagnetic field modes is the local DOS. Modelling the emitter with a dipole moment $\widehat{\mathbf{d}} = d\widehat{\mathbf{e}}$ placed at position **r**, the local DOS or LDOS is defined as

$$\rho_1(\omega,\mathbf{r},\widehat{\mathbf{e}}) = \frac{1}{(2\pi)^3}\sum_n \int_{\mathrm{BZ}} |\widehat{\mathbf{e}}\cdot\mathbf{E}_{n\mathbf{k}}(\mathbf{r})|^2\delta(\omega-\omega_n(\mathbf{k}))\, d\mathbf{k}. \tag{11}$$

The LDOS determines the radiative decay rate of the emitter through Fermi's golden rule as follows [11, 12]:

$$\Gamma = \frac{2\pi}{\hbar^2} d^2 \rho_l(\omega, \mathbf{r}, \widehat{\mathbf{e}}). \tag{12}$$

Thus, the photonic structure entails a modification of the radiative decay rate of an emitter (e.g., an excited atom or molecule, or a quantum dot) as compared to free space or to a homogeneous dielectric medium. In particular, radiative decay is suppressed (decay rate $\Gamma = 0$, or lifetime $\tau = \Gamma^{-1} \rightarrow \infty$) when the emission frequency falls within a complete PBG. The modification of the radiative decay of an emitter is a fundamental quantum electrodynamical effect and it is a basic probe of the modified interaction between radiation and matter. The decay rate can also be enhanced, and the radiative lifetime correspondingly reduced, when the emitter is in resonance with a peak in the local DOS. Such effects are especially pronounced in photonic crystal cavities, which are quasi-zerodimensional systems in which the photonic states are confined in all spatial directions: the enhancement of the radiative decay rate, also known as Purcell effect, will be specifically addressed in Sect. 7. In addition to modifications of the decay rate, the angular pattern of the spontaneous emission is also modified in a photonic crystal. We shall briefly discuss spontaneous emission redistribution at the end of Sect. 4.

3 One-Dimensional Photonic Crystals

The simplest PhC is a structure in which the dielectric function $\varepsilon(x, y, z)$ is periodic along a given direction z and uniform in the others, according to the relation

$$\varepsilon(x, y, z) = \varepsilon(x, y, z + m\Lambda), \tag{13}$$

where m is an integer and Λ is the period, i.e., the length of the unit cell.

In general, one can think about very complicated behaviors of the dielectric function within the unit cell. In practice, the most common 1D PhC is a periodic multilayer having the unit cell composed of two layers with lengths L_1 and L_2 and dielectric functions ε_1 and ε_2, respectively (see Fig. 1). This system is particularly interesting for at least two reasons: (a) from a theoretical point of view, the main properties of PhCs can be easily derived and understood; (b) from an experimental point of view, there are several fabrication techniques that can be utilized to obtain this kind of structure (see also chapters "Polymer Multilayer by Spin Coating", "Co-extruded Multilayer Polymer Films for Photonic Applications", and "Dry Polymerization of Functional Thin Films and Multilayers by Chemical Vapor Deposition").

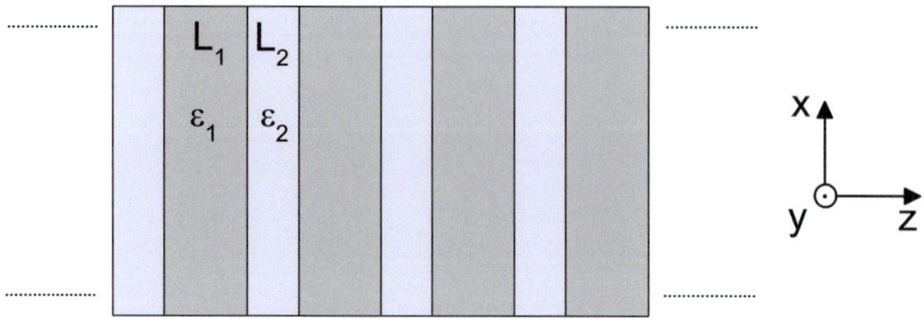

Fig. 1 Sketch of a 1D PhC structure having a unit cell composed of two layers with dielectric function ε_1 and ε_2

As we have seen in the previous section, for a PhC the solution of the Helmholtz equation can be reduced to an eigenvalue problem, where the eigenfunctions describe the electromagnetic field distribution in the multilayer in terms of Bloch waves, and the eigenvalues correspond to the energies of the electromagnetic modes [5].

In a periodic multilayer the electric field takes the form

$$\mathbf{E}(x, z, t) = \mathbf{E}(z) e^{\imath qz} e^{\imath(\kappa x - \omega t)}, \tag{14}$$

where κ is the wave vector component that corresponds to the propagation direction x in the multilayer plane, and $\mathbf{E}(z) = \mathbf{E}(z + m\Lambda)$ is a periodic function. Finally, q is the Bloch vector, and the dispersion relation $\omega = \omega(q)$ can be written in the form of an implicit equation, which is the optical analogue of the well-known Kronig–Penney model for an electron propagating in a 1D periodic potential [13]. In particular, for a bi-layer unit cell, we have

$$\cos(q\Lambda) = \cos(k_{1,z}L_1)\cos(k_{2,z}L_2) - \frac{1}{2}\left(\frac{\alpha_1}{\alpha_2} + \frac{\alpha_2}{\alpha_1}\right)\sin(k_{1,z}L_1)\sin(k_{2,z}L_2), \tag{15}$$

where $\Lambda = L_1 + L_2$ is the period, $k_{i,z} = \sqrt{(\omega/c)^2 \varepsilon_i - \kappa^2}$ is the z-component of the wave vector in the ith layer, and α_i depends on light polarization according to

$$\left.\begin{aligned} \alpha_1 &= \varepsilon_2 k_{1,z} \\ \alpha_2 &= \varepsilon_1 k_{2,z} \end{aligned}\right\} \quad \text{TM modes,}$$

$$\left.\begin{aligned} \alpha_1 &= k_{1,z} \\ \alpha_2 &= k_{2,z} \end{aligned}\right\} \quad \text{TE modes.}$$

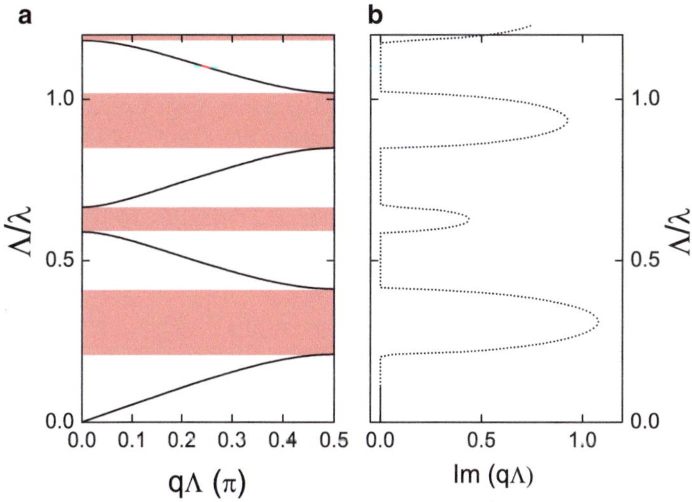

Fig. 2 (**a**) Photonic band structure of a 1D PhC with $L_1/\Lambda = 0.3$, $\varepsilon_1 = 9$, and $\varepsilon_2 = 1$, for light propagating in the direction normal to the multilayer. (**b**) Imaginary part of the Bloch vector within the photonic gap. The results are obtained by solving Eq. (15) numerically

It should be noticed that the solution of (15) in terms of the Bloch vector q can be either purely real, in the case of a mode propagating within the crystal, or complex, for energies within the PBG.

In Fig. 2a we plot the photonic band structure for a periodic multilayer with $L_1/\Lambda = 0.3$, $\varepsilon_1 = 9$, $\varepsilon_2 = 1$ and for light propagating along z (see Fig. 1). As we have seen in the previous section, thanks to scale invariance, the band structure can be shown in dimensionless units for both energy (Λ/λ) and wave vector ($q\Lambda$). Yet, it is important to stress that this makes sense only when the chromatic dispersion of the dielectric function can be neglected [5].

The first main notable feature of this band structure is that, given the relatively large refractive index contrast, the photon dispersion relation is no longer linear despite the absence of chromatic dispersions of the dielectric functions. The dispersion relation depends on the choice of the unit cell composition, and it can be tailored by changing the fraction of a material with respect to the other. Thus one can control phase and group velocities of light propagating in the crystal, for example to increase the light–matter interaction in the proximity of the band-edges when group velocity tends to zero [14], or to achieve phase-matching condition in the case of second harmonic generation or parametric nonlinear processes [15].

A second important characteristic of Fig. 2a is the presence of photonic band gaps (PBGs). In particular, in Fig. 2b we show the imaginary part of the Bloch vector within the PBG: the larger is the imaginary part, the faster is the field exponential decay within the crystal, according to Eq. (14). The capability of inhibiting light propagation in 1D PhCs can be exploited to guide and confine light, taking advantages of the flexibility in engineering the width and the position of the PBGs by varying the PhC geometry (see chapters “Strong Coupling in Organic and Hybrid-Semiconductor Microcavity Structures”, “One-Dimensional Photonic Crystals for Light Management in Organic Solar Cells”, “New Sensing Strategies Based on Surface Modes in Photonic Crystals”, and “Hybrid-Organic Photonic Structures for Light Emission Modification”).

In this respect, given two materials of refractive indices n_1 and n_2, the largest first-order gap centered at the wavelength λ_0 is obtained for the so-called $\lambda/4$-condition, when the thicknesses of the layers are chosen according to

$$L_i = \frac{\lambda_0}{4n_i}. \tag{16}$$

This situation corresponds to a particular case of the Bragg law, and it guarantees the maximum destructive interference of waves reflected at the interfaces between the layers of the structure.

Equation (16) is typically used to design DBRs, which can be understood as 1D PhC having a finite number of periods. In particular, in the case of small refractive index contrast, i.e $\Delta n = |n_1 - n_2| \ll n$, with $n = (n_1 + n_2)/2$ the average refractive index, Eq. (15) can be solved analytically near by the center of the photonic gap at $\omega_s = \pi c/(n\Lambda)$, and it is possible to give an expression of the photonic gap width in terms of the refractive index contrast:

$$\Delta\omega_{gap} = \frac{4}{\pi}\omega_s \frac{|n_2 - n_1|}{n_1 + n_2} \approx \frac{2}{\pi}\omega_s \frac{\Delta n}{n}. \tag{17}$$

Moreover, by taking $\omega = \omega_s$ and solving Eq. (15) in terms of the complex Bloch vector we obtain

$$q\Lambda = \pi + i\frac{\Delta n}{n}, \tag{18}$$

which shows that at the center of the PBG the electromagnetic field in the 1D PhC decays exponentially with an attenuation length $n\Lambda/\Delta n$. From these results one can derive an approximate expression of the DBR reflectivity at normal incidence as a function of the number N of periods:

$$R(\omega_s) \approx 1 - 4\left(\frac{n_1}{n_2}\right)^{2N}, \tag{19}$$

where we have assumed $n_2 > n_1$. It should be noticed that, since the decay is exponential in N, high reflectivity can be obtained even with a small refractive index contrast, like in the case of fiber Bragg gratings [16].

So far we have been considering examples in which the electromagnetic wave propagates in the direction normal to the multilayer plane (i.e., $\kappa = 0$). Yet Eq. (15) can

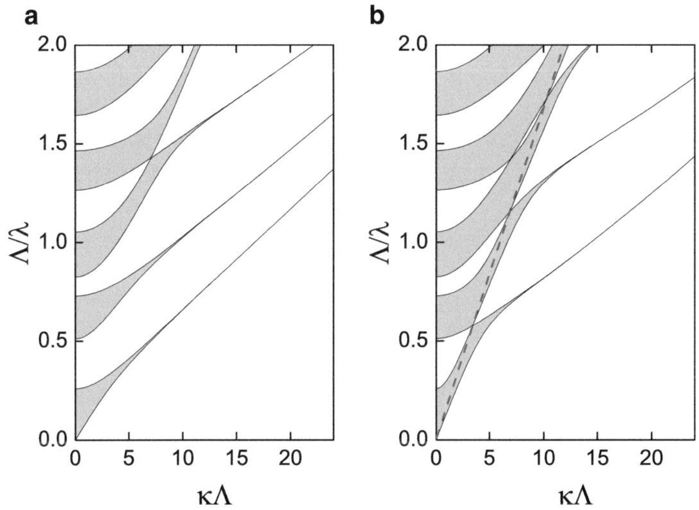

Fig. 3 Photonic band gap dispersion as a function of κ for TE (**a**) and TM (**b**) polarized light. The structure is a 1D PhC with $L_1/\Lambda = 0.3, \varepsilon_1 = 9$, and $\varepsilon_2 = 1$. The *dashed light* in (**b**) indicates the points corresponding to the Brewster angle

also be used at different propagation angles. In this case, even in the presence of isotropic materials, the structure response depends on the light polarization, and the solutions of Eq. (15) can be labelled (transverse-electric) TE or (transverse-magnetic) TM. In Fig. 3 we show the PBG at $q = 0$ as a function of κ for both polarizations. There are a few notable features of the dispersions that help once more to clarify the origin of the PBG, which is simply an interference effect between waves scattered at each interface.

Naturally, the results of this interference depend on the phase shift acquired by the waves in their propagation in each layer, that is $k_{i,z}L_i$. The increase of the PBG energy as a function of κ is directly linked to this phase shift, as $k_{i,z} = \sqrt{(\omega/c)^2\varepsilon_i - \kappa^2}$. In practice, for larger κ, the same phase shift occurs at higher energies. It is worthy noticing that this effect is qualitatively independent of the polarization and that dispersion of the PBG follows the same trend for TE and TM polarization.

The second important aspect of the wave interference at the origin of the PBG is related to the amplitude of the light reflected at each multilayer interface. We have already seen that at normal incidence this is simply related to the refractive index contrast: the larger is Δn the wider is the PBG that can be obtained. Yet, when light propagates at finite κ, the amount of reflected light depends also on its polarization according to the Fresnel's coefficients. This causes a different behavior of TE and TM PBGs as a function of κ. On the one hand, in Fig. 3a the TE PBG width usually increases with κ as the amplitude of TE reflection coefficient at the interface between two media increases monotonically with the incident angle (i.e., with κ). On the other hand, when light is TM polarized, the corresponding Fresnel's coefficient has a zero at the Brewster angle $\theta_B = \arctan(n_1/n_2)$. Thus, the PBG dispersion is more complicated and, when κ corresponds to the Brewster angle, the TM PBG is closed as light can propagate through the multilayer without any back reflection.

While the main direct application of 1D PhC structures are the Bragg mirrors known as DBRs, the concept of 1D periodic structures is also frequently encountered in waveguide optics in the context, e.g., of fiber Bragg gratings, distributed feedback lasers, etc. Such topics are usually treated in Photonics textbooks [13].

4 Planar Microcavities

A sketch of a planar microcavity is shown in Fig. 4. It is composed of a layer of thickness L_c embedded between two dielectric mirrors, typically DBRs. This structure can be considered as the PhC version of the well-known Fabry–Pérot resonator with the transmittance taking the form

$$T(\omega) = \frac{(1 - R(\omega))^2}{1 + R(\omega)^2 - 2R(\omega)\cos(\delta(w))} \tag{20}$$

with

$$\delta(\omega) = 2k_{c,z}(\omega)L_c + 2\phi(\omega), \tag{21}$$

where $R(\omega)$ is mirror reflectance, $k_{c,z}(\omega) = (\omega/c)n_c \cos\theta_c$ is the z component of the wave vector, n_c is the refractive index of the cavity layer, and θ_c depends on the angle of incidence through Snell's law. Finally, $\phi(\omega)$ is the phase shift associated with the reflection at the interface with the dielectric mirror. The system is resonant at specific frequencies ω_m when

$$\delta(\omega_m) = 2\pi m. \tag{22}$$

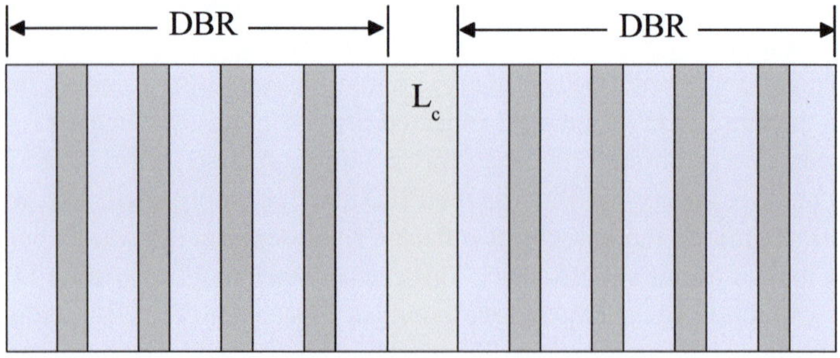

Fig. 4 Sketch of a planar microcavity composed of a layer with length L_c and refractive index n_c embedded between two DBRs

It should be noticed that, unlike the case of metallic mirrors, the phase shift $\phi(\omega)$ of a DBR can be strongly dependent on frequency [17–19]. In the case of periodic mirrors and in the limit $R \approx 1$ (i.e., for sufficiently large N) $\phi(\omega)$ depends only on the period composition and the refractive index n_c of the cavity layer, while it is independent of N [17, 18]. Thus, in a microcavity the resonance frequencies of the cavity modes are usually a complicated function of the structure parameters and of the incidence angle. Analytic expressions for the resonance frequencies are possible only when the cavity mode is close to the center of the stop band [19, 20], where $\phi(\omega)$ can be approximated by a linear function of frequency.

Likewise a Fabry–Pérot resonator, the transmission coefficient near by the resonant frequency ω_m can be approximated with a Lorentzian curve

$$T(\omega) = \frac{\gamma_m^2/4}{(\omega - \omega_m)^2 + \gamma_m^2/4}, \tag{23}$$

where

$$\gamma_m = \frac{c}{n_c L_c} \frac{1-R}{\sqrt{R}}, \tag{24}$$

which is the FWHM of the resonance and is related to the energy decay time in the cavity by the simple relation $\gamma_m = \tau_m^{-1}$. In particular, one can define the *quality factor*

$$Q_m \equiv \frac{\omega_m}{\gamma_m} = \frac{m\pi\sqrt{R}}{1-R}. \tag{25}$$

The quality factor is 2π times the number of oscillation cycles of the electromagnetic fields inside the cavity before the stored energy decays by a factor $1/e$. The larger is the mirror reflectance R, the larger is quality factor. Yet, it should be noticed that Q is also directly proportional to the cavity length, as the time that light spends in the resonator is directly proportional to its length. While, in general, there is no theoretical limit to the quality factor achievable in a planar cavity, in practice, in microcavities with DBRs made of semiconductor or oxides Q is often the order of 10^4 but Q-factors approaching 2×10^5 have been reported [21]. In organic systems, where the refractive index contrast is smaller and the quality of the interfaces in the DBR is lower, the quality factor is typically of the order of 100 [22–24].

Beside the quality factor, the other important parameters that characterized the light confinement in a resonator is the modal volume, which can be generally defined in three dimensions as

$$V_m = \frac{\int \varepsilon(\mathbf{r})|\mathbf{E}_m(\mathbf{r})|^2 d\mathbf{r}}{\max\left[\varepsilon(\mathbf{r})|\mathbf{E}_m(\mathbf{r})|^2\right]}. \tag{26}$$

Fig. 5 Reflectance of a planar microcavity composed of a layer with length L_c and refractive index n_c embedded between two DBRs with $L_1/\Lambda = 0.3$, $\varepsilon_1 = 9$, and $\varepsilon_2 = 1$

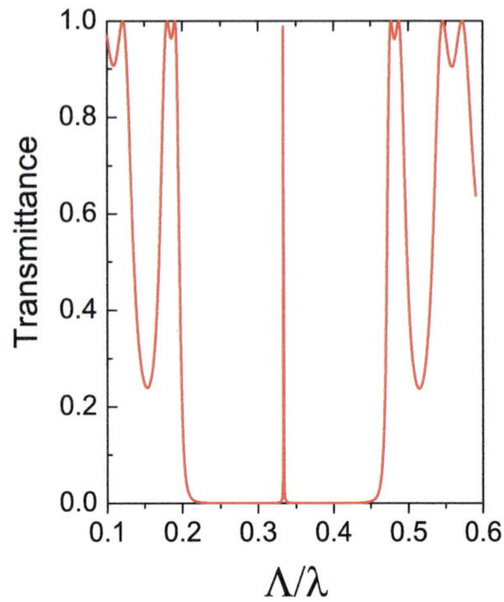

It should be noticed that while the quality factor is essentially a measurement of the light dwelling time in the cavity, the modal volume V_m describes how well the light is confined in space. In other words, the smaller is the modal volume, the tighter the light is localized within the resonator (Fig. 5).

In the case of a planar resonator, we talk about modal length and rewrite Eq. (26) in the 1D case as

$$L_m = \frac{\int \varepsilon(z)|\mathbf{E}_m(z)|^2 dz}{\max\left[\varepsilon(z)|\mathbf{E}_m(z)|^2\right]}. \tag{27}$$

For planar microcavities with dielectric mirrors, the modal length can be as small as (λ/n), where λ is the wavelength of light and n the average refractive index of the structure.

The enhancement of the electromagnetic field intensity inside a resonator is proportional to Q/V. Thus, one wants to obtain structures characterized by a long photon dwelling time and tight confinement within the resonator. In this regard, PhC cavities are among the systems with the highest Q/V, and this explains why they are particularly indicated for those applications that requires a large enhancement of the light–matter interaction (e.g., see chapter "Strong Coupling in Organic and Hybrid-Semiconductor Microcavity Structures").

Planar microcavities yield strong angular redistribution effects in the spontaneous emission of an excited atom, or, generally, a dipolar emitter. Such effects underlie the working principle of the vertical cavity surface emitting laser (VCSEL). The VCSEL is a particular type of semiconductor laser in which the optical beam travels at right angles to the active region (rather than in the plane of the active region, as in conventional edge-emitting diode lasers) thanks to the presence of two Bragg reflectors that form a laser cavity in the vertical direction.

A description of VCSELs can be found, e.g., in [13]. From the fundamental point of view, redistribution of the angular emission has been studied in a number of contexts, e.g., quantum-well excitons [25] or dye molecules [26] in planar microcavities, inorganic [27] or organic light-emitting diodes [28].

5 Two- and Three-Dimensional Photonic Crystals

Two-dimensional photonic crystals have a dielectric constant which is periodic in a plane (xy) and homogeneous in the third (z) direction. The most interesting situation is when light propagates in the xy plane, i.e., for a vertical wavevector $k_z = 0$. The system is invariant under specular reflection with respect to the mirror plane xy, thus the electromagnetic eigenmodes can be classified as even or odd with respect to this mirror symmetry. Even states have nonvanishing field components (E_x, E_y, H_z) and are called H- (or TE-) polarized modes, while odd states have nonzero field components (H_x, H_y, E_z) are called E- (or TM-) polarized. Unlike in the 3D case, there is a strong polarization dependence of the photonic bands.

Most 2D structures that have been studied are based on a square or on a hexagonal Bravais lattice. In general, the most favorable situation for a complete photonic gap in two dimension is to have a structure with the highest possible rotational symmetry: this helps in achieving an overlap between band gaps along different directions of the Brillouin zone. The square Bravais lattice has a fourfold symmetry axis, while the hexagonal lattice has a sixfold rotational axis which is the maximum possible symmetry for periodic Bravais lattices. Higher rotational symmetries are possible for non-periodic structures like quasi-crystals.

In Fig. 6 we show two examples of 2D photonic bands in structures with a full band gap for all directions and polarizations, namely the triangular lattice of air holes in a dielectric medium (Fig. 6a) and the graphite or honeycomb lattice of dielectric pillars in air (Fig. 6b). Both structures are based on the 2D hexagonal Bravais lattice. The triangular lattice of holes has a large gap between the first and the second H-band and a smaller gap between the second and the third E-band, which overlap in the region shown by a dashed area in Fig. 6a. For the assumed value of the dielectric constant ($\varepsilon = 12$), the H-gap is found to open for hole radii $r/a > 0.16$, while the E-gap opens only for hole radii $r/a > 0.4$; therefore, a polarization-independent gap exists only for relatively high values of the air fraction. The graphite lattice of pillars instead has two complete band gaps between higher-lying bands. Notice that the graphite lattice of pillars supports a large gap between the second to third E-bands, which however does not overlap any H-gap.

As a general rule, structures with a connected dielectric lattice (like that of Fig. 6a) tend to have larger gaps for H polarization, while structures with disconnected dielectric pillars (like that of Fig. 6b) tend to support larger E-gaps. This criterion, which can be traced back to Maxwell boundary conditions for the non-vanishing field components, is generally followed for both square and

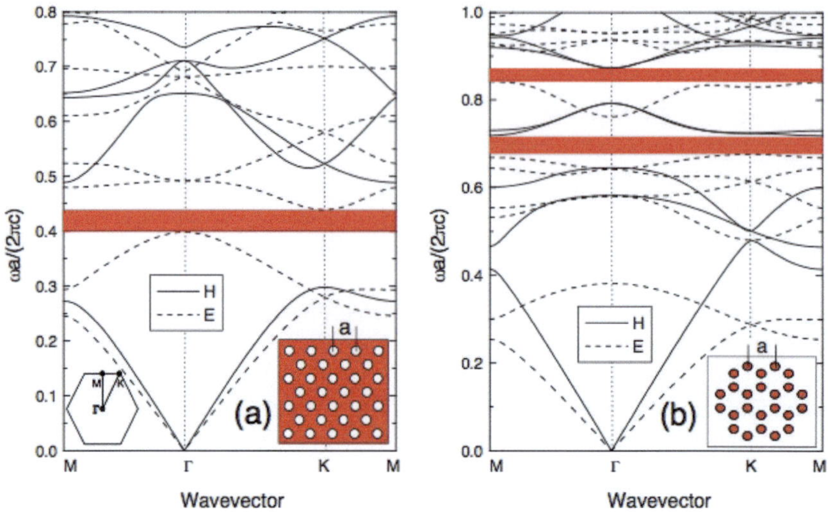

Fig. 6 Dispersion of photonic bands in (**a**) a triangular lattice of air holes with radius $r/a = 0.45$ in a dielectric material and (**b**) a graphite lattice of dielectric pillars with radius $r/a = 0.18$ in air. The dielectric material has $\varepsilon = 12$ in both cases. *Solid (dashed) lines* represent H- (E-) polarized modes: they are even (odd) with respect to a horizontal mirror plane. The *shaded areas* denote the photonic gaps. *Insets*: structures and 2D Brillouin zones

hexagonal Bravais lattices. The overlap of H- and E-gaps is a stronger condition that occurs only in a few cases, those shown in Fig. 6 being the most common examples. In real systems, the photonic gap shown in Fig. 6a is easier to realize, because it is lower in frequency and less sensitive to disorder. For technological applications, a polarization-sensitive gap is often sufficient and the commonly used structure is the triangular lattice of air holes for smaller air fractions than shown in Fig. 6, in order to reduce radiation losses. Actually, 2D photonic structures are often realized in a slab-waveguide geometry and the photonic band dispersion is modified as compared to the pure 2D case, as we discuss in Sect. 6.

2D photonic structures can be defined by a top-down procedure based on lithography and etching. The prototype of a truly 2D photonic crystal is macroporous silicon, which is obtained by wet etching of Si in an electrochemical cell [29]. It should be remembered that the photonic bands like those shown in Fig. 6 refer only to propagation in the xy plane: in a system that can be considered as homogenous in the z direction, the bands retain an out-of-plane dispersion and light is free to propagate in the vertical direction. This mechanism is exploited in a new class of optical fibers, which are known as holey- or photonic crystal fibers (PCF). In the words of their inventor, P. Russell, they represent "a next-generation, radically improved version of a well-established and highly successful technology" [30]. More recently, there have been theoretical as well as experimental results, indicating the possibility of realizing similar structures with organic materials. In particular, the use of polymers for PCF seems to be particularly interesting in the view of tailoring the dispersion relation and the optical birefringence of these structures, as well as adding new functionality by combining polymers and inorganic materials, for example towards the realization of all-fiber nonlinear tunable devices [31–33].

While in this chapter we focus on periodic PhC structures, we should also mention that aperiodic structures known as *photonic quasicrystals* have also been considered as they show many of the properties of conventional photonic crystals. For example, they do display diffraction patterns that result from electromagnetic wave interference. In a sense, photonic quasicrystals share properties of both ordered and disordered photonic structures. For a review of this area see, e.g., [34].

The first photonic structure that has been shown theoretically to possess a complete band gap in three dimensions is the diamond lattice of dielectric spheres in air or air spheres in a dielectric material [35]. The photonic gap opens provided the dielectric constant is large enough and it exists in a wide range of filling factors. Nevertheless, the diamond lattice of dielectric spheres cannot be fabricated in practice, because it cannot be realized by top-down approaches based on lithography and etching, and also it is not a close-packed structure (to be grown with bottom-up, self-assembly approaches).

Other structures possessing a complete photonic gap in three dimensions, that are more amenable to fabrication, have been studied (see also chapters "Organic Opals: Properties and Applications", "Self-Assembling Polymer Photonics", and "Fabrication of Photonic Crystals Using Holographic Lithography"). Two of them are shown in Fig. 7. The structure in Fig. 7a, called the *Yablonovite* [36], is obtained by drilling three sets of cylindrical holes with a triangular pattern, at 35.26° from the surface normal and at 120° with respect to each other. It represents a distortion of the diamond structure of air spheres in a dielectric material, where the surface is a (111) plane of the crystal and the holes correspond to [110] open channels of the diamond structure. The Yablonovite can be realized at infrared and

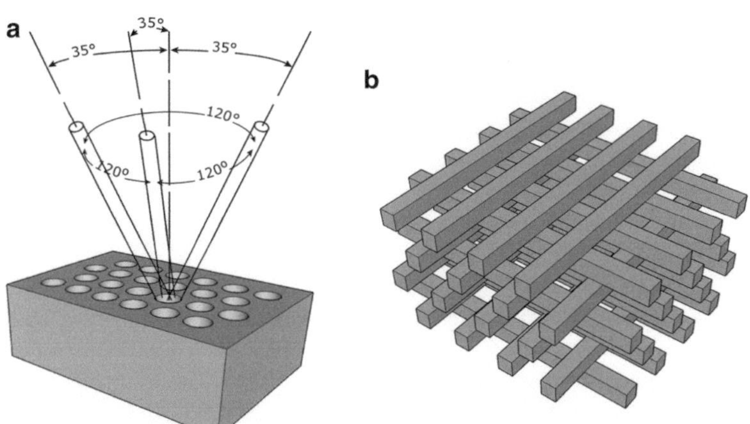

Fig. 7 Schematic plots of 3D photonic crystal structures: (**a**) Yablonovite and (**b**) woodpile

optical wavelengths by deep lithography [37]. The structure in Fig. 7b, often called the *woodpile*, is obtained by stacking four alternate layers of dielectric rods: starting from a first layer, the second layer is rotated by 90deg with respect to the first, and the third layer is shifted by $a/2$ (a is the lattice constant) with respect to the first one [38]. The woodpile structure can be realized by a series of stacking steps [39]. The fabrication procedures of Yablonovite and of woodpile are examples of *top-down* and *bottom-up* approaches, respectively.

Another simple 3D structure with a complete band gap is the fcc lattice of air spheres in a dielectric material: this is called the *inverse opal* structure. The inverse opal is derived from the (direct) opal structure, i.e., an fcc lattice of dielectric spheres in air: indeed, the opal is a natural gem consisting of a close-packed lattice of silica spheres with fcc arrangement. Artificial opals can be obtained by self-assembling of spheres in a colloidal solution. The voids are then filled with a high-index material (Si or TiO_2) and the template is subsequently removed to obtain the inverse structure [40, 41]. With the refractive index of silicon, this procedure allows in principle to realize a photonic crystal with a complete band gap in the optical region. In real systems, however, a small amount of disorder is always present: the photonic gap is not very robust as it is relatively narrow and is formed between higher-lying bands, which are highly sensitive to disorder.

6 Photonic Crystal Slabs

In order to achieve a better control of light propagation, a 2D photonic structure can be embedded in a planar (slab) dielectric waveguide, thereby realizing a photonic crystal slab. A few possible structures are shown in Fig. 8. The theoretically simplest system is the air bridge or self-standing membrane (Fig. 8a), which consists of a patterned dielectric membrane surrounded by air. The air bridge has the highest refractive index contrast between core and cladding. Figure 8b exemplifies the silicon-on-insulator (SOI) structure, consisting of an air/Si/SiO_2 waveguide in which only the Si layer is patterned. Photonic crystal slabs can also be realized on other high-index membranes (GaAs, InP) or on slabs with weak-refractive index contrast like GaAs/AlGaAs or InP/InGaAsP waveguides, in this case both core and cladding need to be patterned in order to maintain the guiding properties of the slab. The fabrication of photonic crystal slabs starts from fabrication of the slab waveguides on a substrate (e.g., SOI wafers), then it requires lithography and dry (reactive-ion) etching steps, followed by removal of the cladding in the case of a self-standing membrane. The first PhC slabs were realized in the nineties on AlGaAs structures [42], while now many PhC slabs are realized in silicon as it is the most suitable material for optical interconnects. Polymer PhC slabs are also being studied [43, 44].

Photonic crystal slabs are characterized by the light-line issue, which is a key feature of dielectric waveguides and is illustrated in Fig. 9. Only photonic modes that lie below the light dispersion of the cladding material(s) in the k-ω plane fulfill

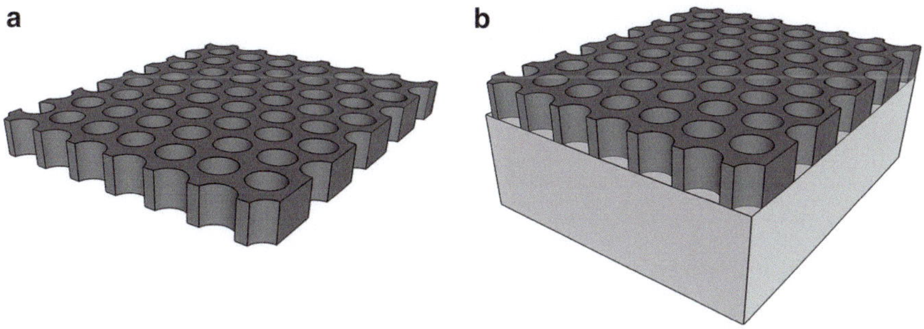

Fig. 8 Schematic photonic crystal slab structures: (**a**) self-standing membrane or air bridge and (**b**) silicon-on-insulator (SOI). In both cases, the high-index slab is patterned with a triangular lattice of air holes

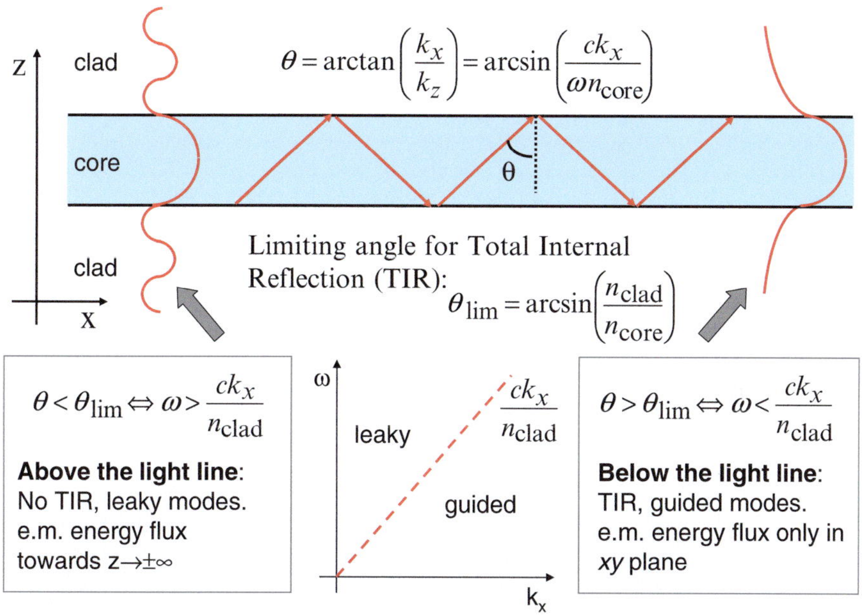

Fig. 9 Schematic illustration of the mechanism of total internal reflection in dielectric slabs, leading to the occurrence of leaky modes above the *light line* and guided modes below the *light line*. The former radiate electromagnetic energy in the far field ($z \to \pm\infty$), while the latter are truly guided in the slab plane

the condition for total internal reflection (TIR) and are therefore truly guided and stationary. On the other hand, modes that lie above the cladding light line(s) are radiative and only quasi-guided. For a specified value of the in-plane wavevector, the truly guided modes lie in a frequency region of discrete spectrum, while quasi-guided modes lie in the region of continuous spectrum and appear as resonances: their linewidth can be related to an imaginary part of the mode frequency.

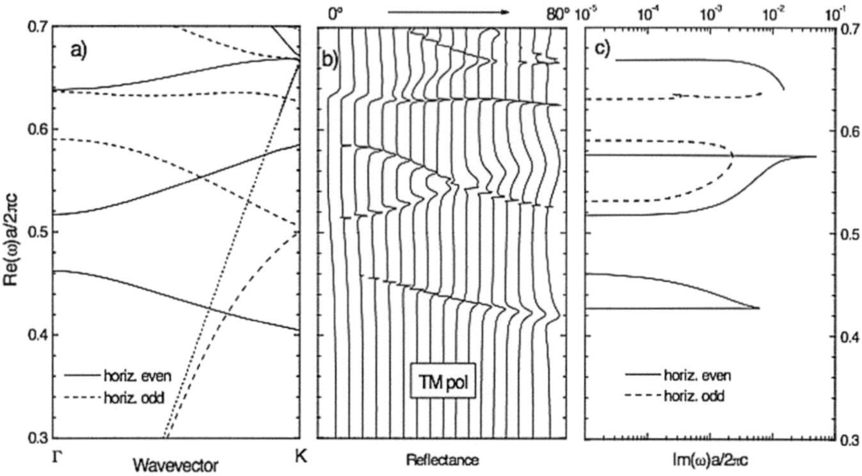

Fig. 10 (**a**) Real part of the frequency, (**b**) specular reflectance from the surface at angles of incidence from 0° to 80° in steps of 5°, and (**c**) imaginary part of the frequency. The results refer to a photonic crystal slab with a triangular lattice of air holes of radius $r/a = 0.3$ patterned in a membrane with core thickness $d/a = 0.3$ and dielectric constant $\varepsilon = 12$. All curves calculated for TM polarized (even) modes with respect to the plane of incidence along the Γ-K orientation. *Solid (dashed) lines* in (**a**) and (**c**) correspond to even (odd) modes with respect to a horizontal symmetry plane. The *dotted line* in (**a**) represents the dispersion of light in air

An example of mode dispersion in a photonic crystal slab is shown in Fig. 10a and it refers to an air bridge patterned with a triangular lattice of holes. Figure 10a shows the real part of the frequency dispersion in the Γ-K direction. The cladding light line (dotted line in the figure) separates the regions of guided and quasi-guided modes. The planar waveguide is monomode up to a frequency $\omega a/(2\pi c) \simeq 0.61$, where a second-order mode appears. Re(ω) is generally larger than for the ideal 2D case due to vertical confinement in the slab waveguide: this effect is particularly pronounced for a strong index-contrast slab. Figure 10b displays the calculated reflectance of a plane wave incident on the surface of the slab, for increasing values of the angle of incidence. Spectral structures appear in reflectance spectra, which mark the excitation of photonic modes and whose evolution as a function of incidence angle corresponds to the $\omega(k)$ dispersion of Fig. 10a: the wavevector parallel to the surface is conserved and it equals $k = (\omega/c)\sin\theta$ (modulo a reciprocal lattice vector). Figure 10c shows the imaginary part of the frequency of quasi-guided modes plotted as a function of the real part. The imaginary part depends strongly on the mode index and on the wavevector and it vanishes when the real part of the dispersion crosses the light line and goes into the guided mode region. It can be seen that an increasing imaginary part in Fig. 10c corresponds to an increasing linewidth of the reflectance structure in Fig. 10b.

It turns out that the complete PBG of the triangular lattice of air holes is suppressed when going to pure 2D (infinitely extended in the vertical direction) to a photonic crystal slab [45]: this follows from vertical confinement in the high-index slab, which modifies the photonic band dispersion of H- and E-modes in a different manner. In particular, E- (or TM-like) modes have a higher blue shift, which reduces the overlap with H- (or TE-like) modes and makes the PBG encounter a second-order guided modes. Even reducing the symmetry of the holes from circular to triangular does not restore a complete PBG [46]. In practice, most PhC slab structures operate in a polarization-dependent manner, as they are based on the TE-like band gap of the triangular lattice of holes.

The imaginary part of the mode frequency leads to an imaginary part of the wavevector through the relation $\mathrm{I}m(k) = \mathrm{I}m(\omega)/v_g$, where v_g is the group velocity. Physically, a propagating quasi-guided mode in an ideal photonic crystal slab is subject to intrinsic radiative losses due to diffraction out of the 2D plane. In real systems, extrinsic diffraction losses related to fabrication aspects (insufficient hole depth, non-vertical profile, roughness of the sidewalls) are present for both guided and quasi-guided modes. The strong intrinsic coupling of quasi-guided PhC slab modes can be used to out-couple radiation from the slab to the far field or, conversely, to in-couple electromagnetic far-field radiation into the slab. An example of out-coupling problem is the enhancement of light emission from a high-index medium like silicon [47], which recently allowed the realization of a PhC-assisted silicon LED [48]. Examples of in-coupling problems are polarization-diversity grating couplers to efficiently couple light from a single-mode optical fiber into a silicon photonic waveguide [49, 50], and light trapping in photovoltaic cells [51]. The flexibility given by the PhC slab parameters is of great help in tailoring the spectral width to the specific problem and in optimizing the coupling efficiency.

7 Defects in Photonic Crystals: Linear Waveguides and Nanocavities

It is well known in electronic structure theory that localized defects introduce impurity states in the gap. An analogous phenomenon takes place in photonic crystals. A common example is a planar defect in a 1D photonic crystal, which is known as a planar (or Fabry–Pérot) microcavity, as we have seen in Sect. 4. Line and point defects in photonic crystals of higher dimensionality also give rise to defect modes within the PBG. The simplest kinds of defects can be treated with the methods introduced for bulk photonic crystals, provided the defect is repeated with supercell periodicity.

A common example of a line defect, or channel waveguide, in a photonic crystal slab is shown in Fig. 11a. The structure consists of a single missing row of holes in the Γ-K direction of the triangular lattice and it is a called a *W1 waveguide*. The corresponding photonic bands are shown in Fig. 11b. The defect modes are found outside the projections of the 2D bands onto the Γ-K direction, which are denoted by shaded regions. The "even" defect mode that falls within the photonic gap starting from the upper bands is called index-guided, because it is laterally confined

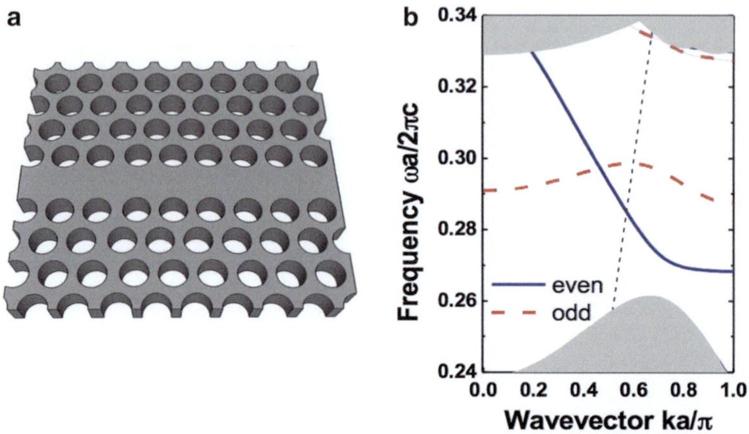

Fig. 11 (**a**) Schematic structure of a W1 waveguide in a photonic crystal slab. (**b**) Dispersion of defect modes of a linear waveguide in a membrane with thickness $d/a = 0.5238$ and dielectric constant $\varepsilon = 12.1$. The holes in the triangular lattice have a radius $r/a = 0.3$. The *shaded regions* correspond to the projected bands of the periodic lattice and the *dashed lines* denote the light dispersion in air

by the dielectric discontinuity between the channel region and the surrounding photonic crystal. The "odd" mode that exists only within the band gap of the 2D lattice is called gap-guided and is laterally confined by the photonic gap, i.e., by the lack of TE-like electromagnetic eigenmodes in the surrounding material. Both index- and gap-guided modes can be seen in Fig. 11b. Notice that the defect modes can be either guided or quasi-guided, according to their frequency with respect to the cladding light line. The portion of the index-guided mode with $k > 0.5\pi/a$, which falls below the light-line dispersion in air, is truly guided and it is the most interesting region for guiding light in PhC slabs with low losses. For treatments of the loss problem in PhC waveguides see, e.g., [52–57]. The linear defect structure in a photonic crystal lends itself to the realization of ultra-compact waveguides with sharp bends, which cannot be obtained with conventional dielectric waveguides based on TIR.

Point defects in photonic crystals with a complete PBG give rise to nanocavities that are able to localize light in all spatial dimensions. The best performing PhC nanocavities are realized in 2D photonic crystal slabs, thanks to the high quality of 2D structures that are fabricated by lithography and etching, combined with efficient confinement in the third direction provided by TIR in the high-index slab. To date, the most suitable structure for high-Q PhC nanocavities is undoubtedly the self-standing silicon slab. A common structure is the L3 nanocavity, realized by removing three holes in the triangular lattice: see Fig. 12a. A milestone in the field was achieved in 2003, when Q-factors up to 45,000 were reported in L3 nanocavities in Si slabs by local geometry optimization of the nearby holes—initially motivated by the principle of "gentle confinement" [58]. Most recently, general optimization of L3 cavity geometry by a genetic algorithm led to the prediction of Q-factors up to

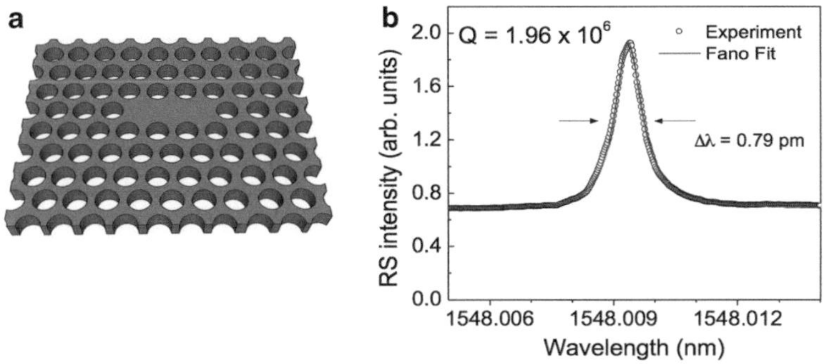

Fig. 12 (**a**) Schematic structure of an L3 cavity a photonic crystal slab. (**b**) Measurement of Q-factor by cross-polarized resonant light scattering and Fano-lineshape fit in a genetically optimized L3 nanocavity in silicon [60]

several millions [59]: the current experimental record is $Q \sim 2 \cdot 10^6$, see [60] and Fig. 12b. Even higher Q-factors can be achieved in cavities obtained by modifying the geometry of a line defect, like heterostructure cavities (where the lattice constant along the defect direction is changed) or width-modulated cavities (where the channel width of the defect is modulated) [61–63]: in both cases, the line-defect mode is confined into a 0D cavity mode. The current record of Q-factor for heterostructure cavities in Si slabs is $Q \simeq 9 \cdot 10^6$ [64]. Another important parameter to characterize nanocavity modes is the modal volume, already defined in Eq. (26). For the most performing PhC slab cavity geometries, like the L3 nanocavity, the modal volume is of the order of $(\lambda/n)^3$, where λ/n is the wavelength of light in the medium. For a detailed review of PhC nanocavities, see, e.g., [65].

Photonic crystal nanocavities lend themselves to a number of fundamental studies of light confinement, light–matter interaction in cavity-QED, quantum and nonlinear processes. Concerning photon confinement, we may mention storage of photons with ns lifetimes [66] with prospective application to optical memories, dynamic switching and adiabatic tuning of cavity modes [65], the enhancement of nonlinear optical processes like second- and third-harmonic generation [67–69]. Among the intriguing phenomena involving radiation–matter interaction with 0D photons, we may mention the control of spontaneous emission lifetime of an excited “atom,” often a quantum-dot excitation [70, 71], ultralow-power nanolasers approaching the regime of low-threshold lasing [72], the strong-coupling (non-perturbative) regime of light–matter interaction [73]. Quantum and nonlinear processes that are especially promising in the context of PhC nanocavities relate to single-photon nonlinearities allowing, e.g., phenomena like photon blockade [74, 75]. These phenomena depend in different ways on the quality factor Q and the mode volume V: while, e.g., the Purcell effect scales as Q/V, other phenomena have a different dependence [65]. In summary, photonic crystal nanocavities represent a most lively area of research where intriguing phenomena following from photon confinement in high-Q, low-volume nanoresonators are only partially explored until now.

We now discuss a highly relevant phenomenon, namely the reduction of spontaneous emission lifetime also known as *Purcell effect* [76]. This can be especially pronounced in high-Q photonic crystal nanocavities where the electromagnetic field is fully confined in all spatial dimensions. Denoting by Γ_n the radiative decay rate of an excited "atom" (generally a two-level system) in a dielectric medium with refractive index n, the decay rate of the same atom interacting with mode m of a photonic nanocavity is modified to $\Gamma_m = F_{\mathrm{P}}\Gamma_n$, where F_{P} is called the *Purcell factor* and is given by

$$F_{\mathrm{P}} = \frac{3}{4\pi^2}\left(\frac{\lambda}{n}\right)^3 \frac{Q_m}{V_m}, \tag{28}$$

where Q_m is the quality factor of mode m and V_m is the mode volume, defined in Eq. (26). The above formula (28) holds under the following conditions: (a) the dipole emitter is placed at the maximum position of the electromagnetic field intensity, (b) the dipole emission frequency is in resonance with the cavity mode, and (c) the transition dipole is parallel to the polarization of the electric field. Under these conditions, if the mode volume is of the order of the cavity mode wavelength cubed, the Purcell factor could in principle become of the order of the cavity Q-factor. In practice, spatial and/or spectral detuning effects usually reduce the ideal Purcell factor to values of the order of 10–100. Still, the Purcell effect represents a fundamental quantum electrodynamical phenomenon and its observation is direct evidence of a profound modification of the radiation–matter interaction when the electromagnetic field is fully confined in zero dimensions. Examples of measured Purcell effect in PhC nanocavities are given in [71] for conventional InGaAs quantum dots in GaAs cavities emitting around 900 nm wavelength, and in [77, 78] for InAs dots in InP or Er^{3+} ions in Si PhC slab cavities emitting at the telecom wavelength $\lambda = 1.54\ \mu$m, respectively.

8 Measuring the Photonic Band Structure

To complete this review, we mention a few experimental methods for measuring the photonic band structure. The frequency window of a PBG can be obtained by transmission (or reflection) measurements. The conceptually simplest method for obtaining the photonic band dispersion in the direction of light propagation is to measure the phase delay $\phi(\omega)$ of an electromagnetic wave through the photonic crystal: if D is the thickness of the sample, the wavevector at any given frequency can be obtained from the relation $\phi(\omega) = k(\omega) \cdot D$. This method has been applied by means of phase-sensitive measurements, first in the microwave spectral region

[79] and later in the near infrared and visible regions [80]. The photonic dispersion in two or three dimensions can also be derived by measuring the beam propagation (i.e., the refraction) of a light beam impinging on the photonic crystal at non-normal incidence [81]. The 2D band dispersion above the light line can be obtained by variable-angle reflectance measurements from the crystal surface [82–84]. The principle is apparent from Fig. 10b: the spectral position of a resonant structure in reflectance yields both the frequency and the parallel wavevector of a photonic mode. The same variable-angle reflectance technique can be applied to 2D and 3D photonic crystals, and in this case it leads to the dispersion of photonic bands in a plane parallel to the crystal surface. The band dispersion below the light line can be obtained by means of attenuated total reflectance (ATR) using a prism: the refractive index of the prism modifies the relation between frequency and parallel wavevector, and it allows reaching wavevector values that are beyond the light line. Finally, the photonic dispersion can also be obtained by angle-resolved luminescence measurements in the far field: this is the photonic analog of angle-resolved photoemission spectroscopy for electronic systems.

Measuring the dispersion of defect modes can be done in different ways. The dispersion of line-defect modes can be measured analyzing Fabry–Pérot fringes in reflection or transmission through a sample of finite size [85]. It can also be obtained by variable-angle reflectance (for the quasi-guided portion of the mode dispersion) or by ATR (for the guided part of the mode) [86, 87], or else by out-coupling with a grating [88]. Nanocavity modes can be probed in the channel-drop filter geometry, i.e., injecting light into a linear waveguide which is side-coupled to the nanocavity [89]. This technique yields the "loaded" Q-factor, which is the inverse sum of intrinsic cavity Q-factor and loading Q-factor due to coupling to the waveguide, and the coupling Q-factor must be disentangled by means of theoretical calculations in order to finally obtain the intrinsic Q. Nanocavity modes can also be probed by photoluminescence when an internal light source is present, e.g., self-assembled quantum dots in the case of III–V photonic crystals [71–73] or light-emitting defects in the case of silicon slabs [90, 91]. In this case, however, reabsorption by the active layer can affect the measured Q-factor, and the measurement must be performed at low temperature when the light source emits only at low T. A powerful technique for measuring the intrinsic cavity Q-factor at room temperature is given by cross-polarized resonant light scattering [92, 93]: a light beam is focused onto the cavity region through a microscope objective, and the reflected beam is collected by the same microscope objective and analyzed in crossed polarization. As a result, the reflected signal is nearly zero (very low background) except when the incident beam is resonant with the cavity mode, as in this case the cross-polarized reflection is resonantly enhanced. This technique can also be extended to nonlinear optics [67, 69] in order to measure, e.g., second- and third-harmonic generation which is strongly increased thanks to the high Q factor and the low modal cavity volume.

9 Conclusions

Photonic crystals allow to control the propagation of light thanks to the periodicity of the dielectric function in real space. This gives rise to Bragg reflections, or multiple interferences, which lead in turn to the formation of a photonic band structure characterized by allowed and forbidden frequency regions. We have reviewed the main principles related to the photonic band structure in 1D, 2D, 3D, and in waveguide-embedded structures (photonic crystal slabs). The latter combine Bragg reflections in the slab plane with TIR in the vertical direction, and result in a powerful approach to achieve a full control of light propagation in 3D.

A complete PBG for all propagation directions and polarizations can be achieved in 3D photonic crystals with proper structures and refractive index. While a complete PBG has been reported in the near infrared, fully 3D photonic structures are difficult to fabricate and it is nearly impossible to introduce line and point defects in a controlled way. On the other hand, photonic crystals slabs can be fabricated with high accuracy by top-down techniques based on lithography and etching, and any kind of defects can be introduced at the level of the lithographic mask in a controlled way. As a result, linear waveguide modes with very low losses and nanocavity modes with ultra-high Q-factors at wavelengths in the telecom range (near infrared) have been realized. The fabrication procedure of silicon photonic crystal slabs is compatible with CMOS technology, and it is very promising in view of integrating photonic and electronic functionalities on the same chip.

The extreme confinement of light in nanocavities with high Q-factor and low mode volume has opened the possibility of studying a number of exciting phenomena in cavity QED, including control of spontaneous emission, various regimes of radiation–matter interaction, nanolasers, enhanced nonlinear phenomena, single-photon nonlinearities. After more than 25 years since its foundation in 1987, the field of photonic crystals has become a wide and consolidated research area, with many emerging hot topics and new discoveries in the directions of pure science and of prospective applications to optical interconnects and to various ultra-low power photonic devices.

References

1. M. Born, E. Wolf, *Principles of Optics*, 7th edn. (Cambridge University Press, Cambridge, 1999)
2. E. Yablonovitch, Phys. Rev. Lett. **58**, 2059 (1987)
3. S. John, Phys. Rev. Lett. **58**, 2486 (1987)
4. E. Yablonovitch, Photonic crystals: Semiconductors of light. Sci. Am. **285**(6), 47–55 (2001)
5. J.D. Joannopoulos, S.G. Johnson, R.D. Meade, J.N. Winn, *Photonic Crystals: Molding the Flow of the Light*, 2nd edn. (Princeton University, Princeton, 2008)
6. K. Sakoda, *Optical Properties of Photonic Crystals*, 2nd edn. Springer Series in Optical Sciences, vol. 80 (Springer, Berlin, 2001)

7. K. Busch, S. Lölkes, R.B. Wehrspohn, H. Föll, *Photonic Crystals Advances in Design, Fabrication and Characterization* (Wiley-VCH, Weinheim, 2004)
8. J.-M. Lourtioz, H. Benisty, V. Berger, J.-M. Gérard, D. Maystre, A. Tchelnokov, *Photonic Crystals* (Springer, Berlin, 2005)
9. C. Weisbuch, H. Benisty, Photonic crystals, in *Progress in Optics*, vol. 49, ed. by E. Wolf (Elsevier, Amsterdam, 2006), pp. 177–313
10. C. Sibilia, T.M. Benson, M. Marciniak, T. Szoplik (eds.), *Photonic Crystals: Physics and Technology* (Springer, Milano, 2008)
11. M. Barth, A. Gruber, F. Cichos, Phys. Rev. B **72**, 085129 (2005)
12. E. Yeganegi, A. Lagendijk, A.P. Mosk, W.L. Vos, Phys. Rev. B **89**, 045123 (2014)
13. A. Yariv, P. Yeh, *Photonics: Optical Electronics in Modern Communications*, 6th edn. Oxford Series in Electrical and Computer Engineering (Oxford University Press, Oxford, 2006)
14. J.M. Bendickson, J.P. Dowling, M. Scalora, Phys. Rev. E **53**, 4107 (1996)
15. M. Scalora, M.J. Bloemer, A.S. Manka, J.P. Dowling, C.M. Bowden, R. Viswanathan, J.W. Haus, Phys. Rev. E **56**, 3166 (1997)
16. K.O. Hill, G. Meltz, J. Lightwave Technol. **15**, 1263 (1997)
17. J.H. Apfel, Appl. Opt. **20**, 1024 (1981)
18. J.H. Apfel, Appl. Opt. **21**, 733 (1982)
19. G. Panzarini, L.C. Andreani, A. Armitage, D. Baxter, M.S. Skolnick, V.N. Astratov, J.S. Roberts, A.V. Kavokin, M.R. Vladimirova, M.A. Kaliteevski, Phys. Rev. B **59**, 5082 (1999)
20. G. Panzarini, L.C. Andreani, A. Armitage, D. Baxter, M.S. Skolnick, V.N. Astratov, J.S. Roberts, A.V. Kavokin, M.R. Vladimirova, M.A. Kaliteevski, Phys. Solid State **41**, 1223 (1999)
21. S. Reitzenstein, C. Hofmann, A. Gorbunov, M. Strauss, S.H. Kwon, C. Schneider, A. Löffler, S. Höfling, M. Kamp, A. Forchel, Appl. Phys. Lett. **90**, 251109 (2007)
22. N. Valappil, M. Luberto, V.M. Menon, I. Zeylikovich, T.K. Gayen, J. Franco, B.B. Das, R.R. Alfano, Photonics Nanostruct. Fundam. Appl. **5**, 184 (2007)
23. L. Frezza, M. Patrini, M. Liscidini, D. Comoretto, J. Phys. Chem. C **115**, 19939 (2011)
24. G. Canazza, F. Scotognella, G. Lanzani, S. De Silvestri, M. Zavelani-Rossi, D. Comoretto, Laser Phys. Lett. **11**, 035804 (2014)
25. G. Björk, S. Machida, Y. Yamamoto, K. Igeta, Phys. Rev. A **44**, 669 (1991)
26. S. Ciancaleoni, P. Mataloni, O. Jedrkiewicz, F. De Martini, J. Opt. Soc. Am. B **14**, 1556 (1997)
27. C. Dill, R.P. Stanley, U. Oesterle, D. Ochoa, M. Ilegems, Appl. Phys. Lett. **73**, 3812 (1998)
28. F.S. Juang, L.H. Laih, C.J. Lin, Y.J. Hsu, Jpn. J. Appl. Phys. **41**, 2787 (2002)
29. J. Schilling, R.B. Wehrspohn, A. Birner, F. Müller, R. Hillebrand, U. Gösele, S.W. Leonard, J.P. Mondia, F. Genereux, H.M. van Driel, P. Kramper, V. Sandoghdar, K. Busch J. Opt. A Pure Appl. Opt. **3**, S121 (2001)
30. P. Russell, Science **299**, 358 (2003)
31. W. Su, S. Lou, H. Zou, B. Han, Infrared Phys. Technol. **63**, 62 (2014)
32. P. Song, Optik **125**, 1330 (2014)
33. C. Markos, I. Kubat, O. Bang, Sci. Rep. **4**, 6057 (2014)
34. Z.V. Vardeny, A. Nahata, A. Agrawal, Nat. Photonics **7**, 177 (2013)
35. K.M. Ho, C.T. Chan, C.M. Soukoulis Phys. Rev. Lett. **65**, 3152 (1990)
36. E. Yablonovitch, T.J. Gmitter, K.M. Leung Phys. Rev. Lett. **67**, 2295 (1991)
37. M. Hermatschweiler, A. Ledermann, G.A. Ozin, M. Wegener, G. von Freymann, Adv. Funct. Mater. **17**, 2273 (2007)
38. K.M. Ho, C.T. Chan, C.M. Soukoulis, R. Biswas, M. Sigalas, Solid State Commun. **89**, 413 (1994)
39. S. Ogawa, M. Imada, S. Yoshimoto, M. Okano, S. Noda, Science **305**, 227 (2004)
40. A. Blanco, E. Chomski, S. Grabtchak, M. Ibisate, S. John, S.W. Leonard, C. Lopez, F. Meseguer, H. Miguez, J.P. Mondia, G.A. Ozin, O. Toader, H.M. van Driel, Nature **405**, 437 (2000)

41. A.F. Koenderink, L. Bechger, H.P. Schriemer, A. Lagendijk, W.L. Vos, Phys. Rev. Lett. **88**, 143903 (2002)
42. T.F. Krauss, R.M. de la Rue, S. Brand, Nature **383**, 699 (1996)
43. J.H. Wuelbern, M. Schmidt, M. Eich, U. Huebner, R. Boucher, F. Marlow, W. Volksen, Appl. Phys. Lett. **91**, 221104 (2007)
44. X. Gan, H. Clevenson, C.-C. Tsai, L. Li, D. Englund, Sci. Rep. **3**, 2145 (2013)
45. L.C. Andreani, M. Agio, IEEE J. Quantum Electron. **38**, 891 (2002)
46. L.C. Andreani, D. Gerace, Phys. Rev. B **73**, 235114 (2006)
47. M. Galli, A. Politi, M. Belotti, D. Gerace, M. Liscidini, M. Patrini, L.C. Andreani, M. Miritello, A. Irrera, F. Priolo, Y. Chen, Appl. Phys. Lett. **88**, 251114 (2006)
48. R. Lo Savio, M. Galli, M. Liscidini, L.C. Andreani, G. Franz, F. Iacona, M. Miritello, A. Irrera, D. Sanfilippo, A. Piana, F. Priolo, Appl. Phys. Lett. **104**, 121107 (2014)
49. L. Carroll, D. Gerace, I. Cristiani, S. Menezo, L.C. Andreani, Opt. Express **21**, 21556 (2013)
50. L. Carroll, D. Gerace, I. Cristiani, L.C. Andreani, Opt. Express **22**, 14769 (2014)
51. L.C. Andreani, A. Bozzola, P. Kowalczewski, M. Liscidini, Sol. Energy Mater. Sol. Cells (2014). doi:10.1016/j.solmat.2014.10.012
52. D. Gerace, L.C. Andreani, Opt. Lett. **29**, 1897 (2004)
53. D. Gerace, L.C. Andreani, Opt. Express **13**, 4939 (2005)
54. L.C. Andreani, D. Gerace, Phys. Status Solidi (b) **244**, 3528 (2007)
55. S. Hughes, L. Ramunno, J.F. Young, J.E. Sipe, Phys. Rev. Lett. **94**, 033903 (2005)
56. M. Patterson, S. Hughes, The disorder problem for slow-light photonic crystal waveguides, in *Optical Properties of Photonic Structures: Interplay of Order and Disorder*, ed. by M.F. Limonov, R.M. De La Rue. (Taylor and Francis, CRC Press, 2012), pp. 88–131
57. S. Mazoyer, J.P. Hugonin, P. Lalanne, Phys. Rev. Lett. **103**, 063903 (2009)
58. Y. Akahane, T. Asano, B.S. Song, S. Noda, Nature **425**, 944 (2003)
59. M. Minkov, V. Savona, Sci. Rep. **4**, 5124 (2014)
60. Y. Lai, S. Pirotta, G. Urbinati, D. Gerace, M. Minkov, V. Savona, A. Badolato, M. Galli, Appl. Phys. Lett. **104**, 241101 (2014)
61. B.S. Song, S. Noda, T. Asano, Y. Akahane, Nat. Mat. **4**, 207 (2005)
62. M. Notomi, A. Shinya, A. Mitsugi, S. Kuramochi, H.Y. Ryu, Opt. Express **12**, 551 (2004)
63. E. Kuramochi, M. Notomi, S. Mitsugi, A. Shinya, T. Tanabe, T. Watanabe, Appl. Phys. Lett. **88**, 041112 (2006)
64. H. Sekoguchi, Y. Takahashi, T. Asano, S. Noda, Opt. Express **22**, 916 (2014)
65. M. Notomi, Rep. Prog. Phys. **73**, 096501 (2010)
66. Y. Takahashi, H. Hagino, Y. Tanaka, B.S. Song, T. Asano, S. Noda, Opt. Express **15**, 17206 (2007)
67. M. Galli, D. Gerace, K. Welna, T.F. Krauss, L. O'Faolain, G. Guizzetti, L.C. Andreani, Opt. Express **18**, 26613 (2010)
68. L.C. Andreani, P. Andrich, M. Galli, D. Gerace, L. O'Faolain, T.F. Krauss, Nonlinear optics in silicon photonic crystal nanocavities, in *Optical Properties of Photonic Structures: Interplay of Order and Disorder*, ed. by M.F. Limonov, R.M. De La Rue (Taylor and Francis, CRC Press, 2012), pp. 361–378
69. F. Priolo, T. Gregorkiewicz, M. Galli, T.F. Krauss, Nat. Nanotechnol. **9**, 19 (2014)
70. M. Fujita, S. Takahashi, Y. Tanaka, T. Asano, S. Noda, Science **308**, 1296 (2006)
71. W.-H. Chang, W.-Y. Chen, H.-S. Chang, T.-P. Hsieh, J.-I. Chyi, T.-M. Hsu, Phys. Rev. Lett. **96**, 117401 (2006)
72. S. Strauf, K. Hennessy, M.T. Rakher, Y.-S. Choi, A. Badolato, L.C. Andreani, E.L. Hu, P.M. Petroff, D. Bouwmeester, Phys. Rev. Lett. **96**, 127404 (2006)
73. K. Hennessy, A. Badolato, M. Winger, D. Gerace, M. Atatüre, S. Gulde, S. Fält, E.L. Hu, A. Imamoglu, Nature **445**, 896 (2007)
74. S. Ferrett, D. Gerace, Phys. Rev. B **85**, 033303 (2012)
75. A. Majumdar, D. Gerace, Phys. Rev. B **87**, 235319 (2013)
76. E.M. Purcell, Phys. Rev. **69**, 681 (1946)

77. M.D. Birowosuto, H. Sumikura, S. Matsuo, H. Taniyama, P.J. van Veldhoven, R. Nötzel, M. Notomi, Sci. Rep. **2**, 321 (2012)
78. R. Lo Savio, M. Miritello, A. Shakoor, P. Cardile, K. Welna, L.C. Andreani, D. Gerace, T.F. Krauss, L. O'Faolain, F. Priolo, M. Galli, Opt. Express **21**, 10278 (2013)
79. E. Yablonovitch, T.J. Gmitter, Phys. Rev. Lett. **63**, 1950 (1989)
80. M. Galli, D. Bajoni, F. Marabelli, L.C. Andreani, L. Pavesi, G. Pucker, Phys. Rev. B **69**, 115107 (2004)
81. M. Notomi, T. Tamamura, Y. Ohtera, O. Hanaizumi, S. Kawakami, Phys. Rev. B **61**, 7165 (2000)
82. V.N. Astratov, D.M. Whittaker, I.S. Culshaw, R.M. Stevenson, M.S. Skolnick, T.F. Krauss, R.M. De La Rue, Phys. Rev. B **60**, R16255 (1999)
83. M. Galli, M. Agio, L.C. Andreani, M. Belotti, G. Guizzetti, F. Marabelli, M. Patrini, P. Bettotti, L. Dal Negro, Z. Gaburro, L. Pavesi, A. Lui, P. Bellutti, Phys. Rev. B **65**, 113111 (2002)
84. M. Galli, D. Bajoni, F. Paleari, M. Patrini, G. Guizzetti, D. Gerace, M. Agio, L.C. Andreani, D. Peyrade, Y. Chen, IEEE J. Sel. Areas Commun. **23**, 1402 (2005)
85. M. Notomi, K. Yamada, A. Shinya, J. Takahashi, C. Takahashi, I. Yokohama, Phys. Rev. Lett. **87**, 253907 (2001)
86. M. Galli, M. Belotti, D. Bajoni, M. Patrini, G. Guizzetti, D. Gerace, M. Agio, L.C. Andreani, Y. Chen, Phys. Rev. B **70**, 081307(R) (2004)
87. M. Galli, D. Bajoni, M. Patrini, G. Guizzetti, D. Gerace, L.C. Andreani, M. Belotti, Y. Chen, Phys. Rev. B **72**, 125322 (2005)
88. N. Le Thomas, R. Houdré, L.H. Frandsen, J. Fage-Pedersen, A.V. Lavrinenko, P.I. Borel, Phys. Rev. B **76**, 035103 (2007)
89. Y. Akahane, T. Asano, B.S. Song, S. Noda, Appl. Phys. Lett. **83**, 1512 (2003)
90. R. Lo Savio, S.L. Portalupi, D. Gerace, A. Shakoor, T.F. Krauss, L. O'Faolain, L.C. Andreani, M. Galli, Appl. Phys. Lett. **98**, 201106 (2011)
91. R. Lo Savio, M. Miritello, A. Shakoor, P. Cardile, K. Welna, L.C. Andreani, D. Gerace, T.F. Krauss, L. OÉaolain, F. Priolo, M. Galli, Opt. Express **21**, 10278 (2013)
92. M.W. McCutcheon, G.W. Rieger, I.W. Cheung, J.F. Young, D. Dalacu, S. Frédérick, P.J. Poole, G.C. Aers, R.L. Williams, Appl. Phys. Lett. **87**, 221110 (2005)
93. M. Galli, S.L. Portalupi, M. Belotti, L.C. Andreani, L. O'Faolain, T.F. Krauss, Appl. Phys. Lett. **94**, 071101 (2009)

Organic Opals: Properties and Applications

Juan F. Galisteo-López, Luz K. Gil, Marta Ibisate, and Cefe López

Abstract The potential of organic materials in the field of photonics, from polymeric to carbonaceous systems, can be enhanced by providing them with a submicrometer structuration. This can strongly affect light–matter interaction within the material, add structural color, or allow for a tailored porosity at the micro and nanoscale. The latter would pave the way for a number of applications ranging from biosensors to lithium-ion batteries. In this sense self-assembled artificial opals have been eagerly explored over the past two decades. In the present chapter we provide a comprehensive account on the state of the art of artificial opals made from organic materials. After introducing the main materials used in the field we describe the properties, both structural and optical, of organic opals which makes them highly relevant from the point of view of applications. Finally, we list a number of potential uses which are being currently explored for these materials in different fields.

Keywords Growth • Self-assembly • Opal like • Structural color • Biology • Synthesis • Tunable reflectors • Functional materials

1 Introduction

The appeal of organic materials in the field of photonics stems from both their optical properties (large transparency window, photoluminescence, non-linear optical properties, etc.) and their mechanical ones (flexibility, etc.) and the possibility of low cost mass production. A paradigm of how many of these properties can be

J.F. Galisteo-López
Instituto de Ciencia de Materiales de Madrid (ICMM-CSIC), c/Sor Juana Inés de la Cruz 3, 28049 Madrid, Spain

Instituto de Ciencia de Materiales de Sevilla (ICMS-CSIC), c/Américo Vespucio 49, 49092 Sevilla, Spain
e-mail: Juan.Galisteo@csic.es

L.K. Gil • M. Ibisate • C. López (✉)
Instituto de Ciencia de Materiales de Madrid (ICMM-CSIC), c/Sor Juana Inés de la Cruz 3, 28049 Madrid, Spain
e-mail: Luzkar12@icmm.csic.es; ibisate@icmm.csic.es; c.lopez@csic.es

D. Comoretto (ed.), *Organic and Hybrid Photonic Crystals*,
DOI 10.1007/978-3-319-16580-6_2

exploited in photonics is certainly the organic light emitting diode. On the other hand, the potential of combining organic and inorganic systems into a hybrid material brings together the best of both worlds. Two examples of the latter are dye-doped silica matrices (which offer a broadband luminescent solid transparent matrix) or flexible polymeric films containing nanometric metal particles (acting as plasmonic resonators able to concentrate the electromagnetic field in sub-λ volumes).

But organic materials are not restricted to polymeric matrices. Carbonaceous materials (CM), defined as a material composed of carbon and hydrogen atoms (though there could be other heteroatoms as nitrogen, oxygen, metals, and so on incorporated into the carbon backbone), are an excellent representative of organic and hybrid material. The plethora of carbonaceous forms and their different attributes (chemical and physical stability, electronic and heat conductivity, allotropes diversity, and the possibility of a rational control of its chemical connectivity and nanostructure) make CM an attractive material in many technological and fundamental research fields. Some examples are separation and purification, gas storage, catalysis support, drug delivery, supercapacitors or Li-ion batteries and photonic materials.

Photonic crystals (PhC), materials where the introduction of a periodic modulation of its refractive index on the order of the wavelength of light allows for a strong control on light generation and transport, have greatly extended the potential of organic and hybrid materials in photonics over the past two decades. Artificial opals [1], the most popular approach to obtain three-dimensional (3D) PhC by means of the self-assembly of submicrometer colloidal particles, represent a clear example of the latter. In this chapter we present an overview of how the integration of organic materials and artificial opals over the past two decades has served both to achieve incredibly complex photonic nanostructures and to boost the functionality of organic materials in fields ranging from photonics to photovoltaics or biology.

The chapter focuses on both materials aspects of organic opals and potential applications being explored nowadays. Within the former we present a list of organic and hybrid materials which are being used as components of artificial opals, as well as processing techniques currently employed to achieve complex nanostructures which permit tailoring light–matter interaction. After introducing the general properties of artificial opals, both structural and optical, we discuss novel applications of organic opals beyond the well-established areas of spontaneous emission modification, chemical and biosensors and photovoltaics.

2 Materials

Organic materials and artificial opals have been linked from the early days of the field of PhC. As a matter of fact, two of the seminal works on artificial opals already followed the two most common approaches to combine opals and organics: using polymeric spheres to fabricate a template which could be further infiltrated with other materials [2] and profiting from the large porosity of opals to infiltrate carbon and produce a highly interconnected organic matrix [3]. Ever since, the range of organic materials employed either as template or backbone has exponentially grown, boosting the potential of organic opals.

2.1 Organic Materials

One of the most extended methods to fabricate artificial opals involves the use of polystyrene (PS) and poly(methyl methacrylate) (PMMA) beads as building blocks [4]. Polymerization processes are involved in the preparation of these beads under different approaches such as precipitation polymerization [5], emulsion polymerization [6], dispersion polymerization [7], seeded polymerization [8], inversion emulsification [9], swelling polymerization, suspension polymerization [10], and modifications on them. The insulating nature and low refractive index of these polymers implies that in most cases they are used as a template to prepare scaffolds which can be further infiltrated with other materials to create more complex structures. An alternative to these conventional polymers is the use of microgel spheres. Poly(*N*-isopropylacrylamide) (PNIPAM) is the most studied water-swellable microgel system [11]. In this polymer water behaves as an appropriate solvent through hydrogen bonding with the amide groups at room temperature. This hydrogen bonding with water is increasingly disrupted with temperature causing water to act as a poorer solvent leading to gradual chain collapse. Opals fabricated with this kind of particles are sensitive to humidity and temperature, due to the polymer's volume phase transition at ~32 °C, so that the particle size decreases at high temperatures [12, 13].

Recently highly monodisperse conducting polymer particles have been prepared using π-conjugated alternating copolymers constituted by fluorene and thiophene units [14]. These particles show enhanced photocarrier lifetimes which makes them very interesting in the preparation of colloidal crystals for optoelectronic applications. On the other hand, conductive polymers can also be introduced in photonic crystals as a constituent of hybrid particles. The incorporation of poly(*p*-phenylenevinylene) (PPV) in mesoporous silica beads yields particles with high fluorescence quantum yield and an improved resistance to the photobleaching compared to PPV [15]. The control in the particle size allows preparing fluorescent opals in which the photonic band gap overlaps with the fluorescence of PPV making them a good playground to study the modification of spontaneous emission.

Beside the above-mentioned approaches, the most widespread strategy to prepare fluorescent organic particles is to dope them with organic dyes or quantum dots. The synthesis of the polymer particles in the presence of dyes allows obtaining not only fluorescent particles over the entire visible range but also multicolor particles through Förster resonance energy transfer (FRET) using several families of dye molecules [16].

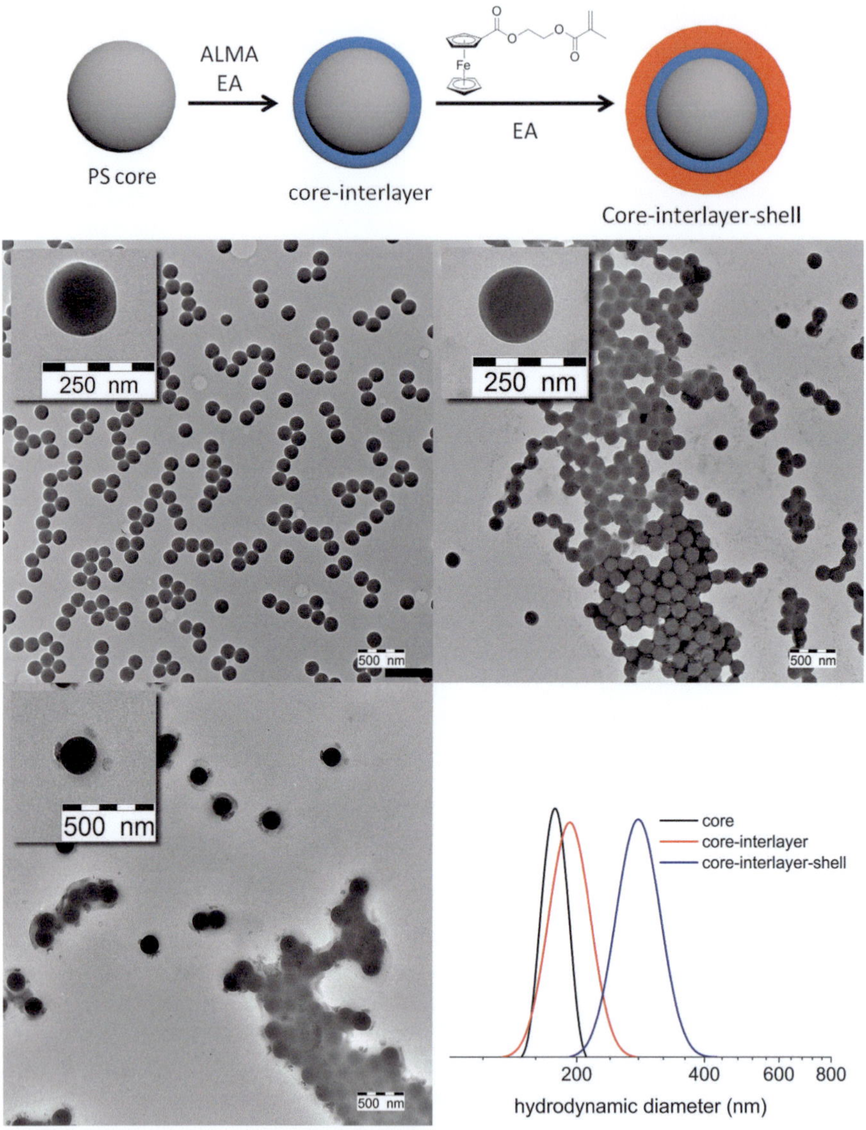

Fig. 1 Scheme of built-up particle with a metallopolymer shell (*top*), TEM images of each stage of the synthesis and the stepwise particle growth obtained by using DLS measurements. Reproduced with permission from [17]

Another way to add functionalities to polymeric nanospheres is the fabrication of core-shell particles. This is usually done in such a way that the core particle provides a high monodisperse particle and the functionality is located on the shell. By using this kind of particles redox- and mechano-chromic response photonic crystals can be fabricated by the melt-shear technique (see Fig. 1). The growth of a metallopolymer shell on PS particles provides this system with a reversible response to redox and mechanical stimuli [17]. The oxidation of the ferrocene moieties induces an increase in the lattice constant of the PhC that affects the opal photonic band gap and which can be optically monitored via spectroscopic studies. The reader can find more information on the engineering of spherical colloids in the chapter by Sparnacci et al.

Other organic materials of interest for optical applications which have been recently incorporated within artificial opals are liquid crystals (LCs) and biopolymers such as DNA. DNA-lipid complexes [18, 19] are extremely attractive for applications in photonics as they are soluble in organic solvents which makes them amenable to be processed as films or fibers. They are transparent over a broad spectral range comprising the visible and unlike most polymers show high thermal stability [20]. DNA-lipid complexes have demonstrated to be a useful host material for dye molecules. Their hierarchic structure allows large dye loading before concentration quenching takes place. This has permitted to observe amplified spontaneous emission in thin films doped with individual dyes and with FRET pairs [21, 22]. These organic complexes can be easily incorporated into ordered and random self-assembled photonic structures [23] which provide an optical feedback environment where resonant random lasing emission has been recently demonstrated [24, 25].

Liquid crystals incorporation in opals makes them thermoresponsive due to the phase change which LCs undergo as the temperature is changed. With inverse opal films based on LCs elastomers and nematic LCs it is possible to modify the phase transition temperature changing the LCs molar ratio thus observing a different behavior in their photonic properties [26]. The thermal response observed for the optimal LCs ratio was a large Bragg peak blue shift of more than 100 nm.

Finally, coordination polymers (CPs) are becoming relevant materials due to their applications in molecular storage, catalysis, sensing, luminescence, electrical conductivity, and magnetism. These applications are determined by the polymer structure, consisting of a coordination compound with repeating coordination entities extending in 1, 2, or 3 dimensions [27]. Porous CP inverse opals are prepared by a coordination replication method. Pseudomorphic replacement process introduces organic elements into a previously fabricated metal oxide inverse opal replicating the structure. These materials show a hierarchical porosity, combining the microporosity of the porous CP and the macroporosity of the inverse opal structure and therefore improving performance for certain applications, as catalysis and molecular storage [28].

2.2 *Carbon-Based Materials*

Organic spheres (particularly PS and PMMA) are acknowledged as ideal building blocks to fabricate 3D ordered arrays with periodicities ranging from a few hundred nanometers to several micrometers, and thus amenable to present their operating wavelength spanning a large spectral range covering from the UV to the NIR. One issue when using such polymeric structures to fabricate composite or inverse structures is that the synthesis of several materials of technological relevance (such as silicon or germanium) involves processes demanding high temperatures for which the polymeric system loses its stability. In this sense, the possibility of using carbonaceous matrices provides the means to overcome these issues [29]. A wide range of carbon spheres (CSs) [30] can be used as building blocks to fabricate the opal templates, which include solid spheres [31], hollow and core-shell structures with diameters ranging from the nano to the mesoscale. From a synthetic point of view this introduces a major challenge: to achieve highly monodisperse and chemically inert carbonaceous nanospheres with excellent size tunability and high thermal and colloidal stability.

Different microstructures and morphologies of CSs can be synthesized via a number of techniques, and varying in terms of process complexity. One route involves carbon decomposition at high temperature and inert atmosphere (e.g., chemical vapor deposition (CVD), arc-discharge, and laser ablation processes). Traditionally, a gaseous carbon precursor is formed by vaporization, sublimation, or atomization of a solid or liquid source. The CSs formation results from the decomposition, radical formation, and aggregation of the precursor molecules under high temperature. The size of the spheres could be tuned by adjusting the process parameters, such as feed rate, composition of the carrier gas, reaction time, and pyrolysis temperatures [32]. However there are some key issues which need to be clarified in order to get CSs rather than other carbon nanostructures. Laser ablation and arc-discharge are not regarded as efficient methods due to their complexity and high cost as well as involving the use of high-purity graphite sources.

Recently, hydrothermal carbonization treatment (HTC) [33] is drawing attention as it allows the fabrication of high purity, homogenous, smooth-surfaced, monodisperse CSs in a one-pot synthesis using sustainable precursors. Using HTC one can obtain core-shell, hollow-shell [34], or hard structure spheres [31] with micrometer tunable size. Further, an sp^2 hybridized backbone and polar oxygenated functionalities groups [35] can be incorporated allowing a wide range of material morphologies and a functional assembly.

Also employing HTC one can obtain functional CM and hybrids via softly, aqueous conditions (<200 °C), often in an autoclave under self-generated pressure, utilizing renewable carbon precursors, for instance from diverse biomass sources (glucose, xylose, maltose, sucrose, amylopectin, starch) and biomass derivatives (hydroxymethylfurfural and furfural) or phenolic compounds, as well as other inorganic structures [35]. Additionally, adjusting the reaction parameters, such as temperature, concentration of precursor, and reaction time one can control the morphology and size of the CS. Another factor affecting the morphology of the CS is the carbon source employed [36]. For instance, starch produces non-spherical particles and glucose is found to be preferential for monodisperse colloidal CS synthesis.

Hollow or carbon-composite spheres can be fabricated either by CVD or HTC using sacrificial sphere templates, including either a hard template as silica or a soft colloidal template such as polystyrene. Here the template could be introduced into the precursor solutions before the main process and it may act as nuclei for the growth of carbonaceous shell. In the HTC case a soft template such as surfactant assembly is used for adequately tuning the hollow CSs [37].

Polymerization–carbonization approaches allow the colloidal spheres formation which, upon stabilization, drying and pyrolysis can be converted into CSs. As an example, resorcinol–formaldehyde resin polymer can be transformed to CS because its highly cross-linked structure allows to retain the spherical morphology in an annealing step [38].

Highly monodisperse CS from phenol–formaldehyde resins have been obtained by a template-free and surfactant-free method [39] and the resulting colloidal spheres have been employed to form three-dimensional periodic structures. However this method involves many synthetic steps making it highly complex and demanding a control over each step. In contrast fewer synthetic steps are needed in the direct pyrolysis of copolymer particles, more precisely monodisperse melamine–formaldehyde microspheres, although certain issues such as the shrinking rate with temperature have to be dealt with [40].

Another issue to consider when using traditional polymer spheres is the undesired crosslinking between neighboring nanospheres, especially when the particle size is below 500 nm. To avoid this phenomenon a hypercrosslinking strategy is used to prepare monodisperse microporous PS nanospheres [41]. An unreactive protecting skin layer of PS is formed in situ allowed by a hydrothermal-assisted carbonization procedure and then carbonaceous nanospheres with a well-developed microporous network are obtained.

Using sugar as a carbon-precursor it is possible to obtain monodisperse colloidal carbon spheres with a diameter ranging from 160 to 400 nm by a two-step HTC strategy where the nucleation and growth process are separated [36]. Hence, a carbon colloid is produced in a first step to serve as seed in a second step of growth.

Similarly, the PS as seed-template followed by a carbon shell formation by HTC allows a hollow-shell structure formation [34]. Features such as low density, large specific surface area, nanostructured wall, as well as a hierarchical hollow structure make CSs a functional material to be used as "processable voids," where they are involved either as bodies or pores. Finally hollow spheres may also be generated by a template-assisted synthesis using either organic or inorganic sacrificial cores.

While the fabrication of carbon spheres to assemble opal templates has drawn a lot of attention over the past few years, efforts have been also directed to obtaining its negative replica: i.e., carbon inverse opals (CIO) (see Fig. 2). Generally speaking, an inverse opal can be also seen as a three-dimensionally ordered macroporous solid with voids interconnected through windows at the points where the spheres from the original scaffold touched. The combination of the wide range of pore sizes together with the existence of an ordered porous structure leads to multifunctional materials which not only find use in optics or photonics but also for applications in chromatography, catalysis and as bioactive materials. These structures can be made from several compounds ranging from oxides, metallic or non-metallic to hybrid organic-silicates and polymers.

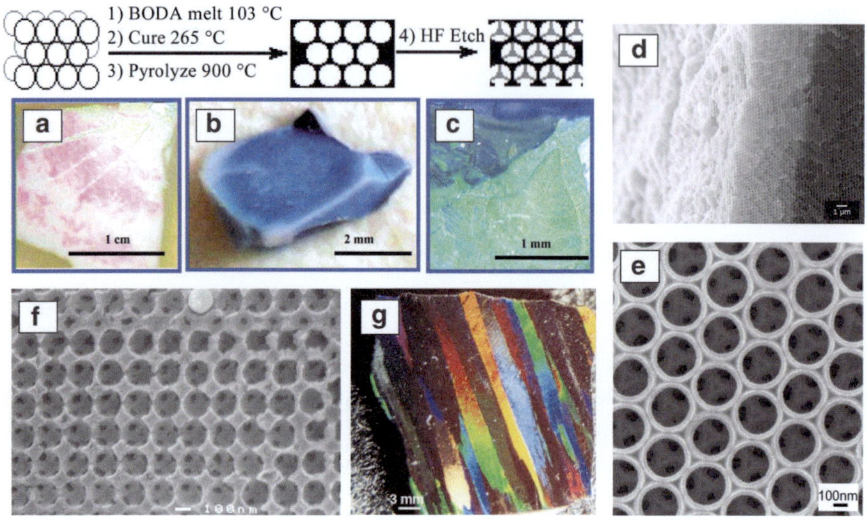

Fig. 2 Different approaches for the fabrication of CIOs. Diagram for CIO produced by in situ polymerization and pyrolysis of BODA monomer using silica template. Optical images show three steps: silica opal (**a**), pyrolyzed carbon/silica composite (**b**), and the resulting CIO (**c**). (**d**) and (**e**) show SEM images of a metal-coated carbon nanocomposite where the carbon scaffold was first coated with hafnium by ALD followed by tungsten coating by CVD. (**d**) shows a cross-section of CIO monolith (*dark layer*) with a coating of hafnium oxide (*bright layer*). (**e**) corresponds to the top side of a tungsten-coated PhC. (**f**) Glassy carbon CIO resulting after carbonization of phenolic resin infiltrated opal. The periodic holes arising from intersphere connections indicate that the volume change on pyrolysis introduces porosity in the glassy carbon. (**g**) Graphite inverse opal with 300 nm air spheres produced by CVD presenting bright opalescence. Reproduced with permission from [92] (**a–c**), [43] (**d**, **e**) and [3] (**f**, **g**)

CIOs result from the removal of the original scaffold that was impregnated with a carbon-based material. Many synthesis techniques can be used ranging from laborious and expensive processes, such as CVD or atomic layer deposition (ALD) to chemical infiltration or other wet chemical methods under mild conditions. The surface chemistry of the sphere templates influences the framework formation providing reactivity and structural tuned features.

In the first example of a CIO dating back to 1998 [3], a silica opal was used as template to form a porous carbon structure which could have dielectric and metallic optical properties, specifically the high in-plane conductivity for the cubic graphite phase leaving a plasmon-defined photonic band gap in the infrared region. The CIO was fabricated following mainly two routes: chemical infiltration of a phenolic resin which was then converted to glassy carbon by phenolic-pyrolyzing and CVD to produce graphitic carbon. This led to a highly conducting photonic structure.

Nowadays, it is known that using hard and soft templates generates microporous, mesoporous, or macroporous carbon with either periodic or disordered pore structures and that employing multiple templating techniques allows a hierarchical pore structuring. Moreover, several attempts have been directed to the synthesis, characterization, and application of carbon-based materials acting as CIO. Further, the carbon pore walls can host nanoparticles [42] that would modify its photonic response.

Certain applications of CIO comprise their use as a scaffold for metal photonic crystals [43] useful for thermophotovoltaic applications. Usually, thermophotovoltaic materials operate at high temperatures and keeping the structure up to at least 2,200 °C in argon is required. Therefore, a carbon framework enhances the thermal stability and then a refractory metal can be deposited. The metallic surface is thicker than the skin depth for light penetration into the metal (>50 nm), hence the light interaction with the scaffold is blocked.

Other materials that have been used as carbon precursors in CIO include polymer resins such as poly(furfuryl alcohol) [44] or phenol–formaldehyde [45]. Also direct carbonization of carbon sources such as pitch, sugar, or cellulose are currently used. The carbonization is usually carried out above 900 °C under inert atmosphere and involves the increase of the amount of carbon atoms that gives the structure a higher aromatic character. In this case the size and/or the type of CM are determining facts for controlling the degree of infiltration of the opal template leading to carbon with uniform spherical pores.

Finally the versatility provided by carbon-based materials is remarkable allowing not only the functionalization of the carbon surface in the CIO as well as CS, but also several coating and infiltration processes which modify the index of refraction [46–48].

3 Properties

An artificial opal is an extremely attractive system both from the point of view of fundamental studies and a wide range of applications. The potential of this type of material arises from its topology, comprising an ordered 3D array of monodisperse spheres. Such nanostructure determines both, an exceptional optical response dictated by multiple Bragg diffraction of light and a perfectly periodic distribution of pores of nanometric dimensions. In what follows the main implications from the point of view of its optical and structural properties will be discussed highlighting aspects specific to organic systems.

3.1 Structural

Contrary to colloidal crystals formed by solutions of charged latexes, which can grow in a wide range of crystalline structures, artificial opals tend to grow forming 3D hexagonal structures (face centered cubic, hexagonal, and random close packed). Furthermore, it is the closest-packed (111) planes that tend to grow parallel to the sample surface independent of the growth method, although growth along the (100) direction may take place at the transition between regions with different number of layers [49]. Growth along directions other than the (111) can also be induced directing it with the use of patterned substrates, the so-called colloidal epitaxy [50].

The presence of this symmetry provides artificial opals with a highly porous structure beneficial for the posterior infiltration of the original scaffold with a different material to create a composite or inverse system (upon the removal of the original mold). The existence of a close packed structure yields the largest packing allowed and therefore a highly interconnected array of voids in the case of inverse opals. In particular two parameters define their porosity: the voids of the inverse structure (which can be controlled via the sphere diameter of the opaline scaffold) and the windows connecting them which can be tuned by creating necks between the spheres (either by physical or chemical means) in the original scaffold. This control over the pore structure in inverse systems is essential for some applications such as Lithium-ion batteries (see below).

Recently evidence has been found that samples grown using the widespread vertical deposition method [51] present a non-close-packed structure with a larger pore volume in between the spheres [52, 53]. Although such results were obtained for inorganic opals, it was also corroborated for polymeric systems [54] and associated with a lattice distortion present in this kind of systems.

The structural properties of other types of organic opals such as carbonaceous systems are fundamental for the applications they are envisaged for. One clear example is that of carbon inverse opals for Lithium-ion batteries where not only the porous structure of the opal but of the carbon itself determines the functionality of the final system as explained below.

3.2 *Optical*

From the point of view of their optical properties, polymeric opals have been the most studied organic self-assembled PhC, as absorption present in carbonaceous systems renders them useless for optical applications. While the interest in the optical response of this kind of systems dates back to over four decades ago [55] it is not until their proposal as PhC that a systematic study in terms of their photonic band structure has been carried out.

Their widespread use is due to the fact that polymeric spheres of different nature (PMMA, PS, etc.) can be fabricated over a large diameter range and with low polydispersity (a few percent), essential to achieve ordered arrays with good optical quality. Although colloidal particles with large diameters can be fabricated from inorganic materials such as SiO_2, their high density makes it difficult to fabricate opals from large (above ca. 500 nm) spheres. This is believed to be due to sedimentation making sample growth too fast in the case of bulk opals, and depopulating the colloidal suspension in the case of vertical deposition, both leading to samples with poor crystalline quality.

One of the main limitations of organic opals is their low refractive index contrast limited to ca. 1.6. While this leads to a weak scattering strength that precludes the existence of broad band gaps, it can be compensated by further infiltration of the sample with other materials having a larger refractive index (see Sect. 4) and the inverted topology. Nevertheless, the possibility to fabricate large area samples with a good crystalline quality and a wide range of sphere diameters has made organic artificial opals an ideal test-bench to probe the optical properties of three-dimensional (3D) photonic crystals.

Samples tend to grow with large crystalline domains and a controlled thickness, which has allowed probing their dispersion relation along high symmetry directions in the surroundings of the pseudogap opening as a consequence of Bragg diffraction by (111) planes parallel to the sample surface [4]. The above-mentioned possibility of growing samples with a wide range of lattice parameters has allowed to explore their optical response in the so-called high energy region [56, 57] where multiple Bragg diffraction by several families of lattice planes makes their dispersion relation more complex. This spectral region is characterized by the presence of low dispersion bands where enhanced light–matter interaction is expected as a consequence of the low group velocity light experiences when coupling to such bands [58]. Recently, in the search for a complete characterization of the optical properties of 3D PhC, Fourier (or back-focal) plane imaging has been applied to the study of organic artificial opals and hybrid metallo-dielectric 2D PhC [59, 60]. This technique has proved valuable in order to retrieve both the dispersion relation of the samples and the equifrequency surfaces determining light refraction within PhC (see Fig. 3).

Other key limitation associated with artificial opals is the presence of residual disorder. While the existence of unwanted structural imperfections (associated with the small albeit ubiquitous polydispersity present in submicrometer spheres) has allowed many studies of the optical response of 3D PhC not permitted in samples

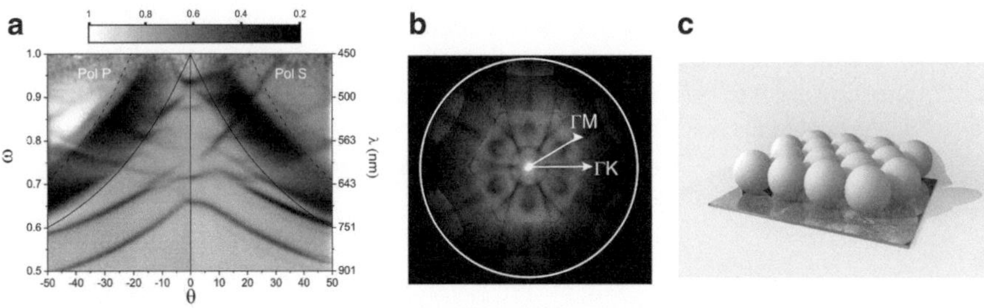

Fig. 3 Optical properties of self-assembled photonic structures. (**a**) Dispersion relations for the two polarizations along the ΓM direction in reciprocal space of a hybrid metallo-dielectric 2D PhC consisting of a hexagonal array of PS spheres deposited on a gold substrate. Results were obtained in reflection configuration by scanning a fiber-coupled spectrophotometer across an enlarged image of the Fourier plane of a microscope objective. (**b**) Equifrequency surface of the same sample retrieved by collecting on a CCD an image of the Fourier plane at a reduced frequency of $\omega = 0.73$. (**c**) Schematic of the hybrid structure

fabricated by other methods, it hampers their use for some applications where losses by diffuse scattering are an issue such as the realization of low-loss waveguiding or high-Q resonators. Nevertheless, the presence of such small defect density renders artificial opals an excellent platform to study the effect of disorder on light transport in 3D PhC described by a combination of Bragg scattering and light diffusion [61]. Further, by using two families of polymeric spheres of different nature but identical diameter, composite opals can be fabricated and, upon further removal of one of the components, 3D PhC with tunable disorder can be achieved [62]. Studying the optical response of these samples has unveiled the dispersive effects, originating in the band structure of the material, in magnitudes describing light propagation in random media such as the scattering mean free path.

4 Processing

The fabrication of organic artificial opals involves self-assembly methods in most cases. In this process polymeric particles form different 3D ordered arrays depending on the particle charge density. Colloidal crystals (CC) arrays are formed from highly concentrated suspensions of charged particles. The addition of photopolymerizable monomers within these CC arrays allows to fabricate a composite CC consisting in a hydrogel matrix which captures the colloidal crystal structure [63]. Different approaches have been used to obtain more dense crystalline structures. One of the first methods used for polymeric particles was centrifugation as their low density prevents the use of the natural sedimentation. Trying to minimize the disorder in the final opal, methods such as confinement [64] and vertical deposition were developed, the latter being the most widely used nowadays (see Fig. 4) [51]. Vertical deposition is based on the evaporation of the solvent of a colloidal suspension forcing the spheres to arrange at the meniscus formed between a vertical substrate, the suspension and air. This method provides opals with high crystalline quality and controlled thickness for a wide range of particle sizes. Crystal thickness increases monotonically with time as solvent evaporates until an equilibrium thickness is reached. This thickness is determined by the meniscus

Fig. 4 (**a**) Schematic of the vertical deposition method. A substrate (*dark grey*) is placed in a colloidal suspension and ordering of the spheres takes place at the meniscus. (**b**) and (**c**) show SEM images of an artificial opal made from a suspension of polystyrene spheres. Scale bars are 4 and 2 μm, respectively

properties which, in turn, depend on temperature, rate of evaporation (ambient humidity and vapor pressure), surface tension of the liquid, and concentration of the colloid.

Polymeric opals tend to grow with a fixed crystalline symmetry (face centered cubic) and mostly with a fixed orientation, having the most compact planes parallel to the sample surface. This is a limitation in terms of the versatility of their optical response, determined by both symmetry and refractive index contrast, but can be circumvented by a number of routes. One is to make use of isotropic plasma etching which can remove most polymeric materials with a nanometric precision. This technique has proved an excellent means to modify the morphology, and its optical response accordingly, of monolayers of polymeric spheres [65, 66] as well as 3D PhC [67].

Due to their high crystalline quality, polymer opals have been used as templates to obtain an ample variety of inverse opals structures [68]. As mentioned above, the low thermal stability of the organic scaffold reduces the number of methods that can be used to infiltrate materials in the structure although it allows to use the temperature as a means to eliminate the scaffold and obtain the replicas.

Diverse techniques can be used to grow materials within the pores of the opaline structure. They can be divided into two different strategies depending on whether the material is synthesized inside the matrix (chemical infiltration) or a previously prepared material is infiltrated in the opal (physical infiltration).

In the physical infiltration case the guest material has to be smaller than the porous size of the matrix. That is why it is used to incorporate dye molecules, quantum dots [69], metallic nanoparticles, etc. [70]. Methods such as dip and spin-coating have been satisfactorily used to obtain, for example, photoluminescent opals [71]. The materials incorporation in previously fabricated opals offers little control over the structure homogeneity and a co-deposition procedure has been used as an alternative method to improve the infiltration process. In this process the opal growth, by dip-coating or vertical deposition, takes place in the presence of the material to be infiltrated [72].

Chemical infiltration requires transporting the reagents into the porous structure where a further reaction takes place. In this way a robust interconnected 3D structure can be grown allowing the fabrication of inverse opals after removing the template.

Wet chemistry methods such as chemical bath deposition [73], sol–gel [74], or electrodeposition [75] allow the growth of a large variety of materials. In the particular case of the electrochemistry deposition different metals [76, 77] and semiconductors can be grown [78] at room temperature. This allows the use of organic opals as templates which are incompatible with some CVD approaches that require high temperatures to obtain high refractive index materials [79]. Nevertheless this drawback can be overcome by protecting the organic opals with an additional inorganic layer at room temperature prior to use in high temperature CVD processes [80, 81].

Room temperature CVD has been used to grow some oxides such as silica [80], and ZnO in PS opals [82] by successive exposure first to tetraethyl orthosilicate or dimethyl zinc and then water vapor (at about nm/s rate). The same procedure is used in the ALD method. In this technique a sequential use of gas phase chemical process allows a highly homogeneous growth of the material within the pores at a much slower rate. In most cases ALD reactions use two precursors. These precursors react with the surface of a material one at a time in a sequential, self-limiting, manner. The slow growth (around 1 angstrom per cycle) that requires several cycles of exposure to the precursors allows a fine control of the degree of opals infiltration [83, 84]. This method is highly satisfactory to infill oxides and nitrides in the opal structure although the higher temperature required for nitride growth excludes the use of organic opals as templates [85].

Beyond the fabrication of artificial opals organic materials can be used to fabricate their inverse replicas. The most general method is the polymerization of organic precursors within colloidal crystal templates. Opals are filled with a liquid monomer, which is polymerized and/or cross-linked by thermal treatment, exposure to UV light, or catalysis. Conventional polymers such as resins [86] or PMMA, PS or PPV inter alia have been used to fabricate organic inverse opals in this way [87, 88]. The polymerization process also allows the fabrication of more complex polymers. For instance, combining nematic and elastomer LCs thermoresponsive inverse opals can be fabricated [26].

It is possible to introduce the polymers in the opal structure by filling colloidal crystals with solutions of preformed polymers. In this manner ferroelectric and conducting polymers have been infiltrated into silica opals [89, 90]. In a similar way DNA-lipid complexes dissolved in ethanol have been introduced in the opal structure during the self-assembly process [23].

A stepwise layer-by-layer liquid growth approach was employed to grow CPs in PS opals. Through immersing the opals in a metal salt solution and ligand solution separately and repeatedly, CPs were grown in the structure. The selective etching of PS with THF allows obtaining CPs inverse opals [91].

When considering CM for photonic structures the main issue is the limited availability of suitable and easily processed carbon precursors. Broadly speaking, the conversion of polymers into CM is hindered by their thermal decomposition. Additionally, phenolic resin carbon-precursors display large shrinkage in carbon-phase during pyrolysis, demanding multistep processes and extremely long curing times.

To overcome the large shrinkage issues in CIO fabrication during pyrolysis, some approaches include a rapid in situ polymerization and pyrolysis of bis-*ortho*-diynyl arene monomer. This process yields a conserved dimension structure of the original template [92]. In this case the carbon precursor must exhibit an excellent compatibility with the opal template and a complete polymerization.

5 Applications of Organic Opals

Their unique structural and optical properties have granted organic artificial opals their presence in many applications since the early stages of the field of PhC. Initial proposals were focused on the spatial and spectral redistribution of the density of photonic states within these periodic arrays to modify light emission of internal sources (15th, 17th, and 21st chapters are related to this subject), enhance light–matter interaction for applications such as solar-cells, or benefit from structural color changes under external stimuli to use them as sensors (see 19th chapter). But recent years have seen a renewed interest in self-assembled organic PhC both for their optical response and for their porous structure. In this section we highlight some of the directions which are being more actively explored nowadays.

5.1 Templating

Beyond their use as photonic materials, organic artificial opals may serve as templates to further fabricate more complicated optical structures. While the most popular approaches within the photonic community are those of composite and inverse opals, where the original organic scaffold may be further infiltrated with another (organic or inorganic) material with different optical properties, a plethora of approaches has emerged over the past decades to employ these self-assembled structures as cost-effective large area templates.

Introduced over three decades ago [93], colloidal lithography is still actively investigated nowadays as a means to achieve nanostructures with ever more complicated morphologies. In this procedure, a monolayer of colloidal spheres acts as either deposition or etching mask to lithographically pattern an underlying layer. By depositing a material on top of the monolayer and further removing the colloidal mask, either by chemical or physical methods, different structures with nanometer control can be fabricated. Depending on the deposition technique, the modification of the colloidal mask (i.e., by using oxygen plasma etching) or the orientation between mask and deposition direction, structures of different morphology can be achieved. A posterior treatment of the grown structures can further fine-tune their shape.

Such control at the nanoscale of the lithographically fabricated structures becomes particularly relevant for the case of plasmonic materials, where the geometry of metallic nanostructures strongly determines their optical response. The potential of colloidal lithography to tailor the optical response of plasmonic nanostructures deposited over large areas has long been explored [94] as well as their use in different fields such as sensing or surface enhanced Raman scattering

(SERS). Over the last years variations of the colloidal lithography technique have been proposed which introduce the possibility of fabricating plasmonic structures only accessible via expensive and time demanding approaches such as focused ion beam or electron-beam lithography. One example is hole-mask lithography [95] where the colloidal array is used as the template to generate a metallic mask of nano-holes through which further metal evaporation is conducted and different structures with tailored optical response such as split-ring resonators can be fabricated [96].

While SiO_2 spheres, produced over a large diameter range and with low polydispersity, can also be used, polymeric ones present a number of advantages. These include the possibility of fabricating large area surfaces with spheres up to several micrometers in size and the ability to fine-tune the colloidal mask employing oxygen plasma etching.

Recently there have been several advances in using the 2D periodicity of monolayers of organic colloidal particles to generate 3D ordering of photonic structures. Some examples include the use of 2D arrays as a mask to produce silicon 3D PhC by sequential passivation reactive ion etching [97] or the use of 2D colloidal monolayers as a phase mask (see Fig. 5) [98]. Also the scattering pattern of individual polystyrene colloids has been used to produce micrometric structures with complex 3D nanostructure [99].

While the potential of the different versions of colloidal lithography listed above is evident in the fabrication of photonic systems of distinct nature, their true applicability in mass production processes will demand an improvement in the optical quality of the starting 2D array. The latter, although capable to reach cm^2 dimensions, lacks a long-range ordering in most occasions and tends to present different types of imperfections (grain boundaries, dislocations, etc.) which may limit some applications.

5.2 *Photonics*

Structural color arising from its photonic band structure is intrinsic to artificial opals and the starting point for many applications. As mentioned at the beginning of this chapter such color depends on both the exact symmetry of the lattice and the refractive indices of the materials comprising the artificial opal. If either structural or refractive index changes can be imparted to the opal, the possibility to change its chromatic properties can open the door to a set of applications, especially if such changes can be carried out in a continuous, controlled, and eventually reversible manner. This tunability, coming from the response of active materials to external stimuli (physical: pressure, temperature, applied voltage, etc. or chemical: pH, ion concentration, etc.), is the basic principle of most sensing applications (see 14th

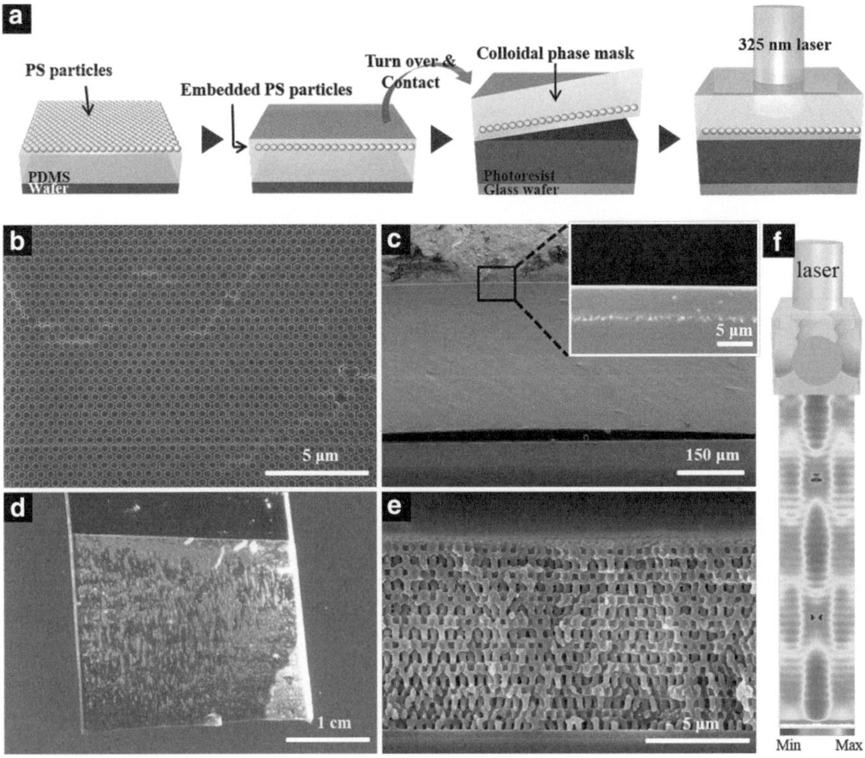

Fig. 5 (**a**) Diagram of the fabrication procedure of a phase mask consisting of a 2D array of polymer spheres embedded in a polymeric matrix and further illumination to create a 3D ordered array. (**b**) SEM image of a PS array on a PDMS substrate. (**c**) Cross-sectional image of the phase mask. (**d**) Shows an optical image of the phase mask and (**e**) a detail of the 3D ordered structure created with the mask. (**f**) Shows a simulation of the light intensity profile within the polymer film below the phase mask. Reproduced with permission from [98]

chapter) but it is also essential for other uses of photonic crystals. Here organic and hybrid organic–inorganic materials have played an important role as stimuli-responsive materials.

One example is the use of hybrid organic–inorganic composite opals as display elements for electronic papers. One of the pioneering approaches is that of Photonic-Ink (P-Ink) [100], consisting of an electroactive polymer network embedded in a silica opal which undergoes volume changes under small external voltages. These changes modify the separation between planes of spheres and thus reflection peaks associated with Bragg diffraction can be tuned across the visible spectrum in a continuous manner. Fast response times, low voltage operation and chromatic properties have made these materials competitive with state-of-the-art display technologies [101].

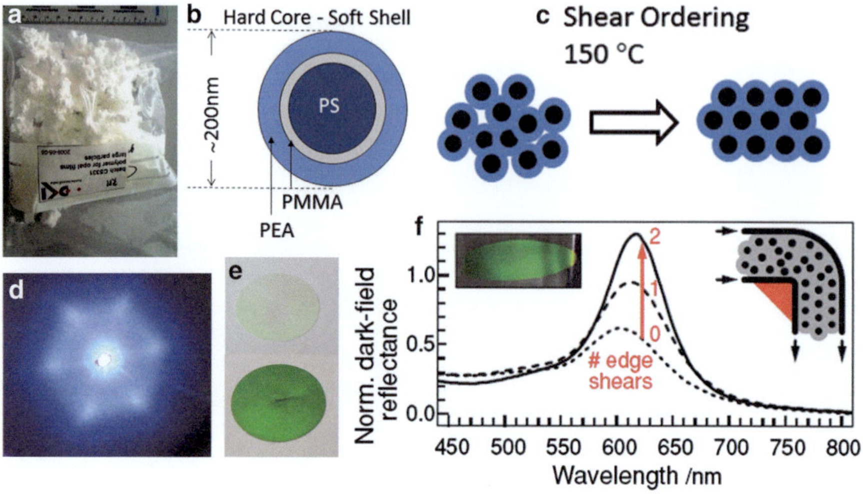

Fig. 6 Fabrication and optical properties of polymer opal films by the combination of melting and shear ordering of core-interlayer-shell organic spheres. (**a**) Shows a bulk batch of as-synthesized spheres and (**b**) a diagram of their composition. (**c**) Schematic of the fabrication procedure. (**d**) Diffraction pattern obtained in transmission with *blue light* for films made from 630 nm diameter spheres. (**e**) Optical photograph of fabricated films showing enhanced color saturation by the addition of 0.05 % weight of carbon nanoparticles. (**f**) *Dark field* reflectance spectra showing increased ordering by edge shearing (*insets* show a typical sample and a schematic of the process). Reproduced with permission from [105]

Other approaches for the application of hybrid artificial opals as display elements have been presented which do not comprise active elements. One example is the infiltration of silica opals with a photocurable polymer which yields free-standing flexible micrometric films showing structural color which can be further patterned using standard photolithographic techniques [102]. Also, PS artificial opals in the shape of domes with tens of micrometers dimensions can be fabricated by means of ink-jet printing technique and used as pixels for nearly 180° viewing angle displays [103]. Structural color can find use beyond display elements, for instance as dye-free pigments to be incorporated in transparent industrial coatings [104]. One necessary step towards any application is that of amenability to mass production. Recently large-scale fabrication of hundreds to thousands of micrometers thick polymer opal films or fibers has been demonstrated through a procedure combining melting and shear ordering of core-shell submicrometer polymeric colloids (see Fig. 6) [105]. Further, the combination with small (ca. 0.05 % wt.) amounts of carbon nanospheres helps increasing the color saturation of the structures by absorption of scattered light [106]. Large-scale production together with stretchability and durability of the samples may bring these systems into applications such as decorative materials or as textiles. Recently conductive opal films were obtained [107] by incorporating single-walled carbon nanotubes into the above samples which could lead the way to applications in optoelectronics.

Other fabrication method which has demonstrated the potential to produce large (wafer-size) samples over short (few minutes) time scales is that of spin-coating. By controlling certain parameters of the coating procedure, such as spin speed and time, organic–inorganic opal composites with different thickness can be fabricated [108]. Additionally, the possibility of controlling the crystal orientation has been demonstrated [109], the main limitation of this technique being the poly-crystalline nature of the fabricated samples, consisting of slightly rotated domains.

Finally, organic artificial opals are being explored as a platform to study all optical switching. While their large third order non-linear susceptibility and ultrafast (femtosecond) response times makes PS, PMMA, or PPV appealing for this use, their low refractive index and the lack of control of self-assembly methods to fabricate high Q cavities or WGs has precluded their application so far. Nevertheless, proofs of principle with PS opals and the modification of their optical response in sub-20 fs times, close to the non-linear response limit, have been presented [110].

5.3 Lithium-Ion Batteries

Certainly, most work on carbon-based opals is being carried out in the field of lithium-ion batteries [111, 112] where carbon is the predominant anode active material in commercial devices. Efforts are being focused at the development of rechargeable batteries with large capacity, high power, and long cycling life. In this direction new types of active materials or modified traditional electrodes are being pursued. CIO have been eagerly explored as porous electrode materials to improve the performance of the batteries.

The porosity of carbonaceous inverse opals can in principle help overcome some of the limitations of graphitic carbon used as anode in commercial Li ion batteries facilitating electrolyte penetration and shortening Li ion diffusion lengths down to the nanometer scale. In one of the first approaches [113] CIOs with large surface area mesoporous walls were fabricated using sol–gel precursors and used as anodes for Li batteries showing good conductivity and electrolyte penetration as well as shortened diffusion path lengths. As a result, improved performance with respect to an unpatterned anode was demonstrated. A CVD approach was alternatively used to fabricate CIO anodes which also presented an enhanced performance [114].

Recent advances in the fabrication of anodes have focused on new fabrication routes to increase the surface area of the electrode [115] or trapping nanoparticles confined within the carbon network in order to improve their capacity [114, 116].

Carbon inverse opals have also found application in other fields such as medicine, where the combination of these structures with Prussian Blue nanoparticles constitute efficient adsorbers for radionuclides in the human body which could outperform commercial products in some cases in terms of their adsorption capacity [117].

5.4 *Biological Applications*

Over the past years the interest in organic and hybrid organic–inorganic artificial opals in the field of biomedical applications has been growing. In this case the biocompatibility and porosity of the material is of fundamental relevance but the requirements (in terms of periodicity, materials, etc.) are quite different from the examples mentioned in the above sections. Some of the more widely explored applications are listed below.

Arguably one of the best known bio-applications of hybrid organic–inorganic self-assembled photonic crystals is that of chemical and biosensors based on hydrogel embedded colloidal crystals or artificial opals. Ever since the first proposal nearly two decades ago [63], hydrogel based PhC have been used as label-free sensors to a large number of biological and chemical stimuli. More details on the working principle as well as some examples can be found on 14th and 18th chapters.

Regenerative medicine is benefitting from the controlled porosity of inverse opals as an appropriate environment for cell growth and differentiation (see [118] and the references therein). In this approach artificial opals made from organic spheres are used as the starting point to then create a highly porous inverse structure using mostly organic and in some cases inorganic or hybrid materials. In this case a porous structure with pores of hundreds of micrometers is needed to allow for cell and nutrient migration. Further, the presence of an ordered array with controlled pore size is essential for reproducibility but also benefits cell migration and differentiation. While clear evidence of superior performance as compared to conventional randomly porous structures has been found, extensive work is being devoted to the topic.

Following different approaches, organic self-assembled PhC have been used over the past few years for DNA detection. In two recent publications, the hybridization of two DNA single strands is monitored in a photonic environment provided by the PhC. In one of them the modified density of states within a photonic crystal is used to control FRET taking place when two dye-labelled single DNA strands hybridize, boosting the sensitivity of the DNA detection two orders of magnitude above conventional fluorophore-based methods (down to 13.5 fM) [119]. On another label-free approach [120], a hydrogel based PhC microcapsule [121] containing a single strand undergoes a volume shrinkage in the presence of the complementary strand. In this system the amount of target DNA can be quantitatively estimated from the spectral shift of the Bragg diffraction peak. The latter approach has been eagerly explored over the past years. While the working principle is identical to other self-assembled PhC based sensors, microcapsules offer several particularities which may render these structures more appropriate for certain applications. These include angle independent response (as their spherical symmetry causes the external viewer to observe (111) planes for every observation geometry) ideal for displays [122] and the possibility of using them as molecularly imprinted self-encoded microcarriers for suspension arrays [123].

6 Outlook

Over the two decades after their proposal as 3D PhC, great advances in the knowledge of the optical and structural properties of organic opals have been made. After the initial stages where the focus was on their use as a test-bench to study light–matter interaction in PhC, an ever growing number of proposals for applications in different fields have appeared. While their use in some areas such as sensing and photovoltaics is already acknowledged, some other fields could also benefit from these materials, from energy storage to medicine or displays.

Most of these applications require the possibility of mass producing samples with reproducible properties and a good crystalline quality. For the case of polymeric opals, advances in fabrication such as the combined use of heating and shear ordering are bringing them from laboratory-sized samples into dimensions like those needed in the industry. Fields such as decoration or displays could be profiting from these advances in recent years. Carbon inverse opals, a material drawing large interest in recent years, are becoming an interesting alternative to plain bulk carbon electrodes in commercial Lithium-ion batteries. While improved performances have been demonstrated, the development of a simple one-pot process for large-scale production is still lacking.

Acknowledgments This work was funded by the EU NoE Nanophotonics for Energy Efficiency (N4E), the Spanish MINECO MAT2012-31659 (SAMAP), and the Comunidad de Madrid project S2009/MAT-1756 (PHAMA). LKG acknowledges COLCIENCIAS for the PhD grant.

References

1. C. López, Adv. Mater. **15**, 1679 (2003)
2. J.E.G.J. Wijnhoven, W.L. Vos, Science **281**, 802 (1998)
3. A.A. Zakhidov, R.H. Baughman, Z. Iqbal, C. Cui, I. Khayrulin, S.O. Dantas, J. Marti, V.G. Ralchenko, Science **282**, 897 (1998)

4. J.F. Galisteo-López, E. Palacios-Lidón, E. Castillo-Martinez, C. López, Phys. Rev. B **68**, 115109 (2003)
5. T. Sowa, T. Watanabe, T. Seguchi, J. Okamoto, J. Polym. Sci. Polym. Chem. Ed. **17**, 111 (1979)
6. I. Piirma (ed.), *Emulsion Polymerization* (Academic, New York, 1982)
7. K.E.J. Barret (ed.), *Dispersion Polymerization in Organic Media* (Wiley, London, 1975)
8. J. Ugelstad, P.C. Mork, H.R. Mfutakamba, E. Soleimani, I. Nordhus, R. Schmid, A. Berge, T. Ellingsen, O. Aune, in *Science and Technology of Colloids*, ed. by G.W. Poehlein, R.H. Ottewill, G.W. Goodwin. NATO ASI Series E, vol. 67 (Martinus Nijhoff, The Hague, 1983), pp. 51–99
9. C. Pichot, C. Graillat, V. Guikhikh, Makromol. Chem. Suppl. 10/11, 199 (1985).
10. E.A. Grulke, Encycl. Polym. Sci. Eng. **19**, 443 (1989)
11. R. Pelton, Adv. Colloid Interface Sci. **85**, 1 (2000)
12. J.G. McGrath, R.D. Bock, J.M. Cathcart, L.A. Lyon, Chem. Mater. **19**, 1584 (2007)
13. Y. Zhang, L. Gao, L. Heng, Q. Wei, H. Yang, Q. Wang, J. Nanosci. Nanotechnol. **13**, 1903 (2013)
14. T. Adachi, L. Tong, J. Kuwabara, T. Kanbara, A. Saeki, S. Seki, Y. Yamamoto, J. Am. Chem. Soc. **135**, 870 (2013)
15. T.L. Kelly, Y. Yamada, C. Schneider, K. Yano, M.O. Wolf, Adv. Funct. Mater. **19**, 3737 (2009)
16. J. Chen, P. Zhang, G. Fang, P. Yi, F. Zeng, S. Wu, J. Phys. Chem. B **116**, 4354 (2012)
17. D. Scheid, C. Lederle, S. Vowinkel, C.G. Schäfer, B. Stühnb, M. Gallei, J. Mater. Chem. C **2**, 2583 (2014)
18. K. Tanaka, Y. Okahata, J. Am. Chem. Soc. **118**, 10679 (1996)
19. L. Wang, J. Yoshida, N. Ogata, Chem. Mater. **13**, 1273 (2001)
20. A.J. Steckl, Nat. Photonics **1**, 3 (2007)
21. Y. Kawabe, L. Wang, S. Horinouchi, N. Ogata, Adv. Mater. **12**, 1281 (2000)
22. M. Ibisate, J.F. Galisteo-López, V. Esteso, C. López, Adv. Opt. Mater. **1**, 651 (2013)
23. M. Leonetti, R. Sapienza, M. Ibisate, C. Conti, C. López, Opt. Lett. **34**, 3764 (2009)
24. P.D. García, M. Ibisate, R. Sapienza, D.S. Wiersma, C. López, Phys. Rev. A **80**, 013833 (2009)
25. J.F. Galisteo-López, M. Ibisate, C. López, J. Phys. Chem. C **118**, 9665 (2014)
26. G. Wu, Y. Jiang, D. Xu, H. Tang, X. Liang, G. Li, Langmuir **27**, 1505 (2011)
27. S. Kitagawa, R. Kitaura, S. Noro, Angew. Chem. Int. Ed. **43**, 2334 (2004)
28. J. Reboul, S. Furukawa, N. Horike, M. Tsotsalas, K. Hirai, H. Uehara, M. Kondo, N. Louvain, O. Sakata, S. Kitagawa, Nat. Mater. **11**, 717 (2012)
29. M.D. Goodman, K.A. Arpin, A. Mihi, N. Tatsuda, K. Yano, P.V. Braun, Adv. Opt. Mater. **1**, 300 (2013)
30. A. Deshmukh, S.D. Mhlanga, N.J. Coville, Mater. Sci. Eng. R Rep. **70**, 1 (2010)
31. Q. Wang, H. Li, L. Chen, X. Huang, Carbon **39**, 2211 (2001)
32. Y. Jin, C. Gao, W. Hsu, Y. Zhu, Carbon N. Y. **43**, 1944 (2005)
33. M.-M. Titirici, M. Antonietti, Chem. Soc. Rev. **39**, 103 (2010)
34. R.J. White, K. Tauer, M. Antonietti, M.-M. Titirici, J. Am. Chem. Soc. **132**, 17360 (2010)
35. J. Ryu, Y.-W. Suh, D.J. Suh, D.J. Ahn, Carbon N. Y. **48**, 1990 (2010)
36. C. Chen, X. Sun, X. Jiang, D. Niu, A. Yu, Z. Liu, J.G. Li, Nanoscale Res. Lett. **4**, 971 (2009)
37. X. Sun, Y. Li, J. Colloid Interface, Science **291**, 7 (2005)
38. J. Liu, S.Z. Qiao, H. Liu, J. Chen, A. Orpe, D. Zhao, G.Q.M. Lu, Angew. Chem. Int. Ed. Engl. **50**, 5947 (2011)
39. J. Zhao, W. Niu, L. Zhang, H. Cai, M. Han, Y. Yuan, S. Majeed, S. Anjum, G. Xu, Macromolecules **46**, 140 (2013)

40. B. Friedel, S. Greulich-Weber, Small **2**, 859 (2006)
41. Y. Ouyang, H. Shi, R. Fu, D. Wu, Sci. Rep. **3**, 1430 (2013)
42. Z. Wang, N.S. Ergang, M.A. Al-Daous, A. Stein, Chem. Mater. **17**, 6805 (2005)
43. P. Nagpal, D.P. Josephson, N.R. Denny, J. DeWilde, D.J. Norris, A. Stein, J. Mater. Chem. **21**, 10836 (2011)
44. Y. Isshiki, M. Nakamura, S. Tabata, K. Dokko, M. Watanabe, Polym. Adv. Technol. **22**, 1254 (2011)
45. K.P. Gierszal, S.B. Yoon, J.-S. Yu, M. Jaroniec, J. Mater. Chem. **16**, 2819 (2006)
46. X. Sun, Y. Li, Angew. Chem. Int. Ed. Engl. **43**, 597 (2004)
47. A. Stein, R. Schroden, Curr. Opin. Solid State Mater. Sci. **5**, 553 (2001)
48. J.F. Galisteo-López, M. Ibisate, R. Sapienza, L.S. Froufe-Pérez, Á. Blanco, C. López, Adv. Mater. **23**, 30 (2011)
49. P.D. García, J.F. Galisteo-López, C. López, Appl. Phys. Lett. **87**, 201109 (2005)
50. A. van Blaaderen, P. Wiltzius, Adv. Mater. **9**, 833 (1997)
51. P. Jiang, J.F. Bertone, K.S. Hwang, V.L. Colvin, Chem. Mater. **11**, 2132 (1999)
52. F. García_Santamaría, P.V. Braun, Appl. Phys. Lett. **90**, 241905 (2007)
53. F. Gallego-Gómez, A. Blanco, V. Canalejas-Tejero, C. López, Small **7**, 1838 (2011)
54. G. Lozano, L.A. Dorado, D. Schinca, R.A. Depine, H. Míguez, Lanmguir **25**, 12860 (2009)
55. P.A. Hiltner, I.M. Krieger, J. Phys. Chem. **73**, 2386 (1968)
56. J.F. Galisteo-López, C. López, Phys. Rev. B **70**, 035108 (2004)
57. H. Míguez, V. Kitaev, G.A. Oin, Appl. Phys. Lett. **84**, 1239 (2004)
58. J.F. Galisteo, M. Galli, A. Balestreri, M. Patrini, L.C. Andreani, C. López, Opt. Exp. **15**, 15342 (2007)
59. J.F. Galisteo-López, M. López-García, A. Blanco, C. López, Langmuir **28**, 9174 (2012)
60. R. Wagner, L. Heerklotz, N. Kortenbruck, F. Cichos, Appl. Phys. Lett. **101**, 081904 (2013)
61. A.F. Koendrink, M. Megens, G. van Soest, W.L. Vos, A. Lagendijk, Phys. Lett. A **268**, 104 (2000)
62. P.D. García, R. Sapienza, L.S. Froufe-Pérez, C. López, Phys. Rev. B **79**, 241109(R) (2009)
63. J.H. Holtz, S.A. Asher, Nature **389**, 829 (1997)
64. S.H. Park, D. Qin, Y. Xia, Adv. Mater. **10**, 1028 (1998)
65. T. Fujimura, T. Tamura, T. Itoh, C. Haginoya, Y. Komori, T. Koda, Appl. Phys. Lett. **78**, 1478 (2001)
66. M. López-García, J.F. Galisteo-López, A. Blanco, C. López, A. García-Martín, Adv. Funct. Mater. **20**, 4338 (2010)
67. G. von Freymann, S. John, V. Kitaev, G.A. Ozin, Adv. Mater. **17**, 1273 (2005)
68. Special Issue on Templated Materials, Chem. Mater. **20**, 599–1190 (2008)
69. P.D. García, Á. Blanco, A. Shavel, N. Gaponik, A. Eychmüller, B. Rodríguez-González, L.M. Liz-Marzán, C. López, Adv. Mater. **18**, 2768 (2006)
70. Y. Tan, W. Qian, S. Ding, Y. Wang, Chem. Mater. **18**, 3385 (2006)
71. I.S. Nikolaev, P. Lodahl, W.L. Vos, J. Phys. Chem. C **112**, 7250 (2008)
72. F. Iskandar, M. Abdullah, H. Yoden, K. Okuyama, J. Appl. Phys. **93**, 9237 (2003)
73. H. Yan, C.F. Blanford, B.T. Holland, W.H. Smyrl, A. Stein, Chem. Mater. **12**, 1134 (2000)
74. B.T. Holland, C.F. Blanford, A. Stein, Science **281**, 538 (1998)
75. P.V. Braun, P. Wiltzius, Nature **402**, 603 (1999)
76. P.N. Bartlett, J.J. Baumberg, P.R. Birkin, M.A. Ghanem, M.C. Netti, Chem. Mater. **14**, 2199 (2002)
77. P.N. Bartlett, J.J. Baumberg, S. Coyle, M.E. Abdelsalam, Faraday Discuss. **125**, 117 (2004)
78. X.D. Meng, R. Al-Salman, J.P. Zhao, N. Borissenko, Y. Li, F. Endres, Angew. Chem. Int. Ed. **48**, 2703 (2009)
79. A. Blanco, E. Chomski, D. Grabtchak, M. Ibisate, S. John, S.W. Leonard, C. López, F. Meseguer, H. Míguez, J.P. Mondia, G.A. Ozin, O. Toader, H.M. van Driel, Nature **405**, 437 (2000)

80. H. Míguez, N. Tétreault, S.M. Yang, V. Kitaev, G.A. Ozin, Adv. Mater. **15**, 597 (2003)
81. A. Blanco, C. López, Adv. Mater. **18**, 1593 (2006)
82. B.H. Juárez, P.D. García, D. Golmayo, A. Blanco, C. López, Adv. Mater. **17**, 2761 (2005)
83. J.S. King, E. Graugnard, C.J. Summers, Adv. Mater. **17**, 1010 (2005)
84. M. Scharrer, X. Wu, A. Yamilov, H. Cao, R.P.H. Chang, Appl. Phys. Lett. **86**, 151113 (2005)
85. A. Rugge, J.S. Becker, R.G. Gordon, S.H. Tolbert, Nanoletter **3**, 1293 (2003)
86. H. Míguez, F.J. Meseguer, C. López, F. López-Tejeira, J. Sánchez-Dehesa, Adv. Mater. **13**, 393 (2001)
87. S.H. Park, Y. Xia, Chem. Mater. **10**, 1745 (1998)
88. M. Deutsch, Y.A. Vlasov, D.J. Norris, Adv. Mater. **12**, 1176 (2000)
89. T.B. Xu, Z.Y. Cheng, Q.M. Zhang, R.H. Baughman, C. Cui, A.A. Zakhidov, J. Su, J. Appl. Phys. **88**, 405 (2000)
90. K. Yoshino, Y. Kawagishi, S. Tatsuhara, H. Kajii, S. Lee, A. Fujii, M. Ozaki, A.A. Zakhidov, Z.V. Vardeny, M. Ishikawa, Microelectron. Eng. **47**, 49 (1999)
91. Y. Wu, F. Li, W. Zhu, J. Cui, C. Tao, C. Lin, P.M. Hannam, G. Li, Angew. Chem. Int. Ed. **50**, 12518 (2011)
92. M.W. Perpall, K.P.U. Perera, J. DiMaio, J. Ballato, S.H. Foulger, D.W. Smith, Langmuir **19**, 7153 (2003)
93. H.W. Deckman, J.H. Dunsmuir, Appl. Phys. Lett. **41**, 377 (1982)
94. C.L. Haynes, R.P. Van Duyne, J. Phys. Chem. B **105**, 5599 (2001)
95. H. Fredriksson, Y. Alaverdyan, A. Dmitriev, C. Langhammer, D.S. Sutherland, M. Züch, B. Kasemo, Adv. Mater. **19**, 4297 (2007)
96. S. Cataldo, J. Zhao, F. Neubrech, B. Frank, C. Zhang, P.V. Braun, H. Giessen, ACS Nano **6**, 979 (2012)
97. A. Vlad, A. Frölich, T. Zebrowski, C.A. Dutu, K. Bush, S. Melinte, M. Wegener, I. Huynen, Adv. Funct. Mater. **23**, 1164 (2013)
98. T.Y. Jeon, H.C. Jeon, S.Y. Lee, T.S. Shim, J.-D. Kwon, S.-G. v, S.-M. Yang, Adv. Mater. **26**, 1422 (2014)
99. X.A. Zhang, J. Elek, C.-H. Chang, ACS Nano **7**, 6212 (2013)
100. A.C. Arsenault, H. Míguez, V. Kitaev, G.A. Ozin, I. Manners, Adv. Mater. **15**, 503 (2003)
101. H. Wang, F. Kerins, U. Kamp, L. Bonifacio, A.C. Arsenault, G.A. Ozin, Inf. Disp. **7–8**, 1–5 (2011)
102. H.S. Lee, T.S. Shim, H. Hwang, S.-M. Yang, S.-H. Kim, Chem. Mater. **25**, 2684 (2013)
103. M. Kuang, J. Wang, B. Bao, F. Li, L. Wang, L. Jiang, Y. Song, Adv. Opt. Mater. **2**, 34 (2013)
104. M. Egen, L. Braun, R. Zentel, K. Tännert, P. Frese, O. Reis, M. Wulf, Macromol. Mater. Eng. **289**, 158 (2004)
105. C.E. Finlayson, J.J. Baumberg, Polym. Int. **62**, 1403 (2013)
106. O.L. Pursiainen, J.J. Baumberg, H. Winkler, B. Viel, P. Spahn, T. Ruhl, Opt. Exp. **15**, 9553 (2007)
107. Y. Imai, C.E. Finlayson, P. Goldberg-Oppenheimer, Q. Zhao, P. Spahn, D.R.R. Snoswell, A.I. Haines, G.P. Hellman, J.J. Baumberg, Soft Matter **8**, 6280 (2013)
108. P. Jiang, M.J. McFarland, J. Am. Chem. Soc. **126**, 13778 (2004)
109. A. Mihi, M. Ocaña, H. Míguez, Adv. Mater. **18**, 2244 (2006)
110. Y. Liu, F. Qin, Z.Y. Wei, Q.B. Meng, D.Z. Zhang, Z.Y. Li, Appl. Phys. Lett. **95**, 1311116 (2009)
111. H. Nishihara, T. Kyotani, Adv. Mater. **24**, 4473 (2012)
112. A.D. Roberts, X. Li, H. Zhang, Chem. Soc. Rev. **43**, 4341 (2014)
113. K.T. Lee, J.C. Lytle, N.S. Ergang, S.M. Oh, A. Stein, Adv. Funct. Mater. **15**, 547 (2005)
114. F. Su, X.S. Zhao, Y. Wang, J. Zeng, Z. Zhou, J.Y. Lee, J. Phys. Chem. B **109**, 20200 (2005)
115. D.-Y. Kang, S.-O. Kim, Y.J. Chae, J.K. Lee, J.H. Moon, Langmuir **29**, 1192 (2013)
116. X. Huang, J. Chen, Z. Lu, H. Yu, Q. Yan, H.H. Hng, Sci. Rep. **3**, 2317 (2013)
117. J.-Y. Hong, W.-K. Oh, K.-Y. Shin, O.S. Kwon, S. Son, J. Jang, Biomaterials **33**, 5056 (2012)
118. Y.S. Zhang, S.-W. Choi, Y. Xia, Soft Matter **9**, 9747 (2013)

119. M. Li, F. He, Q. Liao, J. Liu, L. Xu, L. Jiang, Y. Song, S. Wang, D. Zhu, Angew. Chem. **120**, 7368 (2008)
120. Y. Zhao, X. Zhao, B. Tang, W. Xu, J. Li, J. Hu, Z. Gu, Adv. Funct. Mater. **20**, 976 (2010)
121. J. Wang, J. Zhu, Eur. Polym. J. **49**, 3420 (2013)
122. H. Gu, Y. Zhao, Y. Cheng, Z. Xie, F. Rong, J. Li, B. Wang, D. Fu, Z. Gu, Small **9**, 2266 (2013)
123. Y.-J. Zhao, X.-W. Zhao, J. Hu, J. Li, W.-Y. Xu, Z.-Z. Gu, Angew. Chem. Int. Ed. **48**, 7350 (2009)

Biological Photonic Crystals

Michael Grimann and Thomas Fuhrmann-Lieker

Abstract In this chapter, photonic crystal structures in plants and animals are reviewed. From the perspective of materials science, they can be considered as hybrid nanostructured material systems, comprising inorganic and organic components with appropriate refractive index contrast. General principles of structure formation, the influence of disorder, and possible biological functions are elucidated. Giving examples from the most important and well-investigated taxa, the occurring material combinations and composite structures are presented. Actual trends in biophotonic crystal research that arise interest for applications in materials science are the investigation of photonic glasses and the preparation of natural–artificial hybrid structures with combined properties for the manipulation of light.

Keywords Biological photonic crystals • Structural color • Self-organization • Reflection properties • Diatoms • Birds • Insects • Disorder • Multilayer • Biophotonics • Organic/inorganic composite materials

1 Introduction

For material scientists, it is always useful to look at nature for design principles optimized by a long-term evolutional process and achieved with tailored materials to fulfill their individual task. That is of course also true for photonic crystals since not only in geological minerals like opal but also in the living world, in unicellular organisms, molluscs, arthropods, vertebrates, and plants, periodic structures in the order of the wavelength of light have been realized (Fig. 1). In this chapter we elucidate by few prominent examples the formation principles, the toolbox of material systems, and modern trends in biological photonic crystal research that may help material scientists to adopt them for the realization of artificial and even hybrid device structures.

M. Grimann • T. Fuhrmann-Lieker (✉)
Macromolecular Chemistry and Molecular Materials, Institute of Chemistry and Center for Interdisciplinary Nanostructure Science and Technology, University of Kassel, Heinrich-Plett-Str. 40, 34109 Kassel, Germany
e-mail: michael.grimann@uni-kassel.de; th.fuhrmann@uni-kassel.de

D. Comoretto (ed.), *Organic and Hybrid Photonic Crystals*,
DOI 10.1007/978-3-319-16580-6_3

Fig. 1 Animals with structural colors based on ordered and disordered photonic crystals. (**a**) Sea Mouse *Aphrodita aculeata* (**b**) Abalone *Haliotis* sp. (**c**) Butterfly *Morpho didius*, (**d**) Peacock *Pavo cristatus*. Photo credits: M. Kucki (**a**) [70], I. Lieker (**b**, **d**)

2 Periodic Structures in Biology

2.1 Principles of Periodic Pattern Formation

The question of morphogenesis is one of the great unresolved tasks in developmental biology. D'Arcy Thompson [1] was the first who suggested that pattern formation can be explained by elementary physical laws and therefore mathematical models may be applied. Many of the models invented for the description of periodic pattern formation in various biological and chemical contexts [2–4] can be traced back to the original reaction–diffusion model of Alan Turing [5] who showed that a combination of autocatalytic reactions with unequal diffusion constants of the involved molecules, now usually denoted as morphogenetically active substances, or shortly morphogens, leads to a periodic concentration distribution. The models can convincingly reproduce patterns on various lengthscales, but it proved to be quite difficult to identify the morphogens taking part in a reaction–diffusion system. In fact, periodic stripes that arise at the early embryonal stage of the well-investigated model organism *Drosophila melanogaster* for which a periodic distribution of morphogens as reaction–diffusion eigenmodes under geometrical constraints had been proposed [6] turned out to be defined independently [7–9].

Whereas reaction–diffusion models may apply for larger scale structures, for periodic patterning at a 100 nm–1 μm scale, alternative mechanisms are possible. Similar to reaction–diffusion systems, patterns with periodicity in one dimension may be generated by oscillations in time coupled with deposition steps like the famous Liesegang rings [10]. Internal or external clocks may lead to a layer-by-layer growth of lamellar patterns.

A better order, however, can be achieved if the periodic structure is formed simultaneously. Prerequisite for this kind of periodic morphogenesis is a spontaneous instability which can be relaxed by fluctuations with a distinct periodicity. Such processes are called spinodal due to the characteristic shape of the phase diagram of a binary mixture (Fig. 2a). Within the regime of thermodynamically allowed phase separation, the spinodal line separates the mechanisms of nucleation and growth (metastable regime) from spinodal phase decomposition with characteristic periodic fluctuations (instable regime). Periodic phase separation processes are considered as a model for a number of distinct patterns in organisms with a periodicity of optical wavelengths. Two prominent examples will be given for illustration.

Dufresne et al. compared the photonic nanostructures responsible for the blue color in the feathers of the Eastern Bluebird (*Sialia sialis*) and the Plum-throated Cotinga (*Cotinga maynana*). By the means of small-angle X-ray scattering in

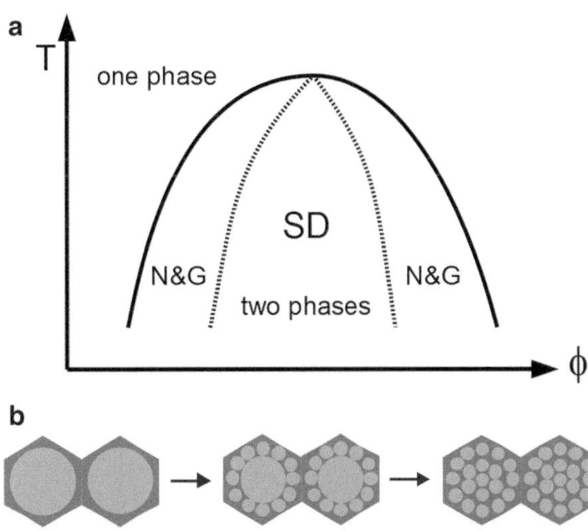

Fig. 2 (**a**) Phase diagram (temperature vs. composition) for a binary mixture. The binodal line (*solid*) separates the homogeneous mixture from the two-phase region, the spinodal line (*dotted*) separates two distinct mechanisms for phase separation: nucleation and growth (N&G) and spinodal decomposition (SD) (**b**) Model proposed by Sumper [12] for pattern formation in diatoms via subsequent phase separation processes involving organic (*light gray*) and aqueous, silica-containing phases (*dark gray*)

addition to electron microscopy they found very different patterns which resemble spinodal decomposition and nucleation-growth patterns, respectively [11].

Sumper proposed that the hierarchical construction of regular patterns in the silica cell wall of diatoms can be attributed to a series of phase separations between an organic phase containing the organic components of the cell wall (i.e., polyamines) and an aqueous phase from which silica is precipitated. Coprecipitation of some amount of the polyamines leads to a new instability and a second level of periodic phase separation with a finer structure (Fig. 2b) [12].

Periodic phase separation may be stabilized by membranes which determine the interface area between hydrophilic and hydrophobic phases. In lipid-oil–water as well as in lipid–water mixtures, thermodynamically stable phases with periodicities in the order of tens of nanometers naturally exist [13]. Even within lipid layers phase separation of the different components leads to periodic patterns, as, for instance, demonstrated for vesicles consisting of sphingomyelin, cholesterol, and di(oleoyl)phosphatidylcholine [14]. In addition, membranes can exhibit mechanical instability by the presence of bending stresses that causes them to buckle, wrinkle, or fold. In this context it should be noted that stroma thylakoid membranes with periodicities in the order of 30 nm [15] are one of the key components of chloroplasts responsible for photosynthesis. Little is known, however, about the self-organized or controlled formation and dynamic reorganization of these and other periodic structures.

2.2 *Biological Function*

For periodic structures in the visible spectral range (which has to be assumed larger than only for the vision of humans) signaling by colors and camouflaging can safely be assumed to be one of the main functions. Other possible biological applications of photonic crystals are less obvious. Why have oysters and mussels iridescent nacre layers in the inner part of their shells? Is there a connection between the intricate photonic patterns of diatoms and their efficient photosynthesis? May photonic crystals hinder UV light entering the cells and causing damage? Definite answers cannot be given yet, but this kind of questions makes biophotonic research an interesting and prospective field.

3 Material Systems in Biological Photonic Crystals

One has to bear in mind that in photonic structures with alternating refractive index, always combinations of different materials are applied. One phase may be, e.g., air or water, the other a biomineral or structural biopolymers. But even in materials denoted as biominerals we intrinsically deal with composite materials of inorganic components with an organic matrix which itself may be composed of several distinct biomolecules. In order to understand the optical properties of biological photonic crystals, one also has to consider this fine structure.

One main feature all studied structures have in common is the absence of complete photonic band gaps due to a low refractive index contrast between the two respective phases. Iridescent colors arise by direction-dependent reflection which change with the angle of observation. However, some structures still produce angle-independent broad-band structural colors due to the incorporation of disorder. Disorder is an intrinsic feature in all biological patterns, ranging from nearly perfect quasi-single-crystalline structures with high long-range order over polycrystalline ones, consisting of differently oriented single-crystalline domains, up to amorphous so-called photonic glasses with only short-range order and completely random structures, which produce color solely by scattering. A plurality of patterns has been realized on a lengthscale capable of manipulating visible and even UV-light covering simple one-dimensional Bragg-like multilayer films as well as complex two- and three-dimensional crystalline lattices. Compared to these genetically controlled and self-organized structures, man-made fabrication of large-scale photonic structures turned out to be still quite challenging, especially for 3D structures.

Presenting nature's choice for photonic materials, we give in this chapter an overview over the main components of biological composites and point out how they are combined to multiscale photonic structures in various species. In Table 1 typical high-dielectric materials found in biological photonic crystals along with their respective refractive indices are compiled. In this context it should be mentioned that one and the same material may exhibit different values in different species due to its specific morphology in the respective organic scaffold. Additionally, the imaginary part of the complex refractive index in some materials is non-zero, either intrinsically or due to incorporated pigments, and therefore the structures show a significant absorption in the measured wavelength range. That is

Table 1 Refractive indices of the main high-dielectric compounds found in biological photonic crystals

Material	Refractive index	Found in
Silica	1.43[a]	Frustule of diatom *Coscinodiscus granii* [16]
$CaCO_3$ (Aragonite)	1.68[b]	Aragonite in direction of *c* axis [17]
Guanine	1.83[c]	Anhydrous guanine in [102] direction [18]
Cellulose	1.41...1.53[b]	Cell walls of soy bean [19]
Chitin	1.55[b] 1.51[b]	Scales of butterfly *Graphium sarpedon* [20] Elytra of beetle *Chrysochroa fulgidissima* [21]
Keratin	1.55[b]	Feather barbules of goose *Anas anas* [20]
Collagen	1.55[c] 1.42[c]	Dried collagen [22] Collagen fibrils in corneal stroma of five species [23]

[a]Value for 632.8 nm
[b]Value for Na_D line (589 nm)
[c]Mean value for the entire visible wavelength range

why the listed values are to be understood as rough estimates. Several material classes are covered, namely minerals (silicate, calcium carbonate), organic crystals (guanine), polysaccharides (cellulose, chitin) and proteins (keratin, collagen).

3.1 Silica

Silicon dioxide or silica occurs mainly in an amorphous hydrated form, $SiO_2 \cdot nH_2O$, with a density around 1.9 g cm^{-3}. Its refractive index strongly depends on the water content and density. With increasing water content the refractive index is decreasing, whereas it increases with increasing density. Silica also forms crystalline phases and many polymorphs are known, among them tridymite, cristobalite, and quartz with increasing densities in that order (around 2.5 g cm^{-3}). Amorphous silica can be found in may organisms, including plants such as grasses, sponges, radiolarians, and diatoms [24].

Diatoms are unicellular algae that incorporate silica into their cell walls as characteristic feature. The cell wall (frustule) consists of two combined halves (thecae) looking like a Petri dish, the upper theca covering the lower one. Each theca consists of a valve and several girdle bands at the circumference of the cell [24, 25]. Typically the thickness of the cell wall is around 1 μm and the cell diameter around 100 μm, but both vary strongly from species to species. *Pennales* are characterized by bilateral symmetry, whereas the *Centrales* exhibit a radially symmetric pattern with a regular array of pores in their cell wall, especially in the genus *Coscinodiscus* (Fig. 3). In *Coscinodiscus granii* and *Coscinodiscus wailesii* the valves consist of a hexagonal array of pores with mean lattice constants of 0.9 and 1.8 μm, respectively. The girdle bands, on the other hand, exhibit a square symmetry with smaller lattice constants of 270 and 310 nm [26]. Because of the higher refractive index of biosilica (1.43) compared to water (1.33) the porous cell

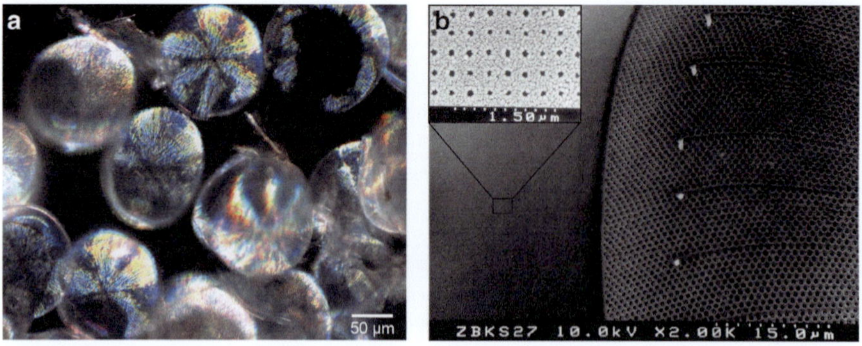

Fig. 3 (**a**) Isolated frustules of *Coscinodiscus granii* in *dark field* illumination. (**b**) SEM micrograph showing the transition region of valve and girdle with a hexagonal and square symmetry (*inset*) respectively. Photo credits: M. Kucki (**a**) [70], S. Landwehr, H. Rühling (**b**)

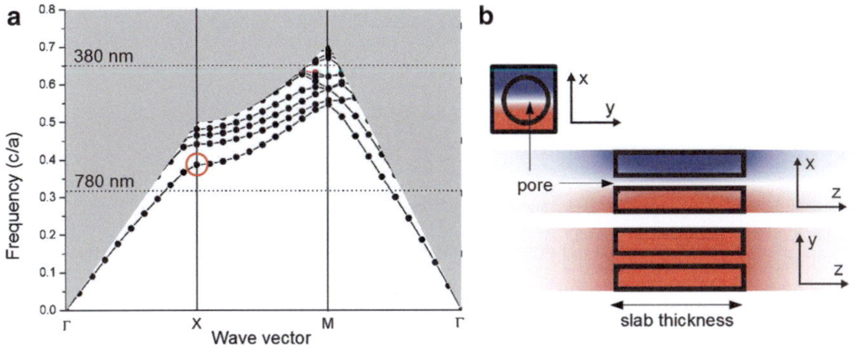

Fig. 4 (**a**) Photonic band diagram of a diatom girdle as slab waveguide in air. Only half of the modes are shown. c: vacuum light velocity, a: lattice constant (250 nm), pore radius 0.36 a, slab thickness 2 a. Indicated with *dotted lines* are the limits of the visible spectrum. (**b**) Electric field profile (in y-polarization) of the lowest mode at the X point (indicated with a *circle* in the band diagram). Three perpendicular cross sections through the unit cell are shown, the sign of the field is indicated by *color*

walls can be described as two-dimensional photonic crystal slab waveguides [16]. As band calculations show (cf. first chapter of this book), in the girdles confined modes are found in the visible range, with resonances in the directions of high symmetry, i.e. the X and M point (Fig. 4). Such modes correspond to stabilized standing waves, as the favored resonator modes in photonic crystal lasers [27]. For the valves resonances are only possible in higher orders for visible light as the structures are bigger. Consequently these modes are coupled to radiating waves. Besides the light guiding properties within the cell wall, the overall shape of the photonic structure leads to a light focusing effect for light incident on the diatom cell. For *C. wailesii* a focus length of about 100 μm was obtained for red light [28]. Whether the cell wall's photonic properties have a biological function supporting the efficient photoharvesting system or are only side-effects of an architecture with good mechanical properties [29] is still in debate and object of current research.

3.2 Calcium Carbonate

Calcium carbonate exists in three different polymorphs, namely calcite, aragonite and vaterite, the latter being less thermodynamically stable. It occurs, for instance, in eggshells, snail shells, and most seashells. Calcite forms trigonal crystals with a density of 2.71 g cm^{-3} [30] and has a uniaxial negative birefringence with an ordinary refractive index of 1.658 and an extraordinary one of 1.486 at 589 nm. In contrast, aragonite has an orthorhombic structure with a density of 2.93 g cm^{-3} [31]. However recent studies reveal a lower symmetry [32]. As a biaxially

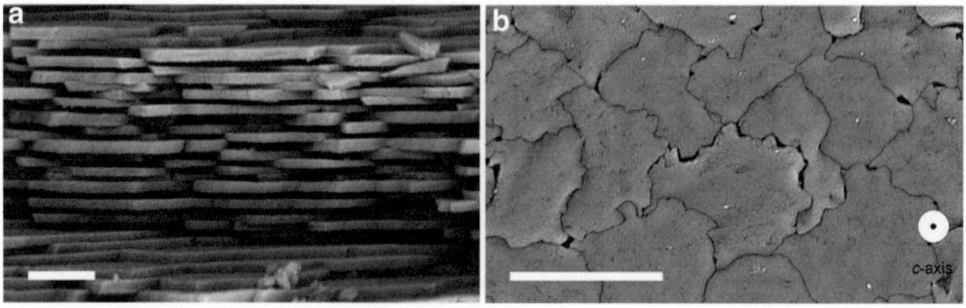

Fig. 5 SEM images of (**a**) cross-section and (**b**) growth surface of nacre of *Haliotis discus hannai*. Scale bars: 6 μm. (Reproduced from J.M. Sun, W.L. Guo, Sci. China Phys. Mech. Astron. 2010, 53, 380)

birefringent material, it exhibits one low and two high refractive indices (1.530, 1.681, 1.685 at 589 nm).

Aragonite is the main component of nacre, which can be found at the inner surfaces of mollusc shells as iridescent Bragg reflector. It may occur in all three most important classes of the phylum *Mollusca*, namely *Bivalvia*, *Gastropoda,* and *Cephalopoda*, with the pearl oysters (*Pinctata*), the abalones (*Haliotis*), and the nautilus (*Nautilus*) as prominent examples, respectively. Nacre is a composite material of layered aragonite platelets, 0.4–0.5 μm thick and 10–20 μm wide, with the crystallographic *c* axis oriented perpendicular to the layers, although some disorder is present (Fig. 5) [33]. The refractive index contrast results from an organic matrix layer consisting of a chitin backbone with a multilayer cladding of hydrophobic proteins that is located between the platelet layer. The chitin and protein chains are arranged perpendicular to each other and parallel to the *a* and *b* axes of aragonite, respectively [34]. Between the inorganic and the protein-containing phases there is an additional layer of acidic, aspartate-rich polypeptides which act as nucleation sites for the calcium carbonate precipitation [35]. For layer thicknesses comparable to the wavelength of visible light this structure, commonly known as "mother of pearl" yields highly iridescent colors, attracting artisans for centuries for the fabrication of decorative gemstones.

3.3 Guanine

Guanine, being one of the four nucleobases in the nucleic acids DNA and RNA, occurs in the living world with great abundance. It is a derivative of purine substituted in the 2- and 6-positions by an amino group and carbonyl-oxygen, respectively. Anhydrous guanine is known to form monoclinic birefringent crystals with refractive indices of 1.83 and 1.48 parallel and perpendicular to its long axis, i.e. the [102] direction [18].

Several fish and spider species, of which two [18] (Fig. 6) will be presented here as examples, incorporate anhydrous guanine crystals in their integument. Silver

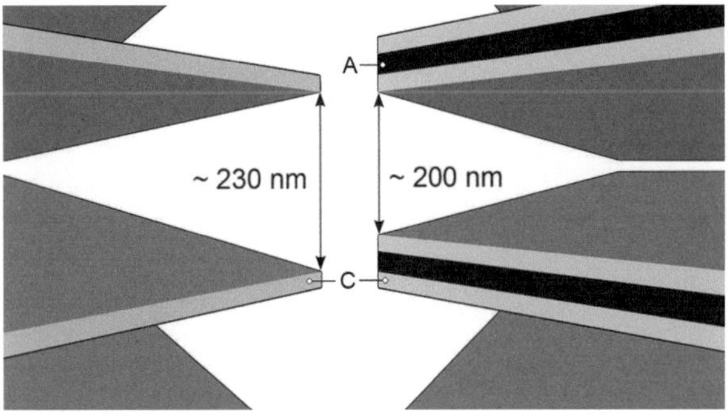

Fig. 6 Schematic of guanine crystals in *Cyprinus carpio* (*left*) versus *Tetragnatha extensa* (*right*). Crystalline guanine phases are depicted by C, amorphous phases by A. Cytoplasm is not shown

colored scales from the skin of Japanese Koi Fish (*Cyprinus carpio*) contain iridophore cells lined up underneath the scales. One cell contains approx. ten guanine crystal plates. They are arranged parallel to each other, having a thickness of 20 nm and being separated by 230 nm-thick cytoplasm layers ($n = 1.33$). The correlation within a single cell is bigger than intercellular, yet a certain long-range order is still present. These slightly disordered multilayers are responsible for the silvery broad-band reflectance, which exhibits a maximum in the green-red spectral region.

The one-dimensional guanine photonic structure in the spider *Tetragnatha extensa* comprises less defined doublet crystal plates with roughly the same thickness but in this case they are separated by 30 nm-thick layers of amorphous guanine. These guanine stacks are in turn separated by cytoplasm with a thickness of 200 nm. Up to ten doublet crystals are located within a cell and no intercellular order is observed here, leading to an even more spectrally uniform reflectance.

3.4 Cellulose

The polysaccharide glucose is the main component in the cell wall of plants. In order to find photonic structures made of this structural polymer, one has therefore to search in the plant kingdom. In comparison with animals, the occurrence of iridescence as hint for photonic crystal behavior is relatively rare. However, there are some prominent examples. When grown in the shade, *Selaginella willdenowii*, a tropical rainforest understory spikemoss, shows blue iridescence in young leaves that can be attributed to a four-layer quarter wave stack in the epidermis [36, 37]. In *Mapania caudata*, the coloration by helicoidal multilayer stacks is additionally supported by the incorporation of silica nanoparticles within the layers [38].

Also in fruits multilayer reflectors can be found. Intensively blue with a characteristic cellular pixelation shine the berries from *Pollia condensata*. Cellulose microfibrils with a helicoidal layer arrangement are the structural base for this coloration. Interestingly, the helicity may vary from cell to cell, selectively reflecting left or right circularly polarized light, respectively [39].

For the interested reader, further examples of iridescent plants can be found in a detailed list compiled by McPherson [40].

3.5 Chitin

After cellulose, chitin is the most common biomaterial. It is a polymer made of *N*-acetyl-D-glycosamine, a derivative of glucose, linked by β-1,4-glycosidic bonds and differs from cellulose by its acetamide-moieties. It is found in the cell walls of fungi, the exoskeletons of arthropods, and in cephalopods. One of the first thoroughly investigated examples of chitin photonic structure analyzed as photonic crystal is the Seamouse (*Aphrodita aculeata*) with its iridescent spines (Fig. 1a). The iridescence is caused by several layers of hexagonally arranged hole cylinders in a chitinous matrix forming a partial band gap [41]. In this section however we will focus on the color-producing structures found in insects like butterflies (*Lepidoptera*) and beetles (*Coleoptera*). The real part of the refractive index of chitin is in the range of 1.5, but due to incorporated pigments like melanin or carotenoides or scaffolding biomolecules like proteins or phospholipids its value may differ greatly (1.40–1.73) [42]. A great morphological variety of chitin structures has been realized, ranging from simple 1D-multilayer reflectors, 2D- and complex three-dimensional photonic structures with astonishing optical properties. And although diffraction gratings with larger periodicities in the micrometer range certainly play a significant role in structural coloration of both butterflies and beetles, they will not be covered by this review as these are not regarded as photonic crystal structures in the strict sense. An excellent review comprising all kinds of structural color mechanisms in beetles can be found in [43].

Many beetles exhibit strongly iridescent features in their exoskeleton. During the formation of the cuticle chitin layers of alternating refractive index are secreted in phases one by another resulting in a one-dimensional Bragg reflector, which gives rise to this structural coloration. They may be formed as epi- [44], exo- [45], or endocuticular [46] multilayers depending on the location within the integument. The resulting color impression is a function of index contrast, layer thickness and spacing, as well as the viewing angle. Less angle-dependent and more broad-band reflecting structures are realized by chirped structures, where layer thickness and spacing gradually change. The resulting appearance of beetles with this kind of structures is metallic, gold, or silver [46]. Moreover, multilayer structures found in *Scarabaeidae* lead to selective reflectance of circularly polarized light [45]. In *Chrysina* (*Plusiotis*) *resplendens* and other scarab beetles, optically anisotropic (birefringent) parallel chitin fibrils are deposited with a progressively rotated orientation above one another, leading to a structure optically similar to a cholesteric liquid crystal [47, 48]. The pitch of the helix defines the circular reflection properties. Interestingly, in contrast to other scarabs that reflect only left polarized light, *C. resplendens* is able to reflect both circular polarizations due to the incorporation of a halfwave plate. A large amount of uric acid (2,6,8-trihydroxy

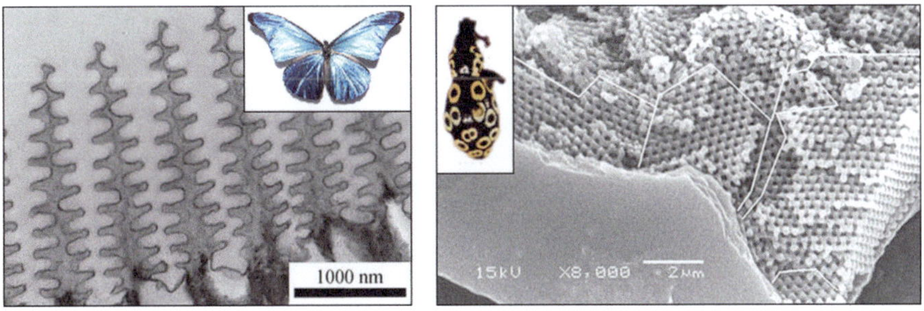

Fig. 7 Chitin structures responsible for color production in *Morpho rhetenor* (*left*) and *Pachyrrhynchus pavonius* (*right*). (Reproduced from G. Liao, H. Zuo, X. Jiang, X. Yang, T. Shi, Front. Mech. Eng. 2012, 7, 394, V.L. Welch, J.P. Vigneron, Opt. Quant. Electron. 2007, 39, 295)

purine) which is present in the matrix in amounts up to 70 % increases the reflectivity by contributing to the birefringence [47].

Three-dimensional photonic structures are found in scales of different weevils and longhorn beetles. The symmetry these structures obey may be either a diamond-like arrangement of spheres (*Lamprocyphus*) [49] or an arrangement similar to opal (Fig. 7), which exhibits a close-packing order (*Pachyrrhynchus*, *Metapocyrtus*) [50, 51]. In *Metapocyrtus* yet an inverse opal structure with hollow spheres in a chitin matrix has been observed. The longhorn beetle *Prosopocera lactator* has scales of greenish-white color which arises from face-centered cubic spheres, which in turn are interconnected by short rods resembling a "ball-and-stick" structure [52]. All mentioned three-dimensional structures produce more or less direction-independent spectrally broad colors due to relatively small coherence lengths. They can be described by the term photonic polycrystal, where many small single-crystalline domains are randomly oriented within the scales.

The morphological variety of photonic structures is even bigger in butterflies, among which the most prominent certainly is the genus *Morpho* with its vivid blue wing scales. The blue color of *Morpho rethenor* originates from "christmas tree"-shaped ridges (Fig. 7) that act as multilayer reflectors [53]. The structural variability of butterfly scales is huge. The lumen beneath the ridges itself may be ordered in one, two, and three dimensions. In the genus *Papilio* it is composed of alternating chitin-air layers, which are irregularly interconnected by small chitin rods, thus forming poorly ordered two-dimensional photonic structures. It has been shown however for *Papilio blumei* that a one-dimensional model, which ignores the interconnections, is capable of describing the resulting color impression [54]. On the contrary, highly ordered polycrystalline three-dimensional structures in the scale lumen of various lycaenid and papilionid species have been identified [55]. All found structures exhibit a cubic single gyroid core-shell network of chitin in air, which leads to elliptically polarized light in reflection. It has been concluded that the ordered smooth endoplasmatic reticulum, which exhibits a bicontinuous cubic symmetry, is involved in the pattern formation. This process seems to be fundamentally different from the one producing 3D biological photonic crystals in beetles, because no gyroid structures are found in beetles and no other symmetries (opal- or diamond-like) have been found in butterflies so far [56].

The photonic crystal structure may be favorably combined with fluorescent materials in order to enhance the optical effect. For *Papilio nireus*, directional emission of fluorescence was demonstrated and attributed to the optical structure comprising a 2D photonic crystal slab and a three-layer distributed Bragg reflector [57].

3.6 Keratin

Keratins are fibrous structural proteins rich in cysteine. Their mechanical properties greatly depend on the degree of cross-linking by disulfide bridges and therefore on their cysteine content. Keratins may be divided into amorphous keratins (without any morphological order), α-keratins and β-keratins named after their molecular conformation as α-helices or β-plated sheets, the former being present in mammalian hard tissues, the latter in, e.g., avian feathers.

The orange beak horn of the King Penguin (*Aptenodytes patagonicus*) shows a selective UV-reflectance around 380 and 400 nm in the dried and living state, respectively. One-dimensional multilayer structures composed mainly of keratin filaments in between juxtaposed membrane folds were found in the upper regions of the beak horn [58]. An average lattice constant of 130 nm was found in dried samples. By estimation of an average refractive index of 1.45 the observed UV-reflectance could be reproduced by computation.

Fibrous two-dimensional photonic structures have been discovered in many bird feather barbules. In the Javanese Peacock (*Pavo muticus*) the colored barbules' cortex consists of melanin rods in a keratin matrix arranged with a square symmetry with distinct spacings for the different resulting vivid color impressions [59]. The melanin rods are arranged parallel to the cortex surface (Fig. 8). The common magpie *Pica pica* exhibits a yellowish-green iridescence in its dark tail feathers and a blue one in the feathers on the wings. The feather barbule's cortex contains cylindrical air tubes distributed in a keratin-melanine matrix with a hexagonal symmetry, in contrast to the square arrangement in the peacock. Interestingly, the lattice spacing in the bluish barbules is larger than in the yellowish-green ones, indicating that in this case a higher stop-band is involved in color production [60].

3.7 Collagen

Collagen is the main structural protein in higher animals and the main component of the connective tissue or extracellular matrix. It consists of three left-handed helices which are connected by hydrogen- and covalent bonds to form a right-handed superhelix. Each helix consists of 200–1,000 amino acids, every third being

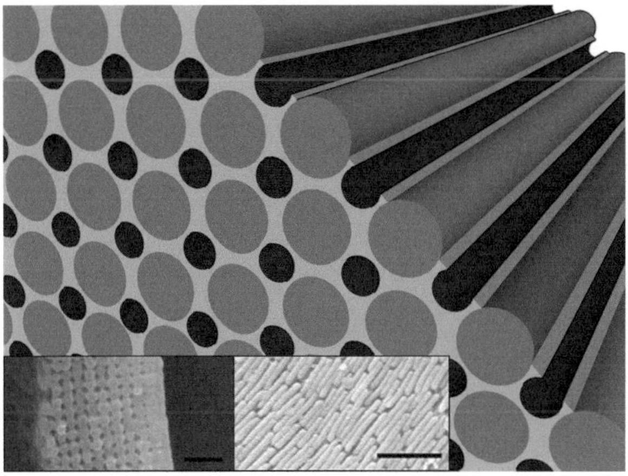

Fig. 8 Schematic of the barbule's cortex of *Pavo muticus*. Melanin rods (*dark gray*) are arranged in a keratin matrix (*light gray*) with a square symmetry. In between there are air voids. The *insets* show SEM images of the transverse (*left*, scale bar: 1 μm) and longitudinal (*right*, scale bar: 500 nm) cross-section respectively. Insets reproduced from J. Zi, B. Dong, T. Zhan, X. Liu in: Bioinspiration: Biological and Medical Physics, Biomedical Engineering (Ed. X.J. Liu), Springer, New York, 2012

glycine. It contains hydroxy-substituted lysins and prolins, which facilitate the formation of the superhelix. Twenty-eight different types of collagen are known so far, type I collagen being most present in skin tissues.

Quasi-ordered two-dimensional arrays of parallel collagen fibers within the dermis of avian and mammalian skin are mainly responsible for color production. The horny epidermis layer of beaks and feet (ramphotheca and podotheca) of more than 30 species of birds with a variety of colors ranging from ultraviolet to yellow hues and blue colored skin of primates (*Mandrillus sphinx*, *Cercopithecus aethiops*) and marsupials (*Marmosa maxicana*, *Caluromys derbianus*) have been studied [61, 62], all showing similar color-producing mechanisms (see also Fig. 11).

4 Recent Trends

The description of biological structural colors as one-, two-, or three-dimensional photonic crystals is only the beginning of a deeper insight into the possibility nature offers us for the design of photonic functionalities. Thus, in this last section, two important trends in the research on biophotonic structures that may show the way into the future will be highlighted: the functionalization of biological structures towards hybrid natural-artificial composites, and the exploration of disorder as a means for coloring without color pigments.

4.1 Hybrid Composites from Biological Photonic Crystals

Additional functionality can be incorporated into a biological photonic structure if the morphogenetic process is manipulated in vivo, or post-functionalizations are done in vitro. Not many species, however, can be cultivated in order to perform in vivo experiments. Remarkable within this context are experiments on biomineralization in abalone sea snails, in which between mantle and shell a substrate is inserted and later, after deposition of nacre, removed again. By this technique, nacre layers could be incorporated into technical multilayer structures, forming hybrid natural–artificial samples [63].

Substantially easier, however, is the cultivation of microalgae, thus we will focus now on diatoms for which many experiments for functionalization have been reported. Templating processes have been developed that allow copying the structure of isolated diatoms into another material system, for instance MgO [64], TiO_2 or even $BaTiO_3$ [65]. However, a more direct way is the addition of precursors structurally similar to silicic acid that replace silica partly in the biomineralized frustules. Al_2O_3 [66], GeO_2 [67], and TiO_2 [68] have been incorporated by this technique.

Similarly, staining with laser dyes can be performed during the cultivation by addition of the appropriate dye. This kind of modification can be used simply as a probe for the cell wall's optical properties as well as for the generation of new light-emitting photonic devices. In vivo-fluorochromation has been shown for various rhodamine [69–71], oxazole [72], and oxadiazole derivatives [73, 74], as diatoms accumulate solved dyes in special vesicles—silica deposition vesicles (SDV)—prior to new cell wall formation during cell division.

Such new biohybrid composite materials obey interesting fluorescence properties. Rhodamine B-stained frustules from *C. granii* exhibit a pronounced directional emission with hexagonal symmetry as could be shown with a conoscopic setup in our laboratory (Fig. 9 cc).

Alternatively, staining has been done after isolation of the frustules by surface chemistry utilizing silanole-functionalized fluorescent dyes [26]. By appropriate surface treatment, other coatings to the frustules can be applied as well, for instance gold [75] or silver [76] nanoparticle layers, TiO_2 [77] and ZnO [78].

Yet another, more fundamental approach to using morphogenetic principles for photonic purposes is the following: It is known which proteins are responsible for the biomineralization of silica in diatoms, and how they can be used for in vitro precipitation of silica in various shapes [79]. A short peptide sequence of these silaffin proteins was mixed into a photopolymer formulation that was patterned to periodic stripes in a holographic setup (cf. tenth chapter of this book). Since the peptide was enriched in defined stripe domains and still active for silica precipitation, a periodic grating of linearly arranged silica spheres was obtained (Fig. 10) [80]. This example may illustrate the power of biocatalysis in producing complex photonic structures by low-temperature processes.

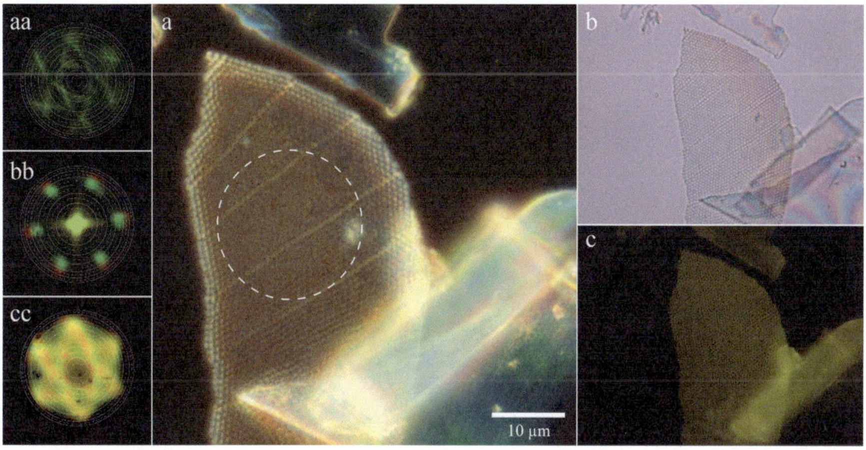

Fig. 9 Orthoscopic images of an isolated, Rhodamine B-stained valve fragment of *C. granii* in (**a**) *dark-field* reflected illumination, (**b**) *bright-field* transmitted illumination, and (**c**) fluorescence. (**aa**), (**bb**), and (**cc**) show the respective conoscopic images. The *dashed circle* in (**a**) indicates the area of which the conoscopic images have been made. In (**aa**) a 544 nm interference filter has been used

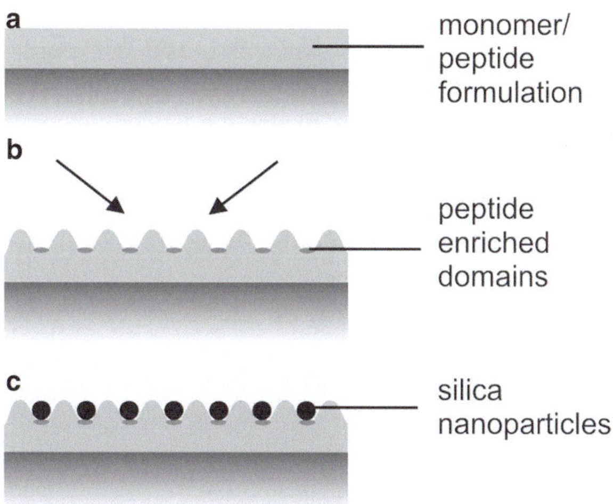

Fig. 10 Principle of holographic patterning with biocatalytic silica deposition after Brott et al. [80]. (**a**) A photomonomer-peptide formulation is deposited onto a glass slide (**b**) After holographic writing and curing, a density/surface relief is grating is formed, confining peptide-enriched domains to stripes. (**c**) The grating is exposed to $Si(OMe)_4$ in phosphate-citrate buffer solution, whereupon silica spheres are deposited along the stripes

4.2 *Photonic Glasses*

That amorphous photonic structures are also useful for the realization of structural colors has been realized recently by many researchers, forming a quickly expanding subfield in biophotonic research. In analogy to the distinction between atomic and molecular crystals and glasses, these mesoscopic structures are called photonic glasses [81]. Especially the colors white and blue often rely on the presence of amorphous structure yet with correlation lengths in the order of visible light (Fig. 11).

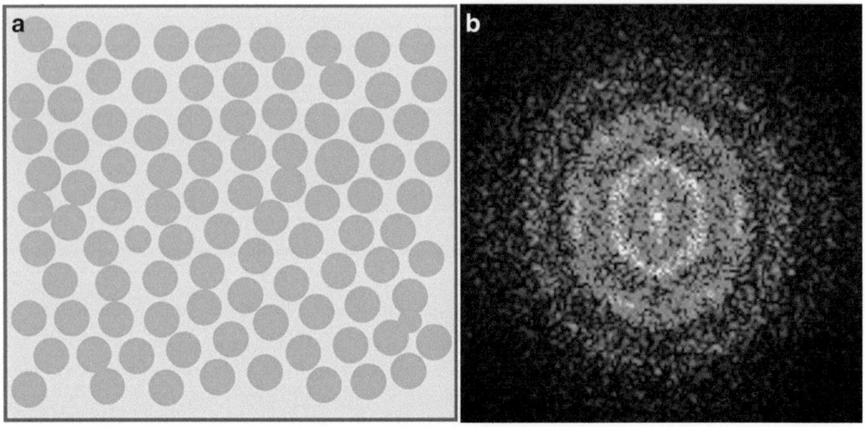

Fig. 11 (**a**) Schematic representation of a collagen array from mammal skin as amorphous photonic glass with near ordering (after [62]). (**b**) A two-dimensional Fourier analysis (from a larger area of the array) reveals *ring-shaped* distributions in the spatial frequency domain demonstrating a substantial nanostructured order

White colors occur frequently in organisms, for a few of which structural details are known. They often order in the first coordination spheres but disorder at larger lengthscales. In the plant kingdom, a well-investigated example is the Edelweiss *Leontopodium nivale* [82]. Parallel fibers with a mean separation of 420 nm are arranged around transparent filaments which are entangled in a disordered way. Here, UV protection is proposed as optical function of the structure. As an example from the animal kingdom, the beetle *Cyphophilus* sp. exhibits a brilliant white in terms of standardized whiteness and brightness values [83]. Using similar cuticular filaments of about 250 nm diameter as in brightly colored weevils, but in a completely disordered arrangement, this example shows the importance of mesoscopic structural order for the resulting optical properties.

Birds can be taken as examples for illustrating how white and blue structural colors are related. Consider the non-iridescent blue plumage of certain birds, like Stellers's Jay (*Cyanocitta stelleri*) [84] or the Indian Blue Peacock (*Pavo cristatus*). Sometimes leucistic forms of these birds occur that lack melanin, which is a black, not blue pigment. Without melanin, the disordered photonic structure gives a white color expression due to scattering in the amorphous structure, but when melanin absorbs one part of the spectrum, the blue color remains. The color impression is both dependent on the pigments and the amorphous order.

A range of colors can be achieved by proper combination of scattering and absorption, as impressively demonstrated in the scales of the Longhorn Beetle *Anoplophora graafi*, each of them with a different color due to different sizes of the random closed packed chitin colloids [85]. In technical applications, the same principle can be applied for the production of deeply colored non-iridescent paints that are only made of white and black colloids [86]—a very astonishing biomimetic story that shows the potential of photonic glasses and the transition from photonic crystal order to disorder.

5 Conclusion

In this chapter, a glance into the world of biological photonic crystals was given. Not all species could be covered, and in this evolving field new species that are familiar to biologists for a long time, are rediscovered from the perspective of photonics every year. But we hope that this compilation of basic principles and important examples will be stimulating for all those looking for biomimetic approaches in "soft matter" photonic crystal research.

References

1. D. Thompson, *On Growth and Form* (Cambridge University Press, Cambridge, 1945) (New ed.)
2. A.J. Koch, H. Meinhardt, Rev. Mod. Phys. **66**, 1481 (1994)
3. S. Kondo, T. Miura, Science **329**, 1616 (2010)
4. G. Nicolis, I. Prigogine, *Self Organization in Nonequilibrium Systems* (Wiley, New York, 1977)
5. A. Turing, Philos. Trans. R. Soc. B **237**, 37 (1952)
6. S.A. Kauffman, *The Origins of Order* (Oxford University Press, Oxford, 1993)
7. L. Wolpert, C. Tickle, *Principles of Development*, 4th edn. (Oxford University Press, Oxford, 2011)
8. M. Akam, Nature **341**, 282 (1989)
9. M.J. Pankratz, H. Jäckle, Trend Genet. **6**, 287 (1990)
10. R.W. Liesegang, Naturwiss. Wochenschr. **11**, 353 (1896)
11. E.R. Dufresne, H. Noh, V. Saranathan et al., Soft Matter **5**, 1792 (2009)
12. M. Sumper, Science **295**, 2430 (2002)
13. K. Larsson, J. Phys. Chem. **93**, 7304 (1989)
14. S. Rozovsky, Y. Kaizuka, J.T. Groves, J. Am. Chem. Soc. **127**, 36 (2005)
15. D. Posselt, G. Nagy, J.J.K. Kirkensgaard et al., Biochim. Biophys. Acta **1817**, 1220 (2012)
16. T. Fuhrmann, S. Landwehr, M. El Rharbi-Kucki, M. Sumper, Appl. Phys. B **78**, 257 (2004)
17. K. Schäfer, E. Lax (eds.), *Landolt-Börnstein – Zahlenwerte und Funktionen*, vol. II-4, 6th edn. (Springer, Berlin, 1961)
18. A. Levy-Lior, E. Shimoni, O. Schwartz et al., Adv. Funct. Mater. **20**, 320 (2010)
19. J.T. Woolley, Plant Physiol. **55**, 172 (1975)
20. H.L. Leertouwer, B.D. Wilts, D.G. Stavenga, Opt. Exp. **19**, 24061 (2011)
21. S. Yoshioka, S. Kinoshita, Phys. Rev. E Stat. Nonlin. Soft Matter Phys. **83**, 051917 (2011)
22. D.M. Maurice, J. Physiol. **136**, 263 (1957)
23. D.W. Leonard, K.M. Meek, Biophys. J. **72**, 1382 (1997)
24. T.L. Simpson, B.E. Volcani (eds.), *Silicon and Siliceous Structures in Biological Systems* (Springer, New York, 1981)
25. J. Picket-Heaps, A.M. Schmid, L.A. Edgar, Prog. Phycol. Res. **7**, 1 (1990)
26. M. Kucki, S. Landwehr, H. Rühling et al., Proc. SPIE **6182**, 6182S1–6182S9 (2006)
27. N. Notomi, H. Suzuki, T. Tamamura, Appl. Phys. Lett. **78**, 1325 (2001)
28. L. de Stefano, I. Rea, I. Rendina et al., Opt. Exp. **15**, 18082 (2007)
29. C.E. Hamm, R. Merkel, O. Springer et al., Nature **421**, 841 (2003)
30. D.L. Graf, Am. Mineral. **46**, 1283 (1961)
31. A. del Negro, L. Ungaretti, Am. Mineral. **56**, 768 (1971)
32. D.J.M. Bevam, E. Rossmanith, D.K. Mylrea et al., Acta Cryst. B **58**, 448 (2002)
33. R.A. Metzler, M. Albrecht, R.M. Olabisi et al., Phys. Rev. Lett. **98**, 268102 (2007)

34. S. Weiner, W. Traub, Philos. Trans. R. Soc. B **304**, 425 (1984)
35. L. Addadi, S. Weiner, Angew. Chem. Int. Ed. Engl. **31**, 153 (1992)
36. D.W. Lee, J.B. Lowry, Nature **254**, 50 (1975)
37. K.R. Thomas, M. Kolle, H.M. Whitney et al., J. R. Soc. Interface **7**, 1699 (2010)
38. G. Strout, S.D. Russell, D.P. Pulsifer, et al., Ann. Botany (2013). doi:10.1093/aob/mct172
39. S. Vignolini, P.J. Rudall, A.V. Rowland et al., Proc. Natl. Acad. Sci. **109**, 15712 (2012)
40. S. McPherson, The Plantsman **9**, 120 (2010)
41. A.R. Parker, R.C. McPhedran, D.R. McKenzie et al., Nature **409**, 36 (2001)
42. M.F. Land, Prog. Biophys. Mol. Biol. **24**, 75 (1972)
43. A.E. Seago, P. Brady, J.-P. Vigneron et al., J. R. Soc. Interface **6**, S165 (2009)
44. T. Hariyama, M. Hironaka, Y. Takaku et al., in *Structural Color in Biological Systems—Principles and Applications*, ed. by S. Kinoshita, S. Yoshioka (Osaka University Press, Osaka, 2005)
45. A.C. Neville, S. Caveney, Biol. Rev. **44**, 531 (1969)
46. A.R. Parker, D.R. McKenzie, M. Large, J. Exp. Biol. **201**, 1307 (1998)
47. S. Caveney, Proc. R. Soc. Lond. B **178**, 205 (1971)
48. D. Goldstein, Proc. SPIE **5888**, 58880T (2005)
49. J.W. Galusha, L.R. Richey, J.S. Gardner et al., Phys. Rev. E **77**(Pt1), 050904 (2008)
50. V. Welch, V. Lousse, O. Deparis et al., Phys. Rev. E **75**, 41919-1 (2007)
51. A. Parker, V. Welch, D. Driver, N. Martini, Nature **426**, 786 (2003)
52. J.F. Colomer, P. Simonis, A. Bay et al., Phys. Rev. E **85**, 011907 (2012)
53. P. Vukusic, J.R. Sambles, C.R. Lawrence, R.J. Wootton, Proc. R. Soc. B **266**, 1403 (1999)
54. M. Lo, C. Lee, Appl. Opt. **53**, A399 (2014)
55. V. Saranathan, C.O. Osuji, S.G.J. Mochrie et al., Proc. Natl. Acad. Sci. U. S. A. **107**, 11676 (2010)
56. B.D. Wilts, K. Michielsen, H. de Raedt et al., Interface Focus **2**, 681 (2012)
57. P. Vukusic, I. Hooper, Science **310**, 1151 (2005)
58. B. Dresp, K. Langley, Anat. Rec. **288A**, 213 (2006)
59. J. Zi, X. Yu, Y. Li et al., Proc. Natl. Acad. Sci. U. S. A. **100**, 12576 (2003)
60. J.P. Vigneron, J.F. Colomer, M. Rassart et al., Phys. Rev. E **73**, 021914 (2006)
61. R.O. Prum, R. Torres, J. Exp. Biol. **206**, 2409 (2003)
62. R.O. Prum, R. Torres, J. Exp. Biol. **207**, 2157 (2004)
63. M. Fritz, A.M. Belcher, M. Radmacher et al., Nature **371**, 49 (1994)
64. H.H. Sandhage, M.B. Dickerson, P.M. Husemann et al., Adv. Mater. **14**, 429 (2002)
65. S. Dudley, T. Kalem, M. Akinc, J. Am. Ceram. Soc. **89**, 2434 (2006)
66. M. Gehlen, L. Beck, G. Calas et al., Geochim. Cosmochim. Acta **66**, 1601 (2002)
67. M.L. Chiappino, F. Azam, B.E. Volcani, Protoplasma **93**, 191 (1977)
68. C. Jeffryes, T. Gutu, J. Jiao et al., ACS Nano **2**, 2103 (2008)
69. C. Li, S. Chu, M. Lee, Protoplasma **151**, 158 (1989)
70. M. Kucki, Biological photonic crystals: diatoms. Dye functionalization of biological silica nanostructures. PhD thesis, University of Kassel, Kassel, 2009
71. M. Kucki, T. Fuhrmann-Lieker, J. R. Soc. Interface **9**, 727 (2012)
72. K. Shimidzu, Y. del Amo, M.A. Brzesinski et al., Chem. Biol. **8**, 1051 (2001)
73. J. Descles, M. Vartanian, A. el Harrak et al., New Phytol. **177**, 822 (2008)
74. V.V. Annenkov, E.N. Danilovtseva, S.N. Zelnsky et al., Anal. Biochem. **407**, 44 (2010)
75. N.L. Rosi, C.S. Thaxton, C.A. Mirkin, Angew. Chem. Int. Ed. **43**, 5500 (2004)
76. M.H. Yu, P. Yuan, J. Thang et al., J. Nanosci. Nanotech. **10**, 612 (2010)
77. D. Losic, G. Triani, P.J. Evans et al., J. Mater. Chem. **16**, 4029 (2006)
78. Y. Cai, K.H. Sandhage, Phys. Stat. Sol. A **202**, R105 (2005)
79. N. Kröger, R. Deutzmann, M. Sumper, Science **286**, 1129 (1999)
80. L.L. Brott, R.R. Naik, D.J. Pikas et al., Nature **413**, 291 (2001)
81. P.D. García, R. Sapienza, C. López, Adv. Mater. **22**, 12 (2010)
82. J.P. Vigneron, M. Rassart, Z. Vértesy et al., Phys. Rev. E **71**, 011906 (2005)
83. P. Vukusic, B. Hallam, J. Noyes, Science **315**, 348 (2007)
84. M.D. Shawkey, G.E. Hill, J. Exp. Biol. **209**, 1245 (2006)
85. B.Q. Dong, X.H. Liu, T.R. Zhan et al., Opt. Exp. **18**, 14430 (2010)
86. Y. Takeoka, S. Yoshioka, A. Takano et al., Angew. Chem. Int. Ed. **52**, 7261 (2013)

Part II
Growth and Properties

Spin-Coated Polymer and Hybrid Multilayers and Microcavities

Francesco Scotognella, Simone Varo, Luigino Criante, Serena Gazzo, Giovanni Manfredi, Robert J. Knarr III, and Davide Comoretto

Abstract Polymer multilayer structures have attracted increasing attention in the recent years because of the straightforward and low-cost techniques that can be used for their fabrication. When the multilayers are composed of a periodical alternation of two materials with different refractive indexes and with layer thicknesses comparable with the wavelength of light, they take the name of distributed Bragg reflectors (DBR). They behave like planar one-dimensional photonic crystals (PhC) and exhibit a photonic band gap (PBG), a spectral region in which photons with suitable energy and wave vector are not allowed to propagate through the crystal. Moreover, within the PBG and at its edges, modifications of radiative photophysical processes occur. The spectral position, efficiency and linewidth of the PBG can be engineered by modifying the layer thicknesses and the refractive indexes of the two materials. While DBRs grown using inorganic materials are well known, polymer and colloidal particle DBRs are receiving a renewed interest due to the possibility to chemically engineer their structural properties and photonic functions; moreover, they can be free-standing and flexible thus being adaptable to any surface. Furthermore, polymers and porous structures can easily embed many other active materials, paving the way to a myriad of applications. In this chapter, we introduce polymer multilayers and planar microcavities fabricated using the spin coating technique, discussing the different materials employed and manufacturing challenges. We will also review different applications that exploit these kinds of photonic structures ranging from lasing to sensing.

F. Scotognella (✉) • S. Varo
Dipartimento di Fisica, Politecnico di Milano, Piazza Leonardo da Vinci 32, 20133 Milan, Italy

Center for Nano Science and Technology@PoliMi, Istituto Italiano di Tecnologia, Via Giovanni Pascoli 70/3, 20133 Milan, Italy
e-mail: francesco.scotognella@polimi.it

L. Criante
Center for Nano Science and Technology@PoliMi, Istituto Italiano di Tecnologia, Via Giovanni Pascoli 70/3, 20133 Milan, Italy

S. Gazzo • G. Manfredi • R.J. Knarr III • D. Comoretto (✉)
Dipartimento di Chimica e Chimica Industriale, Università di Genova, Via Dodecaneso 31, 16146 Genoa, Italy
e-mail: davide.comoretto@unige.it

D. Comoretto (ed.), *Organic and Hybrid Photonic Crystals*,
DOI 10.1007/978-3-319-16580-6_4

Keywords One-dimensional photonic crystals • Distributed Bragg reflector • DBR • Microcavity • Spin coating technique • Photonic crystal laser • Photonic crystal sensor • Photonic crystal switch • DOS manipulation • Flexible photonic crystals • Functional materials @ photonic crystals

1 Introduction

The simplest example of a photonic crystal is a multi-layered structure where layers of materials having different refractive indexes are stacked one on another. In this arrangement, a sequence of two different layers is repeated in the space in cyclic way giving rise to the periodicity needed to create a planar photonic crystal. The name given to this structure sketched in the inset of Fig. 1 is distributed Bragg reflector (DBR). Though basic properties of DBRs were explored long before the

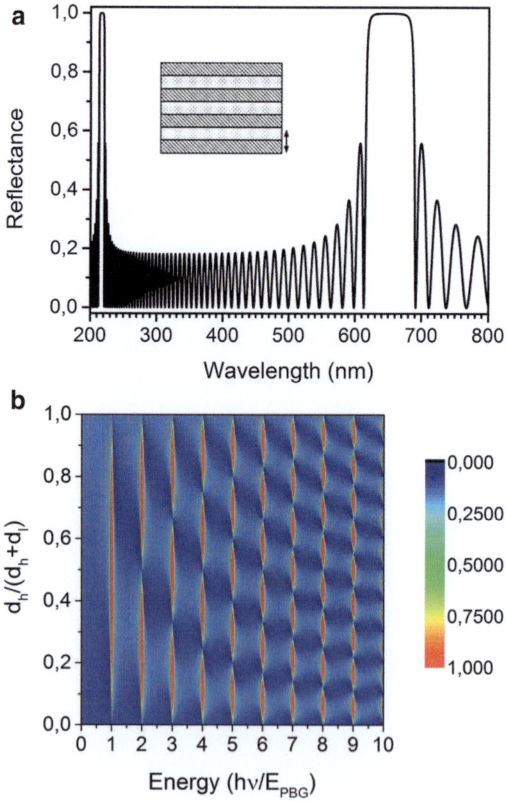

Fig. 1 (**a**) Calculated reflectance spectrum for a DBR of 50 layers of alternated materials having $n = 1.7$ and $n = 1.46$ (similar to that of polyvinylcarbazole and cellulose acetate). In the *inset* the scheme of a DBR is reported. The *arrow* indicates the periodicity of the system. (**b**) Calculated reflectance intensity contour plot for different layers optical path (d_h/d_l, thickness of the high/low refractive index layers) as a function of photon energies

birth of the concept of photonic crystal [1, 2], they remain an important playground to illustrate some of the main properties of PhCs and for their technological applications. Indeed, despite being very simple, this structure is used for many purposes such as sensing, lasing and other photonic devices [3–10]. DBRs theory is discussed in detail in the first chapter of this book. Here we briefly summarize their main properties.

In a DBR, the two alternating layers are composed of transparent materials possessing different refractive indexes. The difference between these values is called dielectric contrast and along with the optical thicknesses of the layers, allows us to tune the PBG spectral features. If exposed to white light, DBRs have peculiar reflectance spectra that appear as the one shown in Fig. 1a. In this example the PBG is observed at about 660 nm producing a frequency range where the reflectance is close to 1. Since at the PBG there are no photonic states allowing light propagation through the crystal and there is no absorption, the photons must be reflected. Together with the first order band gap, higher order gaps can be observed at wavelengths which are integer divisors of the main band gap's one (if we neglect the dispersion of the refractive indexes) (Fig. 1b). The presence of high order gaps in the experimental spectra is a qualitative indication of high optical quality of the system. In the spectral ranges far from the PBG, a progression of fringes is observed in the spectrum background due to interference from light reflected on top and back surfaces, thus providing information on the overall thickness of the structure.

A particular configuration that can be assumed by DBRs is the so-called lambda fourths condition. It is reached when the thickness of the layers composing the crystal gives an optical path that is a quarter of the wavelength of the main band gap (Fig. 1b). Under this condition, a maximum reflectivity is achieved for the first order PBG since $d = \lambda/4n$ provides constructive interference. For the even order gaps ($d = 2k\,\lambda/4n$) the PBG disappears since a destructive interference condition occurs (as it can be seen by the absence of a high reflectivity frequency range that should be around 330 nm in Fig. 1a).

The reflectivity (R) and the width (ΔE) of the band gap (expressed in energy) of a finite DBR with N bilayers can then be calculated as [11] (see also first chapter of this book)

$$R = 1 - 4\left(\frac{n_\mathrm{l}}{n_\mathrm{h}}\right)^{2N} = 1 - 4\left(1 - \frac{\Delta n}{n_\mathrm{h}}\right)^{2N} \tag{1}$$

$$\Delta E = \frac{4E_\mathrm{PBG}}{\pi}\frac{|n_\mathrm{h} - n_\mathrm{l}|}{\mathrm{n_h} + \mathrm{n_l}} = \frac{4E_\mathrm{PBG}}{\pi}\frac{\Delta n}{n_\mathrm{h} + n_\mathrm{l}} \tag{2}$$

where n_h is the high refractive index, n_l is the low refractive index, $\Delta n = n_\mathrm{h} - n_\mathrm{l}$ is the dielectric contrast and E_PBG is the energy at which the band gap is centred.

In order to investigate the photonic band structure of the PhC, it is useful to probe different wave vectors through the use of incidence angle resolved spectroscopy (see also 1st and 11th chapters for additional details). For the 1D system, by increasing the incidence angle, a higher energy shift of the PBG is observed (due to

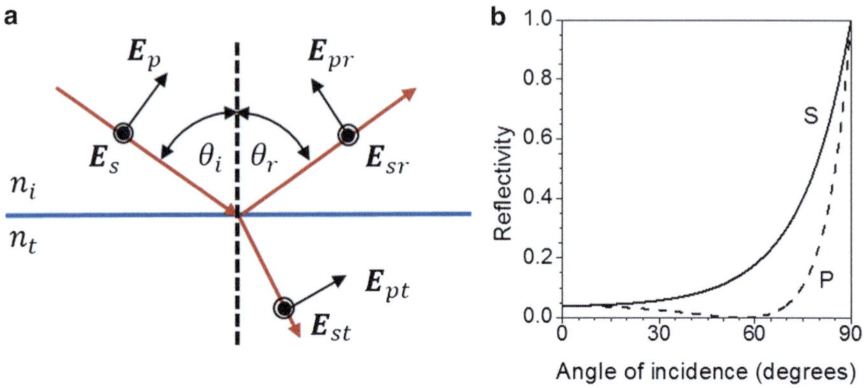

Fig. 2 (**a**) Schematic representation of light polarizations for p and s polarization showing the incident, reflected and transmitted electric fields and angles (θ). (**b**) Dependence of reflectivity on the angle of incidence for S polarized (*solid*) and P polarized (*dashed*) light

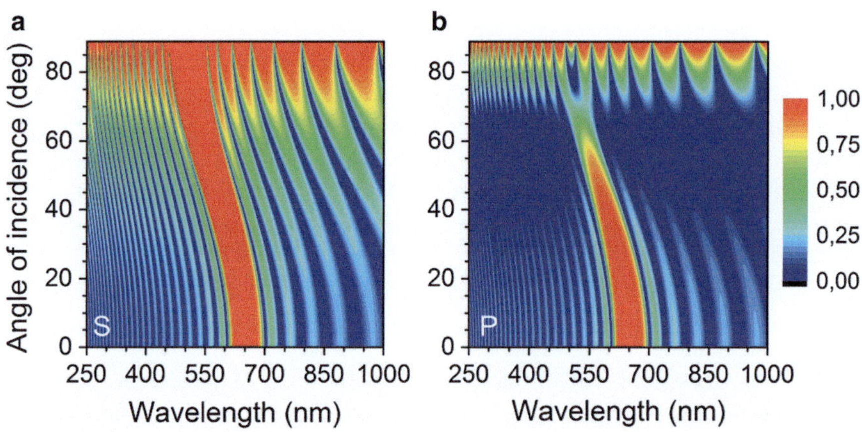

Fig. 3 Contour plots of a DBR reflectance spectra as a function of the incidence angle for S (**a**) and P (**b**) polarization

the change of the optical path). More interestingly, light polarization plays a role in the shape of the band gaps. For S polarized waves (transverse electric, TE), the electric field oscillates only in the plane of the interfaces (i.e. orthogonal to the plane of incidence) while for P polarized (transverse magnetic, TM) waves, the electric field is parallel to the plane of incidence, i.e. the field oscillates both in the plane of the interface and out of it [12] (Fig. 2a).

Since under reflection S and P components have different incidence angle dependence, a very strong polarization dependence is expected also for DBRs response far from the normal incidence condition. The overall angle dependence of transmittance and reflectance spectra is reported as contour plots in Fig. 3 where the reflectance intensity is shown as a colour scale.

The band gap's position (the red band) moves toward smaller wavelengths independently from polarization. However, its width for *S* polarized waves doesn't change significantly upon increasing the incidence angle (Fig. 3a) while for *P* polarized waves, the band gap shrinks in amplitude and spectral width due to the effects of the Brewster's angle (Fig. 2b).

The use of polymers in photonics and optoelectronics is well known [13, 14]. Semiconducting conjugated polymers are widely investigated and exploited as active materials for OLED, photovoltaic cells, transistors and several different devices. Insulating polymers have found few applications in photonics, mainly as passive waveguides/optical fibres or as host for organic and inorganic photoactive guests. The first all-polymer photonic crystal structure proposed has been a DBR prepared by plasma assisted CVD [15, 16]. However, such growth technique was not very developed at the time and only recently seems to find a renewed interest (see Coclite's chapter for more details). In recent years, all-polymer photonic crystal structures have also been widely studied when grown by block-copolymers self-assembly [17–19], co-extrusion [20, 21] and spin coating [3, 5–7, 9, 10, 22–30]. We are not going to discuss all these very stimulating methods since block-copolymers are reported in the sixth chapter while co-extrusion is described in the seventh chapter. There, a detailed review of literature in the field is reported. Here, we limit the discussion to the spin coating technique applied to polymers, nanocrystals and their composites. Nanocomposites have been successfully used to prepare flexible free-standing mirrors by the spin coating technique [29, 30]. Colloidal nanocrystals have also been used to prepare porous DBR [4, 7, 29, 31–34] for different applications that will be discussed later on in this chapter.

2 Spin Coating Multilayer Fabrication

Spin coating has proven to be a reliable, highly reproducible, cost-effective technique for the fabrication of vertical microstructures with polymers or colloids, and is today widely employed both in research applications and in large-scale industrial processes such as microelectronics manufacturing [35]. The technique is conceptually very simple. A planar substrate is coated by dispensing on it an excess of a viscosity controlled solution containing the material that has to be deposited (e.g. a polymer or a nanoparticle dispersed in a suitable solvent). The substrate is then accelerated up to a predetermined rotational speed. Most of the solution is expelled from the surface due to centrifugal force, leaving a film that progressively thins under the combined effect of material outflow and solvent evaporation until it has dried. The technique is very flexible and allows us to obtain a wide range of possible thicknesses varying the solvent employed, the concentration of the material in the polymer solution/colloidal suspension, the rotational speed, the spinning time (if the rotation is stopped before complete drying is reached) as well as the environment conditions such as temperature or humidity that may affect the solvent evaporation rate. However, this flexibility can also be a drawback since it is very difficult to have a firm control over all of the growth conditions.

However, there are some general guidelines that can be followed when using spin coating in order to obtain high quality structures.

Spin Speed

The rotational speed is one of the most important parameters in spin coating. The speed of the substrate modifies the radial force that is applied to the liquid solution and even the characteristics of the air turbulence in the immediate proximity of the substrate. Generally, the maximum speed reached during the procedure determines the thickness of the film that will be obtained. It largely depends upon the balance between the shear force at the borders of the fluid and the evaporation rate. Higher speeds determine lower thicknesses.

Acceleration

Since the solution starts drying before the final speed is reached, the substrate acceleration toward its final rotational speed may influence the properties of the film. Moreover, since the acceleration also induces a non-radial force, it may be used to more uniformly disperse the solution over the surface of the substrate.

Vapour Exhaustion

The solvent evaporation rate during spinning is determined by the solvent properties and by the laboratory environment. Modifying the temperature or the ventilation of the spin chamber can drastically rise the evaporation rate and change spin coating mechanics. Moreover, humidity coupled to high volatile solvents may give rise to "breathing figures" on the surface or even within the film that might affect its transparency [36]. When possible, it is useful to control the parameters of the air surrounding the sample modifying, for example, the vapour exhaustion using close chamber spin coaters.

Despite the simplicity of its qualitative description, it is quite difficult to provide a mathematical, rigorous model to express the final film thickness as a function of the previously mentioned parameters due to the fact that the film formation is essentially a surface phenomenon resulting from an interplay of fluid dynamics, thermodynamics and chemistry of the fluid as well as of the fluid/substrate interface. Many theoretical studies have been conducted for different solvents and polymers but no definitive general laws have been obtained. It is possible however to find general rules that are followed by every spin coating process. The film thickness decreases as the spin speed increases by a power law [37–39]

$$h_l = \frac{kx_0}{\omega^{\alpha}} \tag{3}$$

where k is a constant, x_0 is the initial concentration of the solution, and omega is the angular spinning velocity. The value of the exponent α depends on the details of the process considered but its values are usually around 0.5 [37, 40].

Another aspect to be examined is the film thickness variation during the process. Its value decreases as the substrate is kept spinning by [37]:

$$\frac{dh}{dt} = -\frac{2\rho_s \omega^2}{3\eta_s(t)} \tag{4}$$

where ρ_s is the initial solution density and η_s its viscosity. The change in concentration during spinning affects the hydrodynamics of the process through the power-law dependence of viscosity on concentration [41]. Since there is also a clear dependence on ω^2, the film formation process is faster as the spin speed is higher.

When experimental techniques suitable to measure film thickness are not routinely available, it is difficult to correctly determine the parameters to introduce in theoretical models to obtain reliable predictions. In such cases, spin coating may require several attempts with different parameter sets to establish a reliable fabrication procedure.

Substrate cleaning plays a major role for polymer DBR quality, as well as operating in a sufficiently dust-free environment since micron sized particles or traces present on the substrate before deposition can prove catastrophic for the morphology and the quality of the film, often resulting in defects such as comets or striations. Usually this can be prevented, or at least reduced, by using wet and dry treatment to eliminate organic impurities [42, 43].

The choice of the solvent is important as well, most notably in the fabrication of polymer multilayers and Bragg stacks. In fact, in these structures the layers are built one on another and the solvent used for the first polymer solution (or the liquid phase for the colloidal suspensions) must not dissolve the layer beneath. This condition is called solvents' orthogonality and is required to obtain high optical quality photonic crystals having well-defined interfaces and very low roughnesses. For instance, it has been demonstrated that polystyrene (PS)/cellulose acetate (CA) and polyvinylcarbazole (PVK)/CA DBRs, polymer pairs providing high quality DBR and microcavities, [3, 26] have roughness of about 1 nm [44]. The solvent choice also impacts on the evaporation rate, as well as on the surface tension created between the fluid and the substrate and so on the final morphology. If proven necessary, the addition of surfactants in the solution may prevent aggregation, and improve the overall quality of the film. Alternatively, filtration of the solution before deposition can be carried out to eliminate the presence of aggregates, but a word of caution is needed since this will alter the actual material concentration in the dispensed solution, thus affecting reproducibility.

Selection of material concentration and rotational speed according to the desired final film's thickness are, as stated above, usually to be made by trying different set of parameters and checking results by using a profilometer, a SEM microscope, ellipsometry or interference microscopy. Intuitively, a higher rotational speed or a lower material concentration results in a thinner film. However, concentration changes usually imply modifications of the final film porosity as well: since the average refractive index of a layer is strongly dependent on its porosity, concentration has a tremendous impact on the optical properties of a multilayer, affecting the position of the Bragg peak (changing the optical path by influencing both the average refractive index and final layer thickness) as well as its depth (by changing the index contrast). As such, the choice of concentration can prove to be an extremely powerful tool for tuning optical properties if mastered correctly.

Porosity also has another effect. In spite of solvent orthogonality, if the superficial layer is porous, the solvent of the material used for casting the new layer might reach the underlying film of the same polymer thus dissolving it and inducing interfacial disorder. The effect of this phenomenon is deleterious on the quality of the crystal and in most cases it means that the structure that is being grown must be discarded. DBRs grown by using PVK (refractive index $n = 1.68$) and cellulose acetate (refractive index $n = 1.47$) suffer of this problem. Indeed, the diacetone alcohol (DAA) used to dissolve the low refractive index CA ($n = 1.47$) can percolate through the layers of PVK and dissolve the underlying CA films. To avoid this, every freshly grown PVK layer must undergo a thermal annealing treatment (at about 80 °C) that reduces porosity and makes the PVK film more compact. This also has the additional advantage of eliminating residual solvent and increases the film stability in preparation for further layer coatings [45].

3 Polymer and Hybrid Multilayers Applications

We are now going to discuss different applications of polymer and porous colloidal DBR prepared by spin coating. We start with distributed feedback (DFB) and microcavity lasers then we proceed with photovoltaic devices, electro-optical switches and sensors eventually supporting Bloch surface waves (BSW).

3.1 DFB Laser

Polymer Bragg stacks can be used to fabricate flexible optical devices, for example a fully plastic multilayer laser. An example of such cost-effective, optically pumped, device is reported here, and should provide the reader an insight into the applications of a photonic structure as a way to provide the optical feedback in a lasing system, as well as affecting gain (and consequently the threshold energy density).

DFB laser is one of the most important technological developments in laser physics, currently representing the cornerstone of optical communications by means of wavelength division multiplexing (WDM) that is possible thanks to their narrow band emission that allows the "transmission band" of optical fibres to be divided in several (almost) non-interfering channels. These devices are able to achieve such a narrow band output by modulating the active medium's emission spectrum via a Bragg resonator that gives rise to an interference similar to the one obtained by exploiting a high quality optical cavity.

A flexible polymeric cavity can be made by employing the spin coating technique, alternating, for example, layers of PVK and CA. Due to the compactness of the final structure that prevents infiltration, the active medium (i.e. the molecular species that actually undergoes stimulated emission) must be enclosed in the optical cavity during its fabrication, by adding it to one (or both) of the solutions used to fabricate the multilayer. In our example, Rhodamine 6G (R6G) was blended in the cellulose acetate layer. The multilayer optical structure was modelled to obtain the optimal layers thickness for the full exploitation of the dye emitting properties. For an active material embedded in a cavity with optical path l (refractive index times the thickness), the gain g is proportional to the transit time of the travelling photons, which is determined by their group velocity v_g in the medium. Since in a 1D photonic crystal $v_g = c/n_{eff}$, where n_{eff} is the effective refractive index of the alternating layers, the effective gain g_{eff} can be roughly estimated by considering the increasing of the optical path l, by using the equations

$$\frac{g_{eff}}{g} = \frac{l_{eff}}{l} = \frac{c \times n_{eff}}{v_g} \tag{5}$$

The dielectric periodicity can be modelled, so that both the effective refractive index of the structure and the photon's group velocity can be estimated by considering the thickness of the alternating layers (a and b for PVK and CA, respectively) and their refraction indexes ($n_2 = 1.68$ and $n_1 = 1.47$ at 600 nm for PVK and CA, respectively).

The effective refractive index can be obtained from the Lorentz–Lorenz relationship which gives, in our case, $n_{eff} = 1.505$ [46]. By considering the one-dimensional wave equation for periodic Bloch eigenfunctions [47], it is possible to write an analytical expression for the ratio v_g/c as a function of a, b, n_2 and n_1

$$\frac{v_g}{c} \div \frac{\left(\frac{\alpha}{n_2} + \frac{\beta}{n_1}\right)\sqrt{16n_1^2n_2^2 - \left\{(n_1+n_2)^2\cos\left[2\pi(\alpha+\beta)\right] - (n_1-n_2)^2\cos\left[2\pi(\alpha-\beta)\right]\right\}^2}}{(n_1+n_2)^2(\alpha+\beta)\sin\left[2\pi(\alpha+\beta)\right] - (n_1-n_2)^2(\alpha-\beta)\sin\left[2\pi(\alpha-\beta)\right]} \tag{6}$$

where $\alpha = an_2/\lambda$ and $\beta = bn_1/\lambda$ are dimensionless parameters taking into account the dielectric layers optical thickness in terms of the wavelength λ [48]. To obtain an efficient lasing action, the photonic band-gap (PBG) must be matched with the R6G emission band. Inside this gap, v_g is purely imaginary, and corresponds to an evanescent wave in the lattice as expected for Bragg reflectors [1]. At the band edges the group velocity is real but tends to zero: these are the regions where the enhanced gain provided by the photonic structure is the highest, and where we expect to observe the stimulated emission from the dye.

Figure 4a shows the ratio v_g/c calculated by Eq. (6) (purple line). The thickness parameters have been optimized in order to have one of the photonic band edges in correspondence of R6G gain region. The selected couple of calculated thicknesses is $a = 45$ nm and $b = 140$ nm. Figure 4b instead reports the transmission spectra of a 19 layers system created using similar thicknesses. There, in addition to the first and second order of photonic band gap, the R6G absorption is observed.

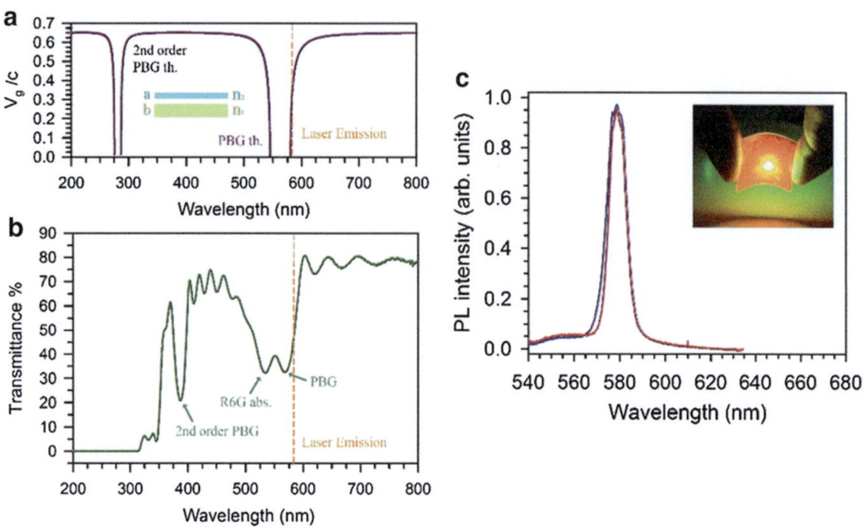

Fig. 4 (**a**) Computed group velocity in the 1D periodic structure as function of the wavelength of travelling photons with $a = 45$ nm, $b = 140$ nm, $n_2 = 1.683$ and $n_1 = 1.475$. (**b**) Transmission spectrum of a 19 layers flexible multilayer sample recorded at normal incidence, showing features originating both from the optical structure and from the absorption spectra of the materials employed. (**c**) Over threshold emission for the flat (*blue line*) and the bended multilayer (*red line*). *Inset*: picture of the fabricated multilayer (Reproduced with permission from [10])

It can be seen from Fig. 4a that the emitter frequency is tuned to spectrally overlap the band gap low energy edge. As previously discussed, being negligible the group velocity, the light matter interaction is enhanced thus increasing the optical gain. The effect also reduces the threshold energy density required for laser operations with respect to a simple polymeric matrix doped with the dye (performed measures showed a reduction of up to an order of magnitude [49]). The device, shown in the inset of Fig. 4c, is very stable in terms of laser peak position, despite any sort of bending. As shown in Fig. 4c, we didn't observe any substantial difference in the laser peak position and width (blue line for the flat sample, red line for the bent sample). This behaviour suggests that the variation of the thickness induced by the stretching is negligible [10].

DFB laser cavities can be built starting from inorganic nanoparticles too [4]. Usually, the main difference in the fabrication process is linked to the high temperature intermediate sintering step required both to improve layers' morphology after deposition and to allow complete drying. The temperature range of several hundreds of degrees is of course devastating for an organic dye that has to be used as active medium, thus imposing insertion in the optical cavity after its fabrication: fortunately, the high degree of porosity attainable in nanoparticle-based Bragg stacks often makes this a straightforward procedure, resulting in a sufficient active medium uniformity throughout the multilayer.

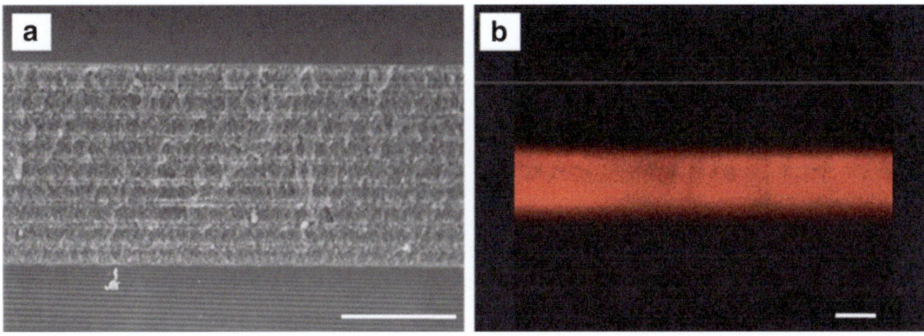

Fig. 5 (**a**) SEM cross section image of a SiO_2/TiO_2 nanoparticle 1D PC (scale bar = 1 μm); (**b**) confocal microscopy image of the dye-loaded nanoparticle 1D PC (scale bar = 1.5 μm) (Reproduced with permission from [4])

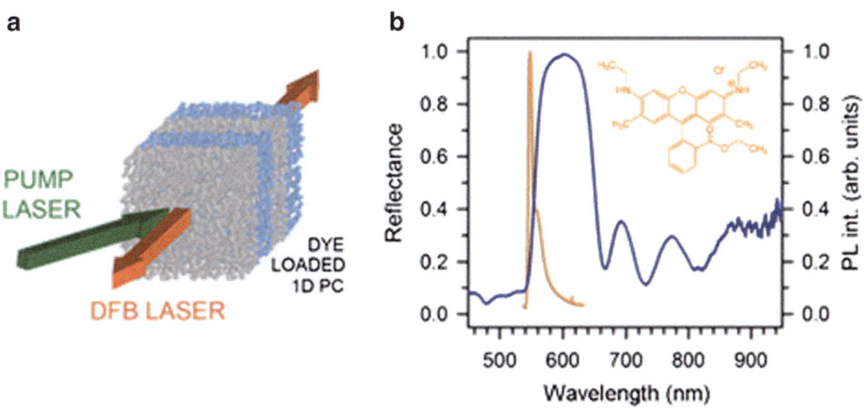

Fig. 6 (**a**) Schematic representation of the sample; (**b**) experimental reflectance spectrum (*blue line*) and photoluminescence (*orange line*) of the dye-loaded nanoparticle 1D PC excited by pulsed laser light at 532 nm. The *inset* shows the molecular structure of the R6G dye (Reproduced with permission from [4])

Hybrid porous DBR made of SiO_2/TiO_2 nanoparticles can be easily infiltrated with R6G solutions. A scanning electron microscopy and a PL confocal picture of such system are reported in Fig. 5a, b.

The experimental reflectance spectrum of a SiO_2/TiO_2 1D PC infiltrated with R6G is provided in Fig. 6b (blue line). It displays a high reflectivity broadband from 550 to 650 nm, with a maximum at about 600 nm. This effect could be entirely ascribed to the periodicity of the dielectric lattice which generates a photonic band gap. The central wavelength of the forbidden band can be calculated by the Bragg–Snell diffraction law:

$$m\lambda = 2(n_h d_h + n_l d_l) \tag{7}$$

where m is the diffraction order, d_l and d_h are the respective thicknesses of the low and high refractive index materials and n_l and n_h are the respective refractive indexes. According to Eq. (7), the stopband or reflected frequency of a 1D PC structure can be manipulated to span the entire range of visible spectral wavelengths by varying the optical thicknesses of the layers comprising the structure; this can be easily achieved with our methodology by controlling the nanoparticle concentration of the dispersion and the spin-rate of the spin coating deposition step. In addition, the intensity of the stop band can be tuned by the modification of the number of bilayers in the photonic crystal according to Eq. (1) (see also first chapter of this book for more details).

By optically pumping the active medium with an off normal laser beam whose power density exceeds the lasing threshold (Fig. 6a), a very narrow emission peak is observed (Fig. 6b, orange line). This corresponds to the spectral region where the emission spectrum of the dye and the blue region of the photonic bandgap overlap. It can be seen that the emission is strongly enhanced by the v_g near 0 (born from a high photonic density of states) as previously described in Fig. 4a.

These two examples of emission enhancement can also be interpreted as an effect of light localization due to the variational theorem applied to PhC [11] (see, for instance, first chapter of this book). At the high energy PBG edge the field is localized within the low dielectric constant material (SiO_2) while at the low energy one it is localized into the high dielectric constant medium (TiO_2). Then, the fluorophore has to be embedded inside the material having refractive index suitable to spectrally tune the band gap edge to its fluorescence. A similar selective localization effect has been also observed for gold nanoparticle doped opals [50]. We notice that in the case of R6G infiltrated colloidal DBR, an increase of the DOS at the low energy PBG edge is expected too. However no PL enhancement in such spectral region is observed. This lack is due to the poor spectral overlap in this region between R6G fluorescence and PBG edge. It is useful to notice that wavelengths falling within the photonic band gap are suppressed and do not give rise to a consistent emission [4].

3.2 All-Polymer Microcavities

Microcavities are an interesting and still developing field of research with applications ranging from laser devices' engineering to quantum computing and strong light–matter interaction (See chapters by D. G. Lidzey et al. and by R. Brückner et al. of this book). A microcavity is essentially a structure formed by two mirrors confining electromagnetic radiation in the region between them (the cavity). The first example of organic microcavity was a hybrid system where a fluorescent conjugated polymer was sandwiched between a metallic mirror and an inorganic DBR [51]. However, metallic mirrors give rise to absorption losses and then the use of dielectric mirrors is much more interesting. As we have already commented in previous sections, polymer multi-layered structures perfectly fit to the aim since they can provide very high reflectivity.

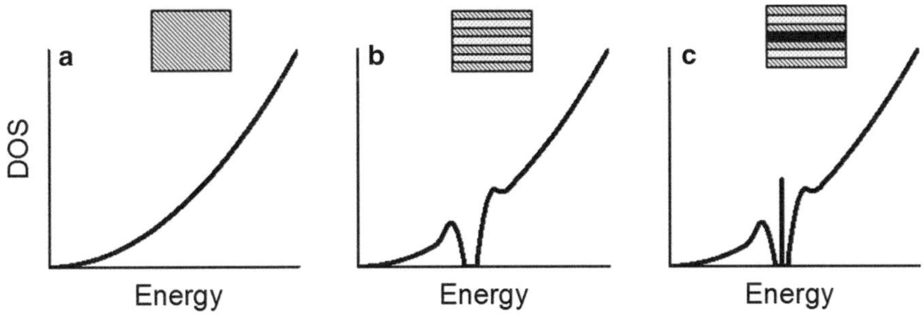

Fig. 7 Sketch of the photonic DOS for (**a**) free photons in a homogeneous medium, (**b**) photons in a PhC (DBR) and (**c**) photons in a microcavity

The planar microcavity can be simply seen as a volume comprised between a pair of Bragg mirrors. However, in a PhC dielectric lattice the cavity plays the role of a periodicity defect. As such, it creates photonic states allowing light propagation through the structure even though they have frequency inside the band gap. In order to better describe the role of the defect, it is useful to consider the photonic Density Of States (DOS), i.e. the number of photonic states available to the system per unit of energy (DOS $\div (\partial E/\partial k)^{-1}$), in order to better understand the properties of such structures [52].

In a homogeneous medium (the effective medium) the DOS has a parabolic photon energy dependence (Fig. 7a). In a PhC (Fig. 7b) the effective medium DOS is modified by the presence of the gap (DOS = 0) and by a relatively small increase at the PBG's edges due to reduction of the group velocity ($v_g \div (\partial E/\partial k)^{-1}$) [50] (see also Eq. (5) and first chapter for a detailed description).

In the case of microcavity, the latter picture is further modified by the presence of sharp peak within the PBG (Fig. 7c). The density of photonic states affects the radiative decay rate ($\tau \div \text{DOS}^{-1}$) of an emitter embedded into the cavity [52]. An acceleration of the radiative decay rate is expected for photons corresponding to the cavity mode. Such modes can be exploited to achieve an extreme narrowing of a dye's emission spectrum. Indeed, the oscillator strength of the emitter for frequencies within the PBG is distributed following the DOS which is squeezed into the cavity mode. The overall effect is a deep change of the fluorescence spectrum with a redistribution of the oscillator strength at different wavelengths associated with specific directional properties.

The transmittance spectrum of a 15 bilayers PVK:CA microcavity embedding a poly[(9,9-di-*n*-octylfluorenyl-2,7-diyl)-alt-(benzo[2,1,3]thiadiazol-4,8-diyl)] (F8BT) layer is reported in Fig. 8a. The spectrum is characterized by the broad absorption feature of the fluorescent material around 450 nm and by the PBG of the DBR structure in the range 530–600 nm. Within this band the cavity mode peak is detected at ~560 nm. The PBG of the DBR and the thickness of the cavity have been engineered in order to be finely tuned on the amplified spontaneous emission peak of F8BT (Fig. 8b) [3]. It is interesting to compare the steady state photoluminescence (PL) spectrum of the microcavity with the corresponding one for the bare

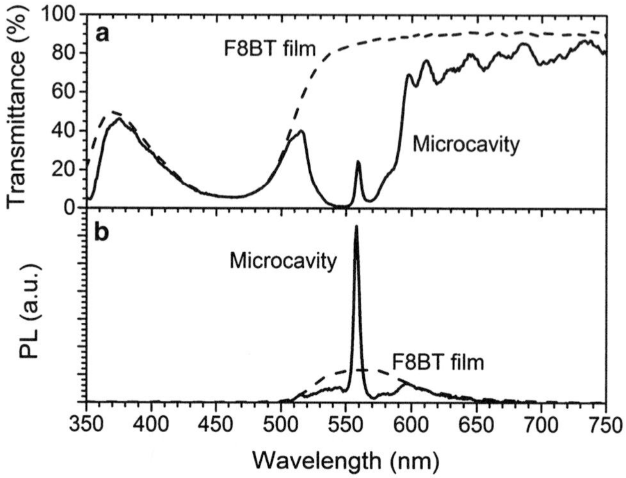

Fig. 8 (**a**) Transmittance spectra for both PVK:CA microcavity with an F8BT defect layer and neat F8BT film. (**b**) Fluorescence spectrum of the PVK:CA microcavity doped with F8BT and of F8BT film

F8BT film (Fig. 8b) The spectral redistribution due to the DOS modification deeply changes the broad PL spectra of the film into a very sharp emission corresponding to the cavity mode where the intensity is dramatically increased. On the other hand, for photon energies within the PBG, the PL emission is inhibited. Notice the very sharp emission ($\Delta E = 6\,\text{nm}$) of the microcavity, which correspond to a relatively large quality factor

$$Q = \frac{E}{\Delta E} \tag{8}$$

It is possible to routinely obtain quality factors of about 150 while the best Q found was 250 ($\Delta E = 2.7$ nm) [3].

Notice that when similar microcavities are prepared with polystyrene instead of PVK, i.e. a material with lower refractive index (1.59 against 1.67), the Q cavity factor is lower and a broader microcavity PL is detected [26].

These structures have been successfully used to achieve lasing action. It has been shown that the microcavity emission is composed of two components having different dependence on the pump fluence. The first one possesses an almost pump power independent width (4 nm). The second one instead shows a clear sharpening with a threshold like behaviour at about $10\,\mu\text{J/cm}^2$ unambiguously indicating the lasing regime [3].

The possibility to prepare by a simple and cheap procedure microcavities providing lasing action at very low threshold is very interesting for photonics since it is combined with mechanical flexibility. Indeed, such microcavities can be peeled off from the substrate and bent or applied to curved surfaces thus providing opportunities unprecedented for their inorganic homologous [3, 26].

3.3 Photonic Photovoltaic Cells

Bragg mirrors can be easily integrated in solar cells in order to improve their efficiency, since they can be used as large bandwidth reflectors at the bottom of the solar cell to reflect the non-absorbed photons back in the device [53]. This topic is more extensively discussed for DBR structures in organic photovoltaic cells in the Martorell' et al. chapter of this book, while additional effects due to photonic structures on photovoltaic cells are also reported in chapters by Lova et al., Marabelli et al, and S. Romanov of this book.

In this paragraph we would like to stress the exploitation of porous Bragg mirrors in dye-sensitized solar cells. In such devices, for example, the porous Bragg mirrors can be placed as active materials in the solar cells, taking advantage of their porosity which allows dye soaking [54]. Alternatively, Heiniger et al. [55] showed that the employment of a Bragg mirror composed by SiO_2 nanoparticle layers and sputtered tin-doped indium oxide (ITO) layers can be exploited in solar cells. Since the silica nanoparticle film is partially infiltrated by the ITO, a continuous conducting network is created inside the Bragg mirror, enabling its use as a replacement for the usual FTO based counter electrode. Moreover, the Bragg mirror, acting as a back-reflector, enhances the photoconductivity of the cell preserving the semi-transparency.

As shown in Fig. 9, a judicious optimization of photoanode thickness and photonic band gap position results in very promising internal photo-current efficiency (IPCE) improvements. Indeed, when the low dye absorbing region is

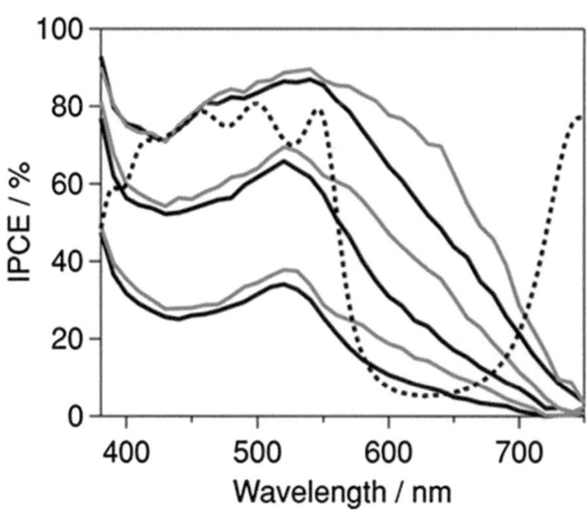

Fig. 9 IPCE for photovoltaic cells, with TiO_2 photoanodes of different thickness, without (*black curve*) and with (*grey curve*) back reflecting porous DBR. The thickness of the photoanodes is 1 μm, 2.8 μm and 9.3 μm, respectively, from bottom to top. The conducting Bragg mirror, that acts also as counter electrode, shows a transmission spectrum that is superimposed to the spectral region where the IPCE increase is expected (*dashed curve*) (Reproduced with permission from [55])

overlapped with the photonic band gap, a remarkable IPCE enhancement is observed, due to back reflection of the light. Remarkably, the enhancement, at $\lambda = 620$ nm, is 82 % for the 9.3 μm thick TiO_2 photoanode, 65 % for the 2.8 μm thick photoanode, and 31 % for the 1 μm thick photoanode, respectively.

3.4 Electro-Optic Switches

As stated previously, the positions of the transmission minima of a 1D photonic crystal are identified as the wavelengths that fulfil the Bragg relation. Infiltration of a molecular species by means of a liquid or gaseous solution in a porous photonic structure causes a change in the value of the effective refractive indexes. This affects the optical path of a hypothetical light beam travelling within the structure and the position of the Bragg peak, due to the different refractive index of the "guest" species [34]. The outcome is very similar if the structure contains a guest that changes its refractive index in response to external stimuli, like a liquid crystal, whose optical properties can be tuned through the application of an electric field. For an extended discussion of the use of liquid crystals in PhC, see chapter by Quan Li et al. of this book.

The porous 1D photonic crystal has been fabricated using spin coating of colloidal suspensions of oxide nanoparticles (e.g. silicon dioxide, zirconium dioxide or titanium dioxide, Fig. 10a). The higher refractive index contrast that can be achieved compared to the ones obtainable using polymeric structures often allows us to achieve a significant optical performance (i.e. a good transmission modulation) with fewer layers [see Eq. (1)].

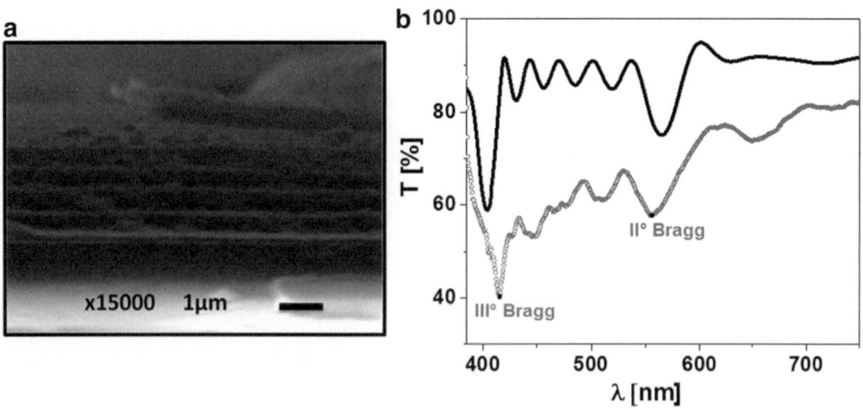

Fig. 10 (**a**) SEM cross section image and (**b**) experimental (*open circle curve*) and simulation (by Transfer Matrix Method, *black line*) transmission spectrum of six bilayer zirconium dioxide/silicon dioxide nanoparticle photonic crystal. The fundamental, second and third orders of the photonic band gap are indicated (Reproduced with permission from [34])

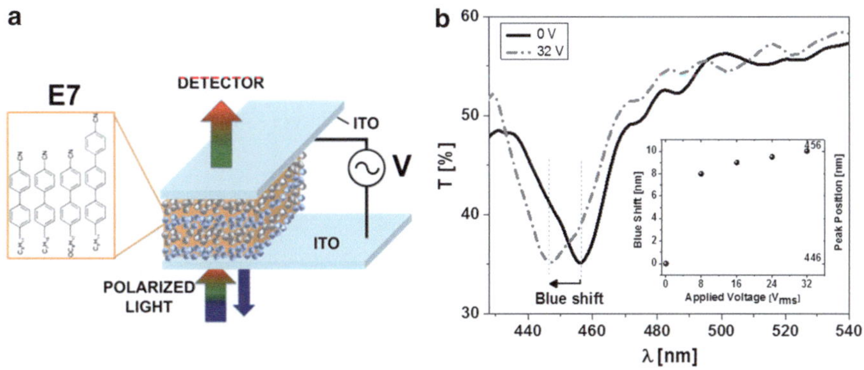

Fig. 11 (**a**) Scheme of the device for photonic band gap tuning applying an electric voltage. (**b**) Transmission spectrum of the liquid crystal doped photonic crystal for different applied voltages (from 0 to 32 V). The blue shift of the third order photonic band gap is displayed. *Inset*: The spectral position of the third order photonic band gap as a function of the applied voltage (Adapted with permission from [34])

Infiltration of a liquid crystal in the structure is a simple and straightforward process which effects are directly observable as a redshift of the Bragg peak due to the guest species filling the pores and increasing the effective refractive index of the multilayer (Fig. 10b). The actual redistribution processes of the liquid crystal mesophases within the porous structure are far from trivial and it is not addressed here.

In order to achieve the refractive index modulation by means of the liquid crystal molecular director rotation, the crystal must be sandwiched between two electrodes so that a suitable electric field bias can be applied to the structure. To simplify the overall architecture and to allow transparency in the visible range, the crystal can be built directly on an ITO substrate, while a second ITO electrode can be added on top of the multilayer after fabrication (Fig. 11a).

Photonic band gap shift in such devices has been demonstrated, as shown in Fig. 11b, with relatively weak applied bias voltages. Thanks to the small thickness of the multilayer, few volts provide quite a strong electric field to drive the liquid crystal mesophase rearrangement thus offering intriguing applications for low-cost optoelectronic switches [56] and displays [57, 58].

3.5 Sensors

We report here some example of 1D PhC sensors. Additional details and examples for PhC structures used as stimuli responsive sensors are reported also in 2nd, 8th, 9th, 14th, 15th, 17th and 18th chapters of this book.

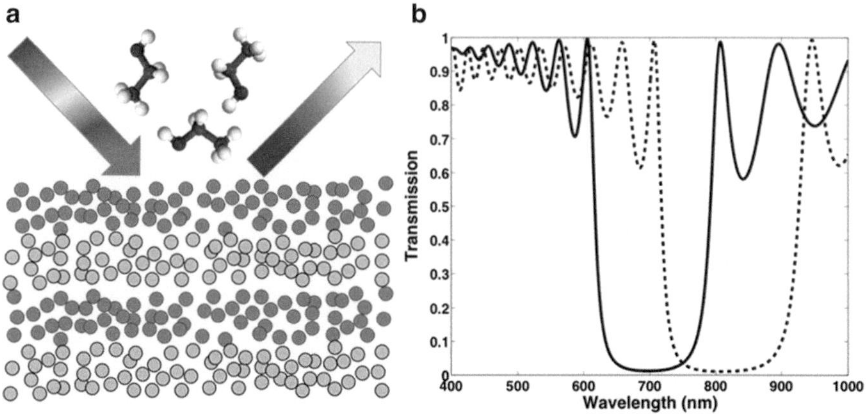

Fig. 12 Schematic representation of a multi-layered photonic crystal architecture illustrating the reflection of incident light (**a**) as well as the optical response upon exposure to external stimuli (**b**)

The porosity of nanoparticle-based and mesoporous Bragg mirrors can be easily exploited for sensing applications, in which the analytes are infiltrated in the photonic crystals, resulting in a change of their optical properties [59].

For example, the fabrication of one-dimensional photonic crystals based on a microporous metal-organic framework material and mesoporous titanium dioxide is proposed for vapour sensing [18]. In this two-component photonic crystal multilayer, zeolitic imidazolate framework (ZIF) ZIF-8 and mesoporous titanium dioxide were selected as different functional components: ZIF-8 for molecular selectivity, mesoporous TiO_2 to ensure high refractive index contrast and to guarantee molecular diffusion within the Bragg stack.

In this structure, schematically shown in Fig. 12a, the photonic band gap position depends not only on the refractive index of the two porous layers (i.e. ZIF-8 and TiO_2), but also on the refractive index of compounds present in the pores of the layers. By filling the pores with an analyte, a band gap shift is observed (as simulated, dashed curves in Fig. 12b). As an example, by permeating pores of the layers with methanol or ethanol vapours (partial pressure, p identical to saturation pressure, p_0), Hinterholzinger et al. [18] recorded a larger shift of the photonic band gap for ethanol (Fig. 12b), which is attributed to more beneficial interactions with ZIF-8 owing to the larger hydrophobicity of ethanol compared to that of methanol.

Figure 13 shows the response of the DBR structure to different alcohols partial pressures (p/p_0). An important finding is the fact that distinctly shaped isotherms are obtained for each of the four analytes (i.e. methanol, ethanol, isobutanol and *tert*-pentanol) over the entire relative pressure range. The characteristic sorption behaviour indicates a high degree of chemical selectivity inherent to the metal-organic-framework-based Bragg mirror, which is especially noticeable at low relative pressures; in fact, at a partial pressure p/p_0 of 0.1 and 0.2, an abrupt increase

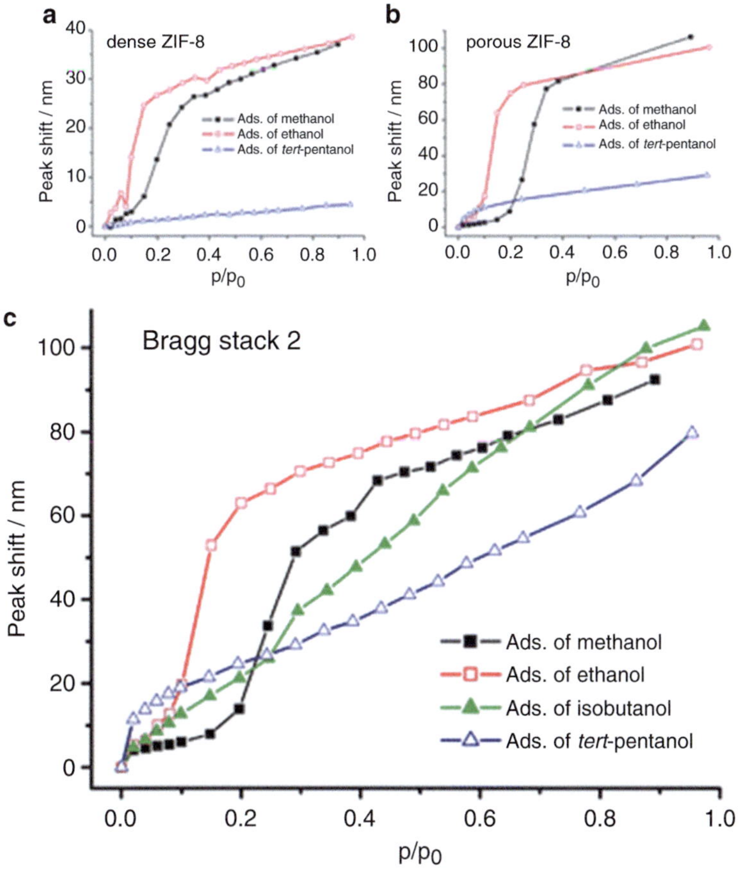

Fig. 13 Optical vapour sorption isotherms demonstrating the adsorption performance of (**a**) dense and (**b**) nanoparticulate ZIF-8 reference samples as well as of (**c**) during exposure to a series of alcohol vapours (Reproduced with permission from [18])

in the isotherms of methanol and ethanol, respectively. Comparison of the shapes of the isotherms for the Bragg mirror (Fig. 13c) and the individual ZIF-8 thin films (Fig. 13a, b) confirms that the optical response is dominated by ZIF-8. This is due to the fact that the analytes with the smaller kinetic diameters, as methanol and ethanol, are more readily adsorbed by the porous ZIF-8 layers. Moreover, while for ethanol and methanol the spectral shift associated with vapour uptake in film and DBR is comparable, for *tert*-pentanol a larger optical shift response is observed for the DBR structure with respect to the ZIF-8 film alone.

The results show that the multi-layered photonic heterostructures are sensitive and selective towards a series of chemically similar solvent vapours. It is thus anticipated that the concept of multilayer heterogeneous photonic structures will provide a versatile platform for future selective, label-free optical sensors.

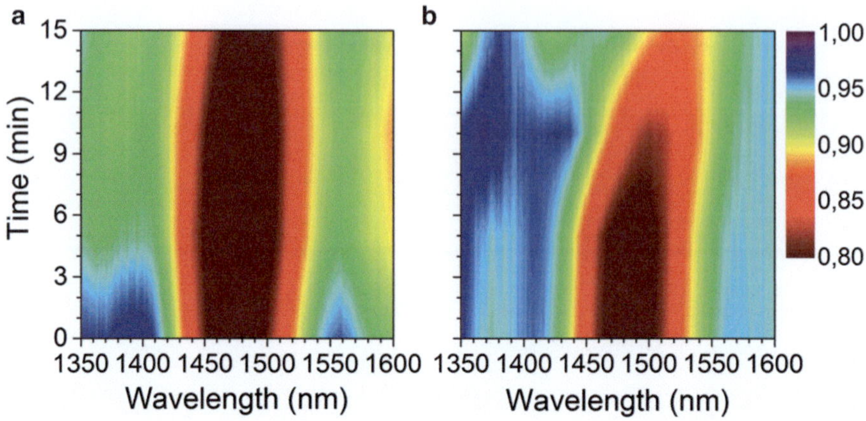

Fig. 14 Transmittance contour plot spectra of (**a**) PS: CA and (**b**) NC: CA DBRs as a function of exposure time to toluene vapours

Another solution to obtain vapour sensible PhC is to use all-polymer photonic crystals. This typically requires the presence either of highly porous matrices [60–63] or of macromolecules undergoing strong swelling effects [17, 64] as building material for the photonic structure. Indeed, photonic crystal sensors generally exploit the modification of refractive index or lattice periodicity due to molecules penetration into the structure [61, 65]. Even if it is well known that penetration of liquid analytes into polymer PhC can produce a strong variation of such parameters [17, 66, 67] (we remind again to see chapters by Furumi and by Quan Li et al. of this book for the role of swelling in porous 3D PhC sensors), gas and vapour sensing might present some issues. Indeed, gas penetration inducing strong thickness and/or refractive index changes can hardly be thought to be effective for standard macromolecules used to prepare all-polymer photonic crystals by spin coating such as polystyrene, polyvinylcarbazole, cellulose acetate and so on. However, it has been recently demonstrated that the addition of a relatively small amount of ZnO nanoparticles to a polystyrene matrix (2.5 % by volume) deeply affect the permeability properties of the nanocomposite (NC) thus inducing a remarkable swelling [8, 9].

In Fig. 14 we report the contour plot of transmittance spectra of 15 bilayers PS: CA and NC:CA DBRs as a function of exposure time to toluene vapours in the spectral region of the first diffraction PBG [8, 9].

For PS:CA DBR, minor spectral changes are observed within the first 15 min of exposure. For NC:CA DBR, a remarkable change of 50 nm is observed using the same exposure conditions (Fig. 14b). Similar effects occur for the higher order PBGs. Such impressive differences in the spectral response of the two structures may be due to PS swelling after exposure to non-polar solvents [68]. The much stronger effect observed for the NC:CA DBR is due to the increased permeability of NC (with respect to bare PS) probably driven by the disorder and free volume effects induced by the ZnO NPs into the amorphous PS host matrix. The low

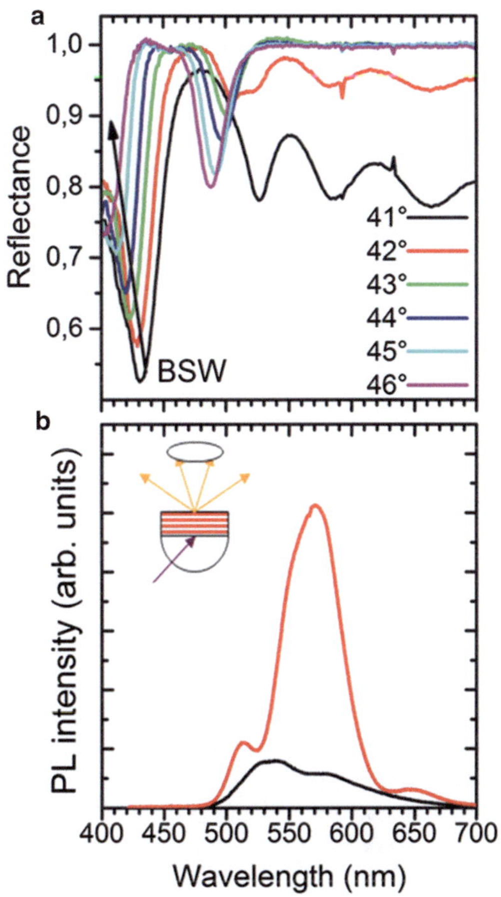

Fig. 15 (**a**) *s*-polarized Attenuated Total Reflectance (ATR) spectra as a function of the incident angle for a CA: PVK DBR covered by an F8BT thin film in the ATR regime; (**b**) Photoluminescence spectra of F8BT film on a glass slide (in *black*) and on top of the DBR on the prism (in *red*). Excitation was obtained with a 405 nm laser impinging from the back with *s*-polarization (scheme of the optical geometry for BSW excitation in the *inset*) [27]

response of standard PS:CA DBR should not instead surprise being PS widely used for packaging thanks to its low permeability to gas and vapours.

Another possible application of all-polymer DBRs as optical sensors is their use in order to excite highly localized waves running at the interfaces between the multilayers and capping layers, i.e. the so-called BSW [27, 69–73]. This item will be described in detail in Chapter by Descrovi et al. where the theory and several examples will be reported; additional comments can be found in chapter by Marabelli et al. where a similar optical configuration is adopted for plasmonic sensors. Here, we would like just to highlight a recent work where spin-coated PVK:CA DBRs are demonstrated for the first time to be suitable to support such excitations [27]. The confinement of BSW at the interface of the DBR with the external environment is given from one side by the total internal reflection condition and from the other side by the PBG. Such waves have to be excited through a prism in order to establish the total internal reflection condition (Fig. 15a). When this happens, the waves travel at the interface where they are very sensitive to the environment. These systems are very promising for sensing since they couple high sensitivity to the lack of optical losses found instead in the analogous structures where the dielectric DBR is replaced by a metallic layer that provides plasmonic excitations [69]. The novelty of the work of Fornasari et al. [27] is double:

1. They demonstrated that even an all-polymer DBR structure, intrinsically less ordered (in particular for the interface sharpness) than inorganic counterparts, does support BSW.
2. They engineered the PhC structure to enhance the excitation of a BSW, which is then used to induce the fluorescence of the well-known semiconducting polymer F8BT.

These results are summarized in Fig. 15a, b. The reflectance spectra as a function of the incidence angle (for angles suitable to the total internal reflection condition) shows, in addition to the PBG structure, a deep minimum shifting from 440 to 400 nm which has been modelled and assigned to a BSW excitation [27]. Notice that, for an angle of about 46°, the BSW is located at 405 nm, i.e. the emission line of a commercial (and cheap) violet laser. If we now replace the white light beam used to probe the optical response of our system with such a laser, the F8BT fluorescence can be strongly excited. The PL intensity is dependent on the incidence angle. In particular, the strongest enhancement of the PL signal is observed when the incidence angle is ~46° i.e. when the BSW is excited at the laser wavelength. In this case the PL excitation is enhanced since it's localized at the interface where the fluorescent polymer is inserted. If we compare the intensity of the PL signal under BSW excitation with the one recorded for a similar F8BT film without the multilayer structure, a tenfold increase of the signal is observed (Fig. 15b).

This result is particularly interesting when this optical setup could be integrated in a fluorescence sensor. In that case, the increase of the PL signal can be used to improve the sensitivity. This seems to be even more interesting when free-standing all-polymer DBRs [3, 8, 9, 26] are used since they can be directly adapted to curved surfaces.

Another interesting sensing application with porous Bragg mirrors is the responsivity to temperature. Thermo-responsive TiO_2/SiO_2 one-dimensional photonic crystals fabricated via sol-gel processing methods have been proposed [74, 75]. They represent a promising class of environmentally responsive nanostructures featuring optically encoded temperature and humidity detection. The thermo-optic response of the layer materials is amplified by their inherent porosity owing to adsorption/desorption of ambient humidity into the mesoporous multilayer structure. Again, the idea is to exploit the spectral shift of the photonic band gap as a function of temperature.

It is interesting to study the hysteresis properties of TiO_2/SiO_2 nanoparticles Bragg mirrors during multiple heating/cooling cycles (Fig. 16), as well as the response and recovery times (≈2–4 s) of the multilayer system during external changes in ambient humidity [74, 75]. The optically detected hysteresis may be

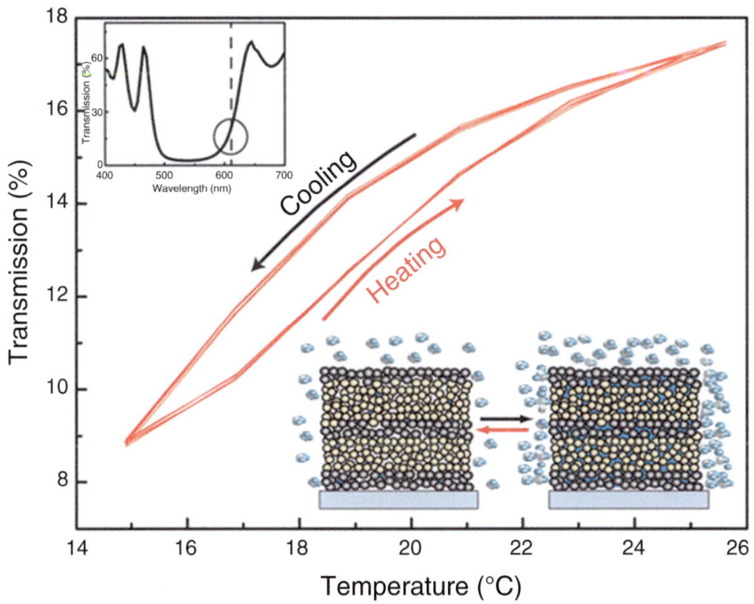

Fig. 16 Heating and cooling curves recorded by monitoring the transmission at 610 nm and 25 % RH between 15 and 25 °C. The stop band of the investigated Bragg mirror is placed in the *upper left corner*. The temperature was cycled five times from 15 to 25 °C in steps of 2 °C and backward with an equilibration time for each temperature of 60 s. The hysteresis may be caused by capillary condensation of ambient humidity into the mesoporous network of the Bragg mirror as is schematically shown in the illustration (*bottom right*) (Reproduced with permission from [75])

caused by capillary condensation of ambient humidity into the mesoporous network of the Bragg mirror during the heating cycles.

Bonifacio et al. [76] proposed a photonic nose, based on an array of photonic crystals with different responsivity to specific analytes. They introduce the photonic nose as a viable and cost-effective approach to address problems relevant to foodstuff monitoring and drinking-water analysis. Detection of low levels of lead in water has been demonstrated as well as the monitoring of storage conditions for fish and meat samples from the vapour headspace composition. They propose the use of a digital colour image system, i.e. a photonic nose, for the optical detection of different analytes [63, 76].

4 Conclusions and Perspectives

In this chapter we described the way to produce polymer and porous colloidal photonic crystals by means of the reliable and cost-effective spin coating technique. Several applications ranging from lasing to sensing, photovoltaics and electro-optical modulation have been reported. We also highlighted the possibility to prepare self-supporting, flexible devices, properties inconceivable for bulk inorganic materials.

We have underlined how active materials' properties (fluorescence, absorption, conductivities and chemical responsivity) can be intensified by exploiting the light localization effects, photonic density of states enhancement and porosity provided by the suitably engineered photonic crystal structures.

Acknowledgments The work in Genova is supported by the Italian Ministry of University and Scientific and Technological Research through the project 2010XLLNM3 (PRIN 2010–2011).

References

1. E. Yablonovitch, Phys. Rev. Lett. **58** (1987)
2. S. John, Phys. Rev. Lett. **58** (1987)
3. G. Canazza et al., Laser Phys. Lett. **11**, 035804 (2014)
4. F. Scotognella et al., Small **5**, 2048 (2009)
5. V.M. Menon et al., Opt. Express **16**, 19535 (2008)
6. L.M. Goldenberg, V. Lisinetskii, S. Schrader, Laser Phys. Lett. **10**, 055808 (2013)
7. S. Colodrero, et al., Adv. Mater. **21** (2009)
8. P. Lova, et al. ACS Photonics, 2, 537 (2015)
9. P. Lova, et al. Phys. Status Solidi C 12, 158 (2015)
10. F. Scotognella et al., Phys. Chem. Chem. Phys. **12**, 337 (2010)
11. J.D. Joannopulos, R.D. Meade, J.N. Win, *Photonic Crystals: Molding the Flow of the Light* (Princeton University Press, Princeton, 1995)
12. E. Hecht, *Optics* (Addison-Wesley, Reading, 2002)
13. G. Lanzani, *Photophysics of Molecular Materials: From Single Molecules to Single Crystals* (Wiley, New York, 2006)
14. P. Samorì, F. Cacialli, *Functional Supramolecular Architectures: For Organic Electronics and Nanotechnology, 2 Volume Set* (Wiley, Weinheim, 2014)
15. A. Convertino et al., Appl. Phys. Lett. **71**, 732 (1997)
16. A. Convertino, A. Valentini, R. Cingolani, Appl. Phys. Lett. **75**, 322 (1999)
17. J.-H. Lee, et al., Adv. Mater. **26** (2014)
18. F.M. Hinterholzinger et al., J. Mater. Chem. **22**, 10356 (2012)
19. C. Park, J. Yoon, E.L. Thomas, Polymer **44**, 6725 (2003)
20. G. Mao, et al., Opt. Mater. Express **1** (2011)
21. H. Song et al., J. Mater. Chem. **19**, 7520 (2009)
22. T. Komikado, S. Yoshida, S. Umegaki, Appl. Phys. Lett. **89**, 061123 (2006)
23. L.M. Goldenberg et al., Opt. Mater. Express **2**, 11 (2012)
24. N.V. Valappil et al., Photonic Nanostruct. **5**, 184 (2007)
25. F. Scotognella et al., Int. J. Photoenergy **2008**, 389034 (2008)
26. L. Frezza et al., J. Phys. Chem. C **115**, 19939 (2011)
27. L. Fornasari et al., Appl. Phys. Lett. **105**, 053303 (2014)
28. A.L. Álvarez, et al., Thin Solid Films **433** (2003)
29. M.E. Calvo et al., Energy Environ. Sci. **4**, 4800 (2011)
30. S. Colodrero et al., Langmuir **24**, 9135 (2008)
31. D.P. Puzzo et al., Nanoletters **9**, 4273 (2009)
32. F. Scotognella et al., Chem. Mater. **23**, 805 (2011)
33. A.C. Arsenault et al., Macromol. Symp. **196**, 63 (2003)

34. L. Criante, F. Scotognella, J. Phys. Chem. C **116**, 21572 (2012)
35. R.K. Yonkoski, D.S. Soane, J. Appl. Phys. **72**, 725 (1992)
36. A. Bolognesi, et al., Langmuir **21** (2005)
37. C.-C. Chang et al., Thin Solid Films **479**, 254 (2005)
38. C.J. Lawrence, Phys. Fluids **31**, 2786 (1988)
39. P. Yimsiri, M.R. Mackley, Chem. Eng. Sci. **61**, 3496 (2006)
40. K. Norrman, A. Ghanbari-Siahkali, N.B. Larsen, Annu. Rep. Prog. Chem. Sect. C Phys. Chem. **101**, 174 (2005)
41. D. Meyerhofer, J. Appl. Phys. **49**, 3993 (1978)
42. J.J. Cras et al., Biosens. Bioelectron. **14**, 683 (1999)
43. L. Henke, N. Nagy, U.J. Krull, Biosens. Bioelectron. **17**, 547 (2002)
44. R.R.K. Unger, C. Czibula, C. Ganser, C. Teichert, G. Jakopic, G. Canazza, S. Gazzo, D. Comoretto, in *Distributed Bragg Reflectors: Morphology of Cellulose Acetate and Polystyrene Multilayer*. 16th International Conference on Transparent Optical Networks (ICTON), Graz (AUT), 6–10 July 2014. IEEE Xplore Digital Library
45. T. Komikado et al., Thin Solid Films **515**, 3887 (2007)
46. M. Montalti, A. Credi, L. Prodi, M.T. Gandolfi, Photophysical properties of organic compounds, in *Handbook of Photochemistry,* 3rd edn. (CRC, Boca Raton, 2006), pp. 83–351
47. V. Morandi et al., Adv. Funct. Mater. **17**, 2779 (2007)
48. M. Born, E. Wolf, *Principles of Optics*, 6th edn. (with corrections) (Pergamon Press, Oxford, 1980)
49. J.P. Dowling, et al., J. Appl. Phys. **75** (1994)
50. V. Morandi et al., J. Phys. Chem. C **112**, 6293 (2008)
51. N. Tessler, G.J. Denton, R.H. Friend, Nature **382**, 695 (1996)
52. M. Barth, A. Gruber, F. Cichos, Phys. Rev. B **72**, 085129 (2005)
53. P. Bermel, et al., Opt. Express **15** (2007)
54. S. Colodrero, et al., J. Phys. Chem. C **113** (2009)
55. L.P. Heiniger et al., Adv. Mater. **25**, 5734 (2013)
56. D.E. Lucchetta, L. Criante, F. Simoni, J. Appl. Phys. **93** (2003)
57. A.C. Arsenault, et al., Nat Photon **1** (2007)
58. D.P. Puzzo et al., Angew. Chem. Int. Ed. **48**, 943 (2009)
59. S.Y. Choi et al., Nano Lett. **6**, 2456 (2006)
60. C.-Y. Kuo, et al., Sens. Actuators B Chem. **124** (2007)
61. M. Ben-Moshe, V.L. Alexeev, S.A. Asher, Anal. Chem. **78** (2006)
62. T. Cassagneau, F. Caruso, Adv. Mater. **14**, 34 (2002)
63. L.D. Bonifacio et al., Adv. Mater. **22**, 1351 (2010)
64. J.-T. Zhang, N. Smith, S.A. Asher, Anal. Chem. **84**, 6416 (2012)
65. X. Xu, A.V. Goponenko, S.A. Asher, J. Am. Chem. Soc. **130** (2008)
66. S.A.A. Michelle M. Ward Muscatello, Adv. Funct. Mater. 18 (2008)
67. S.A. Asher, K.W. Kimble, J.P. Walker, Chem. Mater. **20** (2008)
68. L. Nicolais et al., Polymer **18**, 1137 (1977)
69. M. Liscidini et al., Appl. Phys. Lett. **98**, 121118 (2011)
70. M. Liscidini, J.E. Sipe, Appl. Phys. Lett. **91** (2007)
71. S. Pirotta et al., Appl. Phys. Lett. **104**, 051111 (2014)
72. H. Clevenson et al., Appl. Phys. Lett. **104**, 241108 (2014)
73. R. Badugu et al., Anal. Biochem. **442**, 83 (2013)
74. A.T. Exner, et al., ACS Appl. Mater. Interfaces **5** (2013)
75. I. Pavlichenko, et al., J. Intell. Mat. Syst. Struct. **24**, 2204 (2013)
76. L.D. Bonifacio, G.A. Ozin, A.C. Arsenault, Small **7**, 3153 (2011)

Spherical Colloid Engineering

Katia Sparnacci and Michele Laus

Abstract This chapter provides an overview of the synthesis and functionalization of monodisperse colloidal spheres, a class of colloidal materials that represents the building blocks for the fabrication of photonic crystals. Although the most commonly employed materials are polystyrene and silica, in this contribution we expand the material range through the use of specific synthetic methods. Details about the particle forming mechanism are provided with special emphasis to the control over the size, size distribution, and particle surface characteristics. Colloidal spheres with a solid, hollow, or core shell structure are described, including a variety of organic–inorganic hybrid particles. The incorporation of fluorescent materials into the spherical colloidal particles is also described.

Keywords Colloidal particle synthesis • Monodispersity • Polymerization processes • Particle size control • Particle surface control

1 Introduction

A colloid is a heterogeneous system consisting of a dispersed phase, whose size, at least in one dimension, is in the range from 1 nm to 1 μm, and a dispersion medium [1, 2]. Dispersed phase and dispersion medium can be solid, liquid, or gas and, accordingly, the colloidal systems are usually classified depending on the state of the dispersed phase and the dispersion medium. When the dispersion medium is gas or liquid, the dispersed particles are subjected to intense Brownian motions, i.e. random translational diffusion resulting from particle collisions.

This chapter reports a critical overview of the synthetic approaches to prepare colloidal systems consisting of monodisperse spherical particles, featuring diameters from 20 nm to 1 μm, dispersed in liquids. These particles, under appropriate conditions, can spontaneously self-assemble into macroscopic, highly ordered arrays, ultimately leading to colloidal photonic crystals (CPCs) [3, 4].

K. Sparnacci (✉) • M. Laus
Dipartimento di Scienze e Innovazione Tecnologica (DISIT), Università del Piemonte Orientale "A. Avogadro", Viale T. Michel 11, 15121 Alessandria, Italy
e-mail: katia.sparnacci@mfn.unipmn.it; michele.laus@mfn.unipmn.it

D. Comoretto (ed.), *Organic and Hybrid Photonic Crystals*,
DOI 10.1007/978-3-319-16580-6_5

Gem opals are natural CPCs consisting of monodisperse silica spheres, with average diameter in the range 15–900 nm, ordered in a face centered close-packed structure (FCC) [5]. In a parallel fashion, man-made CPCs are usually called artificial opals [6].

When light interacts with FCC structures, strong diffraction effects from each sphere plane occur. Constructive and destructive interference of diffracted beams determines whether the light is allowed to propagate or not inside the structure. As a result, a photonic band gap (PBG), also referred to as stop band, that is a band of frequencies where light propagation in the photonic crystal is forbidden (the optical analogue of the electronic band gap in semiconductors) is obtained [7]. Photons with the forbidden energies are backward diffracted (i.e., reflected) thus giving rise to the typical reflectance peak (Bragg peak) mainly responsible for the color of such structures. Although CPCs do not possess a complete photonic band gap, yet still they retain great interest for photonics since represent simple and easily affordable elements to build up optical devices. Furthermore, CPCs can be employed as templates for the preparation of several infiltrated structures [8], called inverse opals, using inorganic, organic, metallic, or ceramic components. In this case, when a high refractive index material is employed in the preparation of macroporous inverse opals, a full band gap can be obtained. In fact, the growing of silicon inside the voids of an opal template of close-packed silica spheres, followed by removal of the silica template, allows an inverted opal to be obtained showing a complete photonic band gap in the IR region where silicon is transparent (1.46 μm), and presents a high refractive index (n_D (1.5 μm) = 3.45) [9]. More details in the application of CPCs as template for the fabrication of infiltrated structures can be found in second chapter of this book.

The spectral position of the photonic stop band (λ_B) in CPCs is provided by the Bragg–Snell law [10]:

$$m\lambda_B = 2D\sqrt{n_{eff}^2 - \sin^2\theta} \tag{1}$$

where m is the diffraction order, D is the interplanar spacing, n_{eff} is the effective refractive index, and θ is the incidence angle. The interplanar spacing is determined by the diameter of the colloidal particles (d) and, in the case of a close-packed FCC lattice, is given by:

$$D = \left(\frac{2}{3}\right)^{0.5} d \tag{2}$$

In turn, the particle diameter determines the size of the FCC cell and, then, the spacing between the particle layers, thus determining the spectral position of the PBG [11]. Consequently, by controlling the size of the colloidal particles employed in the self-assembly process, it is possible to finely tune the CPCs optical properties.

In this contest, it is important to stress that size control extends not only to the size of the employed colloidal particles, but also to their size distribution. Highly monodisperse colloidal particles, with size distribution within 2 % of standard deviation [12] must be employed in the preparation of high quality opals. In addition, as the self-assembly process involves anchoring of the particles to the substrate surface as well as lateral interactions among the particles in the opal, special care should be addressed to the design of the particle surface characteristics. In this context, it should be stressed that the particle surface represents a key parameter in determining the optic and mechanical performances of CPCs [4]. Surface functionalization includes both the introduction of functional and/or reactive groups onto the particle surface and the formation of core-shell or multilayer morphology particles. Different methods have been developed to introduce functional groups, as $-NH_2$, $-COOH$, $-OH$, $-SH$, $-SO_3H$, $-Br$ and $-CH=CH_2$, on the surface of organic and inorganic particles. These functional groups can modify the stability and self-assembly properties of the colloidal particles. In addition, the presence of these groups allows easy chemical modification of the particles [13] to obtain photonic crystals responsive to external stimuli for sensing and biosensing [14] applications. Interested readers can find a review on the preparation of gel-immobilized colloidal crystals and their application as biological and chemical sensors in 19th chapter of this book.

Finally, the photonic band gap of CPCs can modify the emissive properties of fluorescent molecules embedded into the photonic crystal when the emission spectrum peak of the fluorophore is matched with the stop band [11]. The feedback mechanism inside the opal can give rise to lasing or enhanced stimulated emission [15], as described in 17th chapter of this book. Consequently, different methods have been developed to load colloidal particles with light emitting dopants.

Within this general frame, a great deal of work has been addressed to the optimization of the chemical methods employed for the synthesis of monodisperse spherical particles with designed size and surface characteristics, made of both organic and inorganic [16–19] materials.

Figure 1 shows the schematic structure of three representative colloidal particles that can be employed as building blocks for the self-assembly of CPCs, namely solid homogeneous functional particle (a), core-shell particle (b), and hollow particle (c).

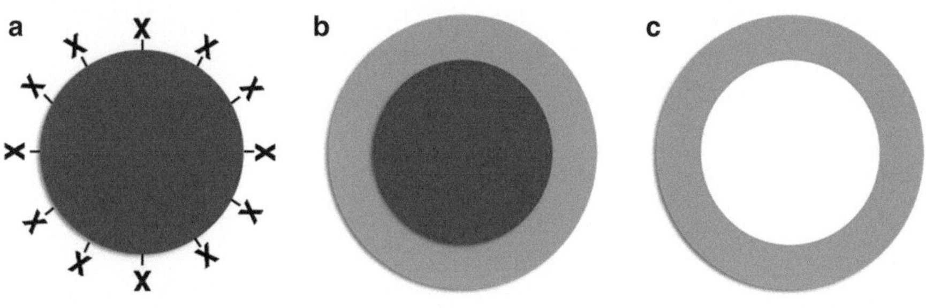

Fig. 1 Schematic structures of three representative colloidal particles: solid homogeneous functional particle (with *X* representing the surface functional groups) (**a**), core-shell particle, (**b**) and hollow particle (**c**)

Section 2 provides an overview of the synthetic methods available for the production of monodisperse spherical colloidal particles, both inorganic and polymeric. Within the former, silica particles are described in detail as they represent the most common inorganic particle system for the preparation of CPCs. Concerning the polymeric particles, several synthetic strategies are described with particular attention to the control over the particle size and the surface characteristics.

Section 3 is devoted to the introduction of specific functionalities or creation of special morphologies in the colloidal particles, thus representing an additional tool to tune the CPCs characteristics. In this regard, two different aspects will be highlighted:

- The production of core-shell and hollow particles with complex structures.
- Incorporation of fluorescent materials, into the colloidal particles, to study the influence of the photonic band gap on the fluorophore emission.

2 Synthetic Methods

A rich variety of chemical methods is available for the production of monodisperse spherical colloidal particles, both inorganic and polymeric, with diameters ranging from a few nanometers to 1 μm [16–19]. A critical overview of these synthetic methods will be given in this section.

Figure 2 shows, as a typical example, scanning electron microscopy (SEM) images of two representative colloidal samples: 220 nm silica particles (Fig. 2a) and 230 nm polystyrene particles (Fig. 2b).

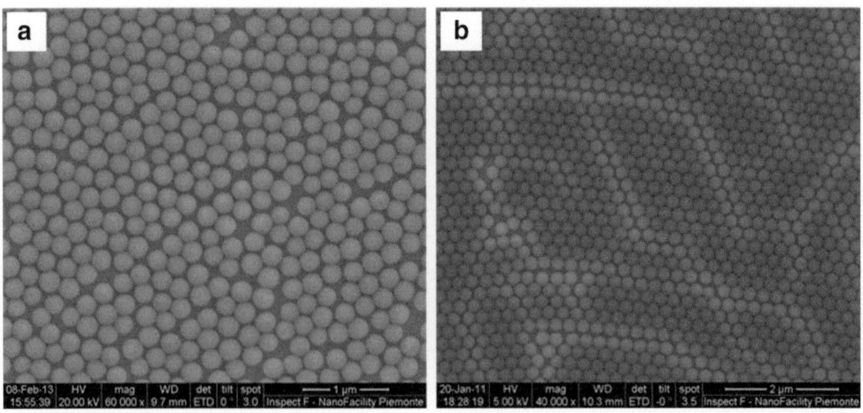

Fig. 2 SEM images of two representative colloidal samples: 220 nm silica particles (**a**) and 230 nm polystyrene particles (**b**)

2.1 *Inorganic Nanoparticles*

Inorganic colloids are attractive because of their high refractive index, high thermal stability, and improved resistance toward organic solvents compared to polymer particles. They are usually obtained via precipitation reactions. According to La Mer's model [20], this process involves two sequential steps: nucleation and nuclei growth. To obtain monodisperse colloidal particles, these two steps must be strictly separated and nucleation should be avoided during the period of growth. Consequently, it is necessary to precisely control the reaction conditions (temperature, pH, ionic strength of the medium, reactant concentrations, and reactant mixing procedure) to obtain a single, short burst of nuclei and then let these nuclei grow uniformly.

Matijevic [19, 21] deeply investigated this process and successfully prepared monodispersed colloidal spheres, cubes, rods, and ellipsoids in a broad range of materials including metal oxides and carbonates.

Recently, Xia and coworkers developed various synthetic methods to prepare uniform colloidal spheres in the mesoscale dimension [18] made up of metals or semiconductor materials. Bottom-up, top-down, and template directed synthesis were employed to produce colloidal spheres with solid, hollow, and core-shell structures with chemical compositions including Se, Bi, Pb, In, Sn, Cd, Pt, Ag_2Se, CdSe, PbS, and TiO_2. In this way, smart colloidal crystals were obtained featuring interesting characteristics, such as thermally switchable bandgaps in case of Ag_2Se, tacking advantage of its reversible polymorphic transition.

2.1.1 Silica Particles

Silica colloid particles represent one of the best characterized inorganic systems. They are generally obtained by the method developed in 1968 by Stöber and Fink [22] which involves the ammonia-catalyzed hydrolysis of tetraethylortosilicate (TEOS) in ethanol at room temperature yielding the precipitation of spherical silica particles whose size could be varied from 50 nm to 2 μm by changing the concentration of the reactants. This method was further investigated and improved by many authors. Bogush [23] developed a correlation between the final particle size and the initial reagent concentrations, with particular emphasis to the determination of the concentration window over which monodispersity is achieved. Van Blaaderen [24] elucidated the mechanism of particle formation and growth. The formation of Stöber particles was demonstrated to start with the hydrolysis and condensation of TEOS, which generate a supersaturated solution of oligomeric precursors. After an induction time, nucleation from this highly supersaturated solution occurs, thus generating small and unstable particles that aggregate until reaching the critical size at which colloidal stability is achieved. Then, aggregation stops and the formed primary particles start growing. The final particle size relies on a balance between nucleation and particle growth, with both processes depending on the relative rates of the condensation and hydrolysis reactions. Figure 3 shows the Stöber process mechanism.

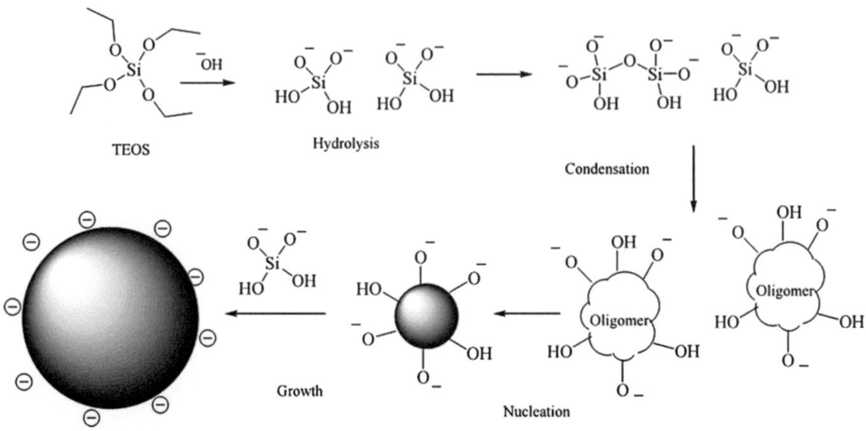

Fig. 3 Mechanism of the Stöber process for the synthesis of monodisperse colloidal silica

The main parameters affecting the reaction rate and the final particle size are the concentration of ammonia and water. Increasing both concentrations results in faster reaction rates and, initially, in an increase in particle size. However, a maximum particle size is obtained corresponding to specific water and ammonia concentrations. A further increase in the concentration of both reactants leads to smaller and even polydisperse particles. This result can be understood taking into account that colloidal stability depends on a balance between the amount of charged groups at the particle surface and the ionic strength of the reaction medium. An increase in water and ammonia concentration increases the surface charge through the dissociation of surface silanol groups, but, at the same time, an increase in the ionic strength of the reaction medium occurs due to the parallel increase of the concentration of various ionic species (NH_4^+ and OH^-).

A further improvement in particle size monodispersity can be achieved by means of a seeded growth technique, a two-step procedure in which preformed silica particles are employed as seeds for the subsequent growth reaction carried out by adding TEOS and water to the particle suspension. Provided that no secondary nucleation occurs, the number of particles remains constant and their size increases [25]. The particle size is then controlled by the amount of added TEOS and the size percentage standard deviation of the final particles decreases as the size increases. In this way, perfect spherical silica particles, with size polydispersity scaling as the inverse of the radius, were obtained [24].

Recently, Santamaria Razo et al. [26] prepared silica particles, with diameters ranging from 160 to 600 nm and less than 3 % of size polydispersity. They employed a one-step procedure based on the Stöber method, and controlled the particles diameters by varying the amount of TEOS. These particles allowed the fabrication of CPCs with a finely tuned and precisely predictable lattice parameters.

The main limitation of the Stöber process remains in preparation of small particles. Particles with sizes below ca. 120 nm are very difficult to obtain and generally have very high polydispersity (typically >20 %).

Monodisperse silica particles with size ranging from 30 to 60 nm can be obtained by the water-in-oil microemulsion system, where particles form in inverse micelles stabilized by a suitable surfactant in a non-polar organic solvent [27]. The microemulsion approach yields silica particles with superior average monodispersity compared to the Stöber method, which requires also the use of large amounts of surfactants (typically several time more by weight than silica) and often necessitates of extensive cleaning.

In this context, Yokoi et al. [28] reported the preparation of silica particles with size comprised between 12 and 23 nm employing the basic amino acid L-lysine in place of ammonia in a process analogous to the Stöber synthesis. They obtained particles with excellent size monodispersity, that self-assemble in highly regular CPCs.

Following the same approach, Kitaev et al. [29] developed a facile one-pot method to synthesize monodisperse silica spheres with size ranging from 15 to 200 nm. Small silica seeds were prepared by TEOS hydrolysis in aqueous solution using arginine as a catalyst and size-determining agent. These seeds were subsequently regrown in the same reaction media. The above particles formed regular close-packed arrays suitable for the production of inverse opals.

The surface of the as synthesized silica colloids comprises silanol groups (–Si–OH) that are ionized [30] at pH higher than 7. Consequently, the particle surface is negatively charged. Modification of the surface chemistry of the silica colloids can be performed employing different silane coupling agents. Figure 4 shows some of the common silane coupling agents used to this purpose.

The surface functionalization of the silica particle can be used to enhance and/or control the overall particle characteristics, including the chemical reactivity and the self-assembling properties [31, 32].

Surface modification can be performed either in a two-step process [33], where preformed silica particles are isolated and then further reacted with a silane

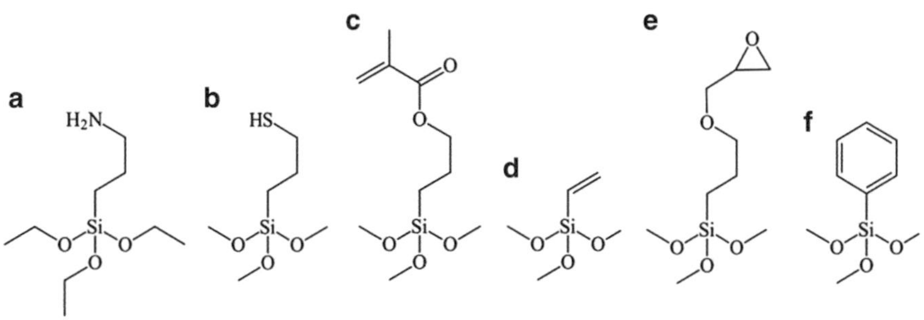

Fig. 4 Chemical structures of some of the most commonly employed silane coupling agents: 3-aminopropyl-triethoxysilane (APTES) (**a**), 3-mercaptopropyl-trimethoxysilane (MPTES) (**b**), 3-methacryloxypropyl-trimethoxysilane (MAPTES) (**c**), vinyl-trimethoxysilane (VTMS) (**d**), 3-glycidyloxypropyl-trimethoxysilane (GPTMS) (**e**), and phenyl-trimethoxysilane (**f**)

coupling agent, or in a one-pot reaction [34], where the functionalized silica spheres are directly obtained.

Wang et al. [35] prepared silica particles with diameters ranging from 400 to 850 nm by the Stöber method and subsequently grafted $-SO_3^-$ groups on their surfaces employing the 2-(4-chlorosulfonylphenyl)ethyl-trimethoxysilane silane coupling agent. These highly charged, monodisperse silica particles readily self-assemble into three-dimensionally ordered crystalline colloidal arrays (CCA) in water suspension. By evaporating water in the CCA, high quality CPCs were obtained with band gap in the near-infrared (NIR) spectral region.

Wu et al. [36] employed a one-step procedure to obtain highly monodisperse vinyl functionalized silica particles with size ranging from 420 to 650 nm and polydispersity below 2 %. The ammonia-catalyzed hydrolysis and condensation reactions of the vinyltriethoxysilane (VTES) organosilane were performed in water. The size of hybrid silica particles can be controlled by varying the dripping speed of the diluted ammonia catalyst, without changing the reagent concentration. These particles were employed to prepare CPCs by a simple gravity sedimentation process and provided ordered structures with good optical properties.

The modification of silica particles surface with silane coupling agents bearing hydrophobic organic groups switches the particle surface nature to hydrophobic. Such hydrophobic particles were generally prepared for the ability to float on the water surface that makes them suitable substrates for assembly methods like floating deposition and Langmuir–Blodget techniques [37].

2.2 Polymeric Particles

Polymer colloids, compared to their inorganic counterparts, offer the advantage of easier tailoring of many properties, such as thermal characteristics and refractive index, by simply changing the starting monomers [38]. The glass transition temperature, which determines the temperature working region of the obtained opals, can be varied between 80 and 160 °C. Additional crosslinking processes [39] allow the preparation of polymer opals stable up to 200 °C. Moreover, CPCs obtained from polymer particles can be easily post-processed for the incorporation of predefined defects [6]. Line and point defects were obtained by E-beam lithography [40] in polymer opals from methacrylates. The introduction of defects by UV lithography process [41] was obtained in photoprocessable poly(*tert*-butylmethacrylate) colloidal particles.

Monodisperse polymer colloidal particles are generally obtained via heterogeneous polymerization processes [42] in which the starting monomer(s) and/or the resulting polymers are in the form of a fine dispersion in an immiscible liquid. The final product is a stable dispersion of polymer particles in a non-solvent, called *polymer latex*. The heterogeneous polymerization techniques can be distinguished on the basis of the following criteria:

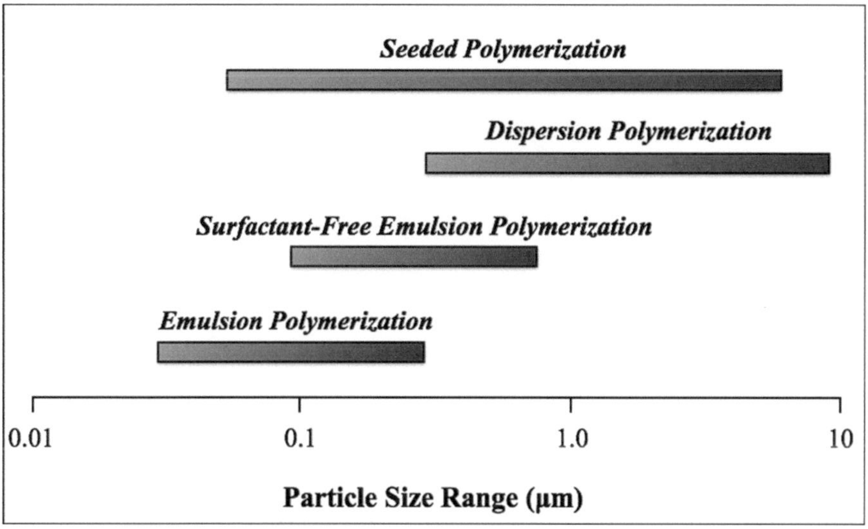

Fig. 5 Particle size range for various heterogeneous polymerization techniques

- Initial state of the polymerization mixture.
- Mechanism of particle formation.
- Shape and size of the final polymer particles.

These processes include emulsion polymerization, surfactant-free emulsion polymerization, dispersion polymerization, and seeded polymerization techniques leading to monodisperse spherical particles with diameters ranging from 50 to 1,000 nm. Figure 5 reports the particle size range corresponding to the above synthetic precesses.

2.2.1 Emulsion Polymerization

In a classical emulsion polymerization [43], a monomer, insoluble or scarcely soluble in the dispersion medium (usually water), is emulsified in it by the aid of a surfactant. The polymerization initiator is water soluble. In this system, the monomer is partly present in the form of droplets (10 µm in diameter or larger) stabilized by the surfactant, and partly in the form of monomer-swollen micelles (5–10 nm in diameter), depending on the surfactant nature and concentration. A small amount of the monomer is also molecularly dissolved in the medium (Fig. 6). For example, also in the case of a water insoluble monomer such as styrene, its solubility at 25 °C is about 0.3 g/l [44], whereas, for the more polar methylmethacrylate, the solubility at the same temperature results about 15 g/l.

It is important to notice that at the beginning of the polymerization, the monomer droplets are much bigger than the micelles and the typical micelles concentration is 10^{18}

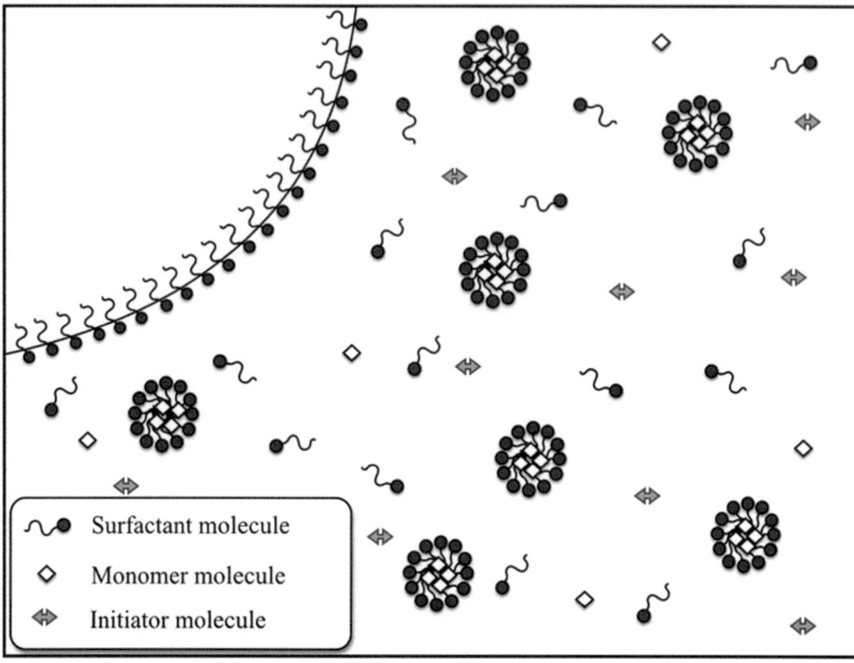

Fig. 6 Initial stage picture of the emulsion polymerization process

per ml, whereas the monomer droplet concentration is 10^{10}–10^{11} per ml. Consequently, the total micelle surface is much greater than the monomer droplets.

The polymerization reaction begins [45] with the decomposition of the initiator in water that generates a burst of primary radicals. These primary radicals polymerize the small amount of monomer present in the aqueous phase becoming oligoradicals that can either enter the monomer-swollen micelles (micellar nucleation) or phase separate in the form of primary particles (homogeneous nucleation) stabilized by the absorption of surfactant molecules from the reaction medium. In either case, the nuclei become the main loci of polymerization and grow, absorbing further oligoradicals and monomer molecules from the droplets, until all the monomer has been consumed. The monomer droplets cannot be the polymerization sites, given the fact that the initiators are insoluble in the monomer, and primary radicals are unlikely to enter the monomer droplets due to their lower surface area than the micelles. The polymerization leads to a polymer latex with particle size in the range 50–500 nm. A polymer particle with a diameter of 100 nm comprises approximately 1,000 polymeric chains, entangled as coils, each chain starting and ending with a functional charged group that originates from the decomposition of the radical initiator.

The size of particles obtained in emulsion polymerization has no direct relationship with the size of the initially formed monomer droplets and micelles. Instead, it is influenced by a number of different factors, including the monomer concentration, the surfactant nature and concentration, the initiator concentration, and the polymerization temperature. Generally, the particle size is directly proportional to the monomer concentration and inversely proportional to the surfactant concentration. Moreover, to obtain monodisperse particles the nucleation time should be extremely short with respect to particle growing time. By properly adjusting the reaction parameters, extremely monodisperse polymer particles can be obtained. Emulsion polymerization was recently employed by Ishii et al. [46] to prepare monodisperse polymer particles, with mean diameters of less than 100 nm, by using an amphoteric initiator and sub-millimolar concentrations of a range of anionic surfactants. Due to their very low size dispersity, these particles form highly regular, close-packed CPCs.

As the surfactant employed in the polymerization reaction remains in the final polymer latex, it is possible to obtain polymer particles with different surface characteristics using cationic, anionic, or non-ionic (steric stabilizer) surfactants. However, since the surfactant is not covalently attached to the particle surface, it can desorb and migrate from particle to particle as soon as they are close to each other in the self-assembly process, thus producing gluing of the colloids and defect formation in the obtained CPCs [38]. To overcome this problem, two techniques have been developed, i.e. the use of a polymerizable surfactant or the elimination of the surfactant from the emulsion polymerization (a process called surfactant-free emulsion polymerization or SFEP).

2.2.2 Emulsion Polymerization with Reactive Surfactants

The use in emulsion polymerization of a reactive surfactant, that is a hydrophilic ionic comonomer, able to stabilize the emulsion and, at the same time, to participate in the polymerization reaction, yields to polymer colloids whose ionic groups are covalently attached to the particle surface [47]. These reactive surfactants are sometimes called surfmers, an acronym formed from the words surfactant and monomers. In these systems, the complex particle forming process mostly involves homogeneous nucleation. The reaction starts in the aqueous phase leading to the formation of water-soluble oligoradicals, rich in the surfmer units, until they reach the limit of solubility and phase separate from the aqueous medium to form primary particles stabilized by the surfmer units located at the nanosphere surface. These particles grow, absorbing oligoradicals and monomer molecules from the system, leading to a polymer colloid in which the nature of the nanosphere surface is dictated by the chemical structure of the surfmer employed. A large variety of surfmers is reported in the literature [48] including anionic surfmers with sulfate or sulfonate head groups [49], cationic surfmers bearing quaternary ammonium groups [50, 51], and non-ionic surfmers [52]. At the end of the polymerization reaction, the structural units from the surfmer are covalently bound to the particle surface, and actively participate to the latex stabilization. Accordingly, the nature of the colloidal particle surface is dictated by the chemical structure of the reactive surfactant employed.

2.2.3 Surfactant Free Emulsion Polymerization

The SFEP is the most frequently employed method to synthesize monodisperse colloid for the preparation of synthetic opals [53]. In this technique, the polymerization is carried out in the same way as in the classical emulsion polymerization, except that no surfactant is added. The first stage of the polymerization involves the decomposition of the initiator in water that generates primary radicals. These radicals polymerize the monomer dissolved in the aqueous phase until they reach a critical chain length at which they are no more soluble and phase separate becoming oligoradical nuclei. Since there is no surfactant in the medium, the nuclei thus formed are not stabilized and collide, forming larger and larger particles. As the particles grow, however, their surface charge density, derived from the charged groups originated from the radical initiator, increases, until they become stabilized by their own electrostatic charges. Thereafter, the process is analogous to the classical emulsion polymerization, with the primary particles becoming the main polymerization loci and growing until all the monomer has been consumed.

The advantage of this method is that both the polymerization rate and the particle number are lower than in the classical emulsion polymerization. Consequently, the nucleation is very short with respect to the particle growth and a lower number of particles are produced thus leading to nearly monodisperse polymeric particles with size ranging from 100 to 700 nm. The size of the obtained particles can be easily tailored by varying the ratio of monomer to water and the ionic strength of the water phase [54, 55]. Moreover, particles with different surface charge can be obtained by employing initiators with negative or positive charges. In addition, the surface charge density of the obtained particles is rather low, and the charged groups are covalently linked to the particle surface, thus preventing the problems associated with the surfactant migration. As a typical example, Fig. 7 reports the chemical structures of two of the most commonly employed initiators, the negatively charged potassium peroxodisulphate (a), and the positively charged α,α′-azodiisobutyramidin dihydrochloride (b).

The SFEP has been studied in detail for styrene [55] and methacrylates [54]. In particular, a systematic investigation of SFEP was performed using three different methacrylates [54] (methyl methacrylate, tert-butyl methacrylate and 2,2,2-trifluoro ethyl methacrylate) to obtain monodisperse spherical particles with diameter ranging from 100 to 500 nm that have been easily crystallized into CPCs of a high optical quality. Waterhouse [56] prepared cationic PMMA colloidal particles, with

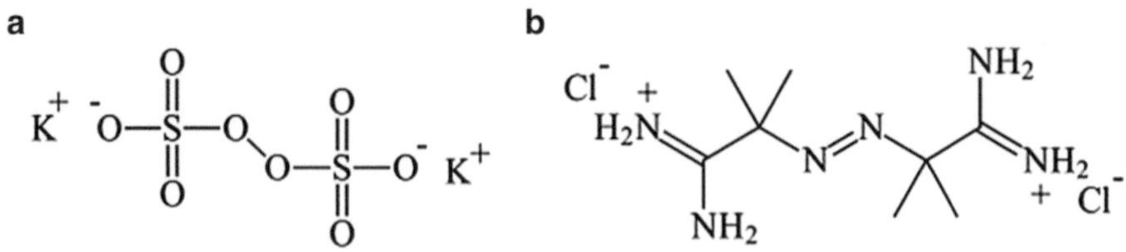

Fig. 7 Chemical structure of potassium peroxodisulphate (**a**), and α,α′-azodiisobutyramidin dihydrochloride (**b**)

diameters ranging from 280 to 415 nm, that have been employed in the preparation of CPCs and inverse opals from silica (SiO_2), titania (TiO_2), or ceria (CeO_2).

Special interest relies on the SFEP process of two or more monomers with a marked difference in their hydrophilic properties, which leads to the synthesis of monodisperse core-shell particles in a single step process as the hydrophobic monomer ends up in the particle core, while the hydrophilic groups create a soft, hydrated shell which exposes the hydrophilic functional groups and ionic charges. Zulian et al. [57] prepared core-shell polymer particles by SFEP of a mixture of styrene and methacrylic acid. The methacrylic acid percentage and the pH of the reaction medium influence both the particle size and the optical properties of the obtained polymer opals. By properly adjusting the reaction parameters, core-shell particles with a size range wide enough to obtain the entire visible spectrum of colors were prepared.

By properly selecting the employed hydrophilic comonomer it is possible to synthesize chemically reactive particles that can be easily post-modified. For example, SFEP of styrene in the presence of a hydrophilic amine-containing monomer resulted in polystyrene particles with amino groups [58] located at the surface.

A modified SFEP method, featuring a slow dropwise addition of the initiator, was employed to obtain highly monodisperse chemically reactive particles, bearing functional surface epoxy groups, with controllable sizes from 383 to 756 nm [59]. CPCs with different iridescent structural colors were fabricated by spin-coating the reactive particles without further purification. Such functional photonic crystals allow for attachment of a variety of stimuli-responsive chemical, that holds much promise for use in sensing and biosensing.

2.2.4 Dispersion Polymerization

Sterically stabilized monodisperse polymer particles can be obtained by dispersion polymerization [60, 61]. In this process, the monomer, the polymeric stabilizer, and the polymerization initiator are dissolved in an organic solvent, which is a non-solvent for the resulting polymer. Accordingly, the reaction mixture is homogeneous at the beginning of the polymerization, which starts in homogeneous solution. When the growing chains reach a critical chain length, they phase separate into primary particles (nucleation) which are sterically stabilized by the presence of some stabilizer chains chemically linked to the polymer particles because of the occurrence of radical transfer reactions. Primary particles so formed are swollen by the monomer and consequently the polymerization reaction proceeds mainly inside the particles, leading to the formation of particles with size in the range 0.5–10 μm. This process is generally employed to obtain, in a single step, polymeric particles larger than 700 nm, that is in a range not accessible by classical or SFEP polymerizations. Thanks to the peculiar particle forming mechanism, at the end of the reaction the polymeric stabilizer is covalently attached to the particle surface.

Consequently the chemical nature of the particle surface can be precisely tailored by carefully choosing the appropriate stabilizer [62].

Chen et al. [63] synthesized monodisperse polystyrene microspheres, with diameters in the range 1.1–5.2 μm, by dispersion polymerization in ethanol/isopropanol media employing poly(acrylic acid) as steric stabilizer. These particles were self-assembled into CPCs with good optical properties.

It is interesting to note that the particles obtained by dispersion polymerization can be rather hydrophobic since they are polymerized in an organic solvent. Similar particles have been employed for the generation of colloid monolayers on water that can be transferred to various substrates [64].

2.2.5 Seeded Polymerization

The synthetic technique so far illustrated for the preparation of polymer colloidal particles involves the formation of nuclei (primary particles) followed by gradual growth of these nuclei to produce the final polymer particles. In all of these systems, preformed polymer particles can be added to the polymerization mixture and these particles act as seed nuclei in the polymerization process, absorbing monomers and oligoradicals from the reaction medium. If the polymerization conditions (number of seed particles and concentration of initiator and stabilizer) are carefully chosen to avoid the formation of new particles, the seed particles grow uniformly, resulting in highly monodisperse particle systems whose final size is controlled by the ratio between the monomer and the added seed particles.

Seeded polymerizations are two-step processes. In the first stage, the seed particles can be prepared either separately or in situ, and in the second stage the monomer is added continuously or stepwise to the reaction vessel, in the presence of the polymerization initiator. Two different particle morphologies can be obtained, depending on the chemical nature of the seed polymer and of the employed shell-forming monomer. When the shell-forming monomer is miscible in the seed-forming polymer, the seed particles are swollen with the added monomer and the polymerization leads to homogeneous particles. On the contrary, when the shell-forming monomer and the seed-forming polymer are immiscible, the polymerization occurs exclusively at the surface of the seeds thus producing core-shell particles.

This technique, initially employed to elucidate the kinetics and mechanism of the emulsion polymerization [65], was subsequently extended by Ugelstad et al. to the preparation monodisperse polymer particles with diameters up to 20 μm, which are less readily obtained by other heterogeneous polymerization techniques [66]. Appropriately selecting the size of the seed particles, this technique can be employed to obtain extremely monodisperse particles and core-shell particles with size extending from 100 nm up to more than 20 μm.

Starting from polytetrafluoroethylene (PTFE) latexes, with particle diameters ranging from 20 to 230 nm, core-shell particles were obtained with particle diameters ranging from 70 to 600 nm, with different shell-forming polymers polystyrene [67], polymethylmethacrylate [68] as well as low T_g polyacrylic copolymers [69]. A very

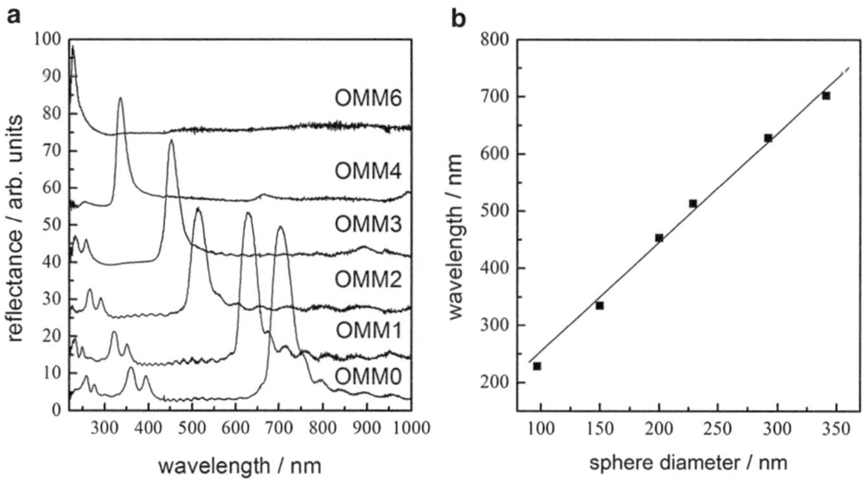

Fig. 8 Reflectance spectra of opals produced by PTFE/PMMA core-shell particles (**a**), and main reflectance peak spectral position as a function of the particle diameter (**b**). Reprinted with permission from [70]

precise control over the particle size is obtained by properly adjusting the ratio between the monomers and the PTFE seeds. In addition, the particle size distribution self-sharpens as the ratio monomer/PTFE increases. These core-shell particles were successfully employed to build 2D and 3D colloidal crystals [70]. In particular, the 2D colloidal crystal leads to very small 2D nanostructuration, useful for the preparation of masks with a combination of nanosphere lithography and reactive ion etching [71]. Moreover, monodisperse PTFE/PMMA core-shell particles, with diameters ranging from 100 to 350 nm, were employed as building blocks in the preparation of 3D colloidal crystals, featuring excellent opal quality [72]. Figure 8a shows the typical reflectance spectra of the obtained opals. The spectral position of the Bragg peak (λ_B) shifts toward shorter wavelength as soon as the particle diameter is reduced (Fig. 8b).

3 Colloidal Particles with Complex Structures and Morphologies for Specific Applications

Currently, there is an intensive effort to develop photonic crystals addressed to optical and sensing technologies [14]. Different chemical strategies can be integrated to include specific functionalities in the colloidal particles. This section focusses on two topics:

- The production of core-shell and hollow particles with complex structures.
- The incorporation of fluorescent materials into the spherical colloidal particles, to study the influence of the photonic band gap on the fluorophore emission.

3.1 Core-Shell and Hollow Particles

The properties of colloidal particles can be further modified by coating them with a thin shell of a different chemical composition with varying thickness.

The formation of shells on colloidal particles dramatically alters the charge, functionality, and reactivity of their surface thus modifying the stability and compatibility of the core particles [73]. Moreover, the formation of core-shell particles allows different materials to be incorporated into the same structure, leading to hybrid, multifunctional colloidal systems.

Polymer core-shell colloids can be obtained by seeded polymerization techniques [74], as previously described. If the shell forming material exhibits a lower T_g value than the core, once a well-ordered assembly of core-shell particles is obtained, thermal annealing allows the shell-forming polymer to soften and form a polymeric film containing a 3D regular periodic array of seed particles [70, 75] (Fig. 9).

Wang et al. [76] synthesized core-shell polymeric particles with a hard PS core and soft shell PMMA/PAA shell by SFEP of the two comonomers. The particles were assembled at 90 °C to obtain colloidal crystal films with a special FCC structure and tough mechanical strength.

The core-shell CPCs have also been employed as a helpful template for the production of special closed-cell inverse opal structures which show tough mechanical properties [77]. These closed-cell structures have been obtained by infiltrating polyimide into the CPCs template prepared using core–shell poly(styrene–methyl methacrylate–acrylic acid) colloidal spheres and subsequent solvent etching of the template. This closed-cell structure shows much better mechanical and thermal properties than those of open-cell structures obtained using polystyrene colloidal spheres as the template.

Inorganic metallodielectric core-shell colloids have received much attention for their photonic behavior. Different synthetic methods were developed to prepare of both metal-core and metal-shell particles.

Metal-core colloids can be obtained by coating silver and gold colloids with a homogeneous silica shell. The critical step in these silica-coating procedures is the transfer of colloids, which are only stable in aqueous solution, to ethanol where the Stöber process is carried out. To this purpose, the metallic colloid must be functionalized with a silane coupling agent, to render the metallic surface vitreophilic, and then coated with a thin layer of silica growth in the aqueous

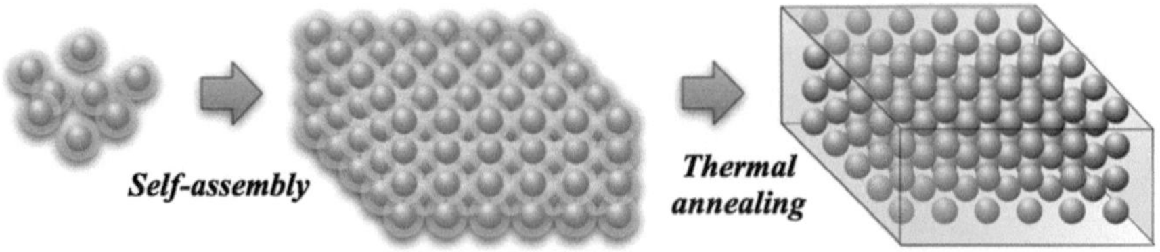

Fig. 9 Scheme for the production of a polymer film containing a 3D regular periodic array of seeds from core-shell particles with low T_g shell material

phase. When the particles are sufficiently stabilized they can be transferred into ethanol for further growth using the Stöber method [78–81].

A faster and more general method to coat various colloidal particles with silica was developed by Graf [82]. This method is based on the absorption of a non-ionic amphiphilic polymer, the poly(vinylpyrrolidone) (PVP), on the particle surface as stabilizer. The stabilized particles could then be transferred to a solution of ammonia in ethanol and directly coated with smooth and homogeneous silica shells of variable thickness using the Stöber method.

Conversely, metal shells can be grown onto silica particles [83]. Small gold nanoclusters were attached to the functionalized surface of colloidal silica particles. Reductive growth and coalescence of these clusters lead to the formation of a gold layer. Variation in the thickness of the gold layer and in the radius of the shell allowed a fine-tuning of the plasmon resonance of the gold shell particles over the whole visible and infrared region of the spectrum. Because of their low polydispersity, these gold shell-silica core particles self-assemble leading to CPCs with lattice constants on the order of visible wavelengths.

Also polymer-inorganic core-shell particles became very popular in the last few decades. They find applications in very diverse areas including catalysis, optics, optoelectronics, and biomedical [84]. Moreover, starting from hybrid core-shell particles, it is possible to remove the cores and to obtain hollow particles consisting of the shell-forming material. The core removal is usually accomplished in a subsequent step, using procedures such as solvent extraction or calcination at high temperature. Xia et al. [85] prepared hybrid core-shell particles by coating the surfaces of monodisperse polystyrene spheres with uniform silica shells. They coated polystyrene particles, bearing functional amino groups on the surface, with uniform shells of amorphous silica using a modified Stöber method. The thickness of the silica coatings was adjusted from tens to several hundreds of nanometers by controlling the concentration of TEOS precursor and/or the growth time. These particles were employed in the preparation of colloidal crystals whose stop bands are located in a spectral region different from that of spherical colloids of a similar size but made of pure polystyrene or silica. Hollow silica particles were obtained by selectively removing the polymer cores via calcination in air at an elevated temperature or by wet etching with toluene.

Wu et al. [86] synthesized monodisperse hollow silica particles by coating polystyrene particles with silica shells in a modified Stöber procedure. By increasing the ammonia concentration, silica shell growth and synchronous dissolution of the polystyrene cores in the same medium were obtained.

3.2 *Fluorescent Particles*

The photonic band gap of CPCs can be used to influence the emissive properties of fluorescent active molecules embedded into the photonic crystal, when the emission spectrum peak of the fluorophore is matched with the stop band. In fact, the absence of propagating EM modes inside the structure gives rise to distinct optical phenomena such as inhibition or attenuation of spontaneous emission of a light emitting material [87]. Various kinds of active media (such as organic dyes, quantum dots, lanthanides, and rare-earth atoms) have been used for investigating stimulated emission from CPCs [88, 89]. Loading of a preformed CPCs with light emitters can be achieved by infiltrating the opal voids with fluorescent materials from solution [90]. However, infiltration may cause the collapse of the photonic structure or the aggregation of the organic dye thus suppressing almost completely the fluorescence [11, 91]. Moreover, in the case of highly fluorescent semiconductor quantum dots, simple infiltration into the voids of colloidal crystals does not protect the quantum dots from oxidation [92]. Therefore, incorporation of these active species within the particles before the CPCs fabrication appears more effective. Within this approach, the particle matrix can also protect the dyes segregated inside the particle, thus increasing their photostability.

Different techniques can be employed to incorporate fluorescent active molecules in either the silica or the polymer particles.

The preparation of fluorescent silica nanoparticles was first reported by Van Blaaderen [93, 94] by condensing a fluorescent dye to a silane coupling agent, the (3-aminopropyl)triethoxysilane, and successively employing the resulting trialkoxysilane derivative in the Stöber synthesis of silica spheres. By optimizing a modular reaction procedure, the composition of the silica spheres was controlled in such a way that the dye molecules can be placed at the surface, in a thin shell inside the particle or distributed through the volume of an inner core. These particles have been crystallized by slow sedimentation onto a patterned surface to obtain bulk photonic crystals. The presence of a fluorescent core allowed the imaging of the colloid formation and the study of buried defect structures by confocal microscopy [95, 96].

Several different light emitting materials can be easily incorporated into silica particles employing this method due to the wide range of amino-reactive markers commercially available [97]. These reactive fluorophores can also be employed to label preformed silica particles containing the appropriate reactive functional groups (generally $-NH_2$ and –SH groups) under mild conditions. The post-functionalization reaction is very useful when expensive fluorophores are employed because of its high incorporation efficiency.

Highly fluorescent semiconductor quantum dots have been included into silica microparticles by exchanging the original stabilizer of the aqueous quantum dot with a silane coupling agent, the 3-mercaptopropyltrimethoxysilane (MPS). The subsequent addition of sodium silicate leads to "raisin bun"-type composite particles that were employed as seeds in the Stöber process to obtain silica spheres with diameters ranging from 100 to 700 nm [98].

Fluorescent CdTe-SiO_2 composite microspheres were prepared by a sol–gel method without the exchange of surface ligands [99]. An aqueous solution of CdTe quantum dots was employed, in replacement of pure water, in a silica seed growth process based on the Stöber method. CdTe quantum dots embedded in the composite microspheres revealed high photostability. Their fluorescence properties were retained due to a confinement effect in the silica matrix.

Fluorescent polymer particles can also be obtained using different approaches depending on the chemical nature of the selected fluorophore. Water soluble fluorescent dyes can be directly incorporated into the polymeric matrix during the polymerization process. In this way, different organic dyes, such as Rhodamine B, Coumarin 6, and Coumarin 334, have been incorporated in monodisperse PMMA, particles obtained by SFEP [100].

In a similar way, the perylenediimide (PDI) dye, a highly stable hydrophobic fluorophore, was incorporated into poly(styrene-acrylic acid) particles by dissolving these particles in the monomer phase employed during the SFEP process [101]. The authors suggested that the primary particles, formed by homogeneous nucleation, absorb the PDI dissolved in the styrene organic phase.

Organic fluorophores can be successfully incorporated in the polymeric particles by a swelling/deswelling process of the already synthesized particles. In this case, the dye diffuses into the polymeric particles swollen in an acetone/water mixture and it is subsequently entrapped by fast shrinking of the particles, obtained by diluting the system with ice water [41].

The fluorophore di(2-thienyl)-2,1,3-benzothiadiazole (Fig. 10b) was incorporated into PMMA particles by a swelling and deswelling process of the already synthesized colloids [102]. Figure 10a shows the SEM images of the obtained fluorescent particles. These particles were employed as building blocks to grow the corresponding CPC. Figure 10c reports the contour plot of the CPC transmission spectra as a function of the incidence angle. The black line shows the normal incidence reflectance spectrum of the sample to better identify the different spectral features. According to the Bragg–Snell equation [Eq. (1)], the dispersion (dependence on the incidence angle) of the photonic stop band is observed as a blue shift from 830 to almost 650 nm for $\theta = 60°$. The van Hove-like structures show an opposite dispersion with respect to the stop band. The two dispersive features merge for $\theta = 35°$. These results are in full agreement with theoretical models previously reported [103].

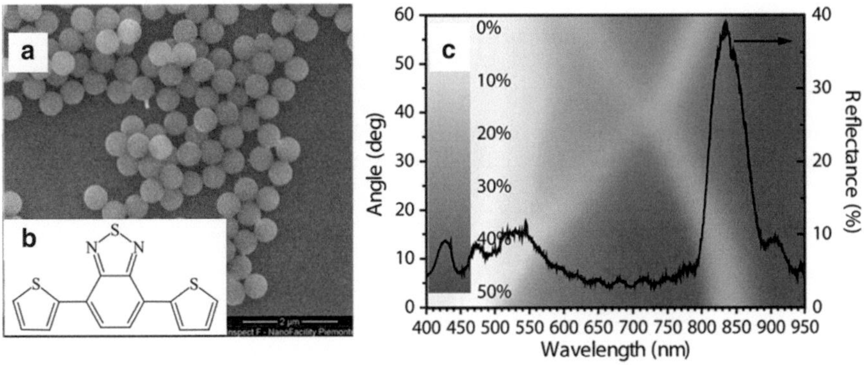

Fig. 10 SEM micrograph of the PMMA particles (**a**) containing the fluorophore di(2-thienyl)-2,1,3-benzothiadiazole (**b**). Contour plot of transmission spectra of 4 F opals as a function of the incidence angle. *Black line* shows the normal incidence reflectance spectrum of the sample in order to better identify the different spectral features (**c**). Reprinted with permission from [102]

Alternatively, a fluorescent comonomer, that is an organic dye that carries a polymerizable group, can be employed during the particle forming process [102]. Several fluorescent comonomers are commercially available or can be easily synthesized. This approach was recently reviewed [104].

As previously described for silica particles, polymer particles with reactive functional groups on their surface can be easily labelled with commercially available reactive fluorophores [105].

The incorporation of semiconductor quantum dots into polymer particles requires a specific procedure. CdSe quantum dots were incorporated into PS/PMMA core-shell particles by a two-stage preparation process. The quantum dots were dispersed in styrene employed as monomer in a modified emulsion process that generates highly fluorescent PS particles. To increase the size and the monodispersity of the PS-based polymer particles, these were subsequently employed as seeds in the polymerization of methyl methacrylate. As a result, monodisperse colloids with size ranging from 200 to 400 nm were obtained and processed to lead high quality CPCs [92].

4 Conclusions

Colloidal particles can spontaneously self-assemble into macroscopic, highly ordered arrays, ultimately leading to colloidal photonic crystals (CPCs). However, the precondition for the preparation of high quality CPCs is the preparation of highly monodisperse colloidal particles.

A variety of synthetic methods have been described in this chapter to make up colloidal systems consisting of monodisperse spherical particles, with diameters ranging from 20 nm to 1 μm. By controlling the chemistry of the particles, the self-assembly process and eventual post-assembly processing steps, an incredible variety of regular structures being organic, inorganic, or hybrid in nature, can be obtained, featuring promising properties for applications including displays, optical devices, photochemistry, and biological sensors.

Acknowledgment This work was partially funded by PRIN 2010–2011 "Materiali Polimerici Nanostrutturati con Strutture Molecolari e Cristalline Mirate."

References

1. W.B. Russel, D.A. Saville, W.R. Schowalter, *Colloidal Dispersions* (Cambridge University Press, Cambridge, 1992)
2. R.J. Hunter, *Introduction to Modern Colloid Science* (Oxford University Press, Oxford, 1993)
3. K.J.M. Bishop, C.E. Wilmer, S. Soh, B.A. Grzybowski, Small **5**, 1600 (2009)
4. O. Kruglova, P.J. Demeyer, K. Zhong, Y. Zhou, K. Clays, Soft Matter **9**, 9072 (2013)
5. J.V. Sanders, Nature **204**, 1151 (1964)

6. B. Lange, N. Metz, M.N. Tahir, F. Fleischhaker, P. Theato, H.C. Schröder, W.E.G. Müller, W. Tremel, R. Zentel, Macromol. Rapid Commun. **28**, 1987 (2007)
7. J.D. Joannopoulos, S.G. Johnson, J.N. Winn, R.D. Meade, *Photonic Crystals: Molding the Flow of Light* (Princeton University Press, Princeton, 2008)
8. B. Lange, F. Fleischhaker, R. Zentel, Macromol. Rapid Commun. **28**, 1291 (2007)
9. A. Blanco, E. Chomski, S. Grabtchak, M. Ibisate, S. John, S.W. Leonard, C. Lopez, F. Meseguer, H. Miguez, J.P. Mondia, G. Ozin, O. Toader, H.M. van Driel, Nature **405**, 437 (2000)
10. D. Comoretto, R. Grassi, F. Marabelli, L.C. Andreani, Mater. Sci. Eng. C **23**, 61 (2003)
11. L. Berti, M. Cucini, F. Di Stasio, D. Comoretto, M. Galli, F. Marabelli, N. Manfredi, C. Marinzi, A. Abbotto, J. Phys. Chem. C **114**, 2403 (2010)
12. M. Allard, E.H. Sargent, Appl. Phys. Lett. **85**, 5887 (2004)
13. H. Kawaguchi, Prog. Polym. Sci. **25**, 1171 (2000)
14. C. Fenzl, T. Hirsch, O.S. Wolfbeis, Angew. Chem. Int. Ed. Engl. **53**, 3318 (2014)
15. B. Lange, F. Fleischhaker, R. Zentel, Phys. Status Solidi **204**, 3618 (2007)
16. Y. Xia, B. Gates, Y. Yin, Y. Lu, Adv. Mater. **12**, 693 (2000)
17. X. Jiang, T. Herricks, Y. Xia, Adv. Mater. **15**, 1205 (2003)
18. U. Jeong, Y. Wang, M. Ibisate, Y. Xia, Adv. Funct. Mater. **15**, 1907 (2005)
19. E. Matijevic, Langmuir **10**, 8 (1994)
20. V.K. La Mer, R.H. Dinegar, J. Am. Chem. Soc. **72**, 4847 (1950)
21. E. Matijevic, Chem. Mater. **5**, 412 (1993)
22. W. Stöber, A. Fink, E. Bohn, J. Colloid Interface Sci. **26**, 62 (1968)
23. G.H. Bogush, M.A. Tracy, C.F. Zukoski, J. Non-Cryst. Solids **104**, 95 (1988)
24. A. Van Blaaderen, J. Van Geest, A. Vrij, J. Colloid Interface Sci. **154**, 481 (1992)
25. S.L. Chen, P. Dong, G.H. Yang, J.J. Yang, J. Colloid Interface Sci. **180**, 237 (1996)
26. D.A. Santamaría Razo, L. Pallavidino, E. Garrone, F. Geobaldo, E. Descrovi, A. Chiodoni, F. Giorgis, J Nanoparticle Res. **10**, 1225 (2008)
27. F.J. Arriagada, K. Osseo-Asare, Colloids Surf. A Physicochem. Eng. Asp. **154**, 311 (1999)
28. T. Yokoi, Y. Sakamoto, O. Terasaki, Y. Kubota, T. Okubo, T. Tatsumi, J. Am. Chem. Soc. **128**, 13664 (2006)
29. K.D. Hartlen, A.P.T. Athanasopoulos, V. Kitaev, Langmuir **24**, 1714 (2008)
30. R.K. Iler, *The Chemistry of Silica: Solubility, Polymerization, Colloid and Surface Properties and Biochemistry of Silica* (Wiley, New York, 1979)
31. Z. Wu, H. Xiang, T. Kim, M.S. Chun, K. Lee, J. Colloid Interface Sci. **304**, 119 (2006)
32. J. Lin, H. Chen, Y. Ji, Y. Zhang, Colloids Surf. A Physicochem. Eng. Asp. **411**, 111 (2012)
33. R.D. Badley, W.T. Ford, F.J. Mc Enroe, R.A. Assinks, Langmuir **6**, 792 (1990)
34. W.Y.D. Yong, Z. Zhang, G. Cristobal, W.S. Chin, Colloids Surf. A Physicochem. Eng. Asp. **460**, 151 (2014)
35. W. Wang, B. Gu, L. Liang, W.A. Hamilton, J. Phys. Chem. B **107**, 12113 (2003)
36. T.S. Deng, J.Y. Zhang, K.T. Zhu, Q.F. Zhang, J.L. Wu, Colloids Surf. A Physicochem. Eng. Asp. **356**, 104 (2010)
37. M. Bardosova, M.E. Pemble, I.M. Povey, R.H. Tredgold, Adv. Mater. **22**, 3104 (2010)
38. K. Busch, S. Lölkes, R.B. Wehrspohn, H. Föll, *Photonic Crystals: Advances in Design, Fabrication, and Characterization* (Wiley, New York, 2006)
39. M. Egen, R. Zentel, Chem. Mater. **14**, 2176 (2002)
40. P. Ferrand, M. Egen, R. Zentel, J. Seekamp, S.G. Romanov, C.M. Sotomayor Torres, Appl. Phys. Lett. **83**, 5289 (2003)
41. B. Lange, R. Zentel, C. Ober, S. Marder, Chem. Mater. **16**, 5286 (2004)
42. R. Arshady, Colloid Polym. Sci. **270**, 717 (1992)
43. P.A. Lovell, M.S. El-Aasser, *Emulsion Polymerization and Emulsion Polymers* (Wiley, New York, 1997)
44. W.H. Lane, Ind. Eng. Chem. Anal. Ed. **18**, 295 (1946)
45. C.S. Chern, Prog. Polym. Sci. **31**, 443 (2006)

46. H. Ishii, M. Ishii, D. Nagao, M. Konno, Polymer **55**, 2772 (2014)
47. M. Borzenkov, O. Hevus, *Surface Active Monomers: Synthesis, Properties, and Application* (Springer, Heidelberg, 2014)
48. A. Guyot, Adv. Colloid Interface Sci. **108–109**, 3 (2004)
49. H.A.S. Schoonbrood, M. Asua, Macromolecules **9297**, 6034 (1997)
50. D. Cochin, F. Candau, R. Zana, Macromolecules **26**, 5755 (1993)
51. M. Laus, K. Sparnacci, M. Lelli, R. Vannini, L. Tondelli, J. Polym. Sci. Part A Polym. Chem. Ed. **38**, 1110 (2000)
52. J.F. Morizur, D.J. Irvine, J.J. Rawlins, L.J. Mathias, Macromolecules **40**, 8938 (2007)
53. J.W. Goodwin, J. Hearn, C.C. Ho, R.H. Ottewill, Colloid Polym. Sci. **252**, 464 (1974)
54. M. Egen, R. Zentel, Macromol. Chem. Phys. **205**, 1479 (2004)
55. J.W. Goodwin, R.H. Ottewill, R. Pelton, G. Vianello, D.E. Yates, Br. Polym. J. **10**, 173 (1978)
56. G.I.N. Waterhouse, M.R. Waterland, Polyhedron **26**, 356 (2007)
57. L. Zulian, E. Emilitri, G. Scavia, C. Botta, M. Colombo, S. Destri, ACS Appl. Mater. Interfaces **4**, 6071 (2012)
58. F. Sauzedde, F. Ganachaud, A. Elaissari, C. Pichot, J. Appl. Polym. Sci. **65**, 2331 (1997)
59. Z. Luo, C. Zou, S. Syed, L.A. Syarbaini, G. Chen, Colloid Polym. Sci. **290**, 141 (2012)
60. K.P. Lok, C.K. Ober, Can. J. Chem. **63**, 209 (1985)
61. M. Laus, M. Lelli, A. Casagrande, J. Polym. Sci. Part A Polym. Chem. Ed. **35**, 681 (1997)
62. K. Sparnacci, M. Laus, L. Tondelli, L. Magnani, C. Bernardi, Macromol. Chem. Phys. **203**, 1364 (2002)
63. C.W. Chen, C.Y. Chen, Colloid Polym. Sci. **287**, 1377 (2009)
64. M. Egen, R. Voss, B. Griesebock, R. Zentel, Chem. Mater. **15**, 3786 (2003)
65. F.K. Hansen, J. Ugelstad, J. Polym. Sci. Part A Polym. Chem. **17**, 3033 (1979)
66. J. Ugelstad, A. Berge, T. Ellingsen, R. Schmid, T.N. Nilsen, P.C. Mørk, P. Stenstad, E. Hornes, Ø. Olsvik, Prog. Polym. Sci. **17**, 87 (1992)
67. E. Giani, K. Sparnacci, M. Laus, G. Palamone, V. Kapeliouchko, V. Arcella, Macromolecules **36**, 4360 (2003)
68. V. Kapeliouchko, G. Palamone, T. Poggio, G. Zuccheri, R. Passeri, K. Sparnacci, D. Antonioli, S. Deregibus, M. Laus, J. Polym. Sci. **47**, 2928 (2009)
69. K. Sparnacci, D. Antonioli, S. Deregibus, M. Laus, T. Poggio, V. Kapeliouchko, G. Palamone, G. Zuccheri, R. Passeri, Macromolecules **42**, 3518 (2009)
70. K. Sparnacci, D. Antonioli, S. Deregibus, M. Laus, G. Zuccheri, L. Boarino, N. De Leo, D. Comoretto, J. Nanomater. **2012**, 1 (2012)
71. L. Boarino, E. Enrico, N. De Leo, F. Celegato, P. Tiberto, K. Sparnacci, M. Laus, Phys. Status Solidi **208**, 1403 (2011)
72. D. Antonioli, S. Deregibus, G. Panzarasa, K. Sparnacci, M. Laus, L. Berti, L. Frezza, M. Gambini, L. Boarino, E. Enrico, D. Comoretto, Polym. Int. **61**, 1294 (2012)
73. D.V. Goia, E. Matijevic, New J. Chem. **22**, 1203 (1998)
74. H.H. Pham, E. Kumacheva, Macromol. Symp. **192**, 191 (2003)
75. X. He, Y. Thomann, R.J. Leyrer, J. Rieger, Polym. Bull. **57**, 785 (2006)
76. J. Wang, Y. Wen, H. Ge, Z. Sun, Y. Zheng, Y. Song, L. Jiang, Macromol. Chem. Phys. **207**, 596 (2006)
77. X. Chen, L. Wang, Y. Wen, Y. Zhang, J. Wang, Y. Song, L. Jiang, D. Zhu, J. Mater. Chem. **18**, 2262 (2008)
78. S.R. Hall, S.A. Davis, S. Mann, Langmuir **16**, 1454 (2000)
79. L. Yang, M.P. Seah, I.S. Gilmore, R.J.H. Morris, M.G. Dowsett, L. Boarino, K. Sparnacci, M. Laus, J. Phys. Chem. C **117**, 16042 (2013)
80. F. Garcìa-Santamarìa, V. Salgueirino-Maceira, C. Lòpez, L.M. Liz-Marzan, Langmuir **18**, 4519 (2002)
81. Z. Chen, Z.L. Wang, P. Zhan, J.H. Zhang, W.Y. Zhang, H.T. Wang, N.B. Ming, Langmuir **20**, 3042 (2004)

82. C. Graf, D.L.J. Vossen, A. Imhof, A. van Blaaderen, Langmuir **19**, 6693 (2003)
83. C. Graf, A. Van Blaaderen, Langmuir **18**, 524 (2002)
84. M.A. Hood, M. Mari, R. Muñoz-Espí, Materials **7**, 4057 (2014)
85. Y. Lu, J. McLellan, Y. Xia, Langmuir **20**, 3464 (2004)
86. Z. Deng, M. Chen, S. Zhou, B. You, L. Wu, Langmuir **22**, 6403 (2006)
87. A. Blanco, C. López, R. Mayoral, H. Míguez, F. Meseguer, A. Mifsud, J. Herrero, Appl. Phys. Lett. **73**, 1781 (1998)
88. Z. Yang, J. Zhou, X. Huang, G. Yang, Q. Xie, L. Sun, B. Li, L. Li, Chem. Phys. Lett. **455**, 55 (2008)
89. R. Withnall, M.I. Martinez-Rubio, G.R. Fern, T.G. Ireland, J. Silver, J. Opt. A Pure Appl. Opt. **5**, S81 (2003)
90. F. Di Stasio, L. Berti, S.O. McDonnell, V. Robbiano, H.L. Anderson, D. Comoretto, F. Cacialli, APL Mater. **1**, 042116 (2013)
91. I.S. Nikolaev, P. Lodahl, W.L. Vos, J. Phys. Chem. C **112**, 7250 (2008)
92. F. Fleischhaker, R. Zentel, Chem. Mater. **17**, 1346 (2005)
93. A. Van Blaaderen, A. Vrij, Langmuir **8**, 2921 (1992)
94. N.A.M. Verhaegh, A. Van Blaaderen, Langmuir **10**, 1427 (1994)
95. A. Van Blaaderen, R. Ruel, P. Wiltzius, Nature **385**, 321 (1997)
96. A. Van Blaaderen, A. Imhof, W. Hage, A. Vrij, Langmuir **8**, 1514 (1992)
97. G.T. Hermanson, *Bioconjugate Techniques* (Elsevier, Amsterdam, 2008)
98. A.L. Rogach, D. Nagesha, J.W. Ostrander, M. Giersig, N.A. Kotov, Chem. Mater. **12**, 2676 (2000)
99. K. Han, Z. Xiang, Z. Zhao, C. Wang, M. Li, H. Zhang, X. Yan, J. Zhang, B. Yang, Colloids Surf. A Physicochem. Eng. Asp. **280**, 169 (2006)
100. S.G. Romanov, T. Maka, C.M. Sotomayor Torres, M. Muller, R. Zentel, Appl. Phys. Lett. **75**, 1057 (1999)
101. A. Diacon, E. Rusen, A. Mocanu, P. Hudhomme, C. Cincu, Langmuir **27**, 7464 (2011)
102. D. Comoretto, V. Robbiano, G. Canazza, L. Boarino, G. Panzarasa, M. Laus, K. Sparnacci, Polym. Compos. **34**, 1443 (2013)
103. E. Pavarini, L.C. Andreani, C. Soci, M. Galli, F. Marabelli, D. Comoretto, Phys. Rev. B **72**, 045102 (2005)
104. A.M. Breul, M.D. Hager, U.S. Schubert, Chem. Soc. Rev. **42**, 5366 (2013)
105. M.S. Falzarano, C. Passarelli, E. Bassi, M. Fabris, D. Perrone, P. Sabatelli, N.M. Maraldi, R. Selvatici, P. Bonaldo, K. Sparnacci, M. Laus, P. Braghetta, P. Rimessi, A. Ferlini, Biomed Res. Int. **2013**, 527418 (2013)

Self-Assembling Polymer Photonics

Andrew J. Parnell and J. Patrick A. Fairclough

Abstract Photonic materials offer the prospect of control over light and consequently the possibility of new and exciting devices based on the manipulation of light. Taking inspiration from the natural world, which readily uses nanostructure to produce structural color, but using synthetic polymeric building blocks, "partial photonic" structures can be created. Polymer based photonic devices based on block copolymers are attractive as they readily self-organize into a variety of structures using well-established design rules. These materials therefore have the potential to be processed with minimal manufacturing input and so enable high volume production of new photonic devices, crucially at a low overall cost.

Keywords Block copolymer • Self-assembly • Lamellar • Bragg reflector • Biomimetic

1 Introduction

Photonic materials are in reality confined to academic laboratories at present, due to the high degree of order and registry between layers needed to make high quality samples. The necessary fabrication techniques frequently use a number of sequential processing steps. These are costly, time consuming and as a consequence have limited the adoption of this interesting and potentially transformative technology. To translate this technology onto an industrial scale requires large reductions in cost and effort. This could be made possible using a self-assembly approach. The natural world readily uses self-assembled structural color for the vivid and diverse colors seen in butterfly wings, bird feathers, and beetle elytra.

The simplest example of a partially photonic structure is that of a dielectric multilayer stack, a "Bragg mirror." This is a series of alternating layers of different refractive indices; importantly, it can be designed to reflect any particular optical

A.J. Parnell
Department of Physics and Astronomy, The University of Sheffield, Sheffield, S3 7RH, UK
e-mail: a.j.parnell@sheffield.ac.uk

J.P.A. Fairclough (✉)
Department of Mechanical Engineering, The University of Sheffield, Sheffield, S3 7HQ, UK
e-mail: p.fairclough@sheffield.ac.uk

D. Comoretto (ed.), *Organic and Hybrid Photonic Crystals*,
DOI 10.1007/978-3-319-16580-6_6

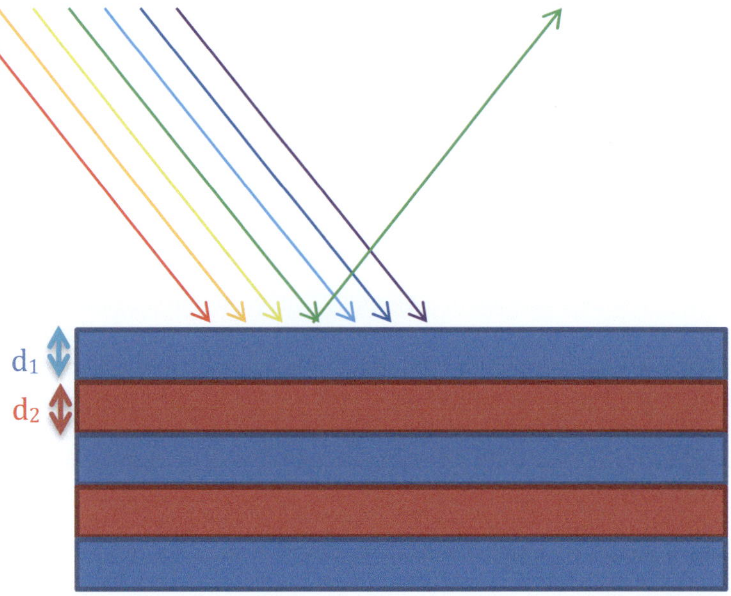

Fig. 1 Schematic of a dielectric stack (a Bragg mirror). The *blue* and *red* colored layers indicate materials of different refractive index (n_1, n_2), these can also be of different layer thicknesses (d_1, d_2)

wavelength. However this cannot deliver a truly 3D photonic material. In a true photonic device the reflected wavelength is a photonic stopband, where the wavelengths of light in the stopband are unable to propagate in the structure, similar to the unallowed states seen in semiconductors band diagrams (Brioullin zones). In Bragg mirror the stopband is only along the stacking direction, when the refractive indices are sufficiently large. In the partial photonic Bragg reflector as shown in Fig. 1, the reflected wavelength depends upon the optical thickness ($n_1d_1 + n_2d_2$) and refractive index of each layer (n_1, n_2), see Fig. 1. Additionally, the width of the wavelength distribution or angular distribution depends on the total number of layers (N). For inorganic materials such as metals it is possible to produce a large difference in refractive index (Δn), consequently good quality Bragg mirrors require only a small number of alternating layers. However, this involves evaporating a sequence of layers (of different refractive index) or some other serial deposition technique using a layer-by-layer approach, (most likely under high vacuum conditions) making the total production costs high.

An example of such a serial technique to create a partially photonic structure, but using polymers, was a study that used spin coating of alternating layers of poly (styrene) and poly(vinylpyrrolidone) to create a distributed Bragg reflector (DBR) [1]. This involved spin coating up to 50 alternating layers of poly(styrene) (PS) and poly(vinylpyrrolidone) (PVP). This technique gave narrow reflection peaks that could be readily controlled by changing the layer thickness. The overall reflection intensity of the structure was improved, by increasing N, the total number of alternating layers. This approach could also be used to create chirped structures, by systematically varying the layer thicknesses. The approach of creating a multilayer stack via spin coating is detailed in the chapter by Scotognella et al.

Another processing approach was to use polymers in the melt phase, as it is possible to coextrude and alternately layer them. In brief the melt was split vertically, separated in the horizontal direction and finally recombined to increase the number of layers [2]. This approach was demonstrated in detail using poly (methyl methacrylate) and poly(styrene), it was also straightforward to use a wide range of other industrial and widely available polymer materials. The reflectance strength of the Bragg mirrors approached 100 % for an extruded structure consisting of 64 polymer layers. This approach is discussed in more depth in the chapter by D. Cavallo et al.

A simpler and lower cost approach is to utilize self-assembly to produce high quality optically active structures. This would rule out the need for expensive vacuum deposition stages, as well as the need for multiple processing steps and, crucially reduce the overall cost. Nature routinely uses self-assembly to form complex and hierarchical order in a biological context, and there are many examples of Bragg mirrors and other complex photonic structures in nature. For instance, the iridescent structures commonly observed in butterflies and birds are created using self-assembly, thus nature can create highly vivid structural colors [3]. In particular some birds' feathers undergo the phase separation mechanism of spinodal decomposition to create non-iridescent structurally colored feathers [4].

A novel class of synthetic materials that display an array of self-organized nanostructured morphologies are block copolymers [5]. The self-organization of these materials arises by linking two chemically distinct polymers, A and B, creating an A–B block copolymer [6]. Conventionally, these two polymers materials when not connected together would macro phase separate into two distinct phases, with dimensions on the micron scale (hence the need to separate your plastic bottles according to the particular polymer for recycling). However due to the physical connection linking them together they are forced to adopt a series of energy minimizing morphologies, on the length scale of the polymer chain. This length scale depends, to a first approximation, on the volume fraction of the constituent units of the block copolymer.

Block copolymers have generated immense interest as materials for a variety of templating and patterning applications [7, 8] due to their ability to self-assemble into an array of 1D, 2D, and 3D periodic structures [9]. They have length scales and morphologies that can be readily controlled through judicious choice of the polymer molecular weight and volume fractions of the two components [10]. In Fig. 2 the block copolymer is represented as a linear chain with a blue fraction of a polymer A component and a red fraction of polymer B component. There are a number of possible block copolymer structures, including amongst others; spheres, cylinders, lamellae and gyroid, these are all shown in Fig. 2. The main advantages of block copolymers over conventional inorganic systems are their large area fabrication potential and low overall cost, they can be manufactured on the

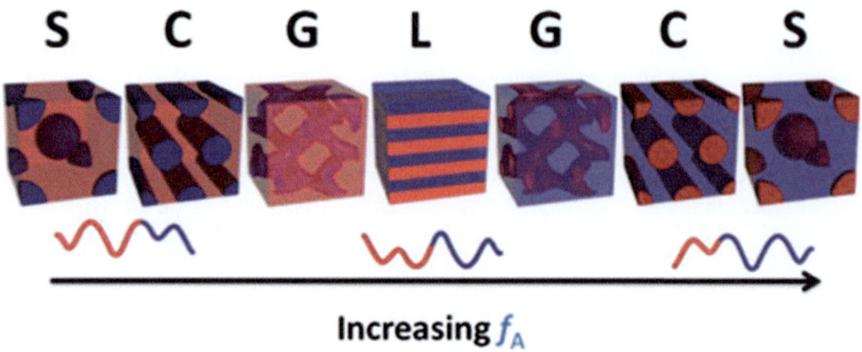

Fig. 2 Schematics of the most commonly observed nanostructures adopted by coil–coil block copolymers, where *S* are spheres, *C* cylinders, *G* gyroid, and *L* lamellae. The *blue block* indicates polymer A (with volume fraction, f_A) and the *red block* polymer B. Reprinted with permission from [10]) Journal of Polymer Science Part B: Polymer Physics (2011) with permission from Wiley

kiloton scale. Kraton™[1] is a styrene-butadiene block copolymer that is routinely used in the soles of shoes as a thermoplastic elastomer, and has found many other applications from barrier materials in medicine (they are free from plasticizers and volatile organic compounds) through to pavements (sidewalks). A number of authors have highlighted the potential of block copolymers as being of interest as partial photonic structures. They are suitable for optical devices due to the high level of ordering and low cost, a number of papers further outlining these attributes can be found elsewhere [9, 11].

In the manufacture of a photonic structure solid films are generally preferred, as a result most workers have concentrated on the solid state properties of these systems where the film is first cast from a solvent and the solvent is then removed. In general this leads to a broadening of the transmission window as the structures dry. This is primarily because the drying process is difficult to control, particularly for large areas, and small differences in solvent gradients make the production of uniform films challenging, consequently non-uniform drying leads to buckling of layers and defects in the final dry film structure. The casting of these films can also be a lengthy process taking hours to days depending on the solvent used. Less volatile solvents such as cumene give the best "annealed" results. In practice time-consuming processing stages such as these could prohibit their use on a commercial basis. The principal challenges that remain to be overcome with these materials are: firstly the ability to reliably and easily control the length scale over a wide wavelength range, second to preserve the narrow transmission peaks during fixing, and thirdly optimizing these materials for production in displays and photonics devices over large areas.

[1] Kraton™ is a registered trademark of Kraton Performance Polymers Inc., Houston, Texas, USA.

1.1 Pure Block Copolymers

One of the fundamental problems with these materials is that the periodicity ($d_1 + d_2$) and thereby the photonic properties of these self-assembled nanostructures are determined by the intrinsic polymer properties. These are controlled principally by the molecular weight, M_w, and the volume fraction, f, and thirdly by the interactions between the blocks A and B of the copolymer. This interaction strength is characterized by the product of the Flory–Huggins interaction parameter, χ_{AB} per monomer, and the total number of segments $N_A + N_B$. The maturity and understanding of block copolymers has led to well-developed design rules that make it relatively routine to produce a particular morphology with a specific periodicity [12]. However, synthesizing one particular diblock copolymer for each specific nanostructure (and color) makes these materials commercially less attractive.

The chemical structures of the polymers discussed in this chapter are displayed in Fig. 3.

1.2 Block Copolymers and Homopolymer

Research has already shown that for lamellar block copolymer structures the domains can be swollen by the addition of homopolymers [13]. However this approach is severely limited by the unbinding process; after the lamellae have swollen by around 20 %, there is an increase in the distribution of lamellar thickness

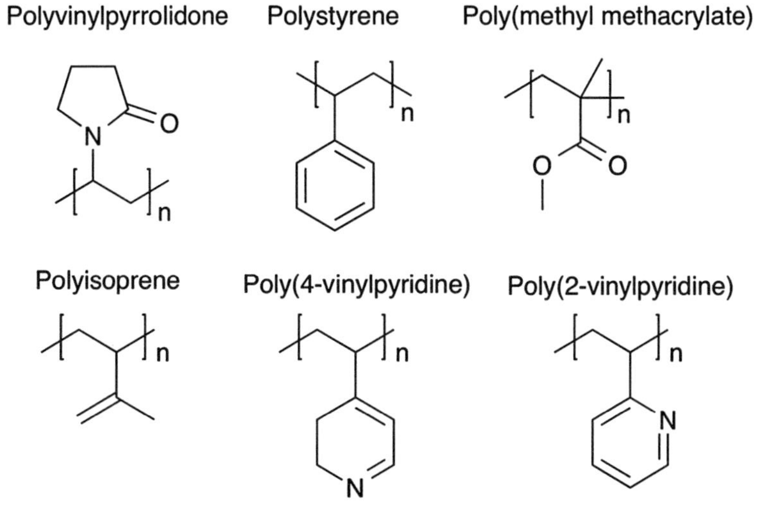

Fig. 3 The chemical structures of the polymers discussed in this chapter

which causes the transmission peak to become broadened. The films made in this study had reflection peak widths of $\Delta\lambda/\lambda \sim 0.15–0.25$.

Parnell et al. [14] used ultra small angle X-ray scattering (USAXS) coupled with UV–VIS spectroscopy to directly relate the block copolymer morphology and domain size to the optical reflectance properties. They examined a series of polystyrene-polyisoprene block copolymer solutions blended with PS and PI homopolymer and saw how the reflectance peak could be tuned over a reasonable wavelength range (~100 nm) and gave narrow peak widths of $\Delta\lambda/\lambda \sim 0.03$. Upon the addition of additional homopolymer macrophase separation takes place, as the homopolymer is unable to be fully accommodated solely in the block copolymer domains.

1.3 Blend of Block Copolymers

Work by Hashimoto et al. [15] showed experimentally how mixing of two block copolymers with dissimilar molecular weights could lead to a homogeneous morphology if the ratio of the molecular weights was less than ~5. This was also confirmed using self-consistent field theory calculations [16]. A study from 2011 examined how two such symmetric high molecular weight diblock copolymers with a 50:50 volume ratio of poly(styrene):poly(isoprene) (PS:PI) are able to be blended together to create a homogeneous structure [17]. The high molecular weights of these block copolymers (500 KDa and 1.5 MDa) means that they form alternating PS and PI domains with repeat periods ranging from 100 to 300 nm depending on the ratio of mixing of the two block copolymers [18].

A series of mixtures using a high molecular weight block copolymer and a low molecular weight block copolymer at various different mixing ratios were created. To order these mixtures, low frequency large amplitude oscillatory shear (LAOS) was used (both linear and rotational shear give high quality ordering). Further work studied the critical shear needed to orient and order the lamellar domains and also performed quenching of non-equilibrium states and orientations of these block copolymers [19].

The nanostructure of these block copolymer mixtures was examined in detail using small angle X-ray scattering (SAXS) at the Diamond Light Source in Oxfordshire (United Kingdom). This technique involves scattering X-rays in a transmission geometry from the ordered block copolymer layers and gives information about the size, morphology, and quality of the ordering in these layers, as such it is a powerful tool to understand the internal structure of nanoscale materials. Figure 4 shows how the mixing of a two lamellar structure forming block copolymers gives rise to an intermediate lamellar structure. These SAXS studies confirmed that there was only one lamellar domain and not the two, which one may expect to see if they had undergone phase separation into two separate lamellae.

The optical properties of the various block copolymer mixtures gave iridescent partially photonic structures that were highly tuneable and gave well-defined and highly periodic one-dimensional Bragg stack structure that was a compositional

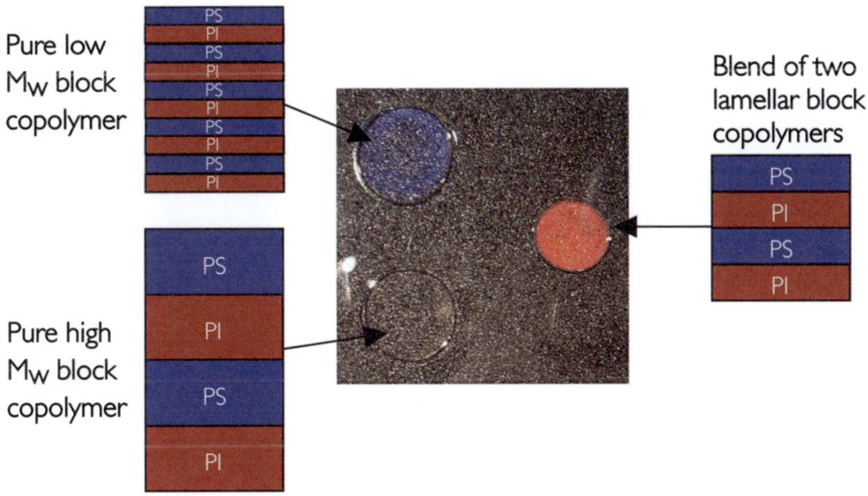

Fig. 4 A pictorial representation of how blending two lamellar forming block copolymers together can be used to produce an intermediate lamellar domain size based on the composition of the mixture. Reproduced from [17] with permission from The Royal Society of Chemistry

average of the two pure lamellar diblock periodicities. The lamellar period depends on the composition of the blend and gives a photonic structure that is easily tuneable in the wavelength range spanning the ultraviolet through the visible and into the infra-red region (λ_{peak} 400–850 nm) with the use of just two diblock copolymers.

Figure 5a shows an image (seen under reflected light) of the shear ordered blend series described above. The compositional change by blending the two diblock copolymer solutions shows a controlled gradual shift in the color of the shear ordered layers with the mixture composition. The intrinsic color of the mixtures in Fig. 5 is due to the internal nanoscale structure of the sheared blend solutions reflecting a narrow and well-defined distribution of wavelengths. From left to right the color in changes from colorless (UV reflector) through blue to red and back to colorless (IR reflector). This progression is due to the addition of increasing amounts of high molecular weight block copolymer. Although colorless at the extremes of blend composition, these sheared solutions have periodic 1D Bragg structures. The lack of visible color at the extremities is due to these structures reflecting in the IR and UV parts of the spectrum. This study on polymer photonics has opened up a simple and cost-effective way of achieving a highly selective and narrow photonic structure that can be blended on demand for an array of applications. This overcomes a significant problem in block copolymer photonics, namely that each individual block copolymer has its own intrinsic wavelength, necessitating separate synthesis for each particular wavelength. The ability to tune the wavelength over the entire visible wavelength range by the use of only two block copolymers opens up opportunities for low cost optical devices.

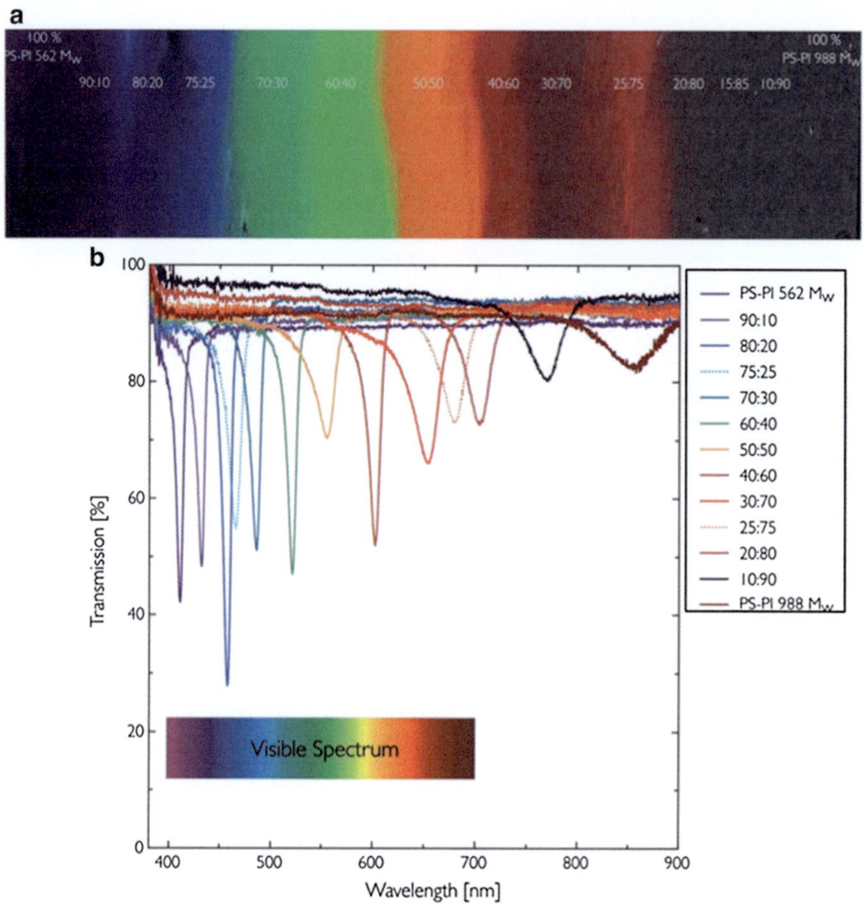

Fig. 5 The optical properties of the block copolymer blend layers. (**a**) Photograph of the blends seen under reflected light and (**b**) the optical properties of the layers in transmission showing that they are highly reflective for a very narrow region. Reproduced from [17] with permission from The Royal Society of Chemistry

1.4 Ionic Liquid Swelling

As mentioned briefly earlier (Sect. 1), the addition of a high boiling point solvent such as cumene can lead to swollen block copolymer films that are well ordered. The problem is that the residual solvent is still a volatile compound and it will over time evaporate, albeit over a long period of time, and so may require some form of encapsulation. A better solution would be to have a system where the solvent remained in the block copolymer and the composite structure remained the same as when it was first made, so that the photonic properties are static as a function of time. Such a system has recently been demonstrated for a poly(styrene)–poly (2-vinylpyridine) block copolymer swollen by an ionic liquid [20].

The films were prepared by spin coating a thin film of the block copolymer followed by 1 day of solvent vapor annealing in a tetrahydrofuran/chloroform mixture (50:50 by volume). A 7:3 molar mixture of imidazole and bis(trifluoromethanesulfonyl) imide was drop cast onto the solvent ordered film and heated to 40 °C for 2 h. After which time the P2VP layer had swollen and the films exhibited strong reflection peaks due to the expanded structure. In contrast the PS domains remain unswollen, as the ionic liquid is a non-solvent for PS. Importantly the study demonstrated that these composite films had stable swollen structures at ambient temperature and humidity, that remained stable for over 100 days. The actuation of colloidal crystal gel films using ionic liquids is discussed in the chapter by T. Kanai, which demonstrates stable swollen photonic systems.

1.5 Comb and Bottlebrush Block Copolymers

A comb polymer is a copolymer whereby one polymer forms the backbone and side chains are grafted or grown from this central chain. Block copolymers have been created as comb–linear systems [21] or comb–comb systems [22].

Kosonen et al. created a hybrid comb structure from a polystyrene-poly (4-vinylpyridine) dodecylbenzenesulphonic acid (DBSA) complex [21] were able to create a hybrid comb structure. In their analysis they conclude that the DBSA chains are bonded to the pyridines by protonation and hydrogen bonding, in effect acting as plasticizers. Structures of these materials had domain sizes up to 140 nm, examples of which are shown in Fig. 6. TEM and AFM clearly show lamellar morphology. These structures also exhibited reflection peaks, the position of which was controllable depending on the degree of complexation with DBSA.

The work by Runge and Bowden [22] used a two-step polymer synthesis to create a series of comb block copolymers with molecular weights over 1 M g mol^{-1}. Synthesizing this would be challenging for a single process linear chain and would require the need for living anionic polymerization, as slow growth rate and chain termination would be inevitable for most polymerization routes, prohibiting such high molecular weight. This method increases the range of chemistries that can be accessed by using the approach. The two-step process also requires much less stringent conditions as compared with the anionic methods used to create the polymers in Parnell's work. Conventionally the high viscosity and associated difficulty in processing a linear polymer of this high molecular weight would also prove challenging. Bottle brush or comb copolymers do not suffer from this effect as they have much fewer chain entanglements. This facilitates the creation of low viscosity inks that may be suitable for inkjet, gravure or intaglio printing, all widely used industrially at high volumes.

The two-step method used ring-opening metathesis polymerization (ROMP) to synthesize the backbone chain and atom transfer radical polymerization (ATRP) [23] to synthesize the arms of the comb. The principal advantage of ATRP is the ability to polymerize a wide variety class of monomers. ROMP is a synthesis route

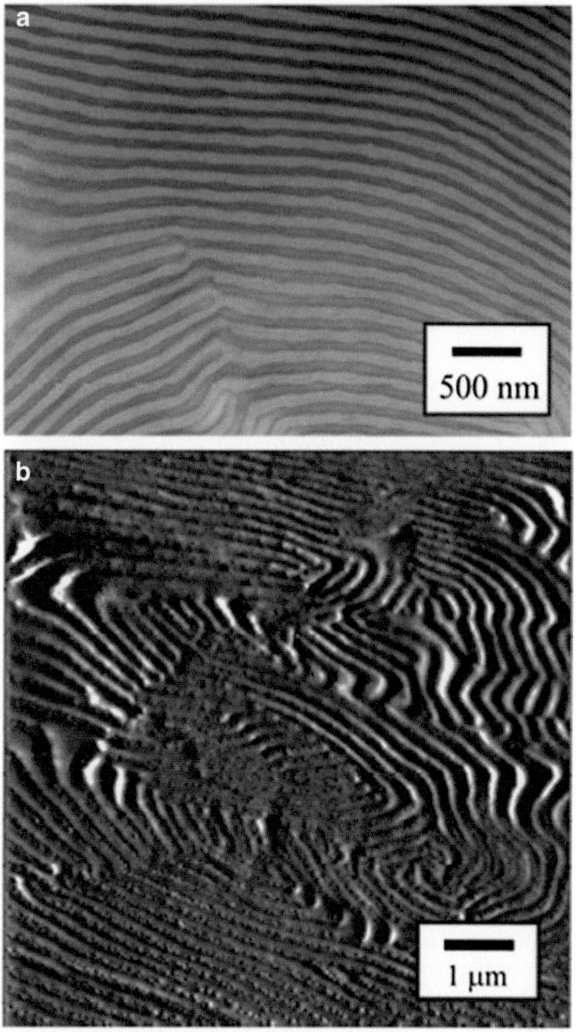

Fig. 6 (**a**) TEM micrograph of a high molecular weight complex PS-block-P4VP(DBSA)$_{2.0}$ illustrating the lamellar structure with a long period of around 140 nm. The P4VP(DBSA)$_{2.0}$ phase shows *dark* in the image due to I_2 staining. Thicknesses of PS and P4VP(DBSA)$_{2.0}$ lamellae are almost equal, which qualitatively agrees with the P4VP(DBSA)$_{2.0}$ weight fraction $w = 0.60$. (**b**) AFM-phase image of PS-block-P4VP(DBSA)$_{2.0}$ illustrating the lamellar structure. The image is taken from [21]. With kind permission of The European Physical Journal (EPJ)

that uses strained cyclic monomers to produce linear polymer chains. Analysis of solid films of these materials showed a series of morphologies from cylinders, spheres, and lamellae, controlled by adjusting the volume fraction in the polymer backbone. The block copolymer films had domain sizes ranging from 130 to 160 nm, the majority produced blue reflecting films after casting from methylene chloride solvent and thermal annealing. An example of the colored copolymer films is shown in Fig. 7, where the solvent has been evaporated to leave a thin layer on the surface of a glass vial. The sensing ability of these films was demonstrated, as upon exposure to methylene chloride a color shift was measured as the structure swells.

Similar work followed later by Grubbs et al. [24, 25]. This involved the use of lactide and styrene based blocks, where they synthesized a series of copolymers with

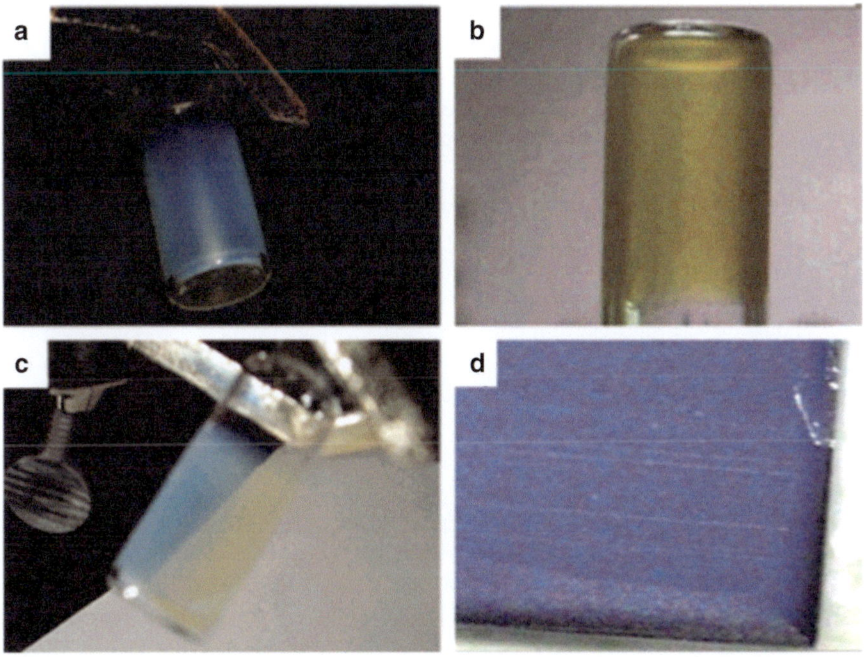

Fig. 7 Showing a coating of the brush copolymer on the inside of a glass vial under reflected light (**a**) transmitted light (**b**) and side by side comparison (**c**) along with a coating on a silicon wafer (**d**). Reprinted with permission from (Runge MB, Bowden NB (2007). Synthesis of high molecular weight comb block copolymers and their assembly into ordered morphologies in the solid state. Journal of the American Chemical Society) (Taken from [22]). Copyright (2007) American Chemical Society

increasing molecular weight and saw a linear increase in the peak position of the reflectance, as the domain size increased. The second study used the idea of blending as mentioned earlier [17] to remove the need to synthesize multiple block copolymers and again using the blending of two copolymers to span the entire optical wavelength.

1.6 Responsive Polymer Gels

An approach that bypasses the need to synthesize a block copolymer with a high molecular weight is to instead use a responsive polymer gel, where the degree of swelling alters the optical properties and gives a high degree of control. With this method it is possible to create a system tuneable over a wide wavelength range, but has severe limitations due to the lengthy processing and annealing stages required [26]. Firstly a comparatively low molecular weight block copolymer of polystyrene–poly(2-vinylpyridine) was cast and annealed in the presence of chloroform vapor to give an ordered lamellar structure, this was subsequently quaternized and crosslinked. The initial 1–3 μm films were then swollen to around 15–20 μm depending on the concentration of ammonium chloride. The quaternized poly (2-vinylpyridine) domain is the variable domain that is being swollen, this induces a change in the overall period and hence color of the reflected light. Interestingly these swollen layers showed a high number of reflection peak orders allowing for a high degree of tuning.

However, there are two competing influences in this case, at low solvent content there is the expected increase in domain spacing and hence a shift to the red end of the spectrum. On further solvent addition the Flory–Huggins interaction parameter is reduced and this leads to a decrease in the domain spacing, leading to a shift to the blue end of the visible spectrum [27]. A major drawback to their use is however the relative insensitivity to trace amounts of solvent vapor and the fact that they tend to be poorly selective with regard to different solvents.

2 Enhancing Refractive Index

Many authors have reported the use of block copolymers with solvent or metal nano-particulate diluents to increase the reflected wavelength into the optical region, by increasing the refractive index difference between the layers. In the Maxwell–Garnett theory, particles, much smaller than the wavelength of light can be used to increase the refractive index of a phase, as the total refractive index is an average of the particles and the matrix. A major issue with this work is the difficulty in assembling the particles in just a single phase. The early theoretical work on block copolymers and nanoparticles [28, 29] showed how the size ratio of the particle to polymer and the volume fractions were important in understanding the equilibrium morphologies. This model assumed that the particles had no preference for either phase. However, as many authors sought to localize particles within a given domain to increase the refractive index difference, particles were increasingly chemically modified to increase their selectivity. Kramer et al. [30–32] utilized a facile concentration based approach to selectively functionalized nanoparticles, with low molecular weight PS and P2VP. This directed the location within the PS–P2VP copolymer. Control of the location is based on the relative amounts of each polymer grafted to the nanoparticle surface.

Thompson [33] extended their earlier theoretical work to include the effect of nanoparticle surface functionality. Xu et al. [34] report on a theoretical investigation into the effects of both chain length and grafting density, illustrating how judicious choice of these can, at equilibrium, produce well-controlled placement of nanoparticles at interfaces or in a given block. By using the lessons learnt from both experiment and theory metal nanoparticles can be selectively located in the respective block copolymer domain. However, the relative complexity is reaching a thermodynamic equilibrium in these systems. Judicious use of solvent annealing over extended periods is often required, as is slow drying of the film [30] to facilitate the organization of the system. Thus commercial exploitation of this technology has not appeared as yet.

3 Devices

The development of block copolymer photonic devices is still a relatively new area, but in this section we will detail some of these devices and the new possibilities that they open up. Most of the devices are proof of concept and at low TRL (technology readiness levels) but when taken together they show the breadth and innovation that is possible with these materials.

3.1 *Electric Field Actuation*

The application of an electric field has demonstrated that the lamellar spacing can be altered controllably and reversibly on a very fast timescale, for a low molecular weight polystyrene–polyisoprene solution system [35]. Solutions of the copolymer in tetrahydrofuran (THF) or toluene were used with 57 and 50 % polymer by weight, respectively. A capacitor style arrangement was used with electric field strengths up to 14 kV mm^{-1}. The time response of the systems was faster than the detector acquisition possible for the SAXS measurements (45 ms). Although the overall change in domain spacing was relatively low at 6 %, it would be possible to think of more polarizable systems that would have much greater volume changes when exposed to high electric fields.

Using an applied voltage [36] used a PS–P2VP hydrophobic hydrophilic block copolymer and made a simple electrochemical cell. They confined a solvent vapor ordered lamellar film between electrodes and immersed it in an electrolytic fluid. The concept and biological structure that inspired this system and layout of the device is shown in Fig. 8. Applying a voltage of up to 10 V caused electrochemical oxidation or reduction and collapse of the P2VP layers. This demonstrated an electrically addressable system, which is a much more appealing technologically than one where the solvent needs to be switched.

This study was closely followed by work from a different group [37], using the same PS–P2VP polymer system. In their setup the cell was filled with a mixture of high purity water and ethanol. The hysteresis of their system was assessed by repeated cycling of the voltage between −2.1 V (blue) and +2.5 V (red), showing that the reflectance peak was at the same position after eight cycles of applied voltage. The main drawback noted in this system is the slow timescale for the P2VP to swell ~20 min.

3.2 *Polymer Laser*

A well-ordered polystyrene–polyisoprene lamellar Bragg mirror was used to build an optically pumped surface emitting polymer laser [39]. This was achieved by

Fig. 8 Taking inspiration from the biological world, mimicking the periodic lamellar squid iridophores seen in Loligo pealei (**a**) and the structure responsible. (**b**) An electrochemical cell used to collapse and swell a PS–P2VP lamellar block copolymer film, with the synthetic equivalent as seen by TEM. (**c**) An electrochemical cell used to collapse an swell a PS-P2VP lamellar block copolymer film, with the synthetic equivalent structure (**d**) as seen by TEM. Reprinted from [36] Advanced Materials (2009) with permission from Wiley. *Part b* of the figure is taken from [38] with permission Copyright 1990, Springer

matching the photoluminescence emission of a laser dye (gain medium) to the reflection peak of the block polymer Bragg mirror. The gain medium was placed between the two polymer Bragg stacks, and the bottom surface was coated with an aluminium mirror. The sandwich structure was then excited by a laser and the emission of the light passes back and forth multiple times through the cavity structure, amplification was achieved and lasing was inferred to take place by the narrow emission line width (~1 nm).

An extension of this approach would be to integrate this amplification system with an emissive polymer that could be made to emit via electroluminescence. This would remove the need to optically pump the system and provide the possibility of low cost to all polymer lasers. Polymer lasers are also discussed in chapters by Scotognella et al. and by S. Furumi.

3.3 Temperature Sensitive Systems

Thermochromaticity, the ability to change color under a temperature stimulus has been widely reported in polymer systems. The paper by Seeboth et al. [40] provides a comprehensive review of the current state of the art covering polymers and copolymers. The advantage of block copolymer systems over conventional thermochromics is that the transition is sharp and between two visually distinct phases as opposed to the gradual color change associated with liquid crystal thermochromics.

Valkama et al. [41] report on the use of reversible hydrogen bonding in a complex of polystyrene poly(4 vinylpyridinium methanesulphonate) with 3-*n*-pentadecylphenol (PS–P4VP(MSA)–PDP. The block copolymer has a volume fraction of 86 % PS (238,100 g/mol PS and 49,500 g/mol P4VP). Swelling of the polymer with PDP leads to the formation of a lamellar phase at low temperatures, as the PDP is selective for the smaller P4VP block. The high volume fraction of PS (86 %) suggests that in the pure bulk phase a lamella phase would be highly unusual. There reversible hydrogen bonding of PDP in the system leads to a sharp color change at 125 °C, as the PDP is no longer selective. This system is chemically simple to produce as self-assembly produces the required complex.

Thomas et al. [42] used a PS–PI system swollen with cumene. Here a more gradual change of color is observed primarily due to the domain spacing changing due to the temperature dependence of the Flory–Huggins interaction parameter χ. This approach should be generally applicable to a wide range of block copolymer systems. A prime candidate for temperature responsive material would be a copolymer with poly(*N*-isopropylacrylamide). This particular polymer exhibits a temperature induced collapse in water around 32 °C, again driven by hydrogen bonding. To our knowledge this has yet to be achieved.

4 Conclusions

In the past 15 years it has become clear that block copolymers have demonstrated huge potential as partial photonic systems. Advances in our understanding have shown that we no longer need to rely on synthesizing a multitude of block copolymers. As blending provides this fidelity in size control. New synthesis routes to produce comb copolymers also make these systems processable via conventional industrial printing methods.

Gels with controlled swelling that can be actuated via electrical control also show promise, with the main obstacle being the response time to switch the color, this will be necessary for display technologies, where fast response times are needed. Again comb or bottle brush polymer architecture may also be of use here.

References

1. J. Bailey, J.S. Sharp, Eur. Phys. J. E **33**, 41 (2010)
2. T. Kazmierczak, H. Song, A. Hiltner, E. Baer, Macromol. Rapid Commun. **28**, 2210 (2007)
3. S. Kinoshita, *Structural Colors in the Realm of Nature* (World Scientific, Singapore, 2008)
4. E.R. Dufresne, H. Noh, V. Saranathan, S.G.J. Mochrie, H. Cao, R.O. Prum, Soft Matter **5**, 1792 (2009)
5. F.S. Bates, G.H. Fredrickson, Phys. Today **52**, 32 (1999)
6. N. Hadjichristidis, S. Pispas, G. Floudas, *Block Copolymers: Synthetic Strategies, Physical Properties, and Applications* (Hoboken, New Jersey, 2002)
7. F.F. Lupi, T.J. Giammaria, G. Seguini, F. Vita, O. Francescangeli, K. Sparnacci, D. Antonioli, V. Gianotti, M. Laus, M. Perego, ACS Appl. Mater. Interfaces **6**, 7180 (2014)
8. M. Li, C.K. Ober, Mater. Today **9**, 30 (Hoboken, New Jersey, 2006)
9. Y. Fink, E.L. Thomas, Polymeric Photonic Band Gap Materials. U.S. Patent 0135880 A1, 26 Sept 2002
10. P.D. Topham, A.J. Parnell, R.C. Hiorns, J. Polym. Sci. B **49**, 1131 (2011)
11. A.C. Edrington, A.M. Urbas, P. DeRege, C.X. Chen, T.M. Swager, N. Hadjichristidis, M. Xenidou, L.J. Fetters, J.D. Joannopoulos, Y. Fink, E.L. Thomas, Adv. Mater. **13**, 421 (2001)
12. I.W. Hamley, *The Physics of Block Copolymers* (Oxford University Press, Oxford, 1998)
13. A.M. Urbas, A. Urbas, R. Sharp, Y. Fink, E.L. Thomas, E.L. Thomas, M. Xenidou, L.J. Fetters, Adv. Mater. **12**, 812 (2000)
14. A.J. Parnell, N. Tzokova, A. Pryke, J.R. Howse, O.O. Mykhaylyk, A.J. Ryan, P. Panine, J.P.A. Fairclough, Phys. Chem. Chem. Phys. **13**, 3179 (2011)
15. D. Yamaguchi, T. Hashimoto, Macromolecules **34**, 6495 (2001)
16. M.W. Matsen, J. Chem. Phys. **103**, 3268 (1995)
17. A.J. Parnell, A. Pryke, O.O. Mykhaylyk, J.R. Howse, A.M. Adawi, N.J. Terrill, J.P.A. Fairclough, Soft Matter **7**, 3721 (2011)
18. J.P.A. Fairclough, A. Pryke, N. Tzokova, A.J. Parnell, Copolymer Blends, Light Reflective Films Comprising the Same and Methods for Their Preparation and Use. U.K. Patent GB2482552A, 8 Feb 2012, 2013
19. O.O. Mykhaylyk, A.J. Parnell, A. Pryke, J.P.A. Fairclough, Macromolecules **45**, 5260 (2012)
20. A. Noro, Y. Tomita, Y. Shinohara, Y. Sageshima, J.J. Walish, Y. Matsushita, E.L. Thomas, Macromolecules **47**, 4103 (2014)
21. H. Kosonen, S. Valkama, J. Ruokolainen, M. Torkkelli, R. Serimaa, G.T. Brinke, O. Ikkala, Eur. Phys. J. E **10**, 69 (2003)
22. M.B. Runge, N.B. Bowden, J. Am. Chem. Soc. **129**, 10551 (2007)
23. K. Matyjaszewski, Mater. Today **8**, 26 (2005)
24. G.M. Miyake, V.A. Piunova, R.A. Weitekamp, R.H. Grubbs, Angew. Chem. Int. Ed. **51**, 11246 (2012)
25. B.R. Sveinbjörnsson, R.A. Weitekamp, G.M. Miyake, Y. Xia, H.A. Atwater, R.H. Grubbs, Proc. Natl. Acad. Sci. U. S. A. **109**, 14332 (2012)
26. Y. Kang, J.J. Walish, T. Gorishnyy, E.L. Thomas, Nat. Mater. **6**, 957 (2007)
27. N. Tzokova, Iridescent Coatings from Block Copolymers – Adhesion and Fabrication, n.d.
28. R.L. Thompson, R.B. Thompson, V.V. Ginzburg, M.W. Matsen, A.C. Balazs, Science **292**, 2469 (2001)
29. R.L. Thompson, R.B. Thompson, V.V. Ginzburg, M.W. Matsen, A.C. Balazs, Macromolecules **35**, 1060 (2002)
30. J.J. Chiu, J.J. Chiu, B.J. Kim, B.J. Kim, E.J. Kramer, E.J. Kramer, D.J. Pine, D.J. Pine, J. Am. Chem. Soc. **127**, 5036 (2005)
31. B.J. Kim, J. Bang, C.J. Hawker, J.J. Chiu, D.J. Pine, S.G. Jang, S.-M. Yang, S. Yang, E.J. Kramer, Langmuir **23**, 12693 (2007)
32. J.J. Chiu, B.J. Kim, G.-R. Yi, J. Bang, E.J. Kramer, D.J. Pine, Macromolecules **40**, 3361 (2007)

33. M.W. Matsen, R.B. Thompson, Macromolecules **41**, 1853 (2008)
34. G.-K. Xu, G.-K. Xu, X.-Q. Feng, X.-Q. Feng, S.-W. Yu, S.-W. Yu, Nano Res. **3**, 356 (2010)
35. K. Schmidt, H.G. Schoberth, M. Ruppel, H. Zettl, Nat. Mater. **7**, 142 (2007)
36. J.J. Walish, Y. Kang, R.A. Mickiewicz, E.L. Thomas, Adv. Mater. **21**, 3078 (2009)
37. Y. Lu, H. Xia, G. Zhang, C. Wu, J. Mater. Chem. **19**, 5952 (2009)
38. K.M. Cooper, R.T. Hanlon, R.T. Hanlon, B.U. Budelmann, Cell Tissue Res. **259**, 15 (1990)
39. J. Yoon, W. Lee, E.L. Thomas, Nano Lett. **6**, 2211 (2006)
40. A. Seeboth, D. Lötzch, D. Lötzsch, R. Ruhmann, O. Muehling, Chem. Rev. **114**, 3037 (2014)
41. S. Valkama, H. Kosonen, J. Ruokolainen, T. Haatainen, M. Torkkelli, R. Serimma, G.T. Brinke, O. Ikkala, Nat. Mater. **3**, 872 (2004)
42. J. Yoon, J. Yoon, W. Lee, W. Lee, E.L. Thomas, E.L. Thomas, Macromolecules **41**, 4582 (2008)

Co-extruded Multilayer Polymer Films for Photonic Applications

Dario Cavallo, Han Goossens, and Han E.H. Meijer

Abstract Co-extrusion through a series of layer-multiplying die elements enables the production of polymer films containing tens to thousands of layers with individual layer thicknesses from the micro- to the nanoscale. The use of polymers with different refractive indexes allows to fabricate, with a simple and inexpensive process, narrowband one-dimensional photonic crystals. Tunability of the photonic gap can be achieved by proper choice of the materials or of the dies, varying the contrast in refractive indexes or the number and thickness of the individual layers. The use of elastomeric polymers offers the opportunity to modify the UV–VIS spectral response of the 1-D photonic crystals simply by deformation of the film. Finally, co-extrusion of multi-layer polymeric films revealed to be an enabling technology, which allows the realization of new polymeric optical devices, going from flexible lasers to gradient refractive index lenses.

Keywords Multilayer co-extrusion • Polymer processing • 1-D photonic crystals • Polymeric photonic crystals • Distributed Bragg reflectors • Flexible photonic crystals

---

D. Cavallo (✉)
Department of Chemistry and Industrial Chemistry, University of Genoa, via Dodecaneso 31, Genoa 16146, Italy
e-mail: dario.cavallo@unige.it

H. Goossens
Department of Chemical Engineering and Chemistry, Eindhoven University of Technology, PO Box 513, Eindhoven 5600 MB, The Netherlands
e-mail: j.g.p.goossens@tue.nl

H.E.H. Meijer
Department of Mechanical Engineering, Eindhoven University of Technology, PO Box 513, Eindhoven 5600 MB, The Netherlands
e-mail: H.E.H.Meijer@tue.nl

D. Comoretto (ed.), *Organic and Hybrid Photonic Crystals*,
DOI 10.1007/978-3-319-16580-6_7

1 Introduction

A quick and efficient way of making plastic multilayer structures is by multilayer co-extrusion. Multilayer co-extrusion has developed into an important polymer fabrication process, providing large growth opportunities for the polymer industry. Co-extruded multilayered films and tapes are challenging traditional materials, such as metals, glass, paper, and textiles. The attraction of co-extrusion is both economic and technical. It is a single-step process starting with two or more polymer materials that are simultaneously extruded and shaped by using a single die to form a multilayered sheet or film. Co-extrusion avoids the costs and complexities of conventional multi-step lamination and coating processes, where individual plies must be made separately, primed, coated, and laminated.

A particular kind of co-extrusion is layer-multiplying co-extrusion, by which products that contain layers thinner than what can be made as an individual ply can be manufactured in a continuous process. In fact, layer-multiplying co-extrusion has been used to manufacture films consisting of thousands of layers with individual layer thicknesses of less than 100 nm [1, 2]. It is difficult to imagine another practical method to manufacture these submicron-layered structures.

Multilayers films may be used to add colours or visual effects, screen infrared or ultraviolet radiation, provide barrier properties for packaging applications, etc. As layer-multiplying co-extrusion evolves towards the nanoscale, unique film properties controlled by interfacial or confinement-induced phenomena are discovered [2, 3].

In this chapter the use of multilayer co-extruded films for advanced applications in the field of photonics will be addressed. Firstly, the principles of layer-multiplying co-extrusion will be discussed, paying the greatest attention to processing and design parameters which affect the final properties of the multilayer film, as, for instance, layer thickness uniformity. Afterwards, the basic notions of photonic crystals, in particular 1-D, will be briefly recalled. Finally, the production of multilayered polymer films exhibiting photonic properties and their application in flexible optical devices, such as waveguides or lasers, will be described.

2 Layer-Multiplying Co-extrusion: Operating Principles, Design and Limitations

2.1 Principles

Multilayer co-extrusion is a process in which polymers are extruded and joined together in a feedblock or die to form a single structure with multiple layers. In the layer-multiplying co-extrusion process the "baker's transformation" is applied, in which the flow is successively stretched, cut, and stacked, in the same way that a baker rolls and folds his dough. The baker's transformation can be realized in a continuous process by using static mixing elements, for instance a multiflux static

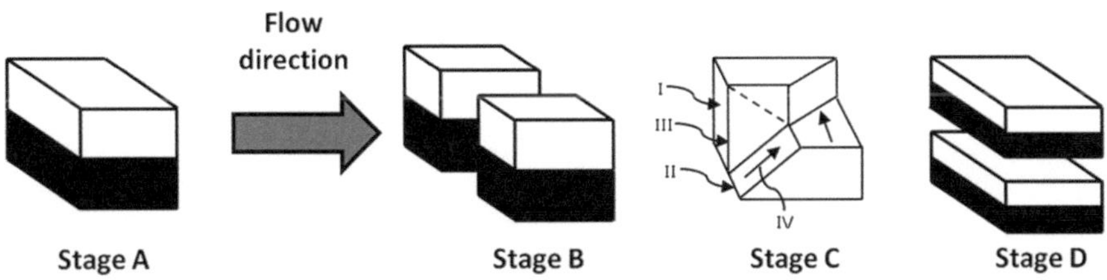

Fig. 1 Operating principle of a mixing element, showing cutting, stretching and stacking of the melt stream. Adapted from [6]

mixer. In the channel of this device, several elements are present, in each of which the flow undergoes a baker's transformation, resulting in an exponential increase in the number of layers [4].

The geometry of the multiflux static mixer was first described by Sluijters [5]. It is a compact design, has a low pressure drop and is easily produced at low costs. The operating principle is demonstrated in Fig. 1, showing the cutting, stretching and stacking of the melt stream inside this element, by using a white and a coloured fluid.

In stage A of the first element, two horizontal input flows (white and black in Fig. 1) are combined to be separated in stage B by a vertical cutting edge (III). The result is two identical strips I and II (see the schematic of stage C). The two-layer assembly of input strip II is accelerated in flow direction by transportation upwards on a ramp (IV), and decelerated by a lateral expansion so that the original width of the assembly before cutting is restored in the top half of the element. The other part of the cut assembly, I, travels downwards on an identical ramp (not shown) and is also allowed to expand laterally. Strip I that travelled downwards takes a position underneath strip II that was transported upwards, and finally the flows are recombined (Stage D). This results in a four-layered stack. In a second element, the number of layers can be doubled to eight and so on. The final number of layers equals $m \cdot 2^n$, where m is the number of layers fed into the first element of the mixer and n is the number of mixing elements used [1, 5, 6].

2.2 *Basic Process Design*

The scheme of a typical setup used to produce microlayered and nanolayered structures in a continuous operation is reported in Fig. 2. While co-extrusion dies are unique, the extruders that are used before the die and the take-away equipment used afterwards, like chill rolls, are the standard equipment that are also used for the manufacture of single-layer films [7].

A two-component co-extrusion system typically consists of two single-screw extruders with, mounted on their exits, two melt metering gear pumps (not shown in

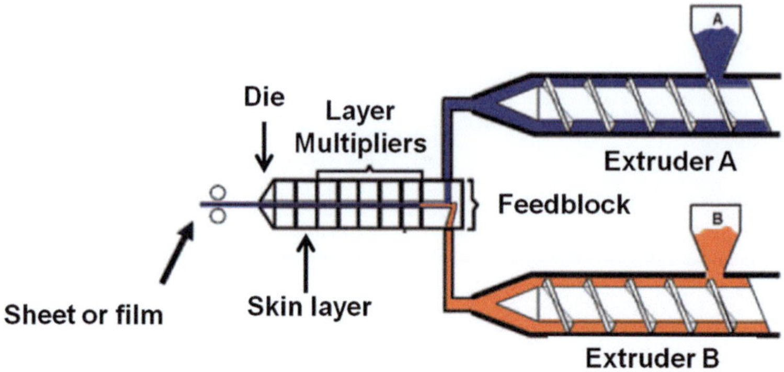

Fig. 2 Two components multilayer co-extrusion system. Adapted with permission from [8]

Fig. 2), an A/B two-layer co-extrusion feedblock, a series of layer-multiplying die elements, two laminating surface extruders and an exit tape or film die.

The inclusion of gear pumps into the co-extrusion system enables an added degree of control over the volume fluxes, and therefore over the relative ratio of the layered polymers as they enter the feedblock. There is no restriction to a 50–50 system composition since any ratio of the two fluxes entering the multiplying system is allowed. Neither is there any restriction to the inflow of two polymers only. Any number of polymers with any layer thickness design and any stack design can be multiplied by using flexible feedblocks to create the starting layered stack. The function of the feedblock is to combine the (typically) two polymer melt streams as two parallel layers. From the feedblock, the two polymer layers enter a series of layer multiplication elements. While the operating principle of these mixing elements just has been discussed it should be noted that there are two types of mixing elements, the (right) E_R and the (left) E_L element. The E_R element is a mixing element in which the flow at the right-hand side of the element is led downwards from top to bottom (and vice versa at the left-hand side), while in the E_L element this is just the other way around directing the flow upwards in the right-hand side of the element and downwards on the left-hand side. Stage C of Fig. 1 shows the E_L element. These two elements are assembled in an alternating fashion to equalize axial resistances over the parallel channels of the elements. Unequal resistances result in pressure differences between the right- and left-hand side channels in the elements causing layer inhomogeneities over the width of the element.

At the exit of the layer multipliers, an additional extruder applies a skin layer on the top and bottom of the multilayered incoming flow. The skin layer could be just a sacrificial material layer used to protect the multilayers in contact with the die surface, where the polymers typically experience the higher shear flows and forces. The importance of these protective layers, preferably made of a third easy-to-peel polymer, to preserve layer uniformity and surface quality will be underlined later on. A typical thickness of this sacrificial layer can reach 20–30 % of the total film thickness.

Finally, the layered polymer system exiting the tape or film die is stretched and cooled, thanks to two cold rolls, and collected. The final thickness of the layered system is set by the draw ratio, which is the ratio between the take-up velocity of the cold rolls and the linear velocity at the die exit.

2.3 Limitations 1: Interface Instabilities

The co-extrusion process has clear advantages over alternative processes: the layering and the formation of the final product is all done in one step, and layers mutually support each other allowing for extreme thinness. However, there are limitations. The two main difficulties that can arise when using the co-extrusion process are interlayer instability and layer thickness non-uniformity, both of which are detrimental for photonic crystals applications. The main causes of these layer defects, and some practical steps to limit them, are discussed in the following paragraphs.

Interfacial instability is a non-steady-state process in which the interface location between layers varies locally in a transient manner. Interface distortion due to flow instabilities can cause thickness non-uniformities in the individual layers, while the thickness of the product still remains constant. If a large-amplitude waveform develops in the flowing multilayer stream within the die, the velocity gradient can carry the crest forward and convert it into a fold. Multiple folding can result in an extremely jumbled, intermixed interface and a film with a marbleized appearance (see Fig. 3a). This type of instability is commonly called a zig-zag instability and it develops in the final die land, which is usually the region of highest shear rate and shear stress as the multilayer melt is shaped to its final dimension before leaving the die. It is believed that there is a critical interfacial shear stress at which interfacial flow instabilities may occur in a given pair of polymers. The zig-zag type of interfacial instability can be reduced or eliminated by increasing the skin layer thickness, thus moving the interface to a lower shear stress. The zig-zag instability can also be reduced by reducing the total extrusion rate, by decreasing the viscosity of the skin layer, or by increasing the die gap. These procedures reduce the interfacial shear stress and a stable flow results [9, 10].

Other types of instabilities may exist, like the wave instability. The wave instability looks like a train of parabolas oriented in the flow direction, each parabola spanning the width of the extruded sheet (see Fig. 3b). It develops inside the die, well ahead of the die land, and the internal die geometry influences both the severity and the pattern.

It has been suggested that this type of instability may be related to the extensional viscosities of the polymers used to constitute the individual layers in the co-extruded structure. Therefore, the die should be designed to spread the layers slowly and at a uniform rate.

One last example of interface instability encountered when processing polymeric structures containing very thin layers is break-up of the layers near the walls.

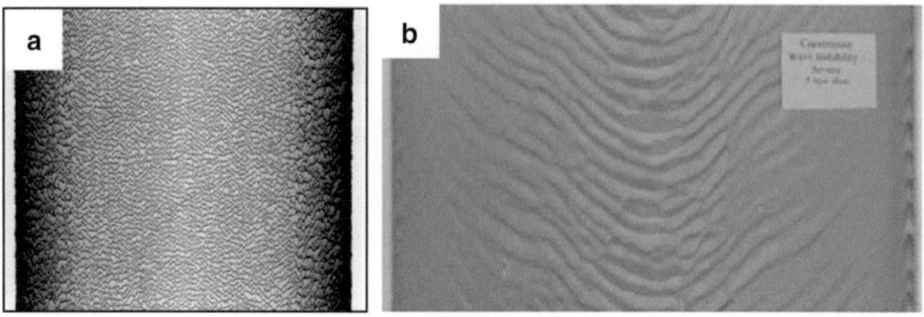

Fig. 3 Interlayer instability patterns: (**a**) zig-zag; (**b**) wave

This break-up is more relevant if relatively low viscous melts are considered and occurs via Rayleigh instabilities [13]. The driving force is provided by the interfacial tension between liquids, which tends to minimize the interfacial area for a given volume of material. In essence, for a jet of fluid immersed in another liquid, small perturbations of the surface shape always exist, although they might be of negligible amplitude. Considering sinusoidal perturbations, some of them will grow the fastest and will eventually pinch the layer into droplets. In the case of two viscous polymers, liquid drop deformation and break-up is described in terms of the capillary number, *Ca*, which is the ratio of the deforming stress $\tau = \eta\dot{\gamma}$ exerted on the drop by the external flow field (with η the viscosity and $\dot{\gamma}$ the shear rate) and the shape-conserving, interfacial stress $\Gamma = \sigma/R$ (with σ the interfacial tension and R the local radius):

$$Ca = \frac{\eta\dot{\gamma}R}{\sigma} \tag{1}$$

For small capillary numbers, the interfacial stress withstands the shear stress and a steady shape and orientation is reached. Above a critical value, $Ca > Ca_{\mathrm{crit}}$, interfacial Rayleigh disturbances grow on the thread and result in its break-up into small drops. Although layers are, as opposed to cylinders, basically stable against Rayleigh distortions, since in stratified systems any disturbance of a flat interface negatively influences the surface-to-volume ratio, in regions of high shear, e.g. close to walls, the flat interface may be distorted so seriously that cylindrical entities that can break up locally occur.

2.4 Limitations 2: Viscous and Viscoelastic Encapsulation

A critical parameter to be taken into account to obtain layer uniformity and continuity is the matching of viscosities of the polymer pair. Poor viscosity matching can result in the low-viscosity polymer encapsulating the high-viscosity one, or in interfacially driven layer break-up. Encapsulation effects can be conveniently

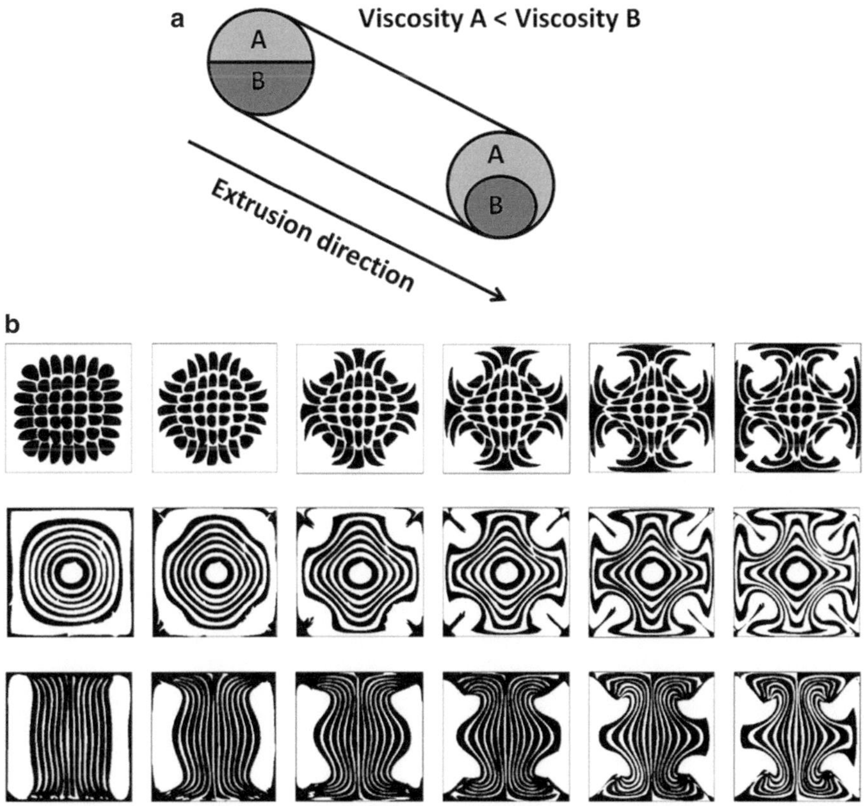

Fig. 4 (**a**) Viscous encapsulation. (**b**) Viscoelastic "encapsulation" even in one-component flows caused by secondary flow-vortices driven by the second normal stress difference. The polymer is *white* and *black* coloured polystyrene and images are shown of cross sections taken at different axial positions along the square channel

described considering a capillary flow of two polymers (see Fig. 4a) [7]. Two polymers introduced side by side into a round tube experience interfacial distortion during flow if the viscosities are mismatched. The lower-viscosity polymer migrates to regions of highest shear (at the wall) and tends to encapsulate the higher-viscosity polymer. It is possible for the low-viscosity polymer to completely encapsulate the higher-viscosity polymer. The degree of interfacial distortion due to viscosity mismatch depends on the extent of viscosity difference, shear rate and residence time [7]. Again, elongational flow influences the encapsulation phenomenon. Moreover, this effect is reported to be more significant with a higher number of layers [11].

Encapsulation effects can also arise from the viscoelasticity of polymer melts. Driven by the second normal stress difference in a flow with non-spherical cross section, e.g. rectangular or square but also general coat hanger die flow channel geometries, secondary flows occur that cause interface deformation. Depending on the initial layer configuration, fantastic patterns result even if a single component fluid flows in a simple channel with constant square cross section, see Fig. 4b. This phenomenon is completely understood since it can be quantitatively predicted, but of course is industrially unwanted, since it disturbs the carefully designed and produced multilayer structure exiting the layer multiplier and entering the coat-hanger die to produce a film [12]. Both optimized die geometries and a proper choice of the polymer used can help suppressing these secondary flow effects.

2.5 *Practical Process Design*

The described effects of interlayer instability, viscous encapsulation, and layer break-up largely compromise the quality of the resulting multilayered films, which is of particular importance when optical applications are concerned. In order to limit these detrimental effects and improve the layered film quality several solutions can be put forward: (1) match the polymer viscosities, (2) optimize the layer-multiplying design and (3) include sacrificial surface layers.

For two given polymer grades, a viscosity match between them can be obtained by a proper choice of the processing temperature. Figure 5 illustrates this concept for co-extrusion of polypropylene and polystyrene [14]. The viscosity is measured by means of a melt flow indexer (MFI), by determining the weight of polymer extruded from a capillary in a given time and under a given load. From the

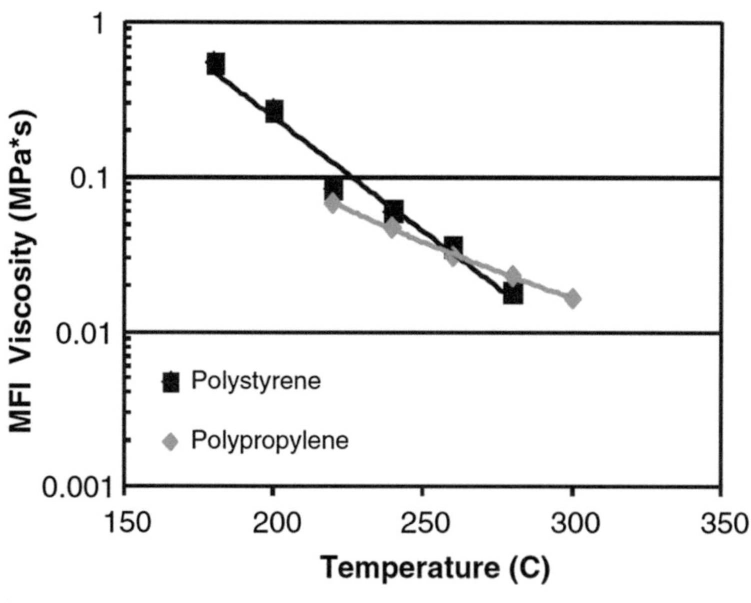

Fig. 5 MFI viscosity of polystyrene/polypropylene as a function of temperature. Adapted with permission from [14]

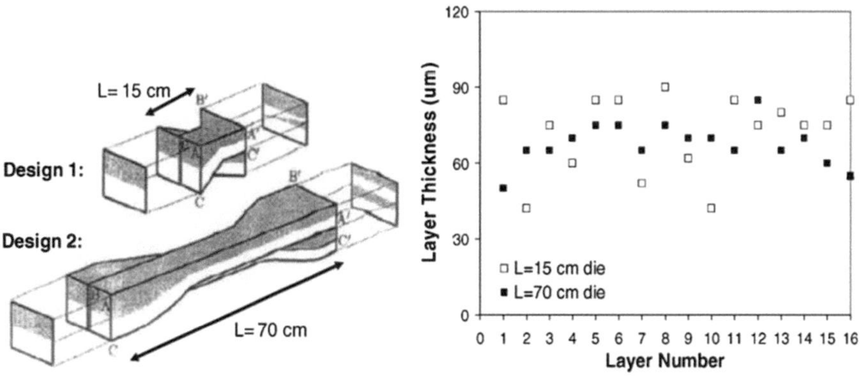

Fig. 6 Scheme of the different layer-multiplying designs (*left*), effect of die length on layer uniformity (*right*). Adapted from [14]

experimental plot of polymer viscosities against temperature, an ideal extrusion temperature of 260 °C for this polymer pair can be deduced. A crude rule of thumb is that polymer viscosities must be matched within a factor of three or four.

The effect of the layer-multiplying die design on layer thickness distribution has been also investigated [4, 8, 14]. Of particular importance is the effect of the layering path length, the higher its value, the more stable the flow field which develops inside the mixing element, eventually leading to better layer thickness homogeneity. This effect is shown in the example of Fig. 6, where two distinct designs, differing in the layering path length L, are compared.

Design I has a sharp spreading transition and short landings, while in Design II transitions are more gradual and landings are longer. The performance of the two designs has been tested by producing 16-layers polycarbonate/poly (methylmethacrylate) films. Cross sections of the films were then cut and the individual layer thicknesses were measured by polarized optical microscopy. The results are shown in Fig. 6 (right). Co-extrusion through dies with longer path length leads to an average layer thickness of 68±8 μm, compared to a much more spread value of 71±17 μm for Design I. Clearly, the improvement in the layer distribution homogeneity occurs at the expense of the compactness of the design.

Improvements in film uniformity and surface quality during film manufacturing using, e.g., a coathanger die, are obtained through the addition of sacrificial surface layers. These protective boundary layers are thicker layers that are added to the surface of the stack to move the thin layers away from the wall of the processing equipment, where the highest stresses are encountered. By moving the very thin layers into a region of the channel with lower stresses, their break-up can be avoided. The sacrificial surface layer can be peeled off from the layered polymer core. Figure 7 illustrates the effect of the sacrificial polymer surfaces on layer uniformity and break-up. As can be noticed, break-up and interfacial instability are observed in the absence of sacrificial layers, while the morphology is more homogeneous when such layers are present.

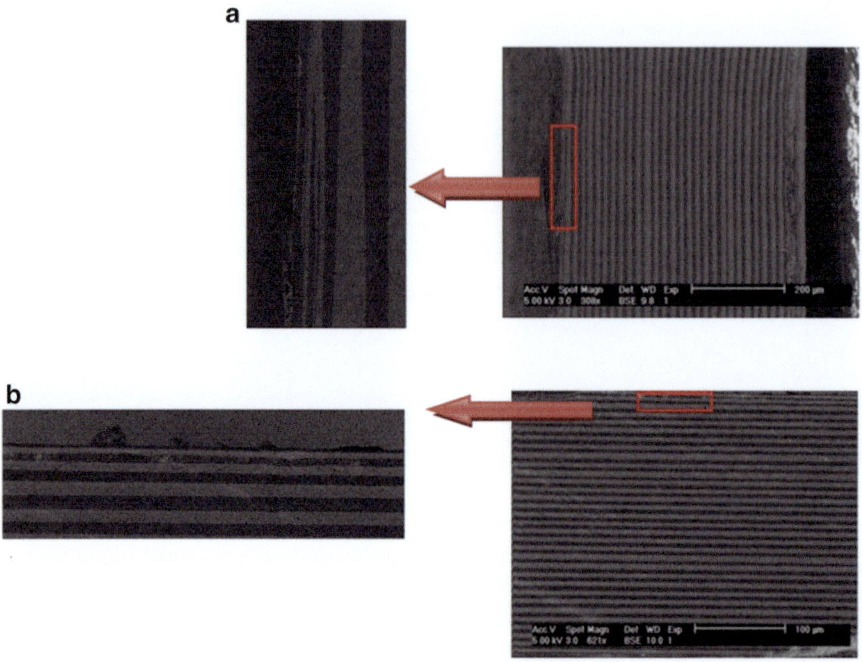

Fig. 7 Scanning electron microscopy images of multilayered films without (**a**) and with (**b**) sacrificial surface layers

3 General Properties of Photonic Crystals

As outlined in the previous section, layer-multiplying co-extrusion is an efficient process to produce 1-D stacks of alternating polymer layers with typical individual thicknesses below the micron range. Such an assembly, if proper conditions in terms of refractive index and layer thickness are met, can act as a 1-D photonic crystals. Before showing the application of co-extrusion to the production of layered optical systems, the main features of photonic crystals will be briefly recalled in the following section.

Photonic crystals are composite materials made of media possessing different refractive indices and arranged in a well-ordered periodical lattice (1-D, 2-D or 3-D), having a periodicity comparable to the wavelength of visible light [15]. Interference phenomena occur inside photonic crystals: light is diffracted by lattice plans and the beams interfere constructively and destructively after the reflection. This complicated combination of diffraction and interference within the dielectric lattice gives rise to a photonic band structure, with allowed and forbidden energy bands for the propagation of photons (light). The situation is analogous to what happens for electron transport in a semi-conductor. Therefore, a photonic band gap is generated: for this energy photons cannot propagate inside the photonic crystal, thus light is backward diffracted giving rise to strong chromatic effects and iridescence [16, 17]. Indeed, the fingerprint of the photonic band gap is a typical peak in the reflectance spectrum of the sample, as well as a dip in the transmission one, occurring at the same energy/wavelength.

The conceptually simplest example of a photonic crystal is a multilayer stack of alternating high and low dielectric constant materials, so that a monodimensional modulation of the refractive index is obtained in the direction perpendicular to the layers plane. Such a 1-D photonic crystal is also indicated as dielectric mirror or distributed Bragg reflector (DBR). As will be shown later in the chapter, this kind of structure can be made by layer-multiplying co-extrusion. Therefore its main characteristics, for what concerns the interaction with light, will be derived in the following paragraphs.

Light wave propagation in 1-D photonic crystals can be described analysing the multiple reflections at the interfaces between low and high refractive index layers. Maximum reflectance condition will arise when, for a set of wavelengths, all reflected beams interfere constructively. The case of a standard crystal can be compared with that of a photonic crystal. The diffraction of X-ray in crystal is described by the well-known Bragg law:

$$m\lambda = 2D\cos(\theta) = 2D\sqrt{1 - \sin^2(\theta)} \tag{2}$$

where θ is the incidence angle (measured from the normal to the crystal surface) and D the interplanar spacing. In a photonic crystal the same picture holds on a different scale since the interplanar spacing, due to the dielectric lattice, is in the order of hundreds of nanometers. Then, diffraction occurs in the visible-near infrared spectral region. However, in a photonic crystal, light is also refracted at the interfaces, as a consequence of the Snell's law. By combining the Bragg and the Snell' laws we obtain the Bragg–Snell law:

$$m\lambda = 2D\sqrt{n_{\mathrm{eff}}^2 - \sin^2(\theta)} \tag{3}$$

The effective refractive index, n_{eff}, is an average index which can be approximately obtained from the refractive indices of the composing media (a and b) weighted over the correspondent volume fractions (f_a and f_b)

$$n_{\mathrm{eff}}^2 = \varepsilon_{\mathrm{eff}} = f_a\varepsilon_a + f_b\varepsilon_b \tag{4}$$

where $\varepsilon_{\mathrm{eff}}$ is the effective dielectric constant and ε_a and ε_b are the dielectric constants of the different materials. A more detailed analysis of the photonic crystals reflection can be found in first chapter of the present book.

A final consideration on the role of the refractive index contrast of the composing media on the photonic crystal properties is useful, in the light of the further description of co-extruded multilayered films. The full width at half maximum (ΔE) of the reflectance peak at the photonic band gap ($E = hc/\lambda$, where λ is provided in Eq. (3)) is a function of the dielectric contrast $n_a - n_b$

$$\Delta E = \frac{4}{\pi} E \frac{|n_a - n_b|}{n_a + n_b} \tag{5}$$

By assuming that the disorder does not play a major role in the photonic crystal, an increase of the dielectric contrast gives rise to a wider photonic band gap [18, 19].

4 Co-extruded Multilayered Polymeric Photonic Crystals

Co-extrusion of nanolayered polymers provides a method for the continuous, low cost fabrication of light-weighted, highly reflective polymer photonic crystals. The first example of melt-processed polymeric photonic crystals was shown by Schrenk et al. [20, 21]. This was a co-extruded multilayered films containing more than 200 alternating layers of polystyrene (PS) and poly(methyl methacrylate) (PMMA) with individual layer thicknesses in the range of 0.06–0.1 μm. The appearance of these films was metallic, with very vivid iridescence.

More recently Hiltner, Baer and co-workers extended the study of multilayered co-extruded films for photonic applications [22]. At first, the effect of multilayer film structure on its performance as 1-D photonic crystal was investigated. The chosen pair of polymer was again polystyrene and poly(methyl methacrylate), due to their good viscosity matching and relatively large difference in refractive index. A refractive index of 1.585 and 1.491 was measured for PS and PMMA, respectively, leading to a difference, Δn, of 0.09. Four and six layer-multiplying elements were used to obtain films with 32 and 128 layers. A sacrificial low-density polyethylene layer was also applied to the outer surfaces of the films. The approximate thickness of the individual polymer layers is around 100 nm. Additionally, two 32 layers films were combined to form a 64 layers stack.

Figure 8 shows an AFM image of a cross section of the 32 layers film, together with the measured layer thickness distribution. An average thickness of 86 nm, with an error of 17 % is obtained. The transmission spectra of the 32 layers film and of the 64 layers stack are reported in Fig. 9. The measured spectral response is

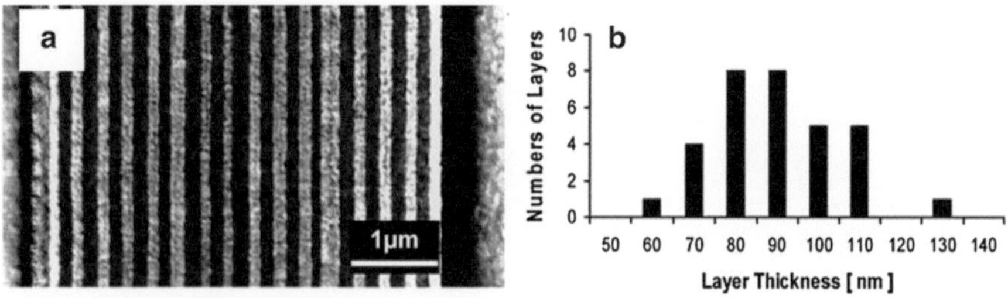

Fig. 8 (**a**) AFM phase image of 32 layers film cross section and (**b**) layer thickness distribution. Adapted with permission from [22]

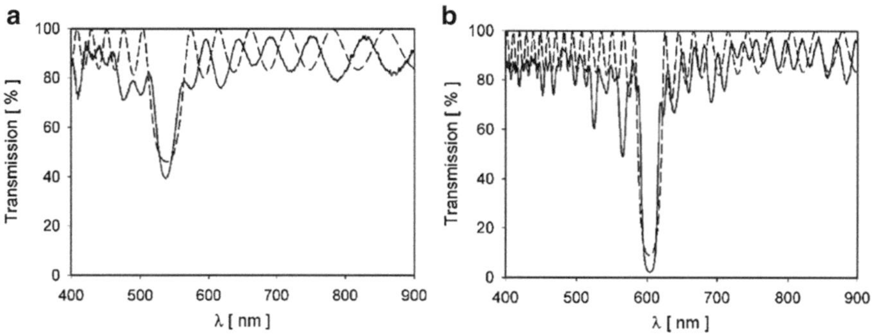

Fig. 9 Comparison between measured (*solid line*) and calculated (*dashed line*) transmission spectra for (**a**) 32 and (**b**) 64 layered stacks. Adapted with permission from [22]

compared with the simulated one, evaluated by the Vasicek matrix method [23], considering layers of equal thickness. As can be seen in Fig. 9a, the agreement between the measured and the calculated spectra of the 32 layers assembly is fairly good, regarding bands intensity, widths and positions. The agreement can be further improved if one considers the effective non-homogeneous distribution of layer thicknesses.

A PS/PMMA stack of 32 layers produces a peak reflection of about 60 % only. In order to obtain close to total reflection an increase in the number of layers is required. Indeed, stacking two 32-layers film resulted in almost 100 % reflection narrow band, at the centre wavelength of the photonic band gap (see Fig. 9b).

A further increase in the number of layers by adding more multipliers to the extrusion process gave rise to a broader layer thickness distribution. As explained in the first section of the chapter, this effect is intrinsic to the co-extrusion process, where layers that are close to the extruder walls experience a different melt flow and shear rate than layers near the centre of the multiplier, and thus become thinner than predicted. This broader layer thickness distribution complicates the interference between forward and backward propagating waves. As a result, the transmission spectra of such films feature a decrease of the reflected intensity of the main band relative to the prediction for uniform layers, and an increase in the intensity of side bands. Finally, as the number and intensity of side bands increase, they start to overlap with the primary band, actually leading to a broadening of the spectrum [22]. Therefore, stacking of films with a lower number of layers is a better procedure than to increase the number of layer-multiplying elements, since it ensures a higher number of layers retaining the narrower distribution characteristic of the films produced with fewer multipliers. In this way, a high, close to total reflection of the primary band can be achieved, together with a low intensity of side bands.

As highlighted in the previous section of this chapter, besides the number and thickness of layers, the refractive index difference between the two media is an important parameter affecting the spectral response of a photonic crystal. Therefore, by altering the polymer pairs used in the multilayer co-extrusion process, the

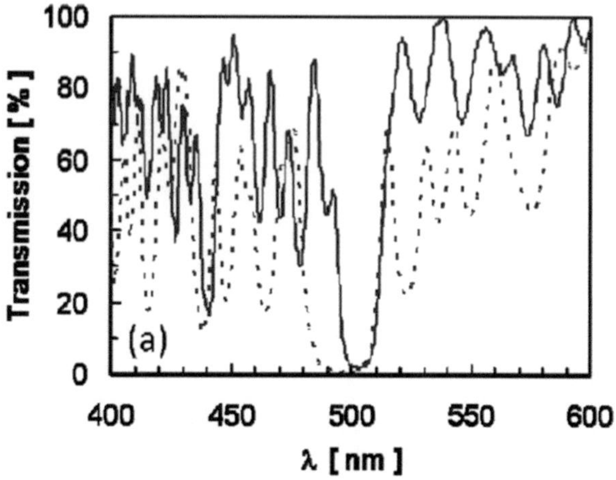

Fig. 10 Transmission spectrum of 128 layers of PS/PMMA (*solid line*) compared with 128 layers of PS/(PVDF-HFP)-PMMA 80–20 wt% blend (*dashed line*). Adapted with permission from [24]

effect of varying Δn on the optical properties of the 1-D photonic crystal can be addressed. Keeping polystyrene ($n = 1.585$) and choosing as second component a blend of poly(methyl methacrylate) with poly(vinylidenefluoride-*co*-hexafluoropropylene) (PVDF-HFP) the value of Δn can be increased. The fluorinated copolymer, containing 15 wt% comonomer, has a very low crystallinity. The crystallization is further suppressed by blending it with 20 wt% PMMA. Thus, an amorphous and transparent miscible blend, with a refractive index of 1.421 is obtained. The value of Δn is then increased to 0.164, relative to the 0.09 of PS/PMMA multilayers films [24]. The effect of Δn on the photonic crystals transmission spectra of a 128-layers assembly can be appreciated in Fig. 10.

In agreement with Eq. (5), the system having a Δn of 0.164 shows a bandwidth of 33 nm for the main reflection peak, against a bandwidth of only 18 nm for the PS/PMMA system ($\Delta n = 0.09$). Moreover, the system having high Δn also exhibits increased reflection from the side bands.

By exploiting the properties of elastomers, tunable photonic crystals can be realized [22]. Elastomeric polymers undergo large and reversible deformation upon the application of small loads. Since their stretching occurs at constant volume, it results in a decrease of thickness. In a 1-D photonic multilayer assembly of elastomers, the decrease of individual layer thickness with deformation leads to a change in the position of the photonic bandgap. Such tunable photonic crystals were produced by co-extruding a 128-layers film made of an elastomeric polyurethane and a poly(amide-*b*-ether). This pair of polymers possesses a good transparency and a reasonable contrast in refractive index ($\Delta n = 0.07$). The transmission spectra of the layered system collected at increasing strain, from 0 to 125 %, are shown in Fig. 11. The spectra consisted of a main strong band and several weaker side bands, as a consequence of layer non-uniformity. The peak reflection is less than 100 %, also due to the relatively small Δn between the two polymers.

Stretching the elastomeric film results in a reversible reflection band shift spanning almost the entire visible range, from 600 to about 400 nm. Visually, this

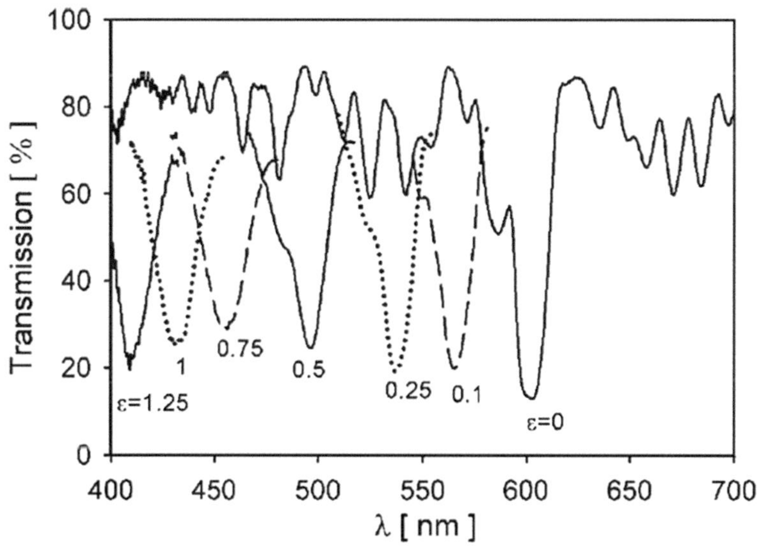

Fig. 11 Transmission spectra of elastomeric photonic crystal at different values of deformation (ε from 0 to 1.25). Adapted with permission from [22]

is seen as a change in the colour from red to blue. The relationship between the applied strain and the peak reflection can be predicted based on geometric modeling of the film thinning during stretching. Reversibly tunable 1-D polymeric photonic crystals represent a novel technology, enabled by the multilayer co-extrusion process, which might find applications in optical strain sensors or tunable reflectors.

5 Optical Applications of Co-extruded Multilayered Photonic Crystals

In the previous section, a simple processing method to obtain multilayered films of different polymer pairs, with properties of a 1-D photonic crystal, has been described. These polymeric photonic crystals are keen to be applied in several fields of optics. The examples of optical waveguides and lasers will be illustrated.

Polymeric photonic crystals waveguides have recently attracted some interest, especially due to their low costs and low processing temperatures, if compared with conventional glass waveguides. The operating principle of photonic crystals waveguide is different from the total internal reflection, which is the mode of wave propagation inside a conventional glass waveguide. Indeed a fibre-shaped waveguide can be made simply by rolling a polymer multilayer film into a tube. In this way, the frequencies which lay in the photonic bandgap of the multilayer structure will be confined to the fibre core and will propagate along its length, their propagation inside the multilayered shell being forbidden. In contrast to the standard total internal reflection waveguides, the core of the fibre can be made of a low-refractive index material, or it can even be hollow.

Bragg fibres are usually fabricated by drawing a macroscopic preform. In turn, the preform can be made by depositing alternating layers of two different polymers on a rotating cladding tube by solvent evaporation. Clearly this process requires the use of two appropriate solvents, which do not cross-solve the polymer during the periodically alternated deposition. Another drawback is the long time required for solvent extraction from the multilayer preform. Alternatively, two different polymer films can be co-rolled around a plastic rod called mandrel. However, with this latter method, it is particularly difficult to avoid the formation of defects, i.e. air microbubbles, inside the multilayer structure. Obviously these defects are detrimental for the waveguide performance. It is apparent that multilayer co-extrusion has the potential to improve the fabrication methods of the Bragg fibre preform.

A Bragg fibre waveguide was made from the polymer pair polystyrene/poly (methyl methacrylate) [25]. A four-layer co-extruded PS/PMMA film was rolled around a PMMA rod (diameter 12 mm, length 30 cm). The original individual layer thickness of PS and PMMA was around 20 μm, while the final thickness of the layers in the photonic structure must be comparable to the wavelength of visible light. Therefore, the preform was drawn 50 times at a temperature of 200 °C, to obtain fibres of final diameter between 300 and 500 μm. This preparation method greatly reduces the number of defects, especially in the optically important first few layers. Electronic micrographs of the Bragg fibre at low and high magnification are shown in Fig. 12. Individual layers in the photonic crystal shell have a thickness of approx. 550 nm. The transmission properties of the obtained fibre revealed that, when using a white-light source, all the spectral components not guided by the photonic bandgap were irradiated within a few centimetres along the fibre length. Only the specific colour guided by the bandgap propagated trough the fibre. The specific transmitted wavelength can be tuned by varying the drawing ratio of the preform, and thus the thickness of the alternating PS/PMMA layers.

Another attractive optical application of polymeric 1-D photonic crystals, due to their intrinsically easy processing and their flexibility, is the realization of all-polymer lasers. Typically, a laser combines three different elements: a gain medium, an optical resonator and a pump source. The gain medium is a material with an electronic structure suitable to amplify light by the process of stimulated emission. An example of a gain medium used in organic lasers is a fluorescent dye. For the gain medium to amplify light, a source of energy is required. This energy can be supplied by the pump either as an electrical current (when feasible) or as light at a different wavelengths, for instance by another laser. The pump source excites enough electrons of the gain medium into higher energy state so that stimulated emission can occur. The optical resonator provides the optical feedback necessary to enable stimulated emission and to control spatial and temporal coherence of the beam. A typical optical resonator is represented by an optical cavity, i.e. a pair of mirrors on either sides of the gain medium. Light bounces back and forth between the mirrors, passing through the gain medium and being amplified

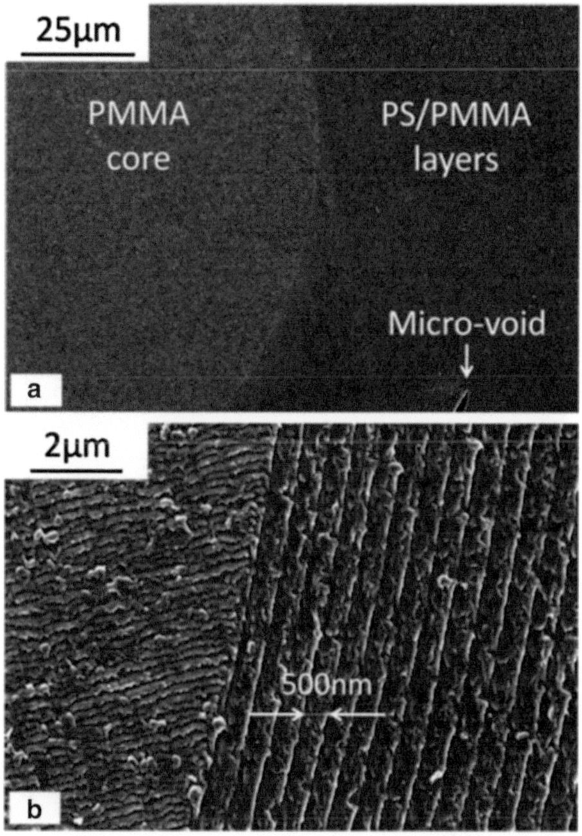

Fig. 12 Scanning electron microscopy details of a Bragg fibre. Reproduced with permission from [25]

each time. The mirrors can also consist of multilayer photonic crystals. More details on the lasing mode of photonic crystals systems can be found in 4th, 16th and 20th chapters of the present book.

Two different designs have been used to produce all-plastic lasers, taking advantage of 1-D polymeric photonic crystals. One configuration is called DBR laser, it consists of a gain medium confined in between two reflective resonators. The two optical resonators were fabricated by co-extruding a 128-layers photonic crystal made of alternated PS and PMMA. The two polymeric Bragg mirrors were sandwiched on either side of a compression-molded PMMA layer, doped with a fluorescent dye molecule at low concentration (10^{-3}–10^{-2} M) [26]. The centre of the photonic band gap was matched with the emission spectra of the dye, by controlling the PS/PMMA layers thickness with co-extrusion and fine-tuning it with biaxial stretching. This is demonstrated in Fig. 13 for the fluorescent organic dye rhodamine 6G perchlorate (R6G). The photonic crystals were designed such that the high-energy edge of the reflection band coincided with the emission maximum of the laser dye.

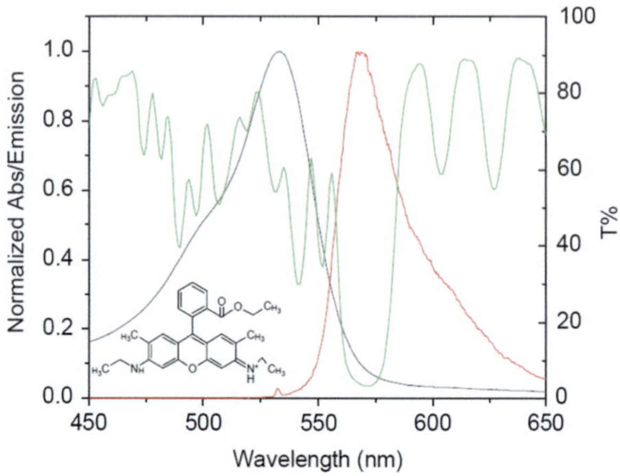

Fig. 13 Reflection band (*green*), absorption (*black*) and emission (*red*) of R6G. The *inset* shows the molecular structure. Reproduced with permission from [26]

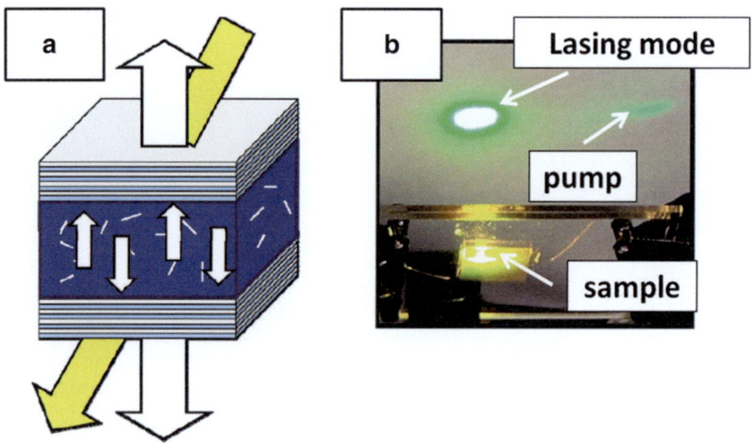

Fig. 14 (**a**) Structure of DBR laser and mechanism of action. The *small white arrows* indicate fluorescence resonating in the structure. The *large yellow arrow* depicts oblique angle optical pumping. The *large white arrow* represents the laser beam. (**b**) Bright emission of R6G DBR laser. Adapted with permission from [24]

The DBR laser was optically pumped by means of a Nd:YAG laser with a tunable optical parametric oscillator, at a wavelength of 530 nm. An oblique angle of the pump beam with respect to the film normal was chosen, to enable the pump light to fully penetrate the DBR mirrors avoiding the reflection band. A scheme of the DBR laser configuration, together with a picture of the laser at work is depicted in Fig. 14. Intense laser emission, stable over long periods of time was observed, above a threshold fluence of 90 μJ/cm^2. The obtained threshold is relatively low, but higher with respect to the best obtained for an all-polymer laser produced by spin-coating (<20 μJ/cm^2) [27].

Also for all-polymer lasers, the use of elastomeric materials enables mechanical tunability for what concerns the emission wavelength. A mechanically tunable DBR laser was realized in a one-step roll-to-roll co-extrusion process by doping the skin layer of the multilayered film with the laser dye R6G [28]. The Bragg reflector is based on two elastomers with low crystallinity: an ethylene-octene copolymer and fluoroelastomeric terpolymer consisting in poly(vinylidene fluoride), hexafluoropropylene and tetrafluoroethylene. Their contrast in refractive indices, Δn, is equal to 0.12. Since the solubility of common fluorescent dyes in the above-mentioned polymers is poor, the co-extruded skin layer was selected to carry the dopant dye. The elastomeric skin layer is a polar terpolymer made of ethylene, acrylic ester and glycidyl methacrylate, which is doped with about 1 wt% of rhodamine 6G perchlorate. The dye-doped skin layer becomes the active gain layer when folded inside the DBR dielectric mirror, i.e. by simply folding the co-extruded film with the skin layer in the middle after the extrusion.

The reflection band of the unstretched assembly is located at 614 nm, as the lasing wavelength. By stretching from 0 to 19 % the photonic band centre shifts to 565 nm. Continuous tuning of the lasing wavelength from red to green was observed by stretching the polymer laser film. The lasing wavelength is a function of both the broad emission spectrum of the R6G dye and the location of the optimal reflection band wavelength of the multilayer mirror. The lasing phenomenon is observed at defect states, i.e., local transmission maxima inside the spectral reflection band [29]. Such defect states can be due to the random variations of the layer thickness of the Bragg reflector or to the intentional introduction of a defect layer in the photonic structure. Defect states not only control the lasing wavelength and spectrum width, but they also lower the lasing threshold and increase the optical conversion efficiency.

A second type of all-plastic laser is the distributed feedback laser (DFB), which consists of a single microlayered 1-D photonic crystal film with active lasing dye in one layer. Thus in DFB lasers, also called "mirrorless" lasers, the feedback mechanism and the gain medium are integrated and distributed throughout the structure. In contrast to the DBR laser, which required laminating 1-D photonic crystal films to a dye-doped active layer, the DFB lasers can be produced in a single step by multilayer co-extrusion [30]. The structure of an all-polymer DFB laser is schematized in Fig. 15a. The feedback leading to stimulated emission arises from interference due to multiple reflections within the structure.

A polymeric DFB laser was produced by co-extruding a film consisting of 128 alternating layers of poly(styrene-*co*-acrylonitrile) (SAN) and a fluoroelastomer terpolymer of vinylidene fluoride, hexafluoropropylene and tetrafluoroethylene. The two polymers have a large difference in the refractive index ($\Delta n = 0.20$), giving rise to vivid reflected light. The laser dye, R6G, was blended in the SAN at a concentration around 1 wt%. The fluoroelastomer terpolymer provided an efficient barrier to the diffusion of the dye molecule, which was then effectively confined in the SAN layers, thus preventing a reduction of its absorption by a dilution effect. Analogously to the all-plastic DBR laser, the photonic bandgap of the structure was finely tuned to match the maximum of the

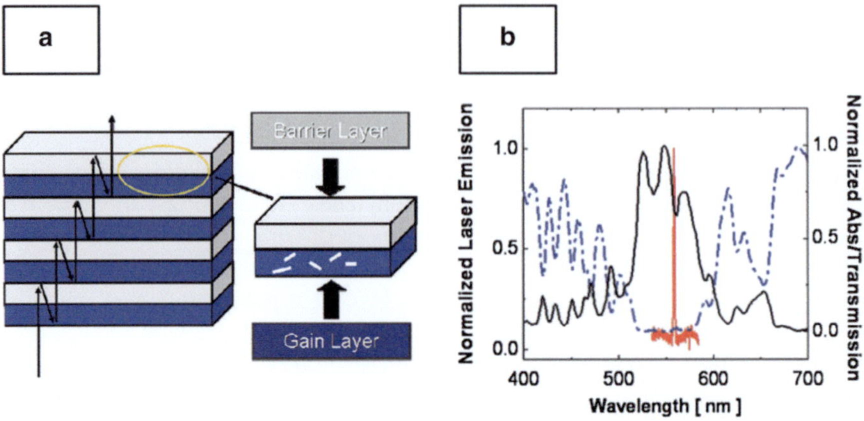

Fig. 15 (**a**) Structure of DFB laser (**b**) lasing spectrum of R6G DFB laser (*solid red line*) superimposed with the transmission spectrum of the multilayer film (*dashed blue line*) and the same transmission spectrum expressed in logarithmic absorbance units (*solid black line*). Adapted with permission from [24, 30]

laser dye emission spectrum. The DFB laser was pumped with a Nd:YAG laser coupled with an optical parametric oscillator at a wavelength of 535 nm. The obtained lasing spectral response is reported in Fig. 15b, together with the transmission spectrum which includes contributions from dye absorption and photonic crystal reflection.

As it occurs in photonic crystal based DBR laser, also a DFB laser shows the lasing mode at defect-like features, near the centre of the reflection band, where absorbance has a relative minimum. The defect modes arise from non-uniform layer thickness distribution, typical for co-extruded polymer multilayer structures.

6 Conclusions

The process of layer-multiplying co-extrusion relies on the principle of the “baker’s transformation”: two melt streams of different polymers are successively cut, spread and stacked. With each passage through one layer-multiplying element, the number of layers is doubled. A total number of hundreds or thousands of alternating layers —with dimensions from the micro- to the nanoscale —can easily be obtained with a simple setup. These highly ordered multilayered structures usually contain defects, such as non-homogeneity of layer thicknesses and layer break-up. By tuning the processing or materials variables the extent of these defects, detrimental for optical applications, can be minimized, and a highly ordered 1-D stack is produced.

Such an assembly of polymer layers, which possess a contrast in their refractive index and a periodicity in the order of visible light, exhibits properties of a 1-D photonic crystals. In particular, the dielectric lattice generates a photonic bandgap, i.e., a region of wavelengths/energies where propagation of light inside the lattice is forbidden. Thus, a close-to-total reflection band appears in the spectrum of the photonic crystal. The exact width and position of the reflectance peak can be finely tuned by playing with layer thickness and the refractive index of the materials. These polymeric 1-D photonic crystals can find several optical applications, from waveguides to lasers.

The process of multilayer co-extrusion is continuous, simple and cheap and it is amenable for the mass-production of large-area nanostructured polymer systems. These polymeric multilayers can find applications in advanced optical devices and provide clear advantages over their inorganic counterparts, considering, for instance, their light-weight and flexibility. Finally, the layer-multiplying co-extrusion process reveals to be an enabling technology, paving the way for the rapid development of novel applications based on polymeric materials.

References

1. C.D. Mueller, S. Nazarenko, T. Ebeling et al., Polym. Eng. Sci. **37**, 355 (1997)
2. R.Y.F. Liu, Y. Jin, A. Hiltner et al., Macromol. Rapid Commun. **24**, 943 (2003)
3. H. Wang, J.K. Keum, A. Hiltner et al., Science **323**, 757 (2009)
4. J.C. Van Der Hoeven, R. Wimberger-Friedl, H.E.H. Meijer, Polym. Eng. Sci. **41**, 32 (2001)
5. R. Sluijters, Mixer. U.S. Patent 3,182,965, 1965
6. R. Wimberger-Friedl, J.W. Mulder, J.G. De Bruin, Method for producing a grid structure. U.S. Patent 6,707,885 B2, 2004
7. J. Dooley, Viscoelastic Flow Effects in Multilayer Polymer Coextrusion. Ph.D. Thesis, Eindhoven University of Technology, The Netherlands, 2002
8. P.C. Lee, J. Dooley, J. Plast. Film Sheeting **29**, 78 (2013)
9. C.D. Han, R. Shetty, Studies on multilayer film coextrusion II. Interfacial instability in flat film coextrusion. Polym. Eng. Sci. **18**(3), 180–186 (1978)
10. W.J. Schrenk, N.L. Bradley, T. Alfrey et al., Polym. Eng. Sci. **18**, 620 (1978)
11. T. Inomoto, T. Kajiwara, J. Soc. Rheol. Jpn. **37**, 91 (2009)
12. P. Yue, C. Zhou, J. Dooley et al., J. Rheol. **52**, 1027 (2008)
13. L. Rayleigh, Philos. Mag. **34**, 145 (1892)
14. M. Ponting, A. Hiltner, E. Baer, Macromol. Symp. **294**, 19 (2010)
15. J.D. Joannopulos, R.D. Meade, J.N. Win, *Photonic Crystals: Molding the Flow of the Light* (Princeton University Press, Princeton, 1995)
16. E. Yablonovitch, Phys. Rev. Lett. **58**, 2059 (1987)
17. S. John, Phys. Rev. Lett. **58**, 2486 (1987)
18. V. Morandi, F. Marabelli, V. Amendola et al., Adv. Funct. Mater. **17**, 2779 (2007)
19. V. Morandi, F. Marabelli, V. Amendola et al., J. Phys. Chem. C **112**, 6293 (2008)
20. T. Alfrey, E.F. Gurnee, W. Schrenk, J. Polym. Eng. Sci. **9**, 400 (1969)
21. J.A. Radford, T. Alfrey, W. Schrenk, J. Polym. Eng. Sci. **13**, 216 (1973)
22. T. Kazmierczak, H. Song, A. Hiltner et al., Macromol. Rapid Commun. **28**, 2210 (2007)
23. A. Vasicek, *Optics of Thin Films*, 1st edn. (North-Holland Publishing Co., Amsterdam, 1960)
24. H. Song, K. Singer, Y. Wu et al., Layered polymeric optical systems using continuous coextrusion. Proc. SPIE **7467**, 74670A (2009)

25. K. Stoeffler, C. Dubois, A. Ajji et al., Polym. Eng. Sci. **50**, 1122 (2010)
26. K.D. Singer, T. Kazmierczak, J. Lott et al., Opt. Exp. **16**, 10358 (2008)
27. G. Canazza, F. Scotognella, G. Lanzani et al., Laser Phys. Lett. **11**, 035804 (2014)
28. G. Mao, J. Andrews, M. Crescimanno et al., Opt. Mater. Exp. **1**, 108 (2011)
29. Y. Wu, K.D. Singer, R.G. Petschek et al., Opt. Exp. **17**, 18038 (2009)
30. H. Song, K. Singer, J. Lott et al., J. Mater. Chem. **19**, 7520 (2009)

Dry Polymerization of Functional Thin Films and Multilayers by Chemical Vapor Deposition

Anna Maria Coclite

Abstract Responsive polymers with electronic and photonic properties have recently gained a widespread interest because of their versatility and cheapness. Chemical vapor deposition (CVD) offers a method to combine the large portfolio of conventional organic polymerization with the advantages of easy thin film processing, uniformity and scalability. CVD polymerization takes place in a mild vacuum environment, the substrate is kept near room temperature and in the complete absence of solvent. This allows to deposit polymers on unconventional delicate and flexible substrates, such as paper or plastic, with control over thickness on large areas. High reflectance distributed Bragg reflectors are based on multilayer structures whose growth can be facilitated by the absence of solvent, hence without risk of swelling or dissolving the underlying material. The conformal coverage of three-dimensional feature is also a unique characteristic of CVD polymers. In this chapter, fundamental background together with successful applications of CVD polymers will be reviewed. New techniques, as initiated CVD and oxidative CVD, will be introduced and it will be emphasized how they allow to obtain polymers with high chemical specificity (i.e., large retention of chemical functionality) at elevated speed. The accent will be on the applications where all-dry-processing is critical, such as deposition of insoluble polymers (e.g., highly cross-linked) for reactive and responsive surfaces, molecular crystals of fluoropolymers, conjugated polymers, composite and multilayer structures.

Keywords Chemical vapor deposition • DBR • Functional materials • Growth • Opal-like • Scale up • Sensing

A.M. Coclite (✉)
Institute of Solid State Physics, Graz University of Technology,
Petersgasse 16, 8010 Graz, Austria
e-mail: anna.coclite@tugraz.at

D. Comoretto (ed.), *Organic and Hybrid Photonic Crystals*,
DOI 10.1007/978-3-319-16580-6_8

1 Introduction

1.1 Overview of Chemical Vapor Deposited (CVD) Polymers

Powerful and innovative organic and hybrid photonic crystals can be fabricated by smart engineering the structure and the combination of properties of the building block molecules [1]. Novel devices obtained combining optical and bio response are under investigation as a new generation of biosensor [2]. This requires fine control on the fabrication process. CVD methods to synthesize polymers thin film can easily fill this niche. The surface quality, composition, and morphology can be engineered thanks to the translation of organic synthesis routes into vacuum-based processes [3]. CVD polymers can be easily processed into thin film, hybrid, and multilayered structures, with high uniformity and scalability [4].

CVD polymerization takes place in a mild vacuum environment, where the monomers are vaporized, and the polymer building blocks are adsorbed on the substrate, which is kept near room temperature [5, 6]. The polymerization occurs on the surface and is completely dry, from vapor phase precursors to thin solid films. The copolymerization of monomers with completely different solubilities is not hindered by the lack of a common solvent and copolymers with random to alternate structure can be obtained without unwanted phase separation [7]. The choice of the organic functionalities drives the desired properties: refraction index [8], surface energy [9], conductivity [10], etc. Some innovative CVD polymerization routes (e.g., initiated CVD and oxidative CVD) have demonstrated full retention of the chemical specificity of the monomers, while this was possible by conventional CVD methods only by sacrificing the polymer growth rate (e.g., with pulsed plasma enhanced chemical deposition) [11, 12].

The absolute absence of solvent allows to deposit polymers on unconventional delicate and flexible substrates, such as paper or plastic, or on substrates, like hydrogels, which would swell upon solvent exposure [13]. Multilayer structures are also easily obtained by just changing the deposition conditions from one layer to the other without the risk of dissolving the underlayer [14]. For comparison, multilayer structures obtained by spin coating and polymer photonics obtained by self-assembly are described in the fourth and sixth chapters, respectively.

In situ monitoring of the polymer growth allows to have a control over the thickness from few nanometers to several micrometers with uniformity over large areas. Deposition rates as high as 1 μm/min have been obtained [15].

Solvent entrained in the polymers is a major cause for polymer properties degradation and device failures. CVD overcomes this problem and results in polymer thin films, which are well adherent, chemically inert, free of leachable, plasticizer, and insoluble. Stability over extended periods of time and resistance to solvent treatments, multiple exposures to heat, humidity, and mechanical wear are fundamental requirements for CVD polymers to be successfully used for every-day applications. In this frame, also durability over a long time frame is extremely important. Insoluble cross-linked and grafted CVD film has shown high resistance to delamination from the underlying substrate and successful results in terms of keeping intact the properties under accelerated test conditions [16].

Engineered three-dimensional shapes are important for optical properties, for example the interstices in opals, the porosity in some photonic crystals, and the embedding of micro- and nano- particles. Opals and porous photonic crystals are extensively discussed in 2nd, 5th, and 11th chapters. CVD polymers can uniquely provide conformal coverage of such features. The topography can be preserved at the micro and nano-scale by adding a film of uniform thickness over every contour of a geometric feature (conformal coverage). The conformality is a direct consequence of the deposition mechanism and the sticking of the molecules on the surface [17]. In contrast, achieving conformality is challenging for solution processing because of the surface tension of the solvents. The application of photonic crystals as sensors, for example, can greatly benefit from the retention of the shape and the large area of the underlying substrate.

2 Fundamentals

2.1 CVD Growth Mechanism and Kinetics

Some of the possible CVD polymerization techniques will be presented in this chapter: plasma enhanced CVD (PECVD), oxidative (oCVD), and initiated CVD (iCVD). Those techniques differ for the deposition mechanism, the type of activation of the film precursor and, of course, on the resulting polymer structure.

In the PECVD process, the film precursors are activated through elastic and inelastic collisions with energetic electrons [18]. The quantity of the reactive species, their energy, and in turn the properties of the plasma are determined by the external parameters, such as the pressure, the applied voltage, the geometry of the reactor, the nature and the flow of gas. A typical parallel plate plasma reactor configuration is sketched in Fig. 1a.

Plasma polymerization can be explained with an activation growth model: [19]

$$\text{Monomer} \xrightarrow{\text{electrons}} \text{Fragments (R, atoms, } \ldots) \tag{1}$$

$$R \rightarrow R_{ads} \tag{2}$$

$$I^+ + S \rightarrow (S^-)^* \tag{3}$$

$$R_{ads} + (S^-)^* \rightarrow R - S \tag{4}$$

The electrons create, through collisions with the molecules in the gas feed, a large number of reactive species: ions (I^+) and radicals (R) (Eq. 1). The gas phase radicals, R, physisorb (Eq. 2) onto the substrate S, since the latter is kept at lower

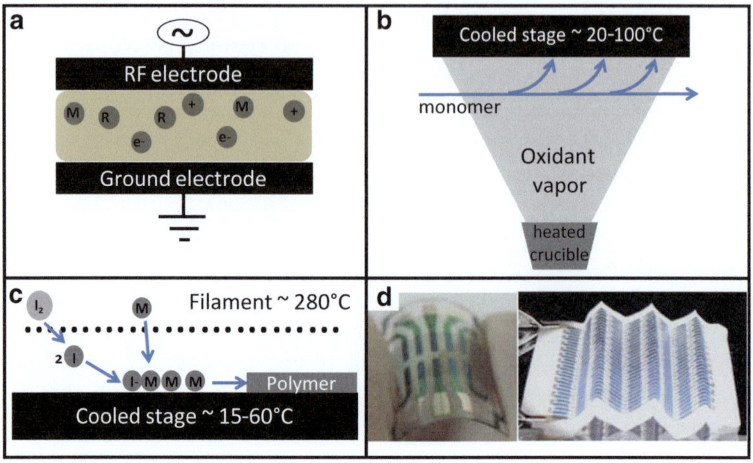

Fig. 1 Schematics of the PECVD (**a**), oCVD (**b**), and iCVD (**c**) processes. M indicates a monomer molecule, R the radicals formed in the process, + the positive ions, e^- the electrons, and I_2 the initiator molecules that get decomposed in 2I by the thermal activation. (**d**) CVD polymers deposited on plastic and paper substrates. (**d**) Adapted with permission from [20]. Copyright 2011, Wiley

temperature than the plasma phase (the adsorbed radicals are indicated with R_{ads} in Eq. 2). At the same time, the surface is activated by ion bombardment (Eq. 3), so activated surface sites $(S^-)^*$ are created, in the form of dangling bonds or impinged ions. Finally, when the physisorbed radicals, during the surface migration, meet an activated site, react with it initiating the formation of the polymer. The final film composition is a result of reactions which takes place on the growing film (deposition of new radicals and formation of new active sites). The resulting chemical structure of plasma polymer can be quite irregular: sometimes plasma polymers can have dangling bonds or uncontrolled cross-linking due to the interaction of the growing polymer with the energetic ions or electrons. The ion bombardment and heterogeneous reactions can lead to the cleavage of some functional groups or water or other oxidized compounds, deteriorating the density of chemical functionalities in the polymer. Pulsed plasma discharges result in less monomer fragmentation and ion bombardment.

The retention of the monomer functionalities can be, instead, easily achieved by oCVD and iCVD processes. The oCVD process is based on step-growth polymerization [21]. In this case the monomer is combined with an oxidizing agent to create dimers of radical cations. Pairs of cation radicals then dimerize and deprotonate. Those react further with the oxidizing agents and with other monomer units, creating longer chains. oCVD follows the same step growth mechanism of oxidative polymerization as in solution. A typical oCVD setup is represented in Fig. 1b. The oxidant source is placed at the bottom and the substrates on the temperature-controlled stage (typically at 25–150 °C), which is on top. Solid and liquid oxidants have been used, e.g. iron(III) chloride, copper(II) chloride, bromine or vanadium oxytrichloride. The vapors of the oxidant are created in the vacuum chamber by heating.

The polymerization by iCVD starts with the thermal decomposition of an initiator species by an array of resistively heated filaments at relatively low temperature (filament temperature, T_{fil}, ~200–400 °C). An initiator is a species containing a labile bond such as azo- or peroxy- linkage, whose thermal decomposition generates free radicals. One example of initiator widely used for its low decomposition temperature is the tert-butylperoxide (TBPO). The initiator radicals react selectively with the unsaturated bonds of the monomer, creating monomer radicals which polymerize. The iCVD concept was first introduced by Lewis et al. [22], who demonstrated that adding as initiator the perfluorooctane sulfonyl fluoride (PFOS), which forms $CF_3(CF_2)_6CF_2$ radicals, dramatically enhances the deposition rate of polytetrafluoroethylene (PTFE). Without the initiator, the polymerization occurs but less rapidly and less controllably because of the thermal decomposition of the monomer and the other background species in the chamber. In terms of polymerization steps, iCVD is very similar to the conventional free radical polymerization, but with the absence of any solvent [5, 6].

$$I_2 \overset{T_{fil}}{\rightarrow} 2I^\bullet \tag{5}$$

$$M \rightarrow (M)_{ads} \tag{6}$$

$$I^\bullet + (M)_{ads} \rightarrow (I - M^\bullet)_{ads} \tag{7}$$

$$(I - M^\bullet)_{ads} + n(M)_{ads} \rightarrow (I - M_n{}^\bullet)_{ads} \tag{8}$$

$$(I - M_n{}^\bullet)_{ads} + R \rightarrow \text{Polymer} \tag{9}$$

The decomposition of the initiator, I_2 (Eq. 5), takes place in the gas phase. $I^\bullet$ indicates the initiator radical. The filament temperature required for such decomposition is too low, to break also the bonds in the monomer (M) structure, therefore the monomer travels intact to the surface where it get adsorbed (M_{ads}, Eq. 6). The low substrate temperature enhances the adsorption on the surface. The generated free initiator radicals ($I^\bullet$) react with the vinyl groups of the monomer molecules absorbed on the surface, creating the primary radicals on the surface $(I - M^\bullet)_{ads}$ (Eq. 7). The propagation step involves the addition of the monomer units to the polymer chains (Eq. 8). Termination of the polymer chains may occur either by addition of a free radical to the end of the chain or by bonding of two growing radicalic polymer chains (both scenarios are indicated with a general R, radical species, in Eq. 9). The quantity of monomer absorbed on the surface can be quantified through the ratio between the monomer partial pressure (P_M) and the saturation pressure (P_{sat}) of the monomer at the substrate temperature. The iCVD growth rate increases with P_M/P_{sat}, suggesting that the rate controlling step for film deposition occurs at the surface. A schematic of the iCVD process is represented in Fig. 1c.

All those methods, PECVD, oCVD, and iCVD, require low substrate temperature and no need for conductive or rigid substrate. Indeed polymers have been deposited also on paper [20], plastics [23], and ionic liquids [24]. Figure 1d shows some examples of paper solar cells, where the conductive polymer was deposited by oCVD.

2.2 *Chemical and Physical Properties*

The great versatility of the organic polymers consists in the possibility of easily tuning the physical properties (e.g., wettability, conductivity) by selecting the correct chemical functionalization. The goal of the CVD polymerization methods is therefore to retain a high percentage of the monomer chemical functionalities. When the activation process goes through the decomposition of the monomer structure, e.g. in plasma processes, the retention of the chemical structure of the monomer becomes a challenging aspect. Methods to limit the damages to the monomer structures involve "soft" conditions: high pressure, low and often pulsed plasma power, resulting, as a drawback, in low deposition rates [25].

In the iCVD and oCVD methods it is possible to achieve high control over the chemical specificity and therefore high structure retention. This aspect can be easily related to the reaction mechanism discussed in the previous section. The monomer does not undergo any fragmentation during the iCVD nor the oCVD processes but selective chemical reactions.

Figure 2a shows the FTIR spectra of the poly(glycidyl methacrylate) (PGMA) film synthesized by iCVD, PECVD and compared with the spectrum of the conventionally polymerized PGMA [26]. The epoxy groups of the PGMA are desirable for many post-deposition functionalization reactions because they can easily react through a ring opening reaction. The FTIR bands in the PGMA film deposited by iCVD are narrow and especially, the adsorption peaks of the epoxide groups (907, 848, and 760 cm^{-1}) are completely retained as in the FTIR spectrum of the solution synthesized PGMA film. In contrast, the FTIR spectrum of the PECVD polymer shows peak broadening and peak intensity loss.

Control over the chemical composition and cross-linking can be used to tune the properties and therefore the functionality of the polymer. Stimuli-responsiveness or swelling can be achieved by a large retention of functional groups, as well as improved mechanical stability and UV sensitivity can be achieved by adding some degree of cross-linking to the structure. The polymer deposited by PECVD shows natural formation of cross-links and covalent bonding of the polymer to the substrate surface, which improves the adhesion and avoid the dissolution of such polymers in solvents. The polymers deposited instead by iCVD require the inclusion of a separate chemical species to the reactor feed, a cross-linker (i.e., a molecule with two or more vinyl bonds) [15].

Graded or homogeneous copolymers can be easily deposited by CVD. This allows to cumulate the properties coming from more than one monomer species.

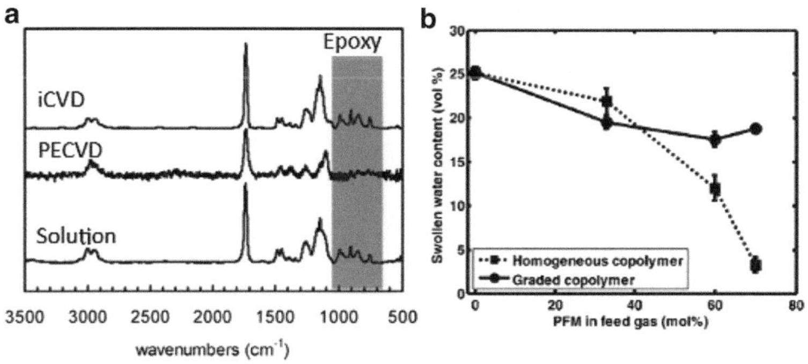

Fig. 2 (**a**) FT-IR spectra of PGMA synthesized by iCVD, PECVD and by conventional synthesis in solutions. The adsorptions of the epoxy groups, which are evidenced in *red* in the figure are narrow and of the same intensity of the polymer deposited by solution phase, while they are broad and less intense in the polymer deposited by PECVD, meaning that in this case some of the groups are cleaved off during the plasma process. (**b**) Volumetric percentage of water in the swollen PHEMA–PFM hydrogels as a function of the PFM percentage in the gas feed and for two copolymer compositions: homogeneous and graded. When the copolymer had a homogenous composition the swelling properties decreased from 25 to 4 % with the PFM content. Instead when the copolymer had a graded composition, i.e. with the PFM nanoconfined in the top 15 nm, the swelling properties showed a much smaller decrease. (**a**) Adapted with permission from [26]. Copyright 2004 American Chemical Society. (**b**) Reproduced with permission from [29]. Copyright 2009 American Chemical Society

The poly(hydroxyethyl methacrylate) (PHEMA) is known as a hydrogel, since its hydrophilic hydroxyl moieties allow the polymer to absorb water in its structure and swell. The swelling properties of PHEMA are used for drug release [27], optical sensors [8], and biodevices [28]. A copolymer of PHEMA and pentafluorophenyl methacrylate (PFM) with graded composition allowed to enhance the surface cell adhesion without altering the swelling properties of the hydrogel [28]. The pentafluorobenzine ring of the PFM monomer was used to covalently bond peptides on the hydrogel surface. The inclusion of PFM in the whole thickness of the copolymer degraded the swelling capabilities of PHEMA, while when the PFM was nanoconfined on the surface (first 15 nm) the high water content of the swollen hydrogel was almost completely retained (Fig. 2b).

2.3 Conformal Coverage and 3D Structures

The ability to encapsulate non-planar substrates with a coating of uniform thickness and composition is called conformality. The surface area and the shape of the substrate material is often very important, e.g. in sensors and optical devices. Conformal coverage of 3D porous photonic crystals like opals (described in 2nd,

5th, 16th, 17th, and 19th chapters) can be highly advantageous to keep the length scale of the features comparable to the light wavelength.

It has been demonstrated that iCVD and oCVD coatings can conformally cover complex geometrical features much better than PECVD or deposition form liquid phase. The latter can be affected by de-wetting, liquid thinning, or surface tension effects, which alter the profile of the coating on the substrate. PECVD processes, instead, depending on the deposition conditions can result in shadowing and line-of-sight profiles: the higher features get coated faster than the lowest. In contrast, the conformality of iCVD and oCVD polymers results from the low sticking probability of the reactants to the surface during a single collision. Sticking coefficients in the range of ca. 0.001–0.01 have been observed for iCVD. In addition, the gas phase diffusion to supply additional monomer to the surface is rapid compared to the rate at which monomer is depleted by the film formation reaction. Low sticking coefficients and high diffusion lead to high conformality.

Figure 3 shows examples of such conformal coatings. A film of poly (3,4-ethylenedioxythiophene) (PEDOT) was deposited by oCVD on newsprint [20]. As it is observable from Fig. 3a the fibers of the newsprint are still visible after deposition, meaning that the oCVD polymer coats them conformally. Figure 3b shows the deposition of a fluorinated polymer over a forest of nanotubes [9].

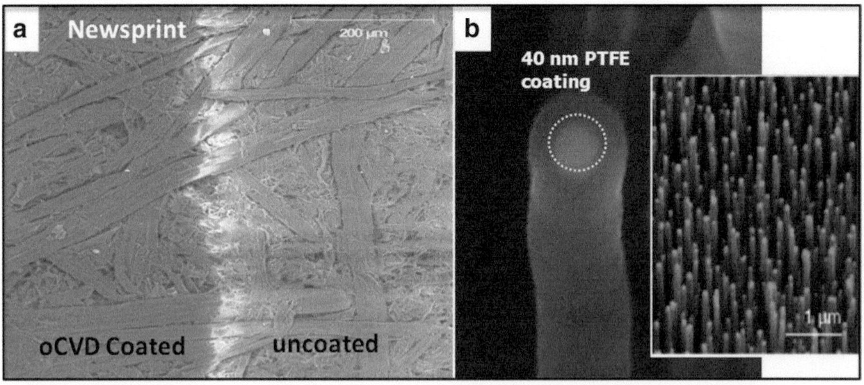

Fig. 3 oCVD and iCVD polymers can be deposited over 3D substrates without altering the underlying topography of the substrate. (**a**) PEDOT polymer deposited on newsprint paper. The *left side* is coated, while the *right side* is uncoated. The fiber structure of the paper is still visible in the coated part. (**b**) Polytetrafluoroethylene (PTFE) coating deposited on a nanotube forest. The coating conformally follows the nanotube shape. (**a**) Reproduced with permission from [20]. Copyright 2011, Wiley. (**b**) Reproduced with permission from [9]. Copyright 2003, American Chemical Society

3 Applications

3.1 Optical Applications: Composite and Multilayers

1D photonic crystals are made of alternating materials with different refractive indexes. Precise layering of different materials is easily achievable by CVD. The dryness of the process represents an enormous advantage because it allows to avoid the risk of dissolving the underlayers with the solvent used for the deposition of the overlapping ones. Cross-linking of the polymer layer is necessary during wet processing to prevent the dissolution, with the drawback of altering the refractive index contrast among the different layers, while this is not required in CVD. The different layers can be easily obtained just changing the deposition conditions from one layer to the other. The implementation of the whole multilayer deposition process in one reactor chamber allows to obtain interfacial smoothness at the nanoscale.

Smooth inorganic films with high refractive index such as silica (SiO_2) or titania (TiO_2) can be obtained by PECVD or Hot Wire CVD (HWCVD). HWCVD takes place in a similar reactor configuration as iCVD and the chemical species are thermally broken by a filament, which differently from iCVD is heated to temperatures above 1,000 °C. The initiator is not involved in the process and the monomer is the species that is thermally decomposed. Different CVD techniques can be combined in a single reactor chamber, to alternate the deposition of inorganic and organic materials in hybrid multilayer structures. Figure 4 shows two examples of multilayers obtained alternating first HWCVD and iCVD (cross section SEM image in Fig. 4d) and then PECVD and iCVD (cross section SEM image in Fig. 4f). All those processes were implemented in the same hardware configuration, without the need of venting the reactor chamber from the deposition of one layer to the other, so minimizing the down time. CVD can also allow to obtain such multilayered structures on large area with great thickness uniformity [30].

Fluorocarbon polymers have low refractive index ($n \sim 1.35$), therefore they are ideal to create large refractive index contrast. Dry deposition methods for fluoropolymers are very attractive because their solvents and surfactants tend to be difficult to find, expensive, and/or persistent in the environment. Wide tunability and high efficiency were obtained on a wide spectral region (300–2,000 nm) by distributed Bragg mirror structure made of fluorinated polymer alternated with inorganic layers by room-temperature ion-beam sputtering [31]. High contrast was also obtained alternating fluorocarbon polymers with other fluorocarbon polymers embedding gold clusters in periodic $CF_x/CF_x(Au)$ layered structures [32]. A bandwidth of 400 nm was obtained, however, with a response time for the complete band shift of about 20 min.

Fast response was obtained in Bragg mirrors obtained alternating seven layers of titania and PHEMA [8]. The inorganic titania layers were deposited by HWCVD. The titania layers were very smooth due to the absence of energetic ions during the HWCVD process, which cause roughening and light scattering. The alternating

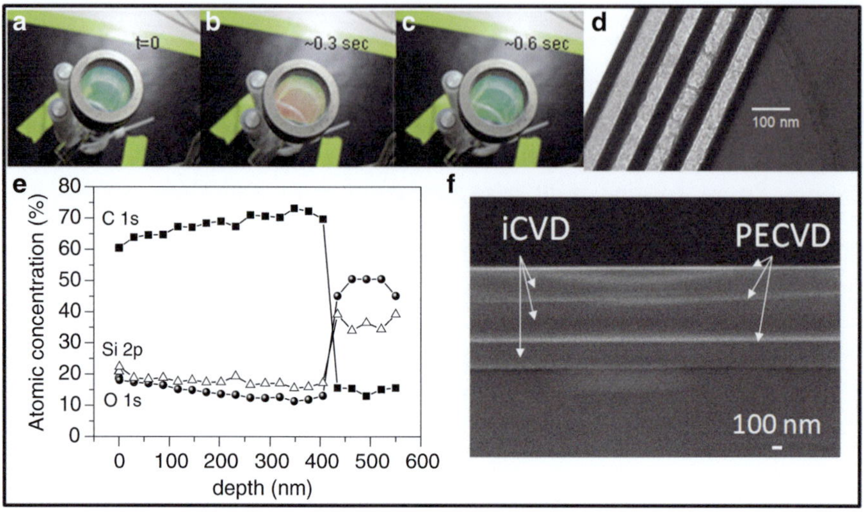

Fig. 4 (**a–d**) Distributed Bragg Mirror obtained alternating seven layers of titania-PHEMA. The PHEMA reversibly swells in water and all the PHEMA layers in the multistack swells of the same percentage in the same time, keeping intact the periodic structure. A fast and reversible swelling-shrinking cycle is shown in the micrographs of the swelling cell containing the distributed Bragg mirror in the dry state, *green phase* $t=0$ s (**a**), after exposure to water vapor, *red phase* t=0.3 s (**b**), and after N_2 purging: returning to dry *green phase* $t=0.6$ s (**c**). SEM in cross section of the multilayer. (**e**, **f**) Multilayer periodic structure obtained alternating organosilicon polymers deposited by iCVD and inorganic silica layer deposited by PECVD. (**e**) XPS depth profile showing the sharp passage from one layer composition to the other. The different layers keep their chemical composition without interdiffusing one into the other. (**f**) SEM in cross section of the multilayers, showing the defined periodic alternating structure. (**a–d**) Reproduced with permission from [8]. Copyright 2008, American Chemical Society. (**e**, **f**) Reproduced with permission from [14]. Copyright 2010, Wiley

PHEMA layers showed fast and reversible swelling in presence of water vapor. Multiple cycles of swelling-shrinking were possible in this structure, each PHEMA layer swelled exactly of the same percentage in the entire multistack. Figure 4a–c shows the photographs of the Bragg reflector in the swelling cell at different time of water vapor exposure. The change from green to red takes place in 0.3 s and after other 0.3 s the process reversed back to green light. Such short response times are typical of biological Bragg reflectors, e.g. in the scales of paradise whiptail the phase shift from blue to red happens in 0.25 s.

Organosilicon polymers were alternated with silica layers by switching from iCVD to PECVD processes in the same reactor chamber [14]. Figure 4e shows the chemical composition profile from the C-rich polymer layer to the C-depleted inorganic layer. The sharp passage from one layer composition to the other, and the visibility of the different layers in the cross section (Fig. 4f) demonstrates that the two materials do not interdiffuse one into the other but maintain the periodic alternation of the layer, fundamental for the achievement of large reflectivity in distributed Bragg mirrors.

3.2 Reactive and Responsive Surfaces

An engineered combination of optical properties and reactive and responsive materials can be used for the new generation of optical devices. Stimuli-responsive nanostructures in 3D-photonic crystals are discussed in the chapter by Li Qaun et al. Butterfly scales have highly selective vapor response; mimicking such structure can have a potential technological application in sensing. Novel sensors or artificial biological crystals can be obtained with a rational design of the chemical and morphological structure of the material. As an example, responsive surfaces can be combined with Bloch Surface Waves (described mainly in 4th and 14th chapters), by depositing it on top of a multilayered photonic crystal. In this case, the responsiveness of the top layer might be amplified by the spatial light confinement provided by the crystal.

Reactive surfaces can be used for the covalent attachment of biomolecules (i.e., peptides or DNA fragments) and nanoparticles on the surfaces. Examples of this type of polymers deposited by CVD are poly(aminostyrene) (PAS) with a high density of functionalizable $-NH_2$ groups and poly(glycidylmethacrylate) (PGMA) with reactive epoxy groups. Figure 5d shows an example of reactive surface functionalized with peptides [28]. The surface was made of a copolymer of PHEMA and PFM, with a graded composition, as explained in Sect. 2.2. The pentafluorobenzine ring of the PFM monomer was used to covalently bond peptides on the hydrogel surface, without altering the swelling capabilities of the PHEMA.

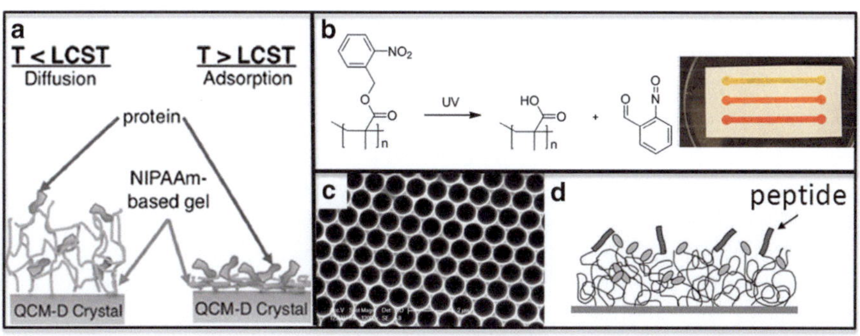

Fig. 5 (**a**) Thermoresponsive properties of the PNIPAAm. The LCST is the lower critical solution temperature, below which the PNIPAAm is in a coil state where proteins can diffuse inside the polymer network. Above the LCST the PNIPAAm is in a globular configuration, on which proteins absorb on the surface. (**b**) Three-dimensional patterning of chromatographic paper. The UV fragmentation reaction (*right*) of the hydrophobic photo-responsive polymer, allowed to obtain hydrophilic channels (*left*, micrograph of dye water in the channels). (**c**) SEM image of periodic nanobowls with 1 μm diameter formed on the surface of a CVD polymer by template colloidal patterning. (**d**) Example of reactive surface with peptides immobilized on the surface to enhance cell attachment and growth. (**a**) Reproduced with permission from [33]. Copyright 2011, American Chemical Society. (**b**) Adapted with permission from [36]. Copyright 2011, Royal Chemical Society. (**c**) Reproduced with permission from [37]. Copyright 2009, American Chemical Society. (**d**) Adapted from [28]. Copyright 2009, Wiley

The combination of swelling and peptide immobilization enhanced the cell attachment and growth on such surface. Biosensors, biological photonic crystals, and more in general biomaterials can be based on such type of structure.

Responsive polymers, i.e. materials that modifying their properties, such as wettability, thickness, stiffness, upon a stimuli (e.g., change in pH, humidity, temperature) can also be included in optical structures to achieve novel combined properties. In Sect. 3.1 an example of distributed Bragg mirror, including a responsive polymer was already shown. The swelling capabilities of the PHEMA hydrogels are due to a large retention of the hydrophilic hydroxyl moieties. Another example of responsive polymer deposited by CVD is the poly(*N*-isopropylacrylamide) (PNIPAAm) whose amide and isopropyl groups induce a temperature-sensitive hydrophobicity. Below the lower critical solution temperature (LCST) the latter polymer is in a swollen state, also called extended coil configuration, in which the amide groups are exposed to the surface, resulting in high hydrophilicity (Fig. 5a). Above the LCST, the polymer is in a globular configuration in which the inter- and intra-chain amide groups are bonded and the hydrophobic isopropyl groups are exposed to the surface. Alf et al. demonstrated that when the PNIPAAm deposited by iCVD is in its swollen state (i.e., below the LCST), the proteins are not adsorbed on the surface but diffuse into the polymer network [33]. Above the LCST, when the polymer is in its globular configuration, the proteins show simple monolayer adsorption.

Patterning can be used to create periodic structures in the order of the wavelength of light. Also bifunctional surfaces can be created with special variations in functionality and chemical composition [34]. Periodic surfaces of alternating hydrophilic and hydrophobic regions have been fabricated using PECVD [35]. Figure 5b shows an example of three-dimensional patterning [36]. The hydrophobic polymer, poly(*o*-nitrobenzyl methacrylate) (PoNBMA) was UV patterned to create a paper-based microfluidic devices. The channels where the UV light decomposed the polymer were filled with liquid. Such microfluidic architectures can be engineered to combine materials with different refractive indexes and obtain waveguides.

Another example of periodic porous structure, obtained by CVD is demonstrated in Fig. 5c. Colloidal templates and conformal CVD deposition were combined for the realization of periodic nanobowls [37]. Polystyrene microspheres of 1 μm diameter were coated with a layer of poly(butylacrylate) and then removed. The CVD polymer was able to withstand the aggressive solvents used for template removal, greatly simplifying the processing steps needed to achieve such patterns. The size of such pattern (1 μm) is not in the visible light wavelength range, but the same approach can be used with smaller spheres with sizes between 400 and 700 nm.

3.3 Molecular Crystals of Fluoropolymers

As mentioned in the Sect. 3.1 fluoropolymers have exceptionally low refractive indexes, therefore their inclusion in distributed Bragg mirrors can be highly advantageous to obtain high contrast and therefore enhanced optical performance [31, 32].

Excellent chemical resistance, high hydrophobicity, low refractive index and dielectric constant, low coefficient of friction, and high thermal stability are typical characteristics of bulk polytetrafluoroethylene (PTFE, Teflon). Many efforts are therefore devoted to the deposition of thin PTFE films, which are spectroscopically indistinguishable from its defect-free, true linear bulk form $(CF_2)_x$. Due to its extremely low solubility in common solvents, it is very difficult to process the PTFE in thin film form. PECVD of teflon-like coatings resulted in structure containing high concentrations of dangling bonds and a variety of bonding environments (i.e., CF_3 and CF, in addition to CF_2) [38]. Films containing high percentages of CF_2 were obtained by using pulsed plasmas, but deposition rates are substantially lower [25]. On the contrary, a structure exclusively made of CF_2 repeating units was successfully obtained by HWCVD [11] and iCVD [22] at much higher growth rates. PTFE films produced by HWCVD technique can reach essentially 100 % CF_2 groups.

Lamellar molecular crystals of fluorocarbon films can be used to add some anisotropic effect to the optical materials [39]. The achievement of control not only over the surface chemistry of the PTFE films but also on their topology can have major implications in technological applications as waveguides. Crystalline fluorocarbon polymers were obtained by Laird et al. [40–42] on single wall carbon nanotubes (SWCNT). The SWCNT induced PTFE crystallization in situ with a particular type of structure, assembled to "nanohybrid shish kebabs." The shishes are the nucleation templates onto which crystalline polymers crystallize with lamellar orientation. The PTFE shows lamellar structures, with a disk shape, spaced on average 13 nm apart and kept strong together by the one-dimensional SWCNT bundles.

Perfluoroacrylates have exceptional properties, including fast polymerization of the unsaturated acrylate group and hydrophobicity for the fluorinated pendant chain. As a result of the $-CF_3$ terminated side groups and comb-like structure, the typical range of surface energies for fluorinated acrylic polymers is very low (5.6–7.8 mN/m). The iCVD of poly(1H,1H,2H,2H-perfluorodecyl acrylate) (PPFDA) has been widely studied either pure [43–45] and copolymerized with other monomers [7, 46, 47]. The side chains with eight perfluorinated carbons, so-called C8, were completely retained during the polymerization giving interesting properties to the iCVD polymer such as hydrophobicity, oleophobicity, crystallinity.

The C8 perfluorinated chains of PPFDA have a natural tendency to crystallize in a smectic B phase, forming a bilayer structure in which the pendant C8 chains are exposed towards themselves and the bulk of the linear polymer is exposed towards the outside. These bilayer lamellae organize in hexagonal structures with spacing of 0.6 nm. The iCVD deposition conditions allow to achieve control over the crystallinity ratio (it is possible by changing the initiator over monomer ratio to go from amorphous coatings to molecular crystals) and over the orientation of the lamellae (Fig. 6). Depending on the filament temperature, different initiation processes take part to the polymerization. At temperatures below 270 °C, the TBPO is decomposed in two tert-butoxy radicals. Those radicals are bulky due to the presence of the isopropyl group. Therefore, they react only with the PFDA molecules that are

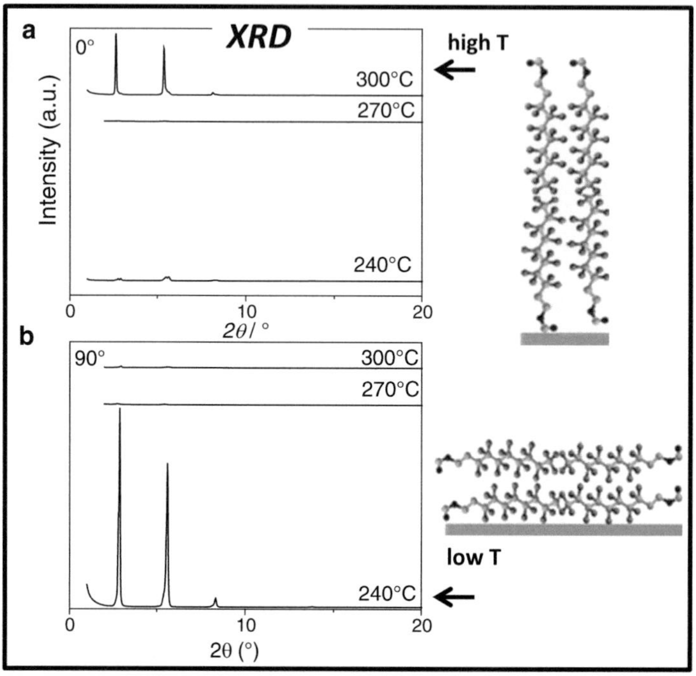

Fig. 6 X-ray diffractogram of PPFDA film deposited at different filament temperatures and measured at 0° (**a**) and 90° (**b**) of the sample tilting angle. When the polymer is deposited at 300 °C the X-ray pattern is visible at 0° and not at 90°, meaning that the chains are oriented perpendicularly to the sample surface. The contrary applies when the polymer is deposited at 240 °C: Since the pattern is visible only at 90° tilting angle, it implies that the polymer chains are laying on the substrate parallel to the surface. Reproduced with permission from [44]. Copyright 2012, Wiley

laying parallel to the substrate, since in this configuration the vinyl groups are more easily reachable.

The XRD patterns in Fig. 6 show that polymerization results in a crystalline thin film in which the chains are all oriented parallel to the substrate, indeed the diffractogram is visible when the sample is tilted at 90° and no peaks are observable when the diffractogram measured at 0°. At filament temperatures higher than 270 °C, the tert-butoxy radicals undergo a second fragmentation, called β-scission, that results in acetone and methyl radicals [48]. Those radicals are much smaller than the tert-butoxy ones, and therefore they can react with the less reachable vinyl bonds of the monomer molecules that are vertically oriented on the substrate. The diffractogram of the polymers deposited at 300 °C shows indeed intense peaks when the sample is at 0° and no peaks when the sample is tilted to 90°.

Further investigations on how to use the control over the orientation of the molecular crystals of PFDA chains are currently being explored. The liquid crystalline-like behavior of iCVD PPFDA can bring to interesting developments in the field of photonic crystals.

3.4 Conjugated Polymers

Since the first discovery of electroluminescence in poly(1,4-phenylene vinylene) (PPV) by Burroughes [49] in 1990, conjugated polymers have attracted great attention also for optoelectronic applications (optical fibers, sensors, solar cells, light-emitting diodes and lasers, see chapter by V. Robbiano et al. for complete description). Moreover, the insertion of a conductive polymer into a photonic crystal structure could be crucial to have electrical control of the optical properties of the crystal.

CVD as a dry deposition method offers many advantages for the deposition of conjugated polymers on a great variety of substrates and with a great variety of useful properties (e.g., flexibility, high conductivity, tunable doping level) [50]. Many conjugated polymers have a highly crystalline nature, that makes them insoluble. Specialty synthesis are then used to add solubilizing substituents to dissolve the monomers or surfactant are added to the polymers, such as poly (styrenesulfonate) (PSS), to stabilize their dispersion in the solvents. The oCVD process has demonstrated to be very successful for the deposition of conjugated polymer for conducting or semiconducting applications [51–53].

One of the most important optoelectronic devices in which an organic material is used is the organic light-emitting diode (OLED). Conjugated polymers can be efficiently used as a hole injection layer to facilitate the charge transfer between the anode and the organic optoelectronic materials. A very promising conjugated polymer used for electrochromic applications is PEDOT. When a voltage is applied to a PEDOT layer, a switch occurs between its oxidized conductive state, in which it appears sky blue, to its reduced insulating state, that is nearly opaque. The electrochromic properties of PEDOT have been used for smart windows and light management devices. PEDOT prepared via electrochemical polymerization is capable of a maximum contrast of 54 %. Fabretto et al. [54] demonstrated that controlling and optimizing the levels of water vapor and surfactant during the CVD of the PEDOT layer resulted in reproducible, highly conductive, high optical switch, PEDOT films. Large-scale electrochromic devices were fabricated through vapor phase deposited PEDOT resulting in both high and consistent device performance [55].

PEDOT films deposited using oCVD have been demonstrated to have electrochromic behavior when cycled between oxidized and reduced states with very high optical contrast and switching speeds. An electrochromic device using oCVD PEDOT is shown in Fig. 7a. Switching speeds of 13 and 8.5 s for light-to-dark and dark-to-light transitions, respectively, were obtained with oCVD PEDOT films about 100 nm thick deposited on ITO/glass [56]. The device was cycled between ± 5 V. The color contrast was 45 % at 566 nm and was 85 % stable over 150 redox cycles. The transition from the light oxidized state to the dark reduced state was gradual and lasted nearly 1 min. The transition from dark to light was

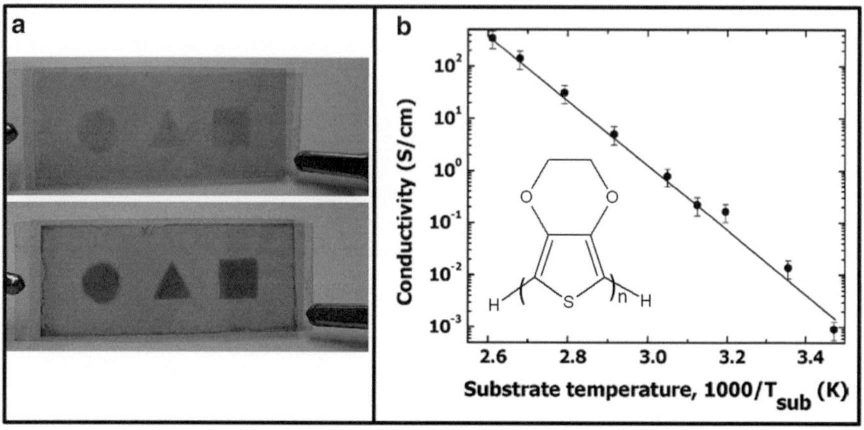

Fig. 7 (**a**) Electrochromic device obtained by depositing oCVD PEDOT on ITO. The oxidized conductive state appears *sky blue* (*up*), while its reduced insulating state, is opaque (*bottom*). (**b**) The deposition conditions strongly influence the performance of the oCVD PEDOT, whose structure is represented in the *bottom left corner*. The conductivity linearly increases with an increase in the substrate temperature. (**a**) Reproduced with permission [56]. Copyright 2007, Elsevier. (**b**) Reproduced with permission from [57]. Copyright 2007, American Chemical Society

much quicker and happened in less than 10 s. Well-established solution-based electrochromic polymers have achieved switching speeds of 0.6 s in similar assembled devices.

The polymer properties can easily be tuned by controlling the oCVD deposition parameters, such as the stage temperature. Figure 7b shows that the conductivity increases linearly from 10^{-3} to 10^2 S/cm when the substrate temperature is increased. The improvement in conductivity is attributed to the longer conjugation length and higher doping levels. The control over the doping level is very important, since higher doping concentration moves the Fermi energy level in the band gap. Im et al. [57] demonstrated that the work function of PEDOT can be varied from 5.1 to 5.4 eV during oCVD by adjusting substrate temperature. The tunability of the work function of oCVD PEDOT films is advantageous for achieving favorable band alignment between dissimilar layers in organic devices, thus lowering the energy barrier to charge transport at the contact interface.

The versatility of the oCVD process to deposit PEDOT on a great variety of substrates has been demonstrated by depositing the oCVD PEDOT electrodes on flexible polyethylene terephthalate (PET) substrate. No significant change in performance was registered upon over 100 compressive flexing cycles. In contrast, the electrical conductivity of commercially available ITO on PET substrates deteriorated rapidly and cracks were visible via optical microscope after repeated flexing.

4 Conclusions

4.1 Commercialization and Conclusion

CVD methods have already been scaled up and commercialized as simple, and low-cost polymer coating technologies [58]. The superior properties demonstrated at the lab-scale can be translated into large reactors, keeping equivalent performances. Gupta et al. [30] used computational fluid dynamic modeling and non-dimensional analysis of convective and diffusive heat and mass transport to show how bench-scale processes and large-scale ones can be run in parallel with similar results. The independency with the substrate material makes CVD compatible for roll-to-roll processing (Fig. 8), which improves the economics of the process, increases the process throughput, and allows for the surface modification of large objects. The low quantity of chemicals involved and the absence of any solvent put CVD among the environment-friendly polymerization methods.

In particular, iCVD and oCVD methods require low energy input, and can achieve high coating conformality to three-dimensional substrates, and high monomer functionality retention. The latter is an established characteristic of the conventional organic synthesis but the possibility of combining it with the advantages of vacuum processing expands the range of applications, where organic polymers can have a fundamental role. In other words, CVD can fill the gap between the versatility of conventional organic synthesis and the requirements of device fabrication.

Fig. 8 Roll-to-roll industrial scale reactor for coating flexible media up to 0.5 m in width. Reproduced with permission from [3]. Copyright 2013, Wiley

The field of organic and hybrid photonic crystals can largely benefit from the CVD methods to deposit thin films in single and multilayered fashions. Distributed Bragg reflectors based on multilayers alternating different polymers or inorganic and organic materials can be easily obtained by CVD. The complete absence of solvent allows to implement insoluble polymers in the multilayer structures (i.e., fluoropolymers, cross-linked polymers, or conjugated ones). It also enables an easy fabrication of the multistack, without the need of finding orthogonal solvents to do not dissolve the underlying layer. Interfacial smoothness between the adjacent layers was obtained due to the processing of the subsequent layers in the same reactor chamber.

Novel polymer compositions can be explored to combine the optical response with other properties, to obtain sensors and new devices. Exotic copolymer compositions can be achieved by mixing monomers with very different solubilities.

The possibility of coating three-dimensional substrate with a uniform thickness can be used to obtain interesting periodic structures, and combine optical properties in microfluidic devices or modify the surface of a periodically porous material with a functional, stimuli-responsive polymer, by keeping intact the porosity.

The inclusion of responsive polymers in photonic crystals can lead to novel sensor devices. Different swelling with different vapors, for example, can lead to different colors. An example of distributed Bragg reflector with fast response to water vapor was obtained alternating PHEMA and titania layers.

Reactive surfaces can be used for subsequent immobilization of nanoparticles or biological molecules. Embedding of gold nanoparticles in a fluorinated polymer structure was demonstrated to be effective for the modification of the refractive index of the pure fluoropolymer.

Anisotropic effects can also be explored by using fluoropolymers with different chain orientation. iCVD allows to control the preferential orientation of the fluorinated pendant chain of the PPFDA from parallel to perpendicular to the substrate surface.

Conjugated polymers are typically difficult to process in a thin film form due to their insoluble crystalline nature. Nevertheless, their electrical properties can be combined to optical ones to obtain electrochromic devices.

The possibility of combining interesting polymeric compositions and a large variety of properties achievable is anticipated to make the field blossom with novel devices and novel solutions.

References

1. V. Robbiano, M. Giordano, C. Martella, F. Di Stasio, D. Chiappe, F. Buatier de Mongeot, D. Comoretto, Adv. Opt. Mater. **1**, 389–396 (2013)
2. J. Clark, G. Lanzani, Nat. Photon. **4**, 438–446 (2010)
3. C. Anna Maria, R.M. Howden, D.C. Borrelli, C.D. Petruczok, Y. Rong, Y. Jose Luis, U. Asli, C. Nan, L. Sunghwan, J. Won Jun, L. Andong, W. Xiaoxue, K.K. Gleason, Adv. Mater. **25**, 5392–5423 (2013)

4. J.L. Yague, A.M. Coclite, C. Petruczok, K.K. Gleason, Macromol. Chem. Phys. **214**, 302–312 (2013)
5. K.K.S. Lau, K.K. Gleason, Macromolecules **39**, 3688–3694 (2006)
6. K.K.S. Lau, K.K. Gleason, Macromolecules **39**, 3695–3703 (2006)
7. S.H. Baxamusa, K.K. Gleason, Adv. Funct. Mater. **19**, 3489–3496 (2009)
8. M. Karaman, S.E. Kooi, K.K. Gleason, Chem. Mater. **20**, 2262–2267 (2008)
9. K.K.S. Lau, J. Bico, K.B.K. Teo, M. Chhowalla, G.A.J. Amaratunga, W.I. Milne, G.H. McKinley, K.K. Gleason, Nano Lett. **3**, 1701–1705 (2003)
10. S.G. Im, E.A. Olivetti, K.K. Gleason, Surf. Coat. Technol. **201**, 9406–9412 (2007)
11. S.J. Limb, C.B. Labelle, K.K. Gleason, D.J. Edell, E.F. Gleason, Appl. Phys. Lett. **68**, 2810–2812 (1996)
12. S.J. Limb, D.J. Edell, E.F. Gleason, K.K. Gleason, J. Appl. Polym. Sci. **67**, 1489–1502 (1998)
13. S.H. Baxamusa, S.G. Im, K.K. Gleason, Phys. Chem. Chem. Phys. **11**, 5227–5240 (2009)
14. A.M. Coclite, G. Ozaydin-Ince, F. Palumbo, A. Milella, K.K. Gleason, Plasma Proc. Polym. **7**, 561–570 (2010)
15. R. Bose, K.K.S. Lau, Biomacromolecules **11**, 2116–2122 (2010)
16. W.S. O'Shaughnessy, S.K. Murthy, D.J. Edell, K.K. Gleason, Biomacromolecules **8**, 2564–2570 (2007)
17. S.H. Baxamusa, K.K. Gleason, Chem. Vap. Depos. **14**, 313–318 (2008)
18. R. d'Agostino, P. Favia, C. Oehr, M.R. Wertheimer, Plasma Proc. Polym. **1**, 11–12 (2004)
19. R. d'Agostino, F. Cramarossa, V. Colaprico, R. Dettole, J. Appl. Phys. **54**, 1284–1288 (1983)
20. M.C. Barr, J.A. Rowehl, R.R. Lunt, J. Xu, A. Wang, C.M. Boyce, S.G. Im, V. Bulovic, K.K. Gleason, Adv. Mater. **23**, 3499–3505 (2011)
21. J.P. Lock, S. Gap Im, K.K. Gleason, Macromolecules **39**, 5326–5329 (2006)
22. H.G.P. Lewis, J.A. Caulfield, K.K. Gleason, Langmuir **17**, 7652–7655 (2001)
23. A.M. Coclite, G. Ozaydin-Ince, R. d'Agostino, K.K. Gleason, Macromolecules **42**, 8138–8145 (2009)
24. R.J. Frank-Finney, L.C. Bradley, M. Gupta, Macromolecules **46**, 6852–6857 (2013)
25. P. Favia, G. Cicala, A. Milella, F. Palumbo, R. Rossini, R. d'Agostino, Surf. Coat Technol. **169**, 609–612 (2003)
26. Y. Mao, K.K. Gleason, Langmuir **20**, 2484–2488 (2004)
27. G. Ozaydin-Ince, K.K. Gleason, M.C. Demirel, Soft Matter **7**, 638–643 (2011)
28. N. Marí-Buye, S. O'Shaughnessy, C. Colominas, C.E. Semino, K.K. Gleason, S. Borròs, Adv. Funct. Mater. **19**, 1276–1286 (2009)
29. L. Montero, S.H. Baxamusa, S. Borros, K.K. Gleason, Chem. Mater. **21**, 399–403 (2009)
30. M. Gupta, K.K. Gleason, Thin Solid Films **515**, 1579–1584 (2006)
31. A. Convertino, A. Valentini, T. Ligonzo, R. Cingolani, Appl. Phys. Lett. **71**, 732–734 (1997)
32. A. Convertino, A. Valentini, R. Cingolani, Appl. Phys. Lett. **75**, 322–324 (1999)
33. M.E. Alf, T.A. Hatton, K.K. Gleason, Langmuir **27**, 10691–10698 (2011)
34. S.H. Baxamusa, L. Montero, S. Borros, K.K. Gleason, Macromol. Rapid Commun. **31**, 735–739 (2010)
35. G.S. Malkov, I.T. Martin, W.B. Schwisow, J.P. Chandler, B.T. Wickes, L.J. Gamble, D.G. Castner, E.R. Fisher, Plasma Proc. Polym. **5**, 129–145 (2008)
36. P.D. Haller, C.A. Flowers, M. Gupta, Soft Matter **7**, 2428–2432 (2011)
37. N.J. Trujillo, S.H. Baxamusa, K.K. Gleason, Chem. Mater. **21**, 742–750 (2009)
38. G. Cicala, A. Milella, E. Palumbo, P. Favia, R. d'Agostino, Diam. Relat. Mater. **12**, 2020–2025 (2003)
39. N. Ouchani, D. Bria, B. Djafari Rouhani, A. Nougaoui, J. Phys. Condens. Matter. **21**, 485401 (2009)
40. L.Y. Li, W.D. Wang, E.D. Laird, C.Y. Li, M. Defaux, D.A. Ivanov, Polymer **52**, 3633–3638 (2011)
41. E.D. Laird, C.Y. Li, Macromolecules **46**, 2877–2891 (2013)

42. E.D. Laird, R.K. Bose, W.D. Wang, K.K.S. Lau, C.Y. Li, Macromol. Rapid Commun. **34**, 251–256 (2013)
43. A.M. Coclite, Y.J. Shi, K.K. Gleason, Adv. Mater. **24**, 4534–4539 (2012)
44. A.M. Coclite, Y.J. Shi, K.K. Gleason, Adv. Funct. Mater. **22**, 2167–2176 (2012)
45. M. Gupta, K.K. Gleason, Langmuir **22**, 10047–10052 (2006)
46. A.M. Coclite, P. Lund, R. Di Mundo, F. Palumbo, Polymer **54**, 24–30 (2013)
47. K.K. Lau, K.K. Gleason, Macromol. Biosci. **7**, 429–434 (2007)
48. G. Ozaydin-Ince, K.K. Gleason, J. Vac. Sci. Technol. A **27**, 1135–1143 (2009)
49. J.H. Burroughes, D.D.C. Bradley, A.R. Brown, R.N. Marks, K. Mackay, R.H. Friend, P.L. Burns, A.B. Holmes, Nature **347**, 539–541 (1990)
50. D. Bhattacharyya, R.M. Howden, D.C. Borrelli, K.K. Gleason, J. Polym. Sci. Polym. Phys. **50**, 1329–1351 (2012)
51. M.C. Barr, R.M. Howden, R.R. Lunt, V. Bulovic, K.K. Gleason, Adv. Energy Mater. **2**, 1404–1409 (2012)
52. D.C. Borrelli, M.C. Barr, V. Bulovic, K.K. Gleason, Sol. Energy Mater. Sol. Cells **99**, 190–196 (2012)
53. D. Bhattacharyya, K.K. Gleason, J. Mater. Chem. **22**, 405–410 (2012)
54. M. Fabretto, K. Zuber, C. Hall, P. Murphy, H.J. Griesser, J. Mater. Chem. **19**, 7871–7878 (2009)
55. M. Fabretto, J.P. Autere, D. Hoglinger, S. Field, P. Murphy, Thin Solid Films **519**, 2544–2549 (2011)
56. J.P. Lock, J.L. Lutkenhaus, N.S. Zacharia, S.G. Im, P.T. Hammond, K.K. Gleason, Synth. Met. **157**, 894–898 (2007)
57. S.G. Im, K.K. Gleason, Macromolecules **40**, 6552–6556 (2007)
58. H.G.P. Lewis, N.P. Bansal, A.J. White, E.S. Handy, Thin Solid Films **517**, 3551–3554 (2009)

Nanoimprint Lithography: Toward Functional Photonic Crystals

Paola Lova and Cesare Soci

Abstract In this chapter we review the use of nanoimprint lithography and its derivative soft-lithography techniques for the fabrication of functional photonic crystals. Nanoimprint is a viable, scalable, and cost-effective solution for large area patterning. While initially it relied primarily on pattern transfer from a rigid mold to a thermally softened polymer by embossing, in the last two decades the process evolved rapidly, giving rise to new technologies that allow direct imprint of functional materials such as conjugated polymers, metals, biological matter, and metal oxides. These advancements generated increasing interest in the use of nanoimprint lithography for the fabrication of photonic structures for light management in optoelectronic devices. After describing standard nanoimprint lithography and its derivative soft-lithography methods, we briefly discuss nanoimprint capabilities and prospects in photonic applications. In particular we review recent implementations of imprinted photonic structures for light management in organic light emitting diodes, solar cells, solid state lasers and sensors.

Keywords Nanoimprint lithography • Soft-lithography • Photonic crystals • Light management

1 Introduction

Photonic crystals are widely used in optoelectronic applications to control light propagation in light emitting diodes [1, 2], as anti-reflecting coatings, diffraction gratings and back reflectors in solar cells [3–5], resonators in lasers [6, 7], and active materials in sensors [8–10]. Notwithstanding many top-down and bottom-up

P. Lova (✉)
Energy Research Institute at NTU (Eri@n) and Interdisciplinary Graduate School, Nanyang Technological University, 21 Nanyang link #03-25, Singapore 637371, Singapore
e-mail: paola001@e.ntu.edu.sg

C. Soci (✉)
Division of Physics and Applied Physics, Centre for Disruptive Photonic Technologies (CDPT) and School of Physical and Mathematical Sciences, Nanyang Technological University, 21 Nanyang link #03-03, Singapore 637371, Singapore
e-mail: csoci@ntu.edu.sg

D. Comoretto (ed.), *Organic and Hybrid Photonic Crystals*,
DOI 10.1007/978-3-319-16580-6_9

fabrication techniques are available [11, 12], concurrent high resolution, low cost, and fast production yield over large areas are hardly achievable with standard lithographic methods. Indeed, bottom-up fabrication methods allow low cost and large area patterning, but high resolution and reproducibility are challenging. On the other hand, top-down technologies yield high resolution and reproducibility, but large area patterning requires large processing time and cost.

In the 1990s, S.Y. Chou developed a new technology, namely nanoimprint lithography (NIL), which allows overcoming these issues [13]. NIL is an embossing tool that relies on the mechanical deformation of a softened resist. Indeed, unlike optical lithography, NIL resolution is not restricted by wave diffraction, scattering, and interference [14, 15]. Moreover, when compared to high-resolution lithography such as electron-beam and extreme ultraviolet lithography, NIL allows patterning large-area nanostructures with lower cost and higher production rate [16].

In the last 20 years, NIL has advanced considerably, overcoming early issues related to resolution [17–19], mold-substrate alignment and overlay [20–23], and polymer flow within the mold features [24–26]. Thanks to these improvements NIL is now widely employed in industrial-scale manufacturing of photonics and microelectronics devices, and it is being considered by the International Technology Roadmap for Semiconductors as a next generation patterning method for manufacturing semiconductor integrated circuits. Commercial NIL systems are available from EV Group [27], Molecular Imprints [28], Nanonex [29], Obducat [30], Smart Equipment Technology [31], NLT [32], and Suss Microtec [33].

In this chapter we will first describe the standard nanoimprint lithography techniques, as well as new "soft lithography" processes that are widely used for the fabrication of functional architectures. We will then review the recent literature on the use of imprinted photonic structures, focusing on the fabrication of active materials for organic light emitting diodes (OLED), organic photovoltaic (OPV) devices, lasers, and sensors.

2 Basic NIL Processes

The basic idea underlying NIL is the replication of features of a patterned mold by embossing monomeric or polymeric resists curable by thermal treatment or ultraviolet exposure. In standard NIL, a thin layer of resist is spun-cast on a substrate and heated above its glass transition temperature (T_g). A prepatterned mold is brought into contact with the coated substrate and pressed on the thermally softened polymer. The polymer resist flows within the mold features and allows pattern transfer. After cooling the mold and the rigid resist are finally separated, revealing the pattern on the latter (Fig. 1a) [13].

Thermal or ultraviolet exposure is used to modify the physical characteristics of polymers or polymer precursor resists during the imprinting process: embossing requires a low viscosity resist to allow deformation, while demolding without affecting the pattern requires high resist mechanical strength, which is acquired

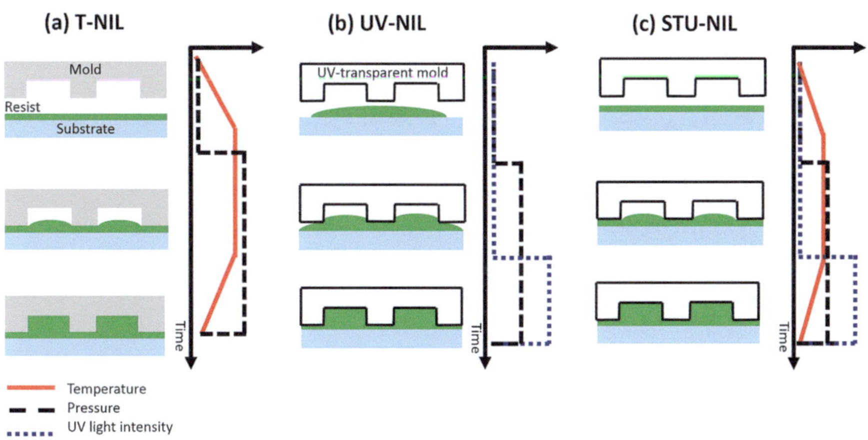

Fig. 1 Thermal (**a**), ultraviolet (**b**), and simultaneous thermal-UV (**c**) NIL processes. The plots display temperature (*red continuous line*), pressure (*black dashed line*), and ultraviolet light intensity (*blue dotted line*) profiles used during the imprinting steps

upon curing [34]. The most common thermoplastic polymers are poly(methyl methacrylate) (PMMA), polycarbonate (PC), polystyrene (PS), polyethylene (PE), polypropylene (PP), polyethylene terephthalate (PET), and siloxane copolymers [15]. These materials cannot be re-molten after curing and are relatively easy to demold. Conversely, methacrylates and epoxides precursors are the most commonly used UV-curable resists. They are hardened during the UV exposure by photo-initiated polymerization [35]. In addition to conventional resists, NIL can also be used for direct imprinting of active and functional materials such as biological materials [36], sol-gel precursors of semiconductor oxides [37], quantum dots [38], metals [39–41], conjugated polymers [42–44], and block copolymers [45].

The imprinting of common thermoplastic resists requires high pressure and hard molds with high hardness, compression and tensile strength, as well as low thermal expansion coefficient and good corrosion resistance to ensure resist deformation and long lifetime. These molds are typically made of metals (e.g., nickel), glasses (e.g., soda-lime and fused silica), crystals (e.g., Si, sapphire, and quartz), or ceramics (e.g., Si_3N_4, SiC, and anodized aluminum oxide) [35, 46] and are fabricated by either top-down lithographic techniques [14, 47] or bottom-up methods such as self-assembly [48] followed by etching and/or metal deposition [49]. On the other hand, the imprinting of low viscous UV-curable resists can be performed with soft polymer molds that allow conformal contact with non-flat and large surfaces. Soft polymer molds have large Young module, global flexibility, and local rigidity: the global flexibility enables uniform contact to large-area substrates, even on uneven surfaces, while the local rigidity minimizes deformation of small imprinted features [50]. Soft molds are usually made by polydimethylsiloxane (PDMS), polyvinyl alcohol (PVA), polyvinyl chloride (PVC), or PMMA [51–53] and are fabricated by replication of a hard mold by NIL itself [49].

The main challenge of NIL is arguably the mold release, due to the high adhesion forces between the stamp and the resist which can damage the imprinted pattern during demolding. To reduce the entity of these forces, anti-sticking molecular monolayers, polymers, metal film or nanoparticles are usually deposited on the mold surface [35].

There are three standard nanoimprinting processes:

Thermal NIL (*T-NIL*), which imprints thermally softened polymers with rigid molds at relatively high pressure. The thermoplastic polymer is heated above its T_g. Once the resist is softened, the mold is brought into conformal contact with the sample substrate and pressed onto it. Figure 1a displays the process steps together with temperature and pressure profiles adopted during the imprint.

Ultraviolet NIL (*UV-NIL*) is used to imprint fluid UV-curable resists. The low viscosity typical of these resists allows room temperature imprinting and relatively low pressure. The resist, which is a polymer precursor, is hardened in situ by UV-irradiation through the transparent mold (Fig. 1b).

Simultaneous Thermal-UV NIL (*STU-NIL*) is used to imprint pre-polymerized resists with hard or soft molds. These resists yield good substrate coverage and lower imprinting temperature with respect to T-NIL so that issues related to thermal expansion and polymers shrinkage are minimized (Fig. 1c).

After demolding, the patterned resist can be dry etched to remove the residual polymer layer or to transfer the pattern to the substrate.

3 Advanced NIL Techniques

The need for fast and low cost fabrication of functional and active materials structures without affecting their optical and semiconducting properties has led to the development of new nanoimprint methods. In particular, advanced NIL and soft-lithography techniques are commonly used for large-scale fabrication of functional photonic crystals.

Roller NIL: Roller-NIL (i.e., roll-to-roll NIL, R2R-NIL, and roll-to-plate NIL, R2P-NIL) was developed by Chou et al. in the late 1990s to achieve high imprinting throughput and large area patterning [54]. In R2R-NIL, a series of rollers coat a moving substrate belt with a resist. Another roller imprints the belt, which is then exposed to UV light to cure the resist (Fig. 2a). R2P-NIL is a variation of this process. Here a substrate covered with the resist is flattened onto a rigid plate which is moved below a roller mold (Fig. 2b) [55, 56]. Roller molds are fabricated by direct patterning of metal cylinders or wrapping flexible molds on the rollers [57–59].

Roller-NIL can reach fabrication speed of ~1 m/min (10^4–10^5 times faster than traditional electron beam lithography) and is currently used in semi-industrial production (see Sects. 4.1 and 4.4) [55].

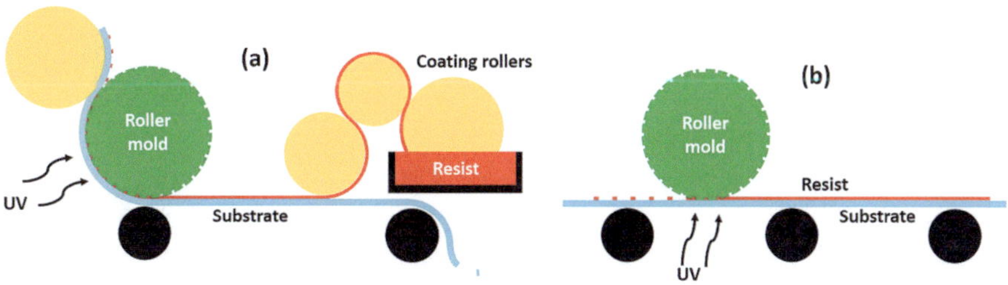

Fig. 2 (**a**) Roll-to-roll and (**b**) roll-to-plate NIL process schemes

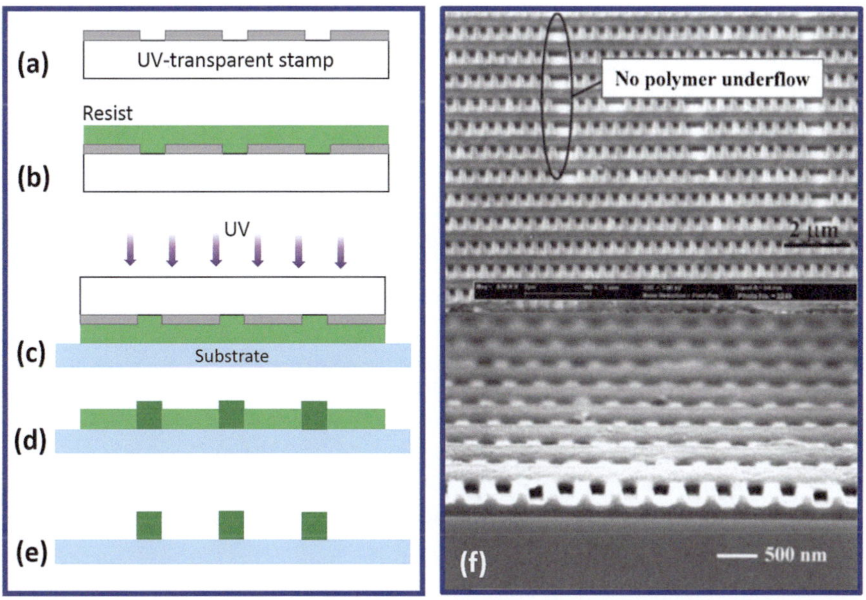

Fig. 3 Reverse UV-NIL process: A metal patterned UV-transparent mold (**a**) is first spun-cast with a resist (**b**) then pressed onto the substrate and exposed to UV light (**c**). The stamp is removed (**d**) and the uncured resist is wet etched to reveal the pattern (**e**). (**f**) Scanning electron microscope (SEM) micrograph of 3D structures fabricated by multiple reverse UV-NIL (Reprinted with permission from [64]. Copyright 2006, American Vacuum Society)

Reverse NIL (R-NIL): In R-NIL, a resist is spun-cast on the mold rather than on the substrate and then transferred on the latter by the imprint process [60–62]. This technique easily allows three-dimensional structures by multiple patterning [63–65]. In the process shown in the left panel of Fig. 3, an UV-curable resist is spun-cast on a patterned metal-quartz stamp. The stamp holding the resist is then pressed onto a substrate and exposed to UV light. After demolding, the unexposed resist is rinsed away with a solvent revealing the grating [64]. Process reiterations allow the growth of patterned multilayers to form 3D structures like the one shown in Fig. 3f.

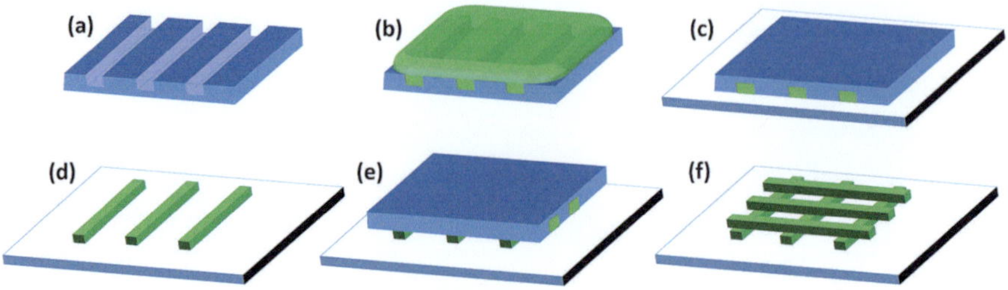

Fig. 4 Schematic of mTM process. An elastomeric mold (**a**) is wetted with the resist (**b**). The excess polymer is removed and the mold is transferred onto a substrate (**c**), the mold is then removed revealing the pattern (**d**). Repetition of the process (**e**) allows 3D structures (**f**)

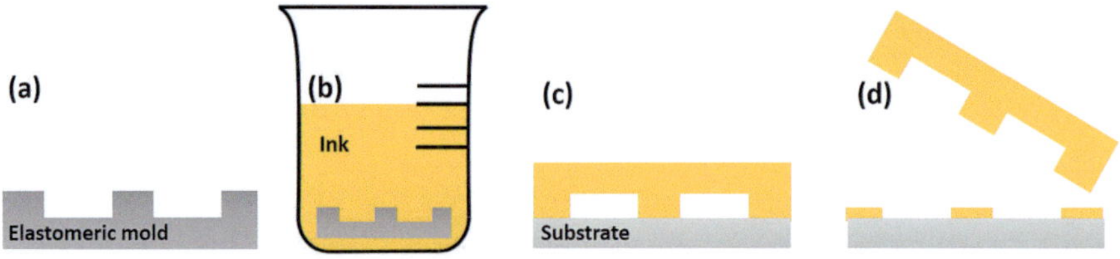

Fig. 5 An elastomeric mold (**a**) is inked with the material to imprint (**b**) and pressed on a substrate (**c**) to release the ink revealing the imprinted pattern (**d**)

Microtransfer molding (mTM): Microtransfer molding is a variation of reverse NIL. In this technique the features of an elastomeric mold are filled with a UV-curable polymer precursor. The excess polymer is removed from the surface using a blading slab. The mold is then pressed onto a surface and cured [66]. Reiterations of the process allow to create 3D structures. The process schematic is shown in Fig. 4.

Microcontact printing (μCP): μCP relies on the property of alkanethiols to form self-assembled monolayers on gold surface [67]. In this technique an elastomeric mold (e.g., silicone, poly (urethane acrylate) (PUA) or PDMS) is inked with an alkanethiol and then pressed on a gold film to transfer the thiol pattern (Fig. 5). Printed alkanethiols are stable enough to be used as etching masks [67]. Examples of the application of μCP are given in Sects. 4.1 and 4.4.

In recent years, μCP was modified to increase pattern homogeneity and to imprint functional materials such as polymers [68], metals [69], nanoparticles [70], proteins [71], lipids [72], and DNA [73] on inorganic and polymer substrates [68]. Among new techniques, *Supramolecolar Contact Printing* uses μCP to immobilize receptor molecules able to selectively physisorb enzymes, proteins, and cells. *Dip Pen Nanolithography* exploits atomic force microscope tips inked with receptor molecules to pattern biological material. *Polymer Pen Imprinting* uses an array of inked polymer tips typically made by PDMS which are brought into contact with the surface to imprint and moved with a piezoelectric system [74]. Another variation of μCP, namely *Lift-up*, consists in the deposition of an active material on the

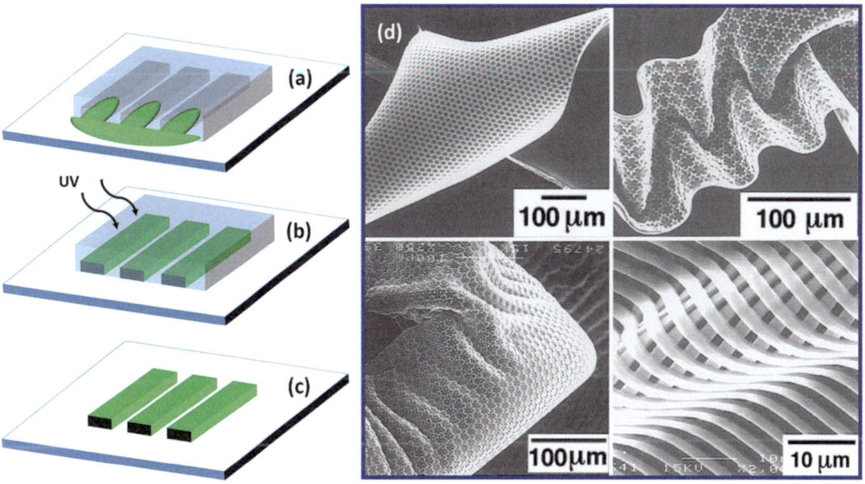

Fig. 6 (**a**)–(**c**) MIMIC process schematics, (**d**) SEM micrograph on resulting free-standing polymer patters (Adapted with permission from [79]. Copyright 1996 American Chemical Society)

substrate on which a soft mold is pressed. The resist in contact with the mold features sticks to it and is removed during the demolding to reveal the pattern [75, 76]. Finally, in *Magnetic Field Assisted μCP* a PDMS layer containing magnetic iron nanoparticles is deposited on the top of the standard PDMS mold and placed into a magnetic field to control the stamp pressure on the substrate during the imprinting. This technique yields high pattern uniformity and homogeneity.

Micromolding in capillaries (MIMIC): In MIMIC an elastomeric mold is placed on a substrate and put in contact with some drop of fluid pre-polymer, polymer solution, or thermally softened polymer (Fig. 6a). The liquid fills the network channels by capillary action and is subsequently cured (Fig. 6b) [77, 78]. MIMIC can yield free-standing film patterns by two procedures. In the first a pattern is formed on a support, which is then etched until complete dissolution. In the second, the pattern in formed between two elastomeric molds, which are then peeled from the free-standing pattern [77]. Free-standing structures resulting from these processes are displayed in Fig. 6d.

Solvent Assisted Micromolding (SAMIM): In SAMIM a good solvent of the resist is applied on the mold surface. As the polymer contacts the wetted mold, a thin layer swells and conforms to the mold pattern [80]. Solvent diffusion and evaporation cause resist solidification. An example of SAMIM is reported in Sect. 4.3.

A recent variation of this process consists in the swelling of the polymer resist with solvent vapors. This method, named *Solvent Vapor Assisted Imprint Lithography (SVAIL)*, was developed to reduce imprinting pressure and temperature which are detrimental for many photoactive polymers and avoid residual polymer layers which require further etching steps [81].

4 Applications

Nanoimprint and its derivative lithographies not only allow low cost, fast production, and large-scale nanofabrication, but also enable capabilities that would be unconceivable otherwise. For instance, it is possible to directly imprint a wide variety of active materials such as photoactive conjugated polymers, gain molecules, metals, and metal oxides without affecting their optical and conductive properties. This remarkable capability has been widely exploited for the production of photonic patterns for light extraction enhancement in OLED, light management in organic solar cell, resonators for organic solid state lasers, and molecular sensors. In the following we review recent results on the fabrication of photonic structures for various device components, such as substrates, active layers and electrodes, and their effect on the device performance.

4.1 Organic Solar Cells

Photoactive conjugated polymers have high extinction coefficient that allow thin film devices processable by roll-to-roll printing [82, 83]. However, their low charge carrier mobility and high recombination rate limit the efficiency of organic photovoltaic devices (see 13th chapter). Thus, roll-to-roll compatible methods for efficiency enhancement are highly researched. In the last decade, NIL became a preferred tool for the fabrication of photonic structures aimed to enhance photovoltaic device performance. PhC have demonstrated absorption enhancement by light diffraction, wave guiding within the photoactive layer, or coupled wave guiding and plasmonic effects achieved structuring the metal electrodes [84, 85]. Recent literature on the imprinting of photonic crystal for light management of the different solar device components is reviewed in the following.

The simplest organic solar cell consists in a film of organic photoactive material sandwiched between two electrodes. A buffer layer can be inserted between the active material and the anode to block electron diffusion toward the latter. From top to bottom, the device structure consists of:

- Glass or polymer substrate;
- Semitransparent anode for positive charges collection, usually made of transparent conducting oxides (e.g., indium tin oxide, ITO);
- Hole injection layer aimed to block electron diffusion to the anode. The most used buffer layer is poly(3,4-ethylenedioxythiophene):polystyrene sulfonate (PEDOT:PSS). At times, it can also double as negative electrode;

- Photoactive material, which consists in an electron donor (i.e., a conjugated polymer) and an electron acceptor (usually a fullerene derivate) layered or blended in a bulk heterojunction (BHJ);
- A metal cathode for negative charge collection.

The following results were achieved by nanoimprinting one of these cell components:

Substrate: The imprinting of periodic 1D and 2D structure in the device substrate aims to diffract light and increase its path within the photoactive layer. Substrate imprinting is the simplest approach to integrate PhCs in organic solar cells. Indeed, commercial UV- and T-NIL can be used. In this regard, 2D structures imprinted by UV-NIL on resists demonstrated increased light absorption [86–88] and device efficiency up to 42 % higher than flat devices [87].

Electrodes: Metal sub-micrometric structures can increase light absorption thanks to light diffraction from the dielectric lattice and surface plasmonic resonance excitation [89–91]. Gold and silver electrodes imprinted with 1D and 2D sub-micrometric periodical patterns have been used as transparent anodes in poly (3-hexylthiophene):phenyl-C61-butyric acid methyl ester (P3HT:PCBM) [92, 93] and copper phthalocyanine:buckminsterfullerene (CuPC:C60) [94] devices. These metal gratings were fabricated by μCP, R2R-NIL [92] (Fig. 7a) and T-NIL associated with metal evaporation [95, 96] on both hard and flexible substrates. The imprinted devices demonstrated 4.4 % maximum power conversion efficiency with enhancement of ~52 % compared to standard ITO electrodes (Table 1).

Hole Injection layer (PEDOT:PSS): Imprinted periodical patterns on the charge injection layer provide two benefits. First, the interface with the photoactive material is increased and charge separation is enhanced. Second, light diffraction and wave guiding increase the light path within the active layer. Nanoimprinting of PEDOT:PSS can indeed increase the device efficiency up to 90 % compared to flat devices [97], but because of its sensitivity to high pressure and temperature [98, 99], soft mold (e.g., PDMS [97, 100, 101] and PUA [102]), and low temperature and pressure processes must be employed (Table 1). Only recently Yang et al. showed that T-NIL of PEDOT:PSS with hard mold can achieve 15 % efficiency enhancement in spite of thermal degradation when the polymer is dehydrated in dry environment for 24 h [103].

Photoactive layer: Active layers are imprinted to achieve three different goals: First, to introduce periodic structure thus light diffraction and/or guided modes in the photoactive layer. Second, to increase the heterointerface between donor and acceptor materials in bilayer geometries. Third, to orientate polymer chains and boost carrier mobility in the donor material.

In bulk heterojunction, the interface between donor and acceptor materials is maximized. On the other hand, phase segregation and low absorption of thin photoactive films limit the efficiency. Since thickening the active layer increases resistance due to high recombination rate and low charge carrier mobility, the

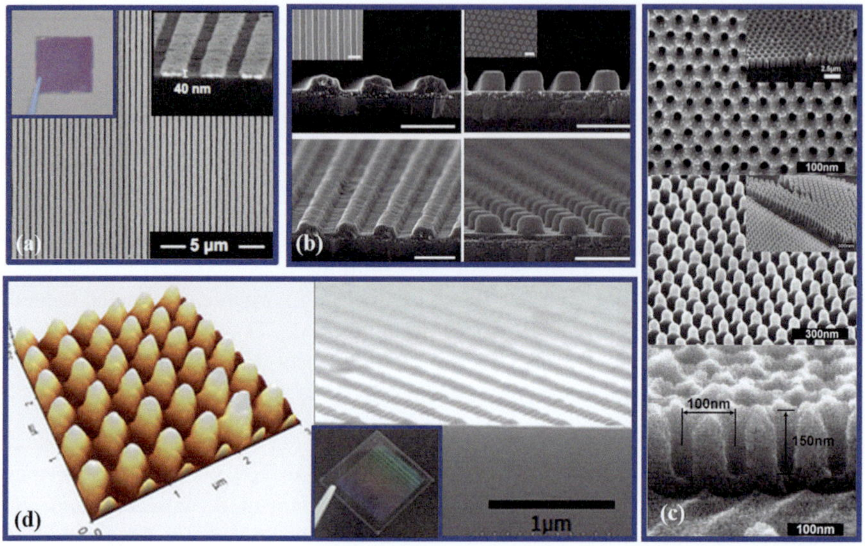

Fig. 7 (**a**) Scanning electron microscope image of a Au grating with period of 280 nm used as electrode and for photonic–plasmonic absorption enhancement. The *inset* shows a photograph of the grating (Adapted with permission from [95]. Copyright 2011 American Chemical Society). (**b**) Imprinted P3HT:PCBM 1D and 2D PhC, scale bars are 500 nm grating (Adapted with permission from [105]. Copyright 2011 American Chemical Society). (**c**) Imprinted P3HT:PCBM bulk heterojunction. grating (Adapted with permission from [117]. Copyright 2010 WILEY-VCH Verlag GmbH & Co. KGaA, Weinheim). (**d**) Atomic force (*left*) and scanning electron (*right*) microscope images of a 2D pattern imprinted on PEDOT:PSS at room temperature (Adapted with permission from [102]. Copyright 2013 Elsevier)

optical thickness can be increased by imprinting periodical patterns. This approach yielded power conversion efficiency increase up to ~15–30 % with respect to flat devices (Table 1) [104–109].

Imprinting of active blends has some drawbacks: Tumbleston et al. showed that T-NIL of region-regular crystalline P3HT:PCBM blend leads PCBM concentration below the optimum in certain area of the device and affect the cell performances [110]. As a consequence, imprinting sub-wavelength patterns in the solely donor to increase donor–acceptor interfacial area seems a more promising strategy. Indeed, sub-wavelength gratings [111–115], nanopillars [116–118], holes [119], and dots array [120] imprinted in the donor polymers yielded efficiency increase up to 200 % (with respect to a flat device with efficiency of 0.82 %) when the size of the imprinted features is comparable or less than the charge diffusion length [119, 121]. On the other hand, when the pattern periodicity is comparable with visible light wavelength, the heterointerface is reduced but light diffraction and guided modes provide larger photoactive layer absorption [122–124] overall increasing efficiency up to 560 % (with respect to a flat device with efficiency of 0.17 %).

Nanoimprinting advantages are not limited to extended donor–acceptor interfacial area and improved light absorption. NIL can control chain alignment orientation

Table 1 Imprinted organic solar cells characteristics and performances

D:A	Junction	Imprinted material	Geometry, pitch (nm)	Method	Efficiency, η (%)			References
					Ref.	PhC	Enhancement	
Substrate								
P3HT:PCBM	BHJ	SU-8	2D, 600–1,200	UV-NIL	2.75	3.92	42	[87]
PCDTBT:$PC_{70}BM$	BHJ	PUA	2D, 265	UV-NIL	3.0	3.9	33	[88]
Electrode								
P3HT:PCBM	BHJ	Au	2D, 200	T-NIL + ME	2.9	4.4	52	[93]
CuPC:C60	DL	Ag	1D, 220	T-NIL + ME	0.96	1.32	35	[94]
P3HT:PCBM	BHJ	Au	1D, 220–420	T-NIL + ME	–	1.55	–	[95]
Hole injection layer								
Pentacene:PTCDI	DL	PEDOT:PSS	1D, 1,200	Soft NIL	0.56	1.60	89	[97]
P3HT:PCBM	BHJ	PEDOT:PSS	1D, 700	Soft-NIL	0.63	0.81	27	[100]
PFSDCNIO:PCBM	BHJ	PEDOT:PSS	1D, 320	MIMIC	0.72	0.96	33	[101]
P3HT:ICBA	BHJ	PEDOT:PSS	2D, 400	Soft-NIL	2.33	2.93	26	[102]
P3HT:PCBM	BHJ	PEDOT:PSS	1D, 140	T-NIL	2.44	2.80	15	[103]
Bulk heterojunction								
PTB7:$PC_{71}BM$	BHJ	BHJ	1D, 700	Soft-NIL	7.20	7.73	7	[91]
P3HT:PCBM	BHJ	BHJ	1D, 500	Soft-NIL	3.56	4.11	18	[104]
P3HT:PCBM	BHJ	BHJ	2D, 400	T-NIL	–	2.91	–	[105]
P3HT:PCBM	BHJ	BHJ	1D ~500	Soft-NIL	3.09	3.68	19	[106]
P3HT:PCBM	BHJ	BHJ	1D, 375	MIMIC	2.39	3.05	28	[109]
P3HT:PCBM	BHJ	BHJ	Random	T-NIL	3.52	4.43	26	[131]
P3HT:PCBM	BHJ	BHJ	1D, 80	T-NIL	2.58	3.37	31	[132]
P3HT:PCBM	BHJ	BHJ	1D-2D, 300:700	Soft-NIL	3.6	4.3	10	[133]
Electron donor polymer								
P3HT:PCBM	DL	P3HT	2D, 100	T-NIL	1.44	2.57	78	[117]

(continued)

Table 1 (continued)

D:A	Junction	Imprinted material	Geometry, pitch (nm)	Method	Efficiency, η (%)			References
					Ref.	PhC	Enhancement	
P3HT:PTCDI	DL	P3HT	1D, 100–200	T-NIL	0.07	0.104	50	[111]
P3HT:C60	DL	P3HT	1D, 140	T-NIL	0.90	1.35	50	[113]
P3HT:PCBM	DL	P3HT	1D-2D, 20–140	R-NIL	2.4	3.1	29	[114]
PCPDTBT:C70	DL	PCPDTBT	1D, 140	T-NIL	2.13	5.52	159	[115]
P3HT:PCBM	DL	P3HT	2D, 100	R-NIL	0.82	2.4	192	[116]
P3HT:PCBM	DL	P3HT	1D-2D, 25–200	SAMIM	1.36	3.25	139	[119]
P3HT:PCBM	DL	P3HT	1D, 200	T-NIL	1.96	2.76	41	[121]
P3HT:C60	DL	P3HT	2D, 100	T-NIL	0.17	1.12	560	[122]
P3HT:PCBM	DL	P3HT	1D,140–560	T-NIL	1.16	3.16	172	[123]
P3HT:EV-BT	DL	P3HT	2D, ~150	T-NIL	0.20	0.30	50	[124]
P3HT:PCBM	DL	P3HT	1D, 360 1,600	SAMIM	1.16	2.77	138	[134]
Electron donor and electron acceptor								
P3HT:PCBM	DL	DL	1D, ~700	T-NIL	–	–	~63	[135]
P3HT:F8BT	DL	P3HT, F8BT	2D, 50–400	SVAIL	0.36	1.85	413	[118]

Efficiency enhancement: percentage enhancement between structured and double layer flat devices unless differently specified
BHJ Bulk heterojunction, *DL* double layer, *EV-BT* 4,7-bis(2-(1-ethylhexyl-4,5-dicyano-imidazol-2-yl) vinyl)benzo[c]1,2,5-thiadiazole, *F8BT* Poly (9,9-dioctylfluorene-alt-benzothiadiazole), *ICBA* indene-C60 bisadduct, *PCPDTBT* Poly[2,6-(4,4-bis-(2-ethylhexyl)-4H-cyclopenta [2,1-b;3,4-b′] dithiophene)-alt-4,7(2,1,3-benzothiadiazole)], *PFSDCNIO* poly[2,7-(9,9-dioctylfluorene)-alt-2- (((4-(diphenylamino)phenyl)thiophen-2-yl) 1,2-dihydro-1-oxoinden-3-ylidene)-malononitrile], *PTB7* Poly({4,8-bis[(2-ethylhexyl)oxy]benzo[1,2-b:4,5-b′]dithiophene-2,6-diyl}{3-fluoro-2-[(2-ethylhexyl)carbonyl] thieno[3,4-b]thiophenediyl}), *PTCDI* perylene tetracarboxylic diimide

Table 2 Recent achievement in nanoimprinted OLED

Imprinted material	Method	Geometry, pitch (nm)	Enhancement[a] (%)					References
			EQE	CE	L	LE	PE	
Substrate and top layer								
Hydrogen silsesquioxane	Low T-NIL	2D, 600	–	17	~160	-	43	[145]
Glass	T-NIL	2D, 200–300	–	–	50	–	–	[142]
PC	Double T-NIL	2D, 200–400	–	180	–	–	'–	[148]
PDMS	Low-T NIL	2D, 10×10^3	–	–	–	~200	–	[138]
UV-resist	UV-NIL	2D, $5*10^4$	–	–	–	70	–	[146]
Resist								
UV-resist	UV-NIL	1D, 500	–	–	–	–	93	[144]
UV-resist	UV-NIL	2D-400	–	6	7		7	[137]
UV-resist	R2R-NIL	2D, 50×10^3	–	–	–	60	–	[139]
PMMA	UV-NIL	2D, 2×10^5	–	–	56	–	–	[140]
Hole injection and active layer								
PEDOT:PSS	mTM	1D-2D, 500	15–25	–	–	–	–	[155]
PEDOT:PSS + MEH-PPV	Soft-NIL	1D, 320	–	35	–	–	–	[157]
PEDOT:PSS + Top layer	Soft-NIL	2D, 250	131		–	–	105	[158]
PEDOT:PSS + Top layer	T-NIL + UV-NIL	2D, 200–400	134	90	–	–	–	[156]

CE current efficiency (cd/A), *L* luminance (cd/m^2), *PE* power efficiency (Im/w), *EQE* external quantum efficiency, LE Light extraction efficiency
[a]Maximum enhancement compared to a flat device reference

in conjugated crystalline [122, 125, 126] and low molecular weight [127] polymers. It has been demonstrated that energetic stabilization [128], nano-confinement [129, 130], and surface interaction with the mold [42] lead to preferential chain orientation with high charge mobility, increasing the solar cell performance [123].

Table 2 summarizes the data presented so far. It is clear that the imprinting of donor–acceptor blend is the worst performing approach with power conversion efficiency enhancement below 25 % with respect to the flat substrate. Slightly better performance is achieved by the imprinted substrate and hole injection layer with enhancement reaching 42 and 200 %, respectively. The highest power conversion efficiencies were achieved by imprinting of the solely donor or acceptor layers, which increased device efficiency by around 500 %.

4.2 Organic Light Emitting Diodes

Although internal quantum efficiency of organic emitters can approach 100 %, waveguiding and total internal reflection due to high refractive indices can prevent photon extraction in OLED. To enhance device performance, 3D light propagation control can be used to forbid in-plane guided mode and to favor out-of-plane light extraction. This can easily be achieved by 2D PhC where the stop band prohibits guided modes in the two in-plane directions, so that light can only propagate out-of-plane (see 15th chapter) [136].

The simplest approach to integrate a photonic structure in OLEDs consists in the imprinting of a resist placed on top of a device [137–140] or on the substrate where the OLED is subsequently grown [141–147]. As an example of this strategy, the left panel of Fig. 8 shows the structure of a devices fabricated on a double patterned substrate. Here T-NIL was used to imprint the two surfaces of a polycarbonate substrate to eliminate waveguiding at the active material–substrate and substrate–air interfaces and to increase light extraction and propagation at high angles [148]. The double pattern improved the device current efficiency by a 2.8 factor and enhanced the flexible OLED luminance by a factor five with respect to the flat structure. The right panel of Fig. 8 shows the comparison between a flat and a patterned OLED consisting in imprinted PS pillars, where a buffer layer was

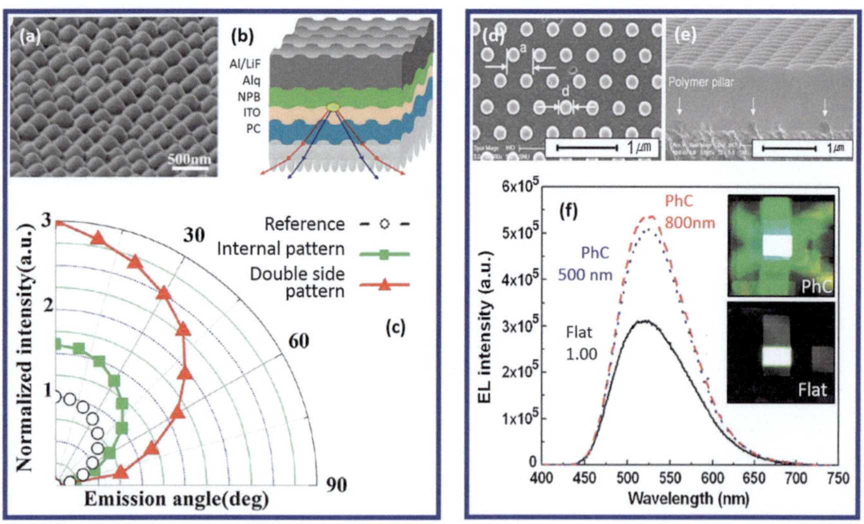

Fig. 8 *Left panel*: (**a**) thermally imprinted polymer substrate, (**b**) double imprinted OLED structure, and (**c**) angle dependence of light emission of the imprinted OLED. (Adapted with permission from [148]. Copyright 2014 American Chemical Society). *Right panel*: SEM micrograph of polystyrene imprinted polymer pillars (**d**) and cross-sectional image of the PhC buffer layer deposited on the pillar (**e**). (**f**) Electroluminescence intensity OLED with and without the PhC. (Adapted with permission from [147]. Copyright 2008 AIP Publishing LLC)

deposited to smooth the photonic structure. The smoothed features enhanced luminescence intensity by 56 % [147].

NIL capacity of direct imprinting functional materials have been widely exploited to control and enhance the spontaneous emission of organic [149, 150] and composite emitters [151–153] and to structure organic electrodes [75, 154, 155] in flexible devices. Zhou et al. showed a double imprinted white OLED made by phosphorescent emitters with a multilayer energy cascade [156]. The device was produced by direct embossing of a PEDOT:PSS layer where the OLED was then fabricated. A second embossing on a UV-curable resist, which was spun-cast on the OLED bottom surface, allowed out-coupling enhancement by a factor 2 and improvement of external quantum efficiency by ~130 %. The same group reported the incorporation of Bragg gratings in both hole transporting (PEDOT:PSS) and emitting (poly(p-phenylene vinylene), MEH-PPV) layers by NIL, achieving 35 % current efficiency enhancement with respect to the flat devices [157]. These results prove that, in addition to high throughput and low costs, NIL can greatly simplify the fabrication process to yield high efficient structure.

Table 2 summarizes recently published achievement attained integrating nanoimprinted photonic structures in OLED active materials, substrates, and top layers. The data show a general enhancement of device performance with maximum gain for devices coupled with imprinted substrates and top layers.

4.3 *Lasers*

Since optically pumped laser emission from organic molecules confined in a photonic structure was demonstrated [159], researchers focused on the development of low cost fabrication processes for dielectric lattices to reduce lasing threshold (see 11th and 17th chapters). Nanoimprint and soft-lithography are the techniques of choice to structure polymer distributed feedback and band edge lasers. So far, NIL has been adopted to imprint periodic structures for lasing in polymer and inorganic substrates, dye doped resists, and gain polymers.

In 1998, Berggren et al. reported one of the first examples of PhC imprinted on flexible polymer for lasing [160]. They achieved low threshold laser (Table 3) by deposition of tris(8-hydroxyquinolinato)aluminum (Alq) doped with 4-dicyanomethylene-2-methyl-6-p-dimethylaminostyryl-4H-pyran (DCM) on a Mylar® substrate where a Bragg grating was previously patterned by T-NIL. So far, lasing from gain molecules and polymers imprinted on either hard [161–164] or flexible polymer [165–167] substrates yielded lasing threshold lower than 20 kW/cm^2.

A simpler but effective approach to lower PhC laser threshold relies on the imprinting of a resist doped with the gain material. Direct imprinting reduces the fabrication process to one single step, thus decreasing material consumption as well as fabrication time and costs. As a result, standard T- and UV-NIL were applied to

Table 3 Characteristic of recently reported nanoimprinted organic laser

λ_{lasing} (nm)	Gain material	Method	Geometry, pitch (nm)	Threshold				References
				Pump pulse (ns)	μJ/cm^2	nJ/pulse	kW/cm^2	
Substrate: DFB								
657	Molecule	T-NIL	1D, 600	2	–	–	12	[160]
522, 558	Polymer	UV-NIL	1D, 360	4	–	–	0.77	[162]
533	Polymer	UV-NIL	1D, 350 main	4	0.06	–	0.015	[163]
533	Polymer	UV-NIL	1D, 355 + 175	4	0.23	–	0.057	[164]
542, 562	Molecule	UV-NIL	1D, 368–384	5–6	1×10^4	–	–	[165]
610, 712	Molecule	T-NIL	1D, 195–450	1	–	14.8	–	[166]
564, 600	Molecule	mTM	1D, 400	0.01	169	–	–	[171]
Active layer: DFB								
580	Polymer	T-NIL	1D-2D, 344	1×10^{-4}	–	32	–	[170]
637	Oligomer	Room-T NIL	1D, 600	3	140	–	–	[176]
638	Polymer	Room-T NIL	1D, 400	3	25	–	–	[178]
564	Polymer	T-NIL	1D, 350	2×10^{-4}	33	–	–	[180]
638	Polymer	SAMIM	2D, 400	1	–	225	–	[181]
Doped resist: DFB								
580	Molecule	T-NIL	1D, 368	10	–	900	–	[169]
618	Molecule	Room T-NIL	1D, 440	6	131	–	–	[168]
562	Molecule	Room-T NIL	1D, 600	10	750	–	–	[179]
602, 614	Molecule	UV-NIL	2D, ~300	16	220	–	–	[172]
Doped resist: BEL								
588, 600	Molecule	T-NIL (Low T)	2D, 500–580	1	300	–	–	[174]
566–600	Molecule	T-NIL (Low T)	2D, 460–540	0.7	630	–	–	[175]
560–618	Molecule	UV-NIL	2D, 364–424	5	100	–	–	[182]

DFB distributed feedback laser [159], *BEL* band edge laser [173]

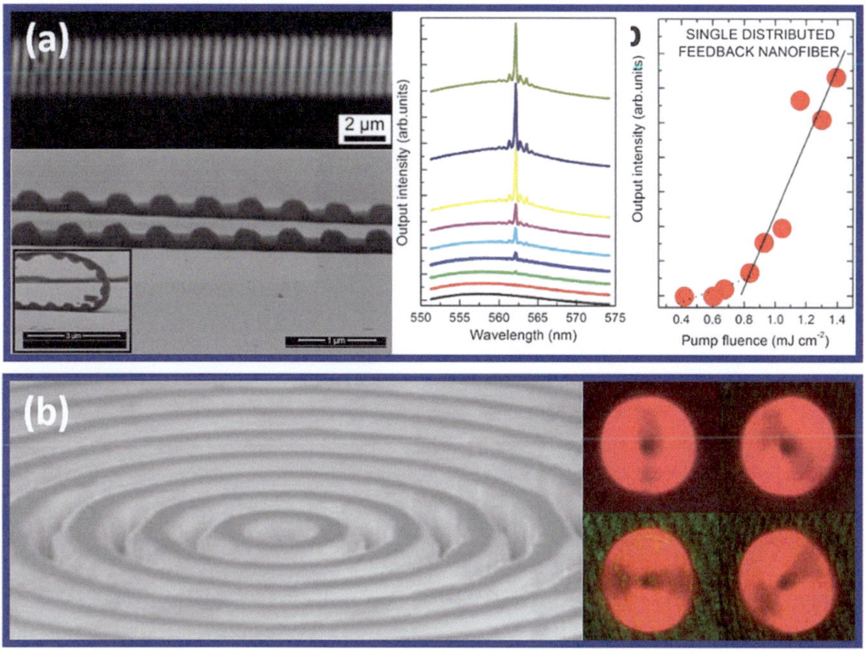

Fig. 9 (**a**) DFB laser fiber SEM micrograph (*left*) and lasing threshold (*right*) (Adapted with permission from [179]. Copyright 2014 WILEY-VCH Verlag GmbH & Co. KGaA, Weinheim). (**b**) Circular R6G DFB laser SEM micrograph and photograph (inset) (Adapted with permission from [168]. Copyright 2007 AIP Publishing LLC)

dye doped resist to obtain low threshold distributed feedback [168–171] and band edge lasers [172–175] (Table 3).

PhC can also be directly imprinted into gain polymers and oligomers. To avoid degradation of optical properties, low temperature imprinting methods such as SAMIM and low temperature NIL of 1D gratings have been proposed (see Table 3). Pisignano et al. demonstrated a new direct patterning of a thiophene based oligomer achieved by prolonged hard mold pressure at room temperature, which does not require thermal nor UV curing [176]. The same group extended the new imprinting method to photoactive polymers achieving thresholds below 10 kW/cm^2 [177, 178].

Figure 9 displays some distributed feedback (DFB) lasers [159] fabricated using different approaches: a thermally imprinted rhodamine 6G (R6G) doped PMMA laser fiber [179] and a circular DFB laser directly imprinted in MEH-PPV doped PMMA [168].

Table 3 summarizes recent advances in imprinted polymer and organic lasers. From the data we see that lower lasing threshold is reported for distributed feedback lasers where the photonic structure is imprinted on the substrate, while the highest threshold is reported for an imprinted distributed feedback laser fiber (Fig. 9a), where the feedback grating in the narrow 1D fiber does not confine light as effectively as the grating that extends over a 2D plane.

4.4 Sensing

There is a steadily growing need for lab-on-a-chip PhC detectors that combine high resolution, sensitivity, and rapid detection in numerous applications such as healthcare, environmental monitoring, and security; however, the development of such detectors is hampered by the lack of low cost materials, fabrication processes, and reproducibility. Research is focusing its effort to the development of new low cost methods for all-plastic disposable devices. In this paragraph we summarize the advance in PhC based sensors fabricated by NIL which promise large-scale production of multi-parameter, label-free sensing platforms.

Standard PhC sensing is generally based on the variation of either effective refractive index or lattice spacing caused by molecules interaction within the crystal [183, 184]. These structures represent the simplest PhC sensors and can be directly imprinted on polymers (see also 2nd and 18th chapters). In 2003 Cunningham et al. reported on a nanoimprinted 1D PhC sensor for the imaging of bimolecular interaction. The sensor was made by a functionalized UV-imprinted epoxy resin and sputtered titanium oxide as low and high index media, respectively [185]. Later on, the same group replaced the epoxy resin with a thermally curable sol-gel silica precursor in order to increase PhC dielectric contrast and its sensitivity [186]. These detectors reported refractive index sensitivity (i.e., PBG spectral shift) for both bulk materials and thin layers contacted with the PhC surface. More recently, the *bird flu* spreading revealed the need for very fast detection of viruses by low cost disposable devices. Endo et al. responded to this demand with self-assembled copolymer imprinted to a 2D PhC on cyclo-olefin substrate and functionalized with H1N1 virus antibody. In their system, an antigen–antibody complex is formed after exposure to the virus. The complex acts as an antireflective coating and reduces PBG reflection intensity (Fig. 10). This device, with sensitivity up to 1 pg/ml antigen in human saliva, opened a new perspective in the monitoring and control of disease spreading [187]. An alternative approach aimed to disease control was provided by the pioneering work of Morhard et al. in year 2000, which reported bacterial detection by antibodies patterned by μCP [188]. *Escherichia coli* antibodies were directly imprinted on a rigid substrate. The selective antibody-bacteria binding gave rise to a diffraction pattern arising from the new cellular PhC.

In the last few years, NIL was also used for the realization of PhC sensor based on lasing. Similar to standard PhC sensor, the interaction between the PhC laser and external molecules affects the PBG features inducing the spectral shift of the laser peak. In 2005, Rose et al. showed high sensitivity to trace of explosive vapors [189]. In their system, a gain semiconducting polymer deposited on a PDMS grating is exposed to di- and trinitrotoluene. Exposure to explosives inhibits lasing by increasing the laser threshold. DFB lasers sensitivity to environmental refractive index variation was also recently demonstrated for both polymer and silica substrates patterned by NIL [190–192]. In these devices a gain material is deposited on a grating to achieve lasing. The device is then exposed to environments with different refractive indices, which induce the spectral shift of the lasing peak.

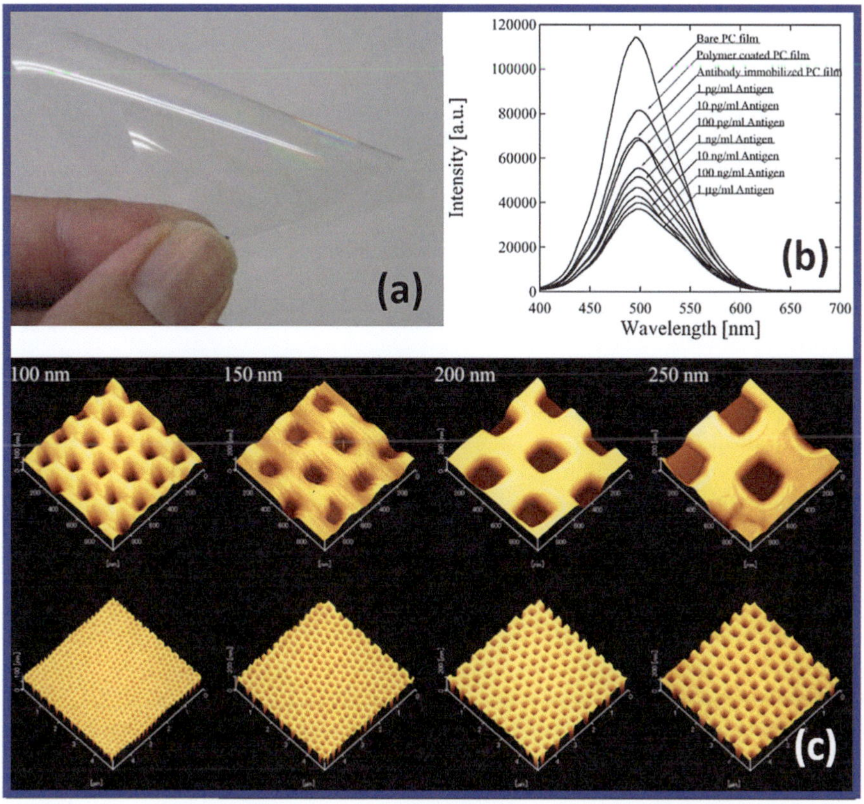

Fig. 10 (**a**) Photograph of Imprinted of 2D grating H1N1 sensor, (**b**) Reflection spectra of the sensor exposed to different H1N1 virus concentrations, and (**c**) atomic force microscopy micrographs of the sensor surface (Adapted with permission from [187]. Copyright 2010 Elsevier)

In 2009 Kristensen et al. reported a functionalized dye doped polymer PhC suitable for the detection of cervical carcinoma cells. In this sensor, the density of carcinoid cells growth on the laser itself generates linear spectral shift of the lasing peak [193]. A more sophisticated sensor geometry was recently reported by Vannhme et al., where an optically pumped DFB laser was connected to a microfluidic channel by a waveguide. The whole structure was produced by T-NIL of PMMA substrates. The imprinted geometry allowed the detection of fluorescent markers exploiting first-order DFB. This was made possible by placing the distributed feedback laser and the measuring site into two different parts of the device [194].

These examples show how NIL is making the production of PhC sensor easier and faster. NIL has already been adopted as a fabrication processes for Klarite^{TM} Surface Enhancement Raman Spectroscopy (SERS) sensing platform commercialized by Renishaw [195]. In a publication dated 2013, the sensing platform was transferred from silicon to plastic platform by roll-to-roll and sheet-level NIL, achieving faster and cheaper mass production together with a eightfold sensitivity improvement to Benzenthiol compared to the silicon sensor [196]. In 2014, a similar roll-to-roll SERS detector allowed sensitivity to traces of ibuprofen (10^{-4} M). In the same work, a functionalized sensor coupled with a Young's interferometer waveguide and a flow cell reported sensitivity down to 1 ppm of formaldehyde [197].

5 Conclusions

Since 1995, when it was just an embossing tool for polymer, the potential of nanoimprint lithography has attracted enormous attention from both the Academia and the Industry. As a result, the technology has evolved rapidly and its capabilities have escalated. Today nanoimprint and the soft lithographies derived from it allow imprinting one- to three-dimensional structures with feature sizes ranging from 5 nm to few hundreds of micrometers in a huge variety of materials, at cost and time scales unconceivable with any other lithographic technique.

In this chapter we introduced standard ultraviolet and thermal nanoimprint lithography and derivative soft lithography used for the fabrication of photonic structures in optoelectronic devices. We reviewed the imprinting of photonic structures in conducting and semiconducting polymer, metal, and metal oxide components of OLED, solar cells, lasers, and sensors, giving few examples of the extreme versatility of NIL for the control of light absorption, confinement, and propagation in optoelectronic devices.

References

1. L. Petti, M. Rippa, F. Capasso, G. Nenna, A. De Girolamo, M.G. Maglione, C. Minarini, Nanotechnology **315206**, 24 (2013)
2. L. Hou, Q. Hou, Y. Mo, J. Peng, Y. Cao, Appl. Phys. Lett. **243504**, 87 (2005)
3. G.-G. Zheng, F.-L. Xian, X.-Y. Li, Chin. Phys. Lett. **054213**, 28 (2011)
4. A. Mihi, M.E. Calvo, J.A. Anta, H. Miguez, J. Phys. Chem. C **13**, 112 (2008)
5. R. Betancur, P. Romero-Gomez, A. Martinez-Otero, X. Elias, M. Maymo, J. Martorell, Nat. Photon. **995**, 7 (2013)
6. L. Frezza, M. Patrini, M. Liscidini, D. Comoretto, J. Phys. Chem. C **19939**, 115 (2011)
7. G. Canazza, F. Scotognella, G. Lanzani, S. De Silvestri, M. Zavelani-Rossi, D. Comoretto, Laser Phys. Lett. **035804**, 11 (2014)
8. S. Hu, Y. Zhao, K. Qin, S.T. Retterer, I.I. Kravchenko, S.M. Weiss, ACS Photon. **590**, 1 (2014)
9. H. Ding, C. Liu, H. Gu, Y. Zhao, B. Wang, Z. Gu, ACS Photon. **121**, 1 (2014)
10. H.S. Lim, J.-H. Lee, J.J. Walish, E.L. Thomas, ACS Nano **8933**, 6 (2012)
11. Y. Xia, B. Gates, Z.Y. Li, Adv. Mater. **409**, 13 (2001)
12. G. von Freymann, V. Kitaev, B.V. Lotsch, G.A. Ozin, Chem. Soc. Rev. **2528**, 42 (2013)
13. S.Y. Chou, P.R. Krauss, P.J. Renstrom, Appl. Phys. Lett. **3114**, 67 (1995)

14. D. Morecroft, J.K.W. Yang, S. Schuster, K.K. Berggren, Q. Xia, W. Wu, R.S. Williams, J. Vac. Sci. Technol. B **2837**, 27 (2009)
15. L.J. Guo, J. Phys. D Appl. Phys. **R123**, 37 (2004)
16. T. Eriksson, S. Yamada, P.V. Krishnan, S. Ramasamy, B. Heidari, Microelectron. Eng. **293**, 88 (2011)
17. S.Y. Chou, P.R. Krauss, P.J. Renstrom, Science **85**, 272 (1996)
18. M.D. Austin, H. Ge, W. Wu, M. Li, Z. Yu, D. Wasserman, S.A. Lyon, S.Y. Chou, Appl. Phys. Lett. **5299**, 84 (2004)
19. G.-Y. Jung, E. Johnston-Halperin, W. Wu, Z. Yu, S.-Y. Wang, W.M. Tong, Z. Li, J.E. Green, B.A. Sheriff, A. Boukai, Y. Bunimovich, J.R. Heath, R.S. Williams, Nano Lett. **351**, 6 (2006)
20. R. Ganesan, J. Dumond, M.S.M. Saifullah, S.H. Lim, H. Hussain, H.Y. Low, ACS Nano **1494**, 6 (2012)
21. S. Takei, T. Ogawa, R. Deschner, C.G. Willson, Microelectron. Eng. **44**, 116 (2014)
22. X. Sun, W. Chen, R. Zhou, W. Chen, J. Zhang, Rev. Sci. Instrum. **125002**, 84 (2013)
23. J. Choi, K. Nordquist, A. Cherala, L. Casoose, K. Gehoski, W.J. Dauksher, S.V. Sreenivasan, D.J. Resnick, Microelectron. Eng. **633**, 78–79 (2005)
24. S.M. Kim, J.H. Kang, W.I. Lee, Polym. Eng. Sci. **209**, 51 (2011)
25. A. Jain, R.T. Bonnecaze, J. Appl. Phys. **234511**, 113 (2013)
26. C. Masclaux, C. Gourgon, C. Perret, S. Labau, Microelectron. Eng. **2048**, 88 (2011)
27. http://www.evgroup.com/en/products/lithography/nanoimprint_systems/. 5 November 2014
28. http://www.molecularimprints.com/. 5 November 2014
29. http://www.nanonex.com/. 5 November 2014
30. http://www.obducat.com/. 5 November 2014
31. http://www.set-sas.fr/en/cat424981-424981-NPS300.html. 5 November 2014
32. http://www.nilt.com/544/nanoimprint-lithography-(nil)-tools. 5 November 2014
33. https://www.suss.com/en/products-solutions/technologies/imprint-lithography.html. 5 November 2014
34. Y. Hirai, J. Photopolym. Sci. Technol. **551**, 18 (2005)
35. L.J. Guo, Adv. Mater. **495**, 19 (2007)
36. M. Elsayed, O.M. Merkel, Nanomedicine **349**, 9 (2014)
37. C. Goh, K.M. Coakley, M.D. McGehee, Nano Lett. **1545**, 5 (2005)
38. M.J. Hampton, J.L. Templeton, J.M. DeSimone, Langmuir **3012**, 26 (2010)
39. K. Felmet, Y.-L. Loo, Y. Sun, Appl. Phys. Lett. **3316**, 85 (2004)
40. S. Buzzi, F. Robin, V. Callegari, J.F. Löffler, Microelectron. Eng. **419**, 85 (2008)
41. B. Radha, S.H. Lim, M.S.M. Saifullah, G.U. Kulkarni, Sci. Rep. **1078**, 3 (2013)
42. M. Aryal, K. Trivedi, W. Hu, ACS Nano **3085**, 3 (2009)
43. L. Jiang, X. Wang, L. Chi, Small **1309**, 7 (2011)
44. Y. Xu, F. Zhang, X. Feng, Small **1338**, 7 (2011)
45. H.-W. Li, W.T.S. Huck, Nano Lett. **1633**, 4 (2004)
46. J. Kouba, M. Kubenz, A. Mai, G. Ropers, W. Eberhardt, B. Loechel, J. Phys. Conf. Ser. **34**, 897 (2006). IOP Publishing
47. X.-Z. Chen, H.-Y. Li, Chin. Phys. Lett. **2830**, 24 (2007)
48. G.H. Jeong, J.K. Park, K.K. Lee, J.H. Jang, C.H. Lee, H.B. Kang, C.W. Yang, S.J. Suh, Microelectron. Eng. **51**, 87 (2010)
49. W. Zhou, *Nanoimprint Lithography: An Enabling Process for Nanofabrication* (Springer, Heidelberg, 2013), p. 33
50. S. Zankovych, T. Hoffmann, J. Seekamp, J.U. Bruch, C.M.S. Torres, Nanotechnology **91**, 12 (2001)
51. L. Hongbo, D. Yucheng, *Nanoimprint Lithography, Lithography* (InTech, Rijeka, 2010)
52. D.-G. Choi, J.-H. Jeong, Y.-S. Sim, E.-S. Lee, W.-S. Kim, B.-S. Bae, Langmuir **9390**, 21 (2005)
53. K. Hyewon, L. Jiyeon, P. Joonhyung, H.L. Hong, Nanotechnology **197**, 17 (2006)
54. H. Tan, A. Gilbertson, S.Y. Chou, J. Vac. Sci. Technol. B **3926**, 16 (1998)

55. S.H. Ahn, L.J. Guo, ACS Nano **2304**, 3 (2009)
56. J. John, Y. Tang, J.P. Rothstein, J.J. Watkins, K.R. Carter, Nanotechnology **505307**, 24 (2013)
57. S. Ahn, J. Cha, H. Myung, S.-M. Kim, S. Kang, Appl. Phys. Lett. **213101**, 89 (2006)
58. T.-C. Huang, J.-T. Wu, S.-Y. Yang, P.-H. Huang, S.-H. Chang, Microelectron. Eng. **615**, 86 (2009)
59. S. Park, K. Choi, G. Kim, J. Lee, Microelectron. Eng. **604**, 86 (2009)
60. X.D. Huang, L.-R. Bao, X. Cheng, L.J. Guo, S.W. Pang, A.F. Yee, J. Vac. Sci. Technol. B **2872**, 20 (2002)
61. V. Reboud, A.Z. Khokhar, B. Sepulveda, D. Dudek, T. Kehoe, J. Cuffe, N. Kehagias, M. Lira-Cantu, N. Gadegaard, V. Grasso, V. Lambertini, C.M. Sotomayor Torres, Nanoscale **3495**, 4 (2012)
62. K.G.A. Tavakkoli, M. Ranjbar, S.N. Piramanayagam, S.K. Wong, W.C. Poh, R. Sbiaa, T.C. Chong, Nanosci. Nanotechnol. Lett. **835**, 4 (2012)
63. K.-S. Han, S.-H. Hong, K.-I. Kim, J.-Y. Cho, K.-W. Choi, H. Lee, Nanotechnology **045304**, 24 (2013)
64. N. Kehagias, V. Reboud, G. Chansin, M. Zelsmann, C. Jeppesen, F. Reuther, C. Schuster, M. Kubenz, G. Gruetzner, C.M. Sotomayor Torres, J. Vac. Sci. Technol. B **3002**, 24 (2006)
65. H. Park, H. Li, X. Cheng, J. Vac. Sci. Technol. B **2325**, 25 (2007)
66. X.-M. Zhao, Y. Xia, G.M. Whitesides, Adv. Mater. **837**, 8 (1996)
67. A. Kumar, G.M. Whitesides, Appl. Phys. Lett. **2002**, 63 (1993)
68. T. Kaufmann, B.J. Ravoo, Polym. Chem. **371**, 1 (2010)
69. C.-P. Chang, P.-C. Wang, Y.-Y. Nian, Y.-M. Liu, M.-D. Ger, J. Taiwan Inst. Chem. Eng. **2755**, 45 (2014)
70. S.-T. Han, Y. Zhou, Z.-X. Xu, L.-B. Huang, X.-B. Yang, V.A.L. Roy, Adv. Mater. **3556**, 24 (2012)
71. C.-J. Pan, H. Qin, Y.-D. Nie, H.-Y. Ding, Colloids Surf. B **18**, 104 (2013)
72. O.A. Nafday, T.W. Lowry, S. Lenhert, Small **1021**, 8 (2012)
73. S.A. Lange, V. Benes, D.P. Kern, J.K.H. Hörber, A. Bernard, Anal. Chem. **1641**, 76 (2004)
74. J. Voskuhl, J. Brinkmann, P. Jonkheijm, Curr. Opin. Chem. Biol. **1**, 18 (2014)
75. T. Granlund, T. Nyberg, L. Stolz Roman, M. Svensson, O. Inganäs, Adv. Mater. **269**, 12 (2000)
76. J. Yao, X. Yan, G. Lu, K. Zhang, X. Chen, L. Jiang, B. Yang, Adv. Mater. **81**, 16 (2004)
77. E. Kim, Y. Xia, G.M. Whitesides, Nature **581**, 376 (1995)
78. E. Kim, Y. Xia, G.M. Whitesides, J. Am. Chem. Soc. **5722**, 118 (1996)
79. Y. Xia, E. Kim, G.M. Whitesides, Chem. Mater. **1558**, 8 (1996)
80. E. King, Y. Xia, X.-M. Zhao, G.M. Whitesides, Adv. Mater. **651**, 9 (1997)
81. N.E. Voicu, S. Ludwigs, E.J.W. Crossland, P. Andrew, U. Steiner, Adv. Mater. **757**, 19 (2007)
82. F.C. Krebs, Sol. Energy Mater. Sol. Cells **465**, 93 (2009)
83. F.C. Krebs, J. Fyenbo, M. Jorgensen, J. Mater. Chem. **8994**, 20 (2010)
84. D.-H. Ko, J.R. Tumbleston, A. Gadisa, M. Aryal, Y. Liu, R. Lopez, E.T. Samulski, J. Mater. Chem. **16293**, 21 (2011)
85. D. Duché, L. Escoubas, J.-J. Simon, P. Torchio, W. Vervisch, F. Flory, Appl. Phys. Lett. **193310**, 92 (2008)
86. L. Yingchi, K. Christoph, G. Abay, A. Mukti, M. Sorin, T.S. Edward, L. Rene, J. Phys. D Appl. Phys. **024008**, 46 (2013)
87. J.-Y. Chen, M.-H. Yu, C.-Y. Chang, Y.-H. Chao, K.W. Sun, C.-S. Hsu, ACS Appl. Mater. Interfaces **6164**, 6 (2014)
88. D.H. Wang, J. Seifter, J.H. Park, D.-G. Choi, A.J. Heeger, Adv. Energy Mater. **1319**, 1319–1322 (2012)
89. T.H. Reilly, J. van de Lagemaat, R.C. Tenent, A.J. Morfa, K.L. Rowlen, Appl. Phys. Lett. **92**, 013504 (2008)

90. N.C. Lindquist, W.A. Luhman, S.-H. Oh, R.J. Holmes, Appl. Phys. Lett. **93**, 123308 (2008)
91. J. You, X. Li, F.-X. Xie, W.E.I. Sha, J.H.W. Kwong, G. Li, W.C.H. Choy, Y. Yang, Adv. Energy Mater. **1203**, 2 (2012)
92. M.-G. Kang, H.J. Park, A. Se Hyun, T. Xu, L.J. Guo, IEEE J. Sel. Top. Quantum Electron. **1807**(16) (2010)
93. S.Y. Chou, W. Ding, Opt. Exp. **A60**, 21 (2013)
94. M.-G. Kang, T. Xu, H.J. Park, X. Luo, L.J. Guo, Adv. Mater. **4378**, 22 (2010)
95. H.J. Park, T. Xu, J.Y. Lee, A. Ledbetter, L.J. Guo, ACS Nano **7055**, 5 (2011)
96. W. Ding, S.Y. Chou, in *Photovoltaic Specialist Conference (PVSC), 2014 I.E. 40th* 2014, p. 2804
97. W.-Y. Chou, J. Chang, C.-T. Yen, F.-C. Tang, H.-L. Cheng, M.-H. Chang, S. Lien-Chung Hsu, J.-S. Chen, Y.-C. Lee, Appl. Phys. Lett. **183108**, 99 (2011)
98. O.P. Dimitriev, D.A. Grinko, Y.V. Noskov, N.A. Ogurtsov, A.A. Pud, Synth. Met. **2237**, 159 (2009)
99. U. Lang, P. Rust, B. Schoberle, J. Dual, Microelectron. Eng. **330**, 86 (2009)
100. J.B. Emah, R.J. Curry, S.R.P. Silva, Appl. Phys. Lett. **103301**, 93 (2008)
101. X. Zhu, W.C.H. Choy, F. Xie, C. Duan, C. Wang, W. He, F. Huang, Y. Cao, Sol. Energy Mater. Sol. Cells **327**, 99 (2012)
102. J.-H. Choi, H.-J. Choi, J.-H. Shin, H.-P. Kim, J. Jang, H. Lee, Org. Electron. **3180**, 14 (2013)
103. Y. Yang, K. Lee, K. Mielczarek, W. Hu, A. Zakhidov, Nanotechnology **485301**, 22 (2011)
104. S.-I. Na, S.-S. Kim, S.-S. Kwon, J. Jo, J. Kim, T. Lee, D.-Y. Kim, Appl. Phys. Lett. **173509**, 91 (2007)
105. D.-H. Ko, J.R. Tumbleston, W. Schenck, R. Lopez, E.T. Samulski, J. Phys. Chem. C **4247**, 115 (2011)
106. X.H. Li, W.E.I. Sha, W.C.H. Choy, D.D.S. Fung, F.X. Xie, J. Phys. Chem. C **7200**, 116 (2012)
107. L. Stolz Roman, O. Inganäs, T. Granlund, T. Nyberg, M. Svensson, M.R. Andersson, J.C. Hummelen, Adv. Mater. **189**, 12 (2000)
108. M. Niggemann, M. Glatthaar, A. Gombert, A. Hinsch, V. Wittwer, Thin Solid Films **619**, 451–452 (2004)
109. Y.-S. Cheng, C. Gau, Sol. Energy Mater. Sol. Cells **566**, 120 (2014). Part B
110. J.R. Tumbleston, A. Gadisa, Y. Liu, B.A. Collins, E.T. Samulski, R. Lopez, H. Ade, ACS Appl. Mater. Interfaces **8225**, 5 (2013)
111. D. Cheyns, K. Vasseur, C. Rolin, J. Genoe, J. Poortmans, P. Heremans, Nanotechnology **424016**, 19 (2008)
112. O. Wiedenmann, A. Abdellah, G. Scarpa, P. Lugli, J. Phys. Conf. Ser. **012115**, 193 (2009)
113. Y. Yang, M. Aryal, K. Mielczarek, W. Hu, A. Zakhidov, J. Vac. Sci. Technol. B **C6M104**, 28 (2010)
114. D.-G. Choi, K.-J. Lee, J.-H. Jeong, D. Hwan Wang, O. Ok Park, J. Hyeok Park, Sol. Energy Mater. Sol. Cells **1**, 109 (2013)
115. Y. Yang, K. Mielczarek, A. Zakhidov, W. Hu, ACS Appl. Mater. Interfaces **19282**, 6 (2014)
116. D. Chen, W. Zhao, T.P. Russell, ACS Nano **1479**, 6 (2012)
117. M. Aryal, F. Buyukserin, K. Mielczarek, X.-M. Zhao, J. Gao, A. Zakhidov, W. Hu, J. Vac. Sci. Technol. B **2562**, 26 (2008)
118. X. He, F. Gao, G. Tu, D. Hasko, S. Hüttner, U. Steiner, N.C. Greenham, R.H. Friend, W.T.S. Huck, Nano Lett. **1302**, 10 (2010)
119. X. He, F. Gao, G. Tu, D.G. Hasko, S. Hüttner, N.C. Greenham, U. Steiner, R.H. Friend, W.T.S. Huck, Adv. Funct. Mater. **139**, 21 (2011)
120. W. Wiedemann, L. Sims, A. Abdellah, A. Exner, R. Meier, K.P. Musselman, J.L. MacManus-Driscoll, P. Müller-Buschbaum, G. Scarpa, P. Lugli, L. Schmidt-Mende, Appl. Phys. Lett. **263109**, 96 (2010)
121. M. Zhou, M. Aryal, K. Mielczarek, A. Zakhidov, W. Hu, J. Vac. Sci. Technol. B **C6M63**, 28 (2010)

122. J.S. Kim, Y. Park, D.Y. Lee, J.H. Lee, J.H. Park, J.K. Kim, K. Cho, Adv. Funct. Mater. **540**, 20 (2010)
123. Y. Yang, K. Mielczarek, M. Aryal, A. Zakhidov, W. Hu, Nanoscale **7576**, 6 (2014)
124. W. Zeng, K.S.L. Chong, H.Y. Low, E.L. Williams, T.L. Tam, A. Sellinger, Thin Solid Films **6833**, 517 (2009)
125. K.M. Coakley, B.S. Srinivasan, J.M. Ziebarth, C. Goh, Y. Liu, M.D. McGehee, Adv. Funct. Mater. **1927**, 15 (2005)
126. Z. Hu, G. Baralia, V. Bayot, J.-F. Gohy, A.M. Jonas, Nano Lett. **1738**, 5 (2005)
127. S.A. Schmid, K.H. Yim, M.H. Chang, Z. Zheng, W.T.S. Huck, R.H. Friend, J.S. Kim, L.M. Herz, Phys. Rev. B **115338**, 77 (2008)
128. H. Hlaing, X. Lu, T. Hofmann, K.G. Yager, C.T. Black, B.M. Ocko, ACS Nano **7532**, 5 (2011)
129. Z. Hu, A.M. Jonas, Soft Matter **21**, 6 (2010)
130. G. Ding, Y. Wu, Y. Weng, W. Zhang, Z. Hu, Macromolecules **8638**, 46 (2013)
131. J.H. Lee, D.W. Kim, H. Jang, J.K. Choi, J. Geng, J.W. Jung, S.C. Yoon, H.-T. Jung, Small **2139**, 5 (2009)
132. B.-C. Chen, Y.-S. Cheng, C. Gau, Y.-C. Lee, Thin Solid Films **384**, 564 (2014)
133. S.-I. Na, S.-S. Kim, J. Jo, S.-H. Oh, J. Kim, D.-Y. Kim, Adv. Funct. Mater. **3956**, 18 (2008)
134. J.Y. Park, N.R. Hendricks, K.R. Carter, Langmuir **11251**, 27 (2011)
135. T. Kohei, T. Kazuhiro, I. Manabu, F. Honoka, N. Naoki, K. Hiroaki, H. Yoshihiko, Jpn. J. Appl. Phys. **06GJ03**, 52 (2013)
136. A. David, H. Benisty, C. Weisbuch, Rep. Prog. Phys. **126501**, 75 (2012)
137. Q.-C. Hsu, J.-J. Hsiao, T.-L. Ho, C.-D. Wu, Microelectron. Eng. **178**, 91 (2012)
138. C.-J. Yang, S.-H. Liu, H.-H. Hsieh, C.-C. Liu, T.-Y. Cho, C.-C. Wu, Appl. Phys. Lett. **253508**, 91 (2007)
139. J.P. Yang, Q.Y. Bao, Z.Q. Xu, Y.Q. Li, J.X. Tang, S. Shen, Appl. Phys. Lett. **223303**, 97 (2010)
140. Wei, M.-K.; Su, I.-L., *Opt. Express* **5777**, 12, (2004)
141. D. Suh, H.H. Lee, J. Vac. Sci. Technol. B **1123**, 22 (2004)
142. K. Ishihara, M. Fujita, I. Matsubara, T. Asano, S. Noda, H. Ohata, A. Hirasawa, H. Nakada, N. Shimoji, Appl. Phys. Lett. **111114**, 90 (2007)
143. K. Trivedi, U.S. Bhansali, B. Gnade, W. Hu, Nanotechnology **405204**, 20 (2009)
144. A.O. Altun, S. Jeon, J. Shim, J.-H. Jeong, D.-G. Choi, K.-D. Kim, J.-H. Choi, S.-W. Lee, E.-S. Lee, H.-D. Park, J.R. Youn, J.-J. Kim, Y.-H. Lee, J.-W. Kang, Org. Electron. **711**, 11 (2010)
145. Y.D. Kim, K.-H. Han, S.-J. Park, J.-B. Kim, J.-H. Shin, J.J. Kim, H. Lee, Opt. Lett. **5901**, 39 (2014)
146. P. Melpignano, V. Biondo, S. Sinesi, M.T. Gale, S. Westenhöfer, M. Murgia, S. Caria, R. Zamboni, Appl. Phys. Lett. **153514**, 88 (2006)
147. S. Jeon, J.-W. Kang, H.-D. Park, J.-J. Kim, J.R. Youn, J. Shim, J.-H. Jeong, D.-G. Choi, K.-D. Kim, A.O. Altun, S.-H. Kim, Y.-H. Lee, Appl. Phys. Lett. **223307**, 92 (2008)
148. Y. Luo, C. Wang, L. Wang, Y. Ding, L. Li, B. Wei, J. Zhang, ACS Appl. Mater. Interfaces **10213**, 6 (2014)
149. J. Choi, D. Kim, P.J. Yoo, H.H. Lee, Adv. Mater. **166**, 17 (2005)
150. J.R. Lawrence, P. Andrew, W.L. Barnes, M. Buck, G.A. Turnbull, I.D.W. Samuel, Appl. Phys. Lett. **1955**, 81 (2002)
151. M. Tamborra, M. Striccoli, M.L. Curri, J.A. Alducin, D. Mecerreyes, J.A. Pomposo, N. Kehagias, V. Reboud, C.M. Sotomayor Torres, A. Agostiano, Small **822**, 3 (2007)
152. V. Reboud, N. Kehagias, C.M. Sotomayor Torres, M. Zelsmann, M. Striccoli, M.L. Curri, A. Agostiano, M. Tamborra, M. Fink, F. Reuther, G. Gruetzner, Appl. Phys. Lett. **011115**, 90 (2007)
153. X. Li, G. Yang, J. Chen, R. Zhang, W. Zhou, Y. Liu, J. Zhang, Q. Wang, Surf. Rev. Lett. **367**, 16 (2009)
154. C. Piliego, M. Mazzeo, B. Cortese, R. Cingolani, G. Gigli, Org. Electron. **401**, 9 (2008)

155. J.M. Ziebarth, A.K. Saafir, S. Fan, M.D. McGehee, Adv. Funct. Mater. **451**, 14 (2004)
156. L. Zhou, Q.-D. Ou, J.-D. Chen, S. Shen, J.-X. Tang, Y.-Q. Li, S.-T. Lee, Sci. Rep. **4040**, 4 (2014)
157. L. Zhou, X. Jiang, Y. Li, A. Shi, J. Chen, Q. Ou, H. Liu, J. Tang, ACS Appl. Mater. Interfaces **18139**, 6 (2014)
158. Q.-D. Ou, L. Zhou, Y.-Q. Li, S. Shen, J.-D. Chen, C. Li, Q.-K. Wang, S.-T. Lee, J.-X. Tang, Adv. Funct. Mater. **24**, 7249–7256 (2014)
159. H. Kogelnik, C.V. Shank, Appl. Phys. Lett. **152**, 18 (1971)
160. M. Berggren, A. Dodabalapur, R.E. Slusher, A. Timko, O. Nalamasu, Appl. Phys. Lett. **410**, 72 (1998)
161. G.A. Turnbull, P. Andrew, M.J. Jory, W.L. Barnes, I.D.W. Samuel, Phys. Rev. B **125122**, 64 (2001)
162. G. Tsiminis, Y. Wang, A.L. Kanibolotsky, A.R. Inigo, P.J. Skabara, I.D.W. Samuel, G.A. Turnbull, Adv. Mater. **2826**, 25 (2013)
163. E.R. Martins, Y. Wang, A.L. Kanibolotsky, P.J. Skabara, G.A. Turnbull, I.D.W. Samuel, Adv. Opt. Mater. **563**, 1 (2013)
164. Y. Wang, G. Tsiminis, A.L. Kanibolotsky, P.J. Skabara, I.D.W. Samuel, G.A. Turnbull, Opt. Exp. **14362**, 21 (2013)
165. C. Vannahme, F. Maier-Flaig, U. Lemmer, A. Kristensen, Lab Chip **2675**, 13 (2013)
166. X. Liu, S. Prinz, H. Besser, W. Pfleging, M. Wissmann, C. Vannahme, M. Guttmann, T. Mappes, S. Koeber, C. Koos, U. Lemmer, Faraday Discuss **174**, 153–164 (2014)
167. M. Punke, T. Woggon, M. Stroisch, B. Ebenhoch, U. Geyer, C. Karnutsch, M. Gerken, U. Lemmer, M. Bruendel, J. Wang, T. Weimann, Proc. SPIE **6659**, 665909 (2007)
168. Y. Chen, Z. Li, Z. Zhang, D. Psaltis, A. Scherer, Appl. Phys. Lett. **051109**, 91 (2007)
169. A. Retolaza, A. Juarros, D. Otaduy, S. Merino, V. Navarro-Fuster, M.G. Ramírez, P.G. Boj, J.A. Quintana, J.M. Villalvilla, M.A. Díaz-García, Microelectron. Eng. **52**, 114 (2014)
170. E.B. Namdas, M. Tong, P. Ledochowitsch, S.R. Mednick, J.D. Yuen, D. Moses, A.J. Heeger, Adv. Mater. **799**, 21 (2009)
171. C. Ge, M. Lu, X. Jian, Y. Tan, B.T. Cunningham, Opt. Exp. **12980**, 18 (2010)
172. B.C. Mads, B. Thomas, L.C.S. Cameron, R.P. Sidsel, M.J. Mette, K. Anders, J. Micromech. Microeng. **115025**, 20 (2010)
173. J.P. Dowling, M. Scalora, M.J. Bloemer, C.M. Bowden, J. Appl. Phys. **1896**, 75 (1994)
174. V. Reboud, J. Romero-Vivas, P. Lovera, N. Kehagias, T. Kehoe, G. Redmond, C.M. Sotomayor Torres, Appl. Phys. Lett. **073101**, 102 (2013)
175. V. Reboud, N. Kehagias, P. Lovera, M. Zelsmann, C. Schuster, F. Reuther, G. Gruetzner, G. Redmond, C.M.S. Torres, Jpn. J. Appl. Phys. **5139**, 47 (2008)
176. D. Pisignano, L. Persano, P. Visconti, R. Cingolani, G. Gigli, G. Barbarella, L. Favaretto, Appl. Phys. Lett. **2545**, 83 (2003)
177. E. Mele, F. Di Benedetto, L. Persano, R. Cingolani, D. Pisignano, Nano Lett. **1915**, 5 (2005)
178. E. Mele, A. Camposeo, R. Stabile, P. Del Carro, F. Di Benedetto, L. Persano, R. Cingolani, D. Pisignano, Appl. Phys. Lett. **131109**, 89 (2006)
179. L. Persano, A. Camposeo, P.D. Carro, V. Fasano, M. Moffa, R. Manco, S. D'Agostino, D. Pisignano, Adv. Mater. **6660**, 26 (2014)
180. M. Salerno, G. Gigli, M. Zavelani-Rossi, S. Perissinotto, G. Lanzani, Appl. Phys. Lett. **111110**, 90 (2007)
181. J.R. Lawrence, G.A. Turnbull, I.D.W. Samuel, Appl. Phys. Lett. **4023**, 82 (2003)
182. F. Bernal Arango, M.B. Christiansen, M. Gersborg-Hansen, A. Kristensen, Appl. Phys. Lett. **223503**, 91 (2007)
183. M. Ben-Moshe, V.L. Alexeev, S.A. Asher, Anal. Chem. **5149**, 78 (2006)
184. X. Xu, A.V. Goponenko, S.A. Asher, J. Am. Chem. Soc. **3113**, 130 (2008)
185. P.Y. Li, L. Bo, J. Gerstenmaier, B.T. Cunningham, in *Sensors, 2003. Proceedings of IEEE 2003,* vol. 1, p. 310
186. I.D. Block, L.L. Chan, B.T. Cunningham, Sens. Actuators B **187**, 120 (2006)

187. T. Endo, S. Ozawa, N. Okuda, Y. Yanagida, S. Tanaka, T. Hatsuzawa, Sens. Actuators B **269**, 148 (2010)
188. F. Morhard, J. Pipper, R. Dahint, M. Grunze, Sens. Actuators B **232**, 70 (2000)
189. A. Rose, Z. Zhu, C.F. Madigan, T.M. Swager, V. Bulovic, Nature **876**, 434 (2005)
190. A.M. Haughey, B. Guilhabert, A.L. Kanibolotsky, P.J. Skabara, G.A. Burley, M.D. Dawson, N. Laurand, Sens. Actuators B **132**, 185 (2013)
191. M. Lu, S.S. Choi, C.J. Wagner, J.G. Eden, B.T. Cunningham, Appl. Phys. Lett. **261502**, 92 (2008)
192. M. Lu, S.S. Choi, U. Irfan, B.T. Cunningham, Appl. Phys. Lett. **111113**, 93 (2008)
193. M.B. Christiansen, J.M. Lopacinska, M.H. Jakobsen, N.A. Mortensen, M. Dufva, A. Kristensen, Opt. Exp. **2722**, 17 (2009)
194. C. Vannahme, S. Klinkhammer, U. Lemmer, T. Mappes, Opt. Express **8179**, 19 (2011)
195. http://www.renishawdiagnostics.com/en/eu-project-delivers-new-possibilities-for-sers-sensors--24329. Retrieved on 15 November 2014
196. S.Z. Oo, R.Y. Chen, S. Siitonen, V. Kontturi, D.A. Eustace, J. Tuominen, S. Aikio, M.D.B. Charlton, Opt. Exp. **18484**, 21 (2013)
197. P. Karioja, J. Hiltunen, S. Aikio, T. Alajoki, J. Tuominen, M. Hiltunen, S. Siitonen, V. Kontturi, K. Böhlen, R. Hauser, in *SPIE Photonics Europe, International Society for Optics and Photonics: 2014*, p. 91410D

Fabrication of Photonic Crystals Using Holographic Lithography

Lijun Wu, Yi Xu, and Kam Sing Wong

Abstract Fabrication of periodical micro- and nano-structures with functional defects is one of the key issues in photonic crystals and metamaterials research fields. Low cost, fast, and simple fabrication techniques capable of producing photonic structures in a large scale are crucial to realize their potential applications. In this chapter, we will firstly provide an overview of recent research activities in the fabrication of photonic crystals especially by holographic lithography technique. Secondly, we will emphasize on discussing the principle, advantage, and experimental processes of the single-prism based holographic lithography method, where the refractive prism compacts and simplifies the optical experimental setup since it can split and recombine multiple beams simultaneously. Specific examples from recent literature will be illustrated. We will also demonstrate how to generate functional defects by modulating the phase of the laser beam in one-, two-, and three-dimensional periodic structures in a single exposure holographic process. Lastly, we will concentrate on discussing the application of the templates fabricated by the holographic lithography technique, such as inversing the templates by infiltrating functional materials, or materials with higher refractive index than polymers to obtain larger photonic band gap.

Keywords 3D photonic crystals • Defects • Growth • Holographic lithography • Nano-/micro-structures • Photonic crystals • Single refraction prism

---

L. Wu (✉)
Guangdong Provincial Key Laboratory of Nanophotonic Functional Materials and Devices, School for Information and Optoelectronic Science and Engineering, South China Normal University, Guangzhou 510006, P.R. China
e-mail: ljwu@scnu.edu.cn

Y. Xu
Department of Electronic Engineering, College of Information Science and Technology, Jinan University, Guangzhou 510632, P.R. China
e-mail: e_chui@qq.com

K.S. Wong (✉)
Department of Physics, Hong Kong University of Science and Technology, Clear Water Bay, Kowloon, Hong Kong, P.R. China
e-mail: phkswong@ust.hk

D. Comoretto (ed.), *Organic and Hybrid Photonic Crystals*,
DOI 10.1007/978-3-319-16580-6_10

1 Introduction

A photonic crystal (PhC) is an artificial periodic dielectric structure that controls photons just as a semiconductor crystal controls electrons. Since its first introduction by Yablonovitch and John in 1987 [1, 2], PhC has attracted enormous interest for the past decades because it offers unprecedented opportunities for the miniaturization and integration of optical devices. PhC also exhibits a variety of new physical phenomena, including the suppression or enhancement of spontaneous emission [3, 4], low-threshold lasing [5], and quantum information processing [6], as have been addressed in 17th and 21st chapters in this book.

One of the most important experimental studies on PhCs has been addressed to the methods for fabricating complex micro/nano-structures, especially three-dimensional (3D) structures in near infrared and visible spectral ranges because they require dielectric unit cells having sub-micron sizes. Various techniques such as electron-beam lithography (EBL) [7], self-assembly [8], multiphoton polymerization (laser direct writing) [9], multilayer stacking of woodpile structures using semiconductor fabrication processes [10], and holographic lithography [11] have been proposed and demonstrated with different levels of success. EBL can control the feature size and shape accurately but it is normally applied to fabricating one-dimensional (1D) or two-dimensional (2D) structures. Obtaining 3D structures requires multiple stacking by precise alignment between different layers. Furthermore, it is subject to high cost and time consuming, which is similar to the laser direct writing method since the structures are generated by scanning the laser beam or electron beam point by point. Self-assembly is cheap and convenient. But it is limited to face-centered-cubic (*fcc*) or hexagonal-close-packed (*hcp*) structures with unavoidable structural defects (for more information, please refer to second and sixth chapters in this book).

Holographic lithography (HL) (or the so-called multi-beam interference lithography) was first implemented by Berger et al. to fabricate two-dimensional (2D) hexagonal periodic patterns in a photosensitive polymer, which subsequently served as an etching mask for a high-index GaAs substrate [12]. This technique was extended by Campbell et al. [11] by introducing an additional laser beam to create 3D structures with *fcc*-like symmetry. In the same year, Shoji et al. published another paper demonstrating a two-step approach to forming a simple hexagonal 3D PhC by superposing a 2D lattice with a 1D reflection grating [13].

Since it is able to produce defect-free nanometer-scale uniform periodic structures over a large area and volume in a single step fabrication, HL technique is a very economical and powerful tool and might hold the key to volume production of photonic structures. Many efforts have been devoted to developing this method to fabricate 2D and 3D micro/nano-structures. Basically, by controlling the beam propagation directions, the number of the interfering beams, the beam intensities, the polarizations of each beam, respectively, periodic patterns of all 14 Bravais lattices [14] can be produced by HL. Furthermore, HL can be applied to generating diamond [15, 16] and woodpile structures [17], which have been intensively studied and highly desired as architectures for 3D PhCs due to their wide photonic bandgaps (PBGs) [18, 19]. In addition, HL can produce compound lattices [20–23], quasicrystals [24–26], and even spiral architectures [27].

There are several ways to obtain multiple beams to form interference pattern in HL technique. In the early stage of applying HL to fabricate periodic micro/nano-structures, the typical and most-often applied method was utilizing two independent optical elements to split and superpose beams [11, 28]. Another scheme was to apply only two interfering beams but expose the sample more than once by rotating the sample step by step [29, 30].

In order to alleviate the burden of adjusting optical setup during the experiment, a single optical element such as a phase mask [31], a reflective optical element [32], or a single prism [33, 34] was used. The single prism strategy has attracted a wide range of interests because of its special properties. Based on this strategy, the splitting and recombining of an incoming laser beam using the same optical element can be enabled simultaneously and thus the complexity of the optical setup can be simplified greatly.

Functional defects in photonic crystals can not only change their physical properties but also offer new possibilities for designing devices with new functionalities [35]. Fabricating periodic micro/nano-structures with designed defects are thus very important for many applications in photonic crystals. Several techniques have been proposed to create functional defects into periodic structures. For example, direct laser writing has been applied to inscribing defects with arbitrary shapes into the precise positions in the PhCs formed by HL [36]. However, this technique requires a complicated second step in which an accuracy alignment is necessary in the manufacturing process. Therefore, being able to create functional defects into periodic structures in one step is becoming another important progress in the development of HL technique.

As the refractive index of the materials sensitive to the UV or visible wavelength range is normally about 1.5, the contrast is not high enough to obtain full PBG. Transferring higher refractive index materials into the templates created by HL is also a very important research direction in fabricating PhCs by HL method.

In this chapter, we will first discuss the principle, advantage, and experimental processes of the single-prism based HL method. Specific examples from recent literature will be addressed. Then we will demonstrate how to generate functional defects by modulating the phase of the laser beam in 1D, 2D, and 3D periodic structures in a single exposure holographic process. Lastly, we will concentrate on discussing the application of the templates fabricated by the HL technique.

2 Multi-Beam Interference

When multiple coherent beams interfere together, the spatial and temporal intensity profile (I) can be determined by the following equation:

$$I\left(\vec{r},t\right) = \left|\vec{E}_1\left(\vec{r},t\right) + \vec{E}_2\left(\vec{r},t\right) + \cdots + \vec{E}_n\left(\vec{r},t\right)\right|^2 \tag{1}$$

$$\vec{E}_i\left(\vec{r},t\right) = A_i\,\vec{\mathrm{e}}_i e^{i\left(\vec{k}_i \bullet \vec{r} - \omega t + \phi_i\right)} \tag{2}$$

where n represents the number of beams involved in the interference. A_i, $\vec{k}_i$, ϕ_i stand for the amplitude, wave vector, and initial phase of the ith plane wave. $\vec{e}_i$ is the unit vector of the electric field and $\vec{r}$ the spatial vector. As the response time of the recording medium (commercial photoresist) is much larger than the oscillation period of the irradiated source, we can thus obtain the time-independent distribution of the interference as following:

$$I(r) = \sum_{i=1:n}\left|\vec{E}_i\right|^2 + \sum_{l\neq m}^{n} A_l A_m \vec{e}_l \cdot \vec{e}_m \cos\left[\left(\vec{k}_l - \vec{k}_m\right)\vec{r} + (\phi_l - \phi_m)\right] \tag{3}$$

As can be seen, the intensity of the resulted interference pattern consists of a constant background part [the first part in Eq. 3] and a modulated cosine part. The contrast of the pattern depends on the parameters of the interfering beams involved. The amplitude, the electric vector, and the phase of each beam could be used to optimize the contrast of the interference. More importantly, the spatial period of the interference pattern is determined by $2\pi/\left(\vec{k}_l - \vec{k}_m\right)$. Therefore, the difference of k vectors could provide a basic framework determining the rotational and translational symmetry of the interference pattern. The polarization and phase of each beam can affect the morphology of the unit cell. For example, these two parameters can be used to break the degeneracy of the compound lattice produced by the subset of different vectors from the multiple beams and engineer the morphology of the unit cell [37]. By arranging multiple beams properly and optimizing the polarization of each beams, all five 2D and all fourteen 3D Bravais lattices with controllable shape of the unit cell can be fabricated [14, 38].

There are several ways to generate multiple beams from one irradiation source. Generally, they can be categorized into amplitude- and wavefront-splitting methodologies.

In the early stage of developing HL technique, the amplitude-splitting method was most often applied to producing multiple beams through the use of dielectric beam splitters [11] or diffractive elements [13]. They are superposed at the exposure area [11] or through another specially designed prism [28], as shown in Fig. 1a. Figure 1b

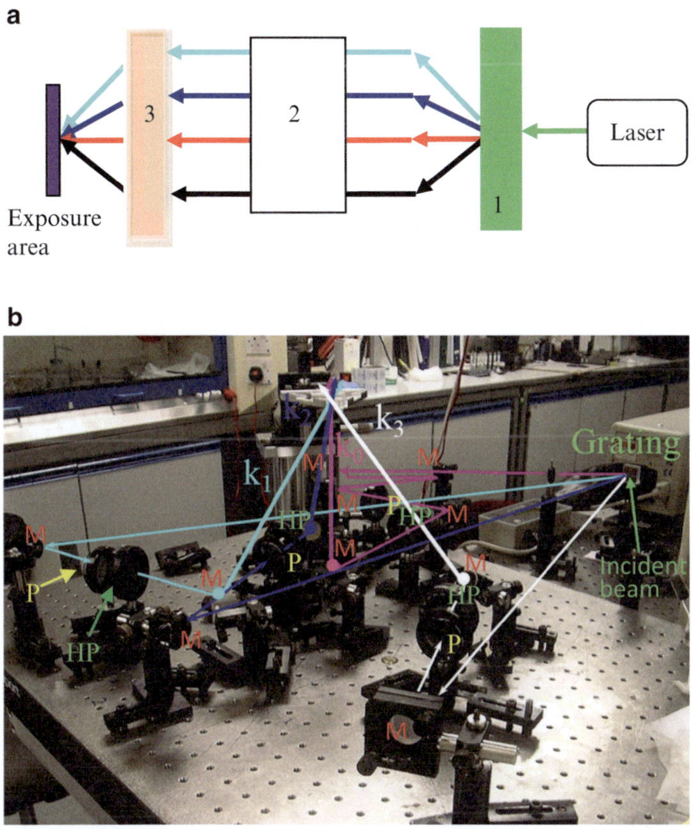

Fig. 1 (**a**) A schematic of the optical setup for four non-coplanar beams. (1) The optical element such as a grating splitting the laser output into multiple beams; (2) The optical elements such as mirrors, half-wave plates, and polarizers steering the beams; (3) The optical elements superposing the beams. (**b**) A typical optical setup for four non-coplanar beams with umbrella-like arrangement. A grating and ten mirrors are used to spatially split and recombine four beams. *M* mirror, *HP* half-wave plate, *P* polarizer

demonstrates a typical optical configuration for four non-coplanar beams with umbrella-like arrangement. From this figure, it can be seen this fabrication strategy can introduce alignment complexity and inaccuracies due to differences in the optical path length and angles among the interfering beams as well as vibrational instabilities.

In order to alleviate the burden of adjusting optical setup and improve the interferometric stability, wavefront-splitting schemes have been developed which divide portions of a single expanded and collimated source beam into multiple beams. The first wavefront-splitting scheme incorporates a Lloyd's mirror to reflect a part of an expanded source to intersect with the transmitted part [39]. However, if one wants to obtain more complex pattern, multiple exposures are required. A recent report shows how to generate three beams with 120° symmetry and produce hexagonal patterns by this method [40].

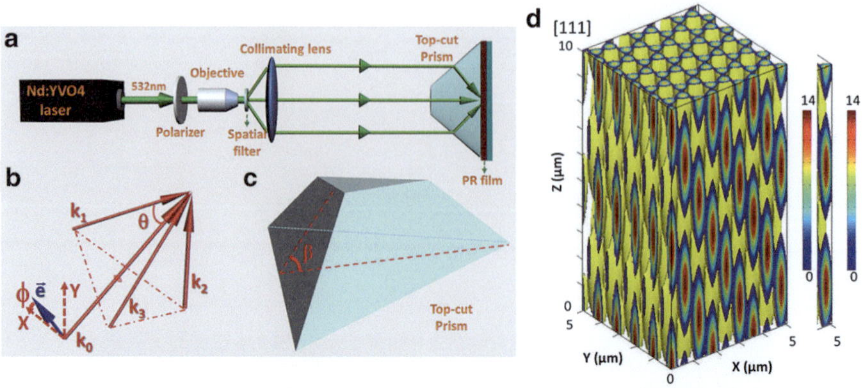

Fig. 2 (**a**) A schematic of the optical setup used for the single-prism based HL. (**b**) The specially designed refracting prism with a cutting angle β determined by the structure. (**c**) A typical geometry for four beams interference, corresponding to a 3D *fcc*-type structure. (**d**) Numerical modeling result of the total exposure in the photoresist using the prism shown in (**c**). The color scale is with arbitrary unit where the amplitudes of $A_1 - A_4$ are set to be 1

A more common and flexible wavefront-splitting strategy is to apply a specially designed prism to split the single expanded and collimated beam into multiple beams [24, 34]. Figure 2a depicts the optical configuration of this strategy. An irradiation laser beam is expanded after entering into an objective and passing a spatial filter. Then it is collimated by a collimating lens. The size of the expanded beam is dependent on the magnification times of the objective and the focal length of the collimating lens. The expanded laser beam shines on the truncated prism and is divided into multiple beams. The same prism is then used to recombine and overlap the generated multiple beams. The interference pattern is determined by the prism. For example, four beams interference with umbrella-like architecture shown in Fig. 2b can be realized by the prism shown in Fig. 2c. Figure 2d presents the simulation result. The shaded regions correspond to all spatial points where the exposure dosage is higher than the polymerized threshold and remain insoluble in the developer.

The single-prism based HL has been proven to be highly stable. Anti-vibration equipment and complicated optical alignment system to adjust the angles between the interfering beams are not required, leading to a very simple and compact optical setup. In the context of mass production, this method is much more practical and robust, and is able to produce PhCs over a larger uniform area than previous demonstrations by two independent-element setups. In addition, this method can be easily extended to generate more complex micro/nano-structures, such as the woodpile lattice [17], by designing the refracting prism properly.

The first demonstration of the single-prism based HL technique fabricating 2D and 3D PhCs is shown in Fig. 3a–c. It can be seen from the SEM pictures that the single-prism based HL is capable of fabricating large area PhCs. Figure 3d presents the transmission and reflection spectrum of the fabricated PhCs. A PBG at around 2.5 μm is clearly visible as a pronounced reflection peak and transmission dip. By changing the irradiation to ultraviolet and designing the refraction prism properly, it

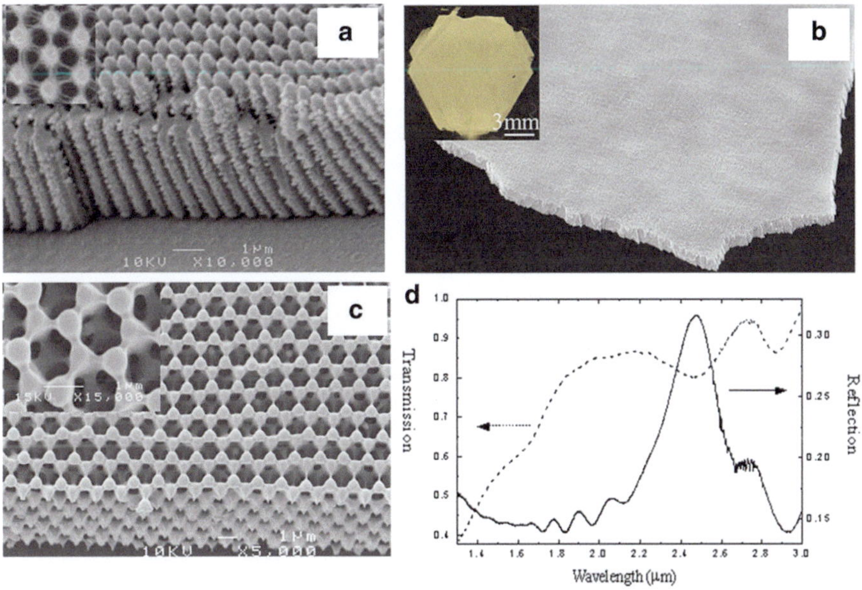

Fig. 3 SEM images of PhCs generated by the prism shown in Fig. 2b. (**a**) 2D hexagonal structures from three side refracted beams. *Top-left inset* is the magnified *top view*. (**b**) and (**c**) 3D *fcc*-type structures from four beams with a small magnification view in (**b**) and a fractured part with (111) in the *top* and its cleavage plane (**c**). A close-up of (111) plane is illustrated in the *top left inset* of (**c**). The *top left* inset in (**b**) is the typical size of the PhCs fabricated. (**d**) Corresponding transmission (*dashed line*) and reflection (*solid line*) spectra of 3D periodic structures shown in (**c**). Reproduced with permission from [34]

is possible to achieve a PBG at the range of the optical communication wavelength (1,460–1,565 nm) [41] and even at visible wavelengths [42] in the (111) direction. The period dimension of the target structure is only limited by the wavelength of the irradiating laser and the resolution of photoresists.

The polarization direction is very important to obtain high contrast interference pattern. It can either be manipulated before the laser beam being expanded [43] or before the refraction prism for each refracted portion [44]. Electromagnetic waves would deflect when they encounter a discontinuous interface of permeability and permittivity. For a plane wave, it can be divided into two types: electric polarization lying in the incident plane and electric polarization normal to such plane. The electric polarization of the first type changes its direction under refraction while the second type keeps its direction. According to Snell's law, two types of incident polarizations have different transmission rates when oblique incident is applied [45]. Therefore, we have to take into account the change of the polarization when the laser beam encounters the prism.

After the demonstration of the capacity for fabricating large area 2D and 3D polymer PhCs using the single refracting prism based HL, numerous advances in experimental fabrication and characterization of micro/nano-structures have been made facilitating by such a simple, compact, and stable setup. These efforts can be classified into two categories: 2D and 3D micro/nano-structures. We will summarize the related recent advances in the following.

2.1 2D Micro/Nano-Structures Fabricated by the Single-Prism Based HL

By blocking the light incident on the top surface of the prism shown in Fig. 2b, we can fabricate large area 2D PhCs. For a prism with sixfold rotational symmetry, we can obtain several types of 2D PhCs [22]. For example, hexagonal PhCs can be fabricated by three symmetrically separated beams with the same intensity, as shown by the simulation and experiment results in Fig. 4a, b. Introducing an auxiliary beam can create hetero-binary structures. The size contrast can be tuned by the intensity ratio between the auxiliary beam and three basic beams, as are shown in Fig. 4c–f. Figure 4g, h demonstrates that honeycomb structures can be realized by using six beams with the same intensity. These results manifest the capacity of fabricating various types of PhCs by the single-prism based HL. By using an obliquely incident laser beam, the shape of the resulted PhCs can be manipulated [46]. Furthermore, tailoring the shape of the prism could produce more sophisticated beam geometries and fabricate more complicated structures. For example, Yang et al. demonstrate that the prism shown at the top of Fig. 5 can

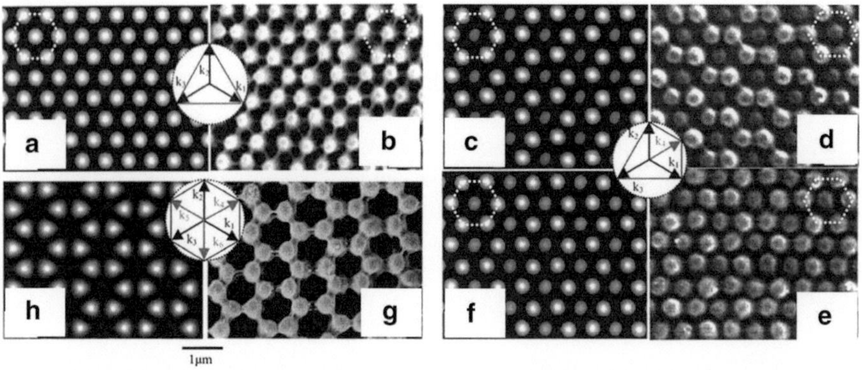

Fig. 4 Calculated intensity distributions on the left-hand side and corresponding SEM images on the right-hand side. The insets in the middle show the beam configurations using in the prism with sixfolded rotation symmetry. (**a**) and (**b**) Regular hexagonal structure from three symmetrically separated beams with the same intensity. (**c**) and (**d**) Hetero-binary structures by introducing an auxiliary beam into the geometry shown in (**a**) and (**b**). The intensity contrast between the auxiliary beam and basic beams is 52 %. The size contrast between the central rod and its six closest neighbors is clearly observed. (**e**) and (**f**) Hetero-binary structures from different intensity contrast 26 %. (**g**) and (**h**) Honeycomb structure by introducing six symmetrically separated beams with the same intensity. Reproduced with permission from [22]

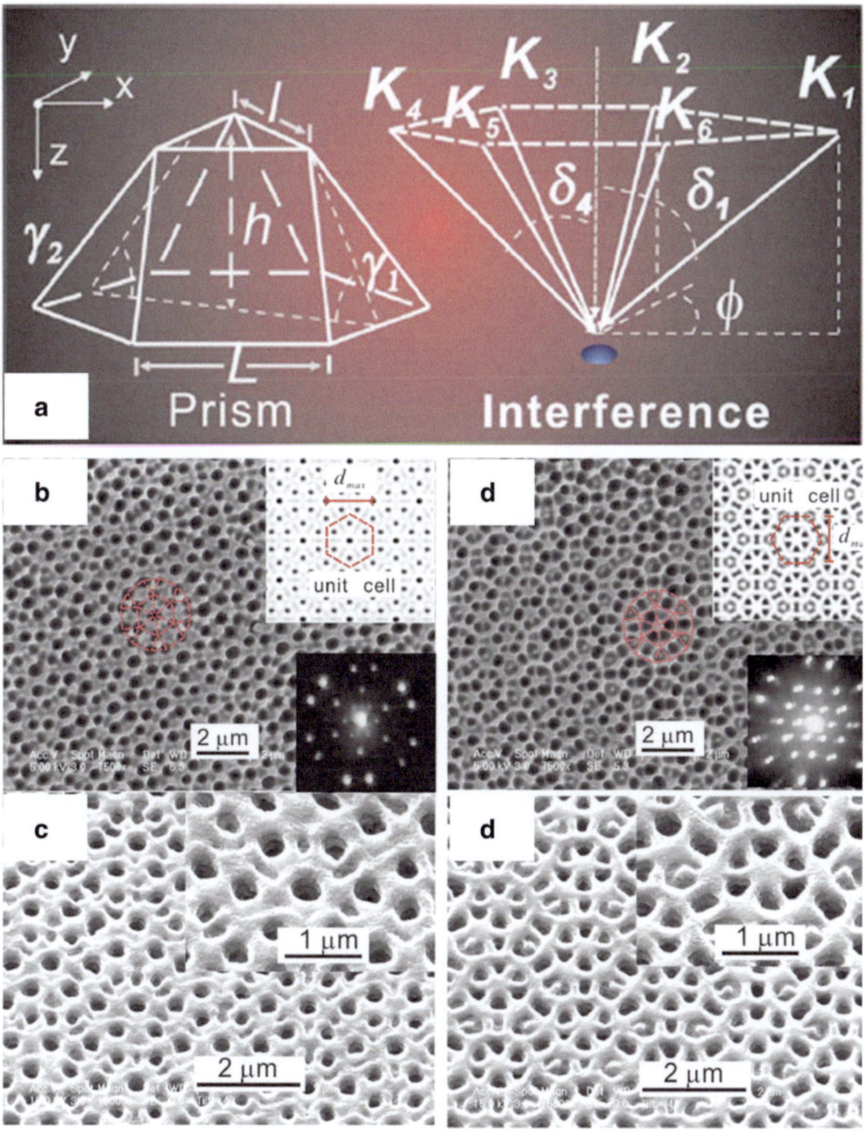

Fig. 5 (**a**) Structure of a hexagonal prism (*left*) producing wavevectors $K_1 - K_6$ (*right*) from the *bottom* of the prism. (**b**) SEM image of a recorded periodic complex PhCs. *Top inset*: Simulation result by $K_1 - K_6$ with phases $\varphi_1 - \varphi_6$ equal to $\varphi + 2q\pi$ in the positive photoresist films. *Bottom inset*: Optical diffraction pattern. (**c**) Oblique view of the SEM image. *Inset*: Magnified view. In each unit cell, there are seven lattice spots (*air holes*). (**d**) Similar to (**a**) except for $K_1 - K_6$ with phases $\varphi_1 = \varphi_3 = \varphi_5 = \varphi + 2q\pi$ but equal to $\varphi_2 = \varphi_4 = \varphi_6 = \varphi + (2q + 1)\pi$. (**e**) Oblique view of the SEM image. *Inset*: Magnified view. In each unit cell, there are six lattice spots (*air holes*). Reproduced with permission from [23]

produce two sets of laser beam with different incident angles respecting to the symmetry axis of the prism [22]. Such beam architecture can be used to produce complex PhCs constructed with a portion of photonic quasicrystals. By manipulating the phases of two sets of laser beams, they proved that the resulted complex PhCs can be further controlled, as are shown in Fig. 5b–e. Details of the beam geometry are outlined in the caption of Fig. 5. These results validate that the single-prism based HL is a stable and powerful technology for the production of large area and uniform 2D micro/nano-structures.

2.2 3D Micro/Nano-Structures Fabricated by Single-Prism Based HL

Fabrication of 3D micro/nano-structures by the amplitude-splitting method becomes challenging due to complicated optical setup as well as inevitable disturbance from environments. It becomes more critical when fabricating complex micro/nano-structures, such as woodpile, icosahedral quasicrystals or spiral structures, where more than five beams are required [27, 44, 47, 48]. Fortunately, the single-prism based HL can release the stable requirement of the optical setup. Various complex micro/nano-structures over large area have been produced by this simple while stable method. For example, Pang et al. have demonstrated that woodpile structure can be fabricated by a specially designed prism as shown in the left inset of Fig. 6, where four side beams and one central beam are constructed by reflecting and dividing the expanded incident beam. Four side beams and the central beam are linear and circular polarized, respectively [47]. Figure 6a–d shows the resulted woodpile structures recorded in SU-8. The structure in Fig. 6b is very close to a diamond structure. Due to unavoidable inhomogeneity in the optical setup which causes the random phase shift between all beams, the resulted large area porous structure would somehow differ from the targeted structure, as shown in Fig. 6c, d. Park et al. introduced a phase shift of two counter propagating side beams relative to the others by using a prism with different lateral dimensions, as shown in the left panel of Fig. 6e, f. The difference of the lateral dimensions of the truncated prism introduces a phase delay for beams 2 and 4 relative to beams 1 and 3, while there is no phase delay between beams 2 and 4, and beams 1 and 3 [16]. They have obtained woodpile structure with correct stacking which can be observed in Fig. 6f.

By shaping the prism and adding other components to the prism, the scope of the single-prism HL method may be further broadened. For example, Xu et al. have demonstrated that the icosahedral quasicrystal can be fabricated using a truncated pentagonal prism which produces seven beams (for fivefold symmetry), as shown in Fig. 7a–c [44]. A mirror with index matching oil is used to obtain the counter-propagation laser beam. The resulted PhCs are shown in Fig. 7e–g with the

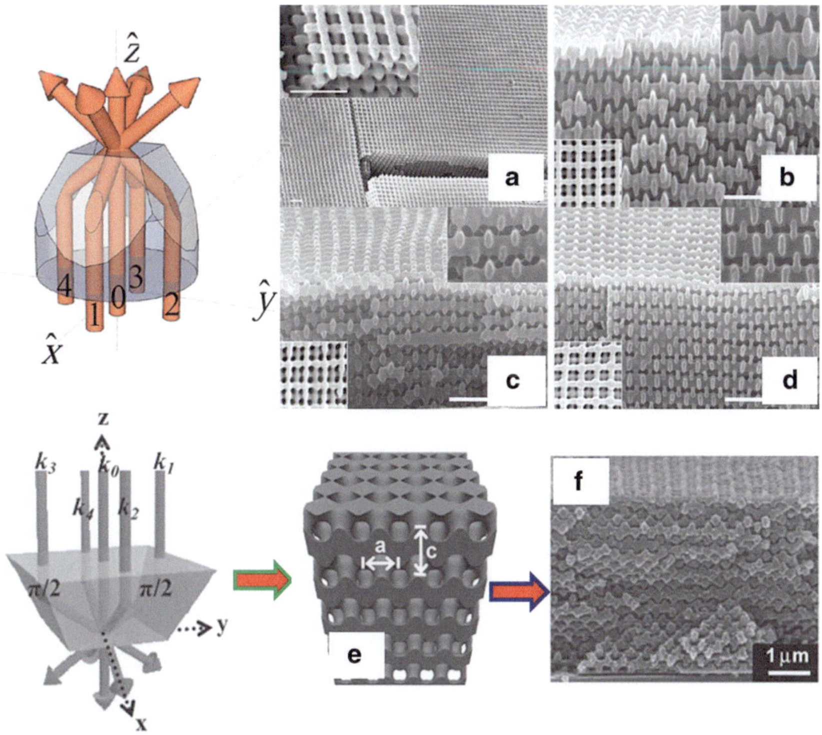

Fig. 6 (4 + 1) beam configuration for the woodpile structures produced by the reflected prisms as shown in the *left panel*. (**a**) A 3D SEM image of the fabricated woodpile structure. The *upper-left inset* shows the zoom-in view of the woodpile structure. (**b**)–(**d**) SEM images for the woodpile structures with lattice ratio (obtained from the *front view insets*) equal to 0.72, 0.88, and 0.82, respectively. (**b**) is very close to the desired diamond structure. (**c**) suffers from slight distortion from (**b**) and (**d**) is the unfavorable result among them. The *upper-right insets* are the zoom-in front views while the *lower-left insets* are the *top views* of the structures. The *white scale bars* are all 1.0 µm. Reproduced with permission from reference [47]. (**e**) shows the simulated iso-intensity distribution of the 3D interference pattern generated by five beams as in the *left panel* with two counterpropagating side beams shifted by $\pi/2$ in phase relative to the others. (**f**) Cross-sectional SEM image of the generated woodpile structures. The lattice constants, a and c are 380 nm and 390 nm, respectively. Reproduced with permission from [17]

simulation results in their left insets. They present fivefold, threefold, and twofold symmetry projections, respectively. By adding a matching prism at the truncated surface of the prism as shown in Figs. 7d and 8, the surface orientation of PhCs could be manipulated [44, 49].

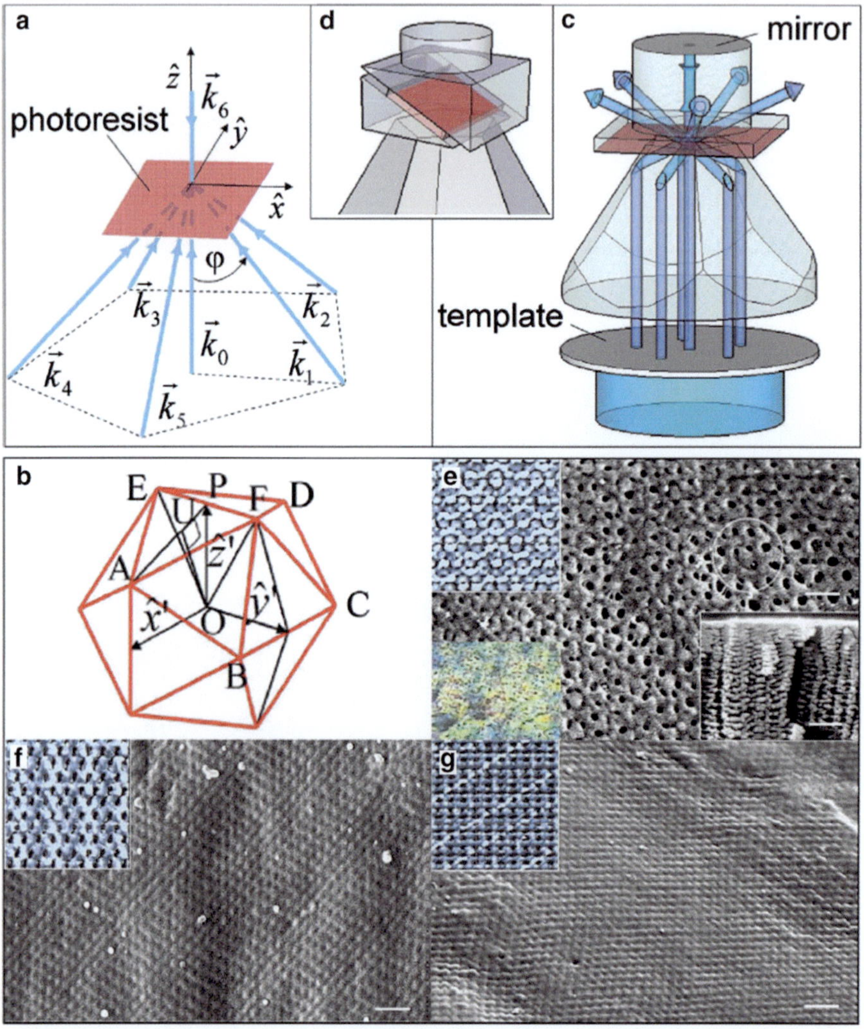

Fig. 7 (**a**) Seven-beam configuration for the icosahedral quasicrystal. (**b**) Icosahedral quasicrystal lattice (*red*) with $\varphi = 63.4°$. (**c**) A truncated pentagonal prism producing seven beams geometry (for fivefold symmetry). (**d**) Setup for obtaining twofold and threefold symmetry projections on the surface of photoresist using a pair of prisms. (**e**) The corresponding SEM image of the fivefold symmetry obtained using the configuration in (**b**). (**f**) SEM image of the threefold symmetry obtained using the configuration in (**c**) with the prism angle 37.4°. (**g**) SEM image of the twofold symmetry obtained using the configuration in (**c**) with the prism angle 31.7°. The *upper-left insets* in (**e**)–(**g**) are the simulated projections of the corresponding symmetries. The circle in (**e**) shows the fivefold symmetry. The *lower-right inset* in (**e**) is the cross-sectional SEM image of the sample. The *lower-left inset*, size 60 μm × 80 μm, in (**e**) is a fivefold normal reflection image. The *scale bars* are all 1 μm. Reproduced with permission from [44]

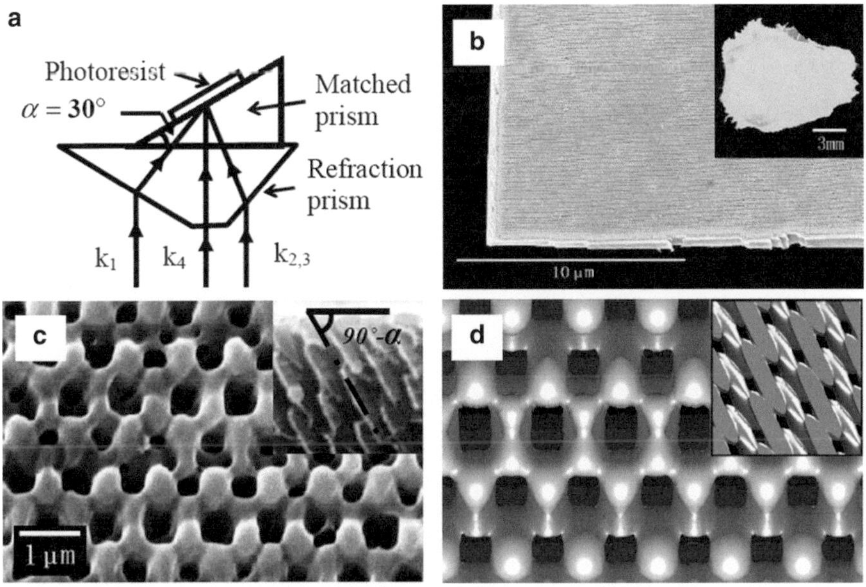

Fig. 8 (**a**) Projection of the beam configuration, reproduced on the k_1 and k_4 plane, to fabricate the $(8\bar{1}\,\bar{1})$-oriented *fcc*-type PhCs. (**b**) and (**c**) show the SEM images of PhC structures generated. The *right-top inset* is the cross-section of the sample, where the *solid line* shows the surface of the sample and the *dash dot line* shows the (111) direction of the sample. (**d**) shows the computer simulation. Reproduced with permission from [49]

3 Multi-Beam Interference Accompanied with Functional Defects

For many applications, accurate introduction of defect such as waveguide or cavity into PhCs is very important; therefore, there has been a paramount need to develop means to incorporate functional defects directly into PhCs, both effectively and accurately [50]. PhCs with defect structures can be made using traditional semiconductor processing technique, but this is a multistep process requiring very precise mask alignment and expensive equipment [51, 52]. Simpler method such as direct EBL to make defect on surface or buried line defect using photolithography and regrowth was demonstrated on opal-based PhCs [53–55]. However, all these techniques suffer from either complex multistep processes and/or lack of control on the precise location of the defect in the lattice.

Multiple-beam HL and single-beam HL such as the method we introduce here have been widely employed to fabricate large size and periodic PhC structures. The major drawback of these HL techniques is that the PhC structure with embedded defect cannot be made simultaneously during the fabrication process. Various techniques exist for introducing functional defects into PhCs generated by HL. One method is to introduce functional defects via laser writing or multiple-exposure [30, 36].

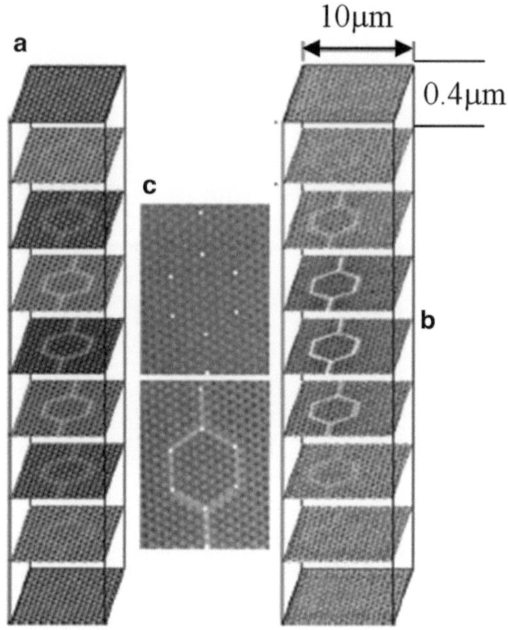

Fig. 9 Latent image of a test device created by sequential holographic and scanning two-photon optical exposure. (**a**) The target structure: the sum of the calculated photoacid densities created in the two-step exposure process. (**b**) Experimentally measured photoacid density: confocal sections are recorded at 0.4 μ intervals. (**c**) Confocal section (*top*) recorded after holographic exposure and used to align the waveguide-writing step and the corresponding section (*bottom*) after waveguide writing. The overlay of *white dots* indicates reference points used to align the test device. Reproduced with permission from [36]

Figure 9 shows the formation of a Mach–Zehnder interferometer by two-photon laser writing into a 3D PhC made by HL [35]. However, this technique introduces a costly and complicated second step to the manufacturing process.

Besides the multi-beam approach, an advanced phase-control HL has also been used to introduce defects into PhCs in a single step [56, 57]. This technique is based on the multi-beam diffraction in a phase mask. As a result, each pixel must be sufficient small to produce effective diffraction, requiring thousands, even millions of pixels for incorporating defects into PhC structures via HL. In this approach, the phase mask needs to have very fine scale in order to produce multiple higher-order diffraction beams. Furthermore, this approach is limited because a particular phase mask is suitable only for one particular structure; and the fabrication of the mask itself may be an extra burden compared with coherent multi-beam interference.

Previous discussion on HL fabrication of PhCs in this chapter is mainly done by manipulating and controlling the polarization, amplitude (A), and wavevector (k) in Eq. (3). One more term we can control is the phase (φ), which gives an additional degree of freedom to design complex optical lattice or periodic lattice embedded with defect structures. Controlling the phase of one or two interfering beams in HL

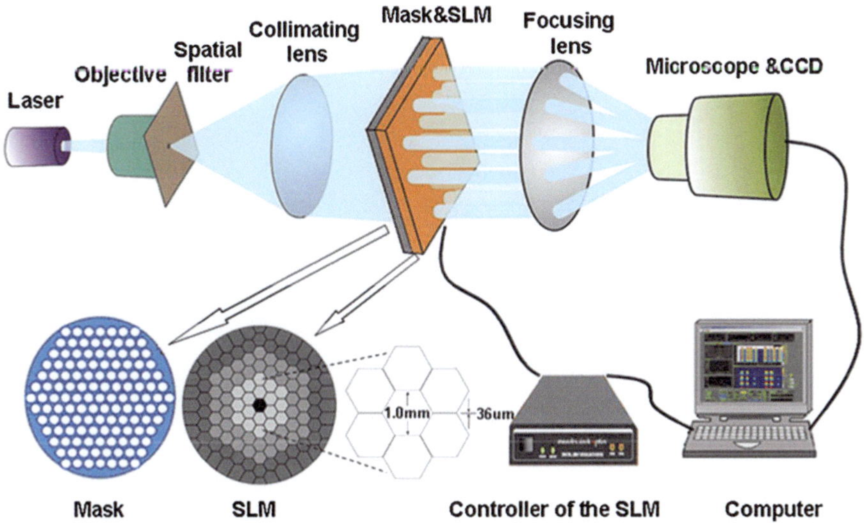

Fig. 10 Experimental setup for adaptive synthesis of intensity distribution with simulated annealing algorithm. Reproduced with permission from [61]

using piezo electric delay, phase retarder, or simply a glass slide to make PhCs or PhCs with line defects has been demonstrated [58–60]. However, for complex PhC structures or defect requiring tens or even hundreds of interfering beamlets, these methods are not practical. One elegant and simple method of simultaneous introduction of functional defects and fabrication of PhCs is using a liquid crystal spatial light modulator (LC-SLM) to dynamically control the phase of individual interfering beamlets in multi-beam HL [61, 62]. The experimental setup for this phase controlled HL is shown in Fig. 10. The main feature of this technique is that the number of beamlets can be increased substantially without compromising the simplicity and the ease of controlling various beams. In combination with genetic algorithm (GA), optimal arrangement for multi-beam interference is obtained to generate a desirable optical lattice, which is applied to producing PhCs with designated functionality [62]. The GA technique [63] is employed to find the minimum number of beamlets and their relative phases for the designated HL interference fringes. Then a mask is made with the correct spatial distribution of interfering light beams as calculated by the GA method and the mask is aligned with LC-SLM so that phase of each beamlet can be controlled independently by the LC-SLM. Activating the LC-CLM pixels to set the appropriate phase at each beamlets will enable one to make the desired optical lattice. Figure 11 shows an embedded waveguide into a triangular 2D PhC lattice created by this phase controlled HL using six interfering beams of various phases. Note that these six interfering beams produce one defect mode for every four lattice periods. The ratio of the defect period to the unit lattice period is determined by the number of interfering beams (i.e., pixels), thus a larger ratio can be produced with a finer LC-SLM with more variable pixels. For example, 20 interfering beams can produce a defect mode in every 8 periods of a 2D square

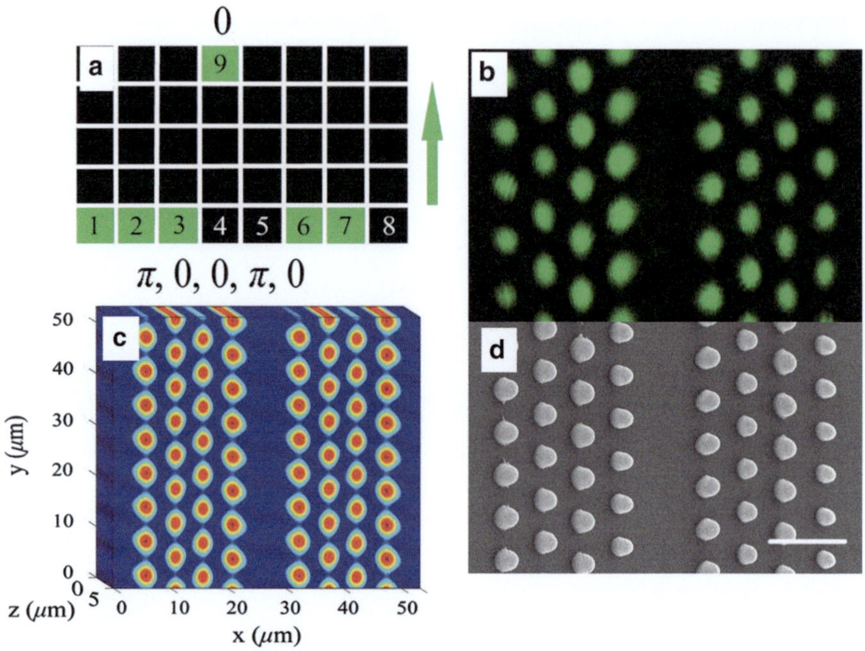

Fig. 11 (**a**) Beam configuration with phases of various beams indicated for formation of line defect in a 2D triangle PhC. (**b**) The simulated light intensity distribution in SU-8 for beam configuration in (**a**). (**c**) The CCD-recorded intensity patterns in air of the optical lattice structure. (**d**) SEM showing the oblique view of the SU-8 structure after exposure of the intensity pattern using a lens of $f = 100$ mm. *Scale bar*: 10 μm. Reproduced with permission from [62]

lattice [62]. For very complex PhCs or defect structures, one needs a large number of interfering beams and thus a more sophisticate method has to be used to find the correct phases for each beam [61]. In this case, a target intensity pattern $I_t(i, j)$ is stored in the computer, the interference intensity pattern ($I_m(i, j)$) is monitored by the CCD and fed back to the computer. The deviation of $I_m(i, j)$ from $I_t(i, j)$ is compared using a fitness evaluation function, an iteration process is used to optimize the phase of each beamlet through LC-SLM until $I_m(i, j)$ matches with $I_t(i, j)$. The detail of the fitness evaluation function and the optimization algorithm used can be found in [61]. Figure 12 shows a zigzag waveguide and a micro-cavity embedded in a 2D PhC fabricated using this technique. An advantage of this adaptive method is that the ideal intensity distribution can be automatically produced even without full knowledge of the SLM response and the phase distortion in the optical system. This so-called self-adaptive annealing algorithm applied to the feedback control to obtain the predetermined intensity distribution is very powerful and will play an important role in future for fabrication of functional photonic devices and microstructures.

In general, introduction of programmable LC-SLM for phase control in multi-beam HL to make PhCs and functional defects has substantially reduced the fabrication complexity. This method has a number of merits over the previous

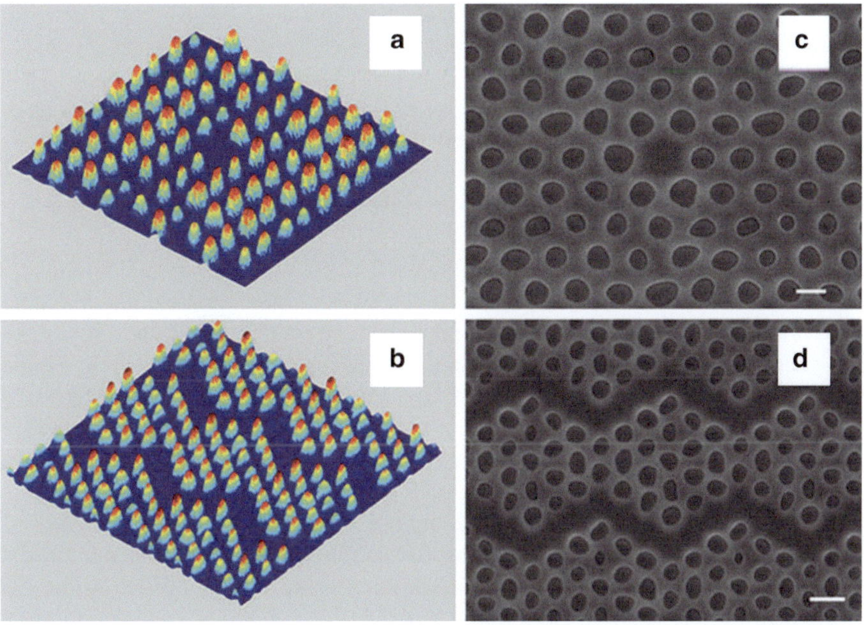

Fig. 12 The final optimized intensity patterns of (**a**) a 2D micro-cavity and (**b**) two bent waveguides. (**c**) (**d**) SEM images of the microstructures of the exposed material. The scale bars in (**c**) and (**d**) are 5 μm and 10 μm, respectively. Reproduced with permission from [61]

employed techniques used to make functional defects in PhCs, for example far fewer beamlets are needed to inset defects and different PhC designs can be quickly tried out using the programmable LC-SLM without the need to go through etching process to make a new phase mask. Indeed, a number of groups now have adopted this LC-SLM for phase control in multi-beam HL to make very complex PhCs and defect structures [64, 65].

4 Recording the Interference Pattern into Photosensitive Materials

To record the interference pattern generated by HL, a film of photosensitive materials is required. The photosensitive materials can be classified into two main categories: positive-tone and negative-tone. For the positive-tone photosensitive medium, the regions exposed to the light intensity higher than the lithographic threshold of the resist are removed in the development process. In the negative-tone resist, they are cross linked and remained. The threshold intensity is determined by the sensitivity of the photoresist and the post-exposure processing. Choosing which type of photosensitive medium is dependent on the application. Positive-tone photoresist such as

AZ1500 and AZ5214 (both from Clariant International Ltd.) has been applied to fabricating diverse metamaterials by the single-prism based HL [66] and 2D hexagonal pattern [67]. The most common photoresist in HL is the negative-tone SU-8 (Microchem Inc.). This photoresist can be as thick as 2 mm and aspect ratios up to 25 have been demonstrated with a standard UV-lithography [68]. It is quite transparent in the near-UV region and highly transparent in the visible region. Thus its properties meet the requirements in HL and we only describe its processing in this chapter.

The commercially available SU-8 photoresist is photosensitive to UV light and consists of three basic components: (1) epoxy Epon SU-8; (2) Gamma-Butyrolactone (GBL) solvent; (3) photoacid generator from the family of triarylium-sulfonium salts. If the irradiation wavelength of the laser is in the visible region, another photoinitiator such as Irgacure 261 (from Ciba Co.) has to be incorporated into the solution to adsorb visible light and produce acid to catalyze the cross linking reaction. More photoinitiators could be found in [67]. The reaction mechanism for cross linking is similar both in the UV and visible range.

The photoresist solution is spun onto a cover glass and the solvent is evaporated by 90–95 °C hard baking to obtain a thick or a thin photoresist resin film, depending on the usage of the structures. During UV exposure, the triaryalsulfonium salts release photoacids in those regions exposed to light intensity higher than the threshold. The subsequent post-exposure bake above the glass transition temperature of SU-8 ($T_g = 50$ °C) can accelerate acid diffusion and induce cationic cross-link of the epoxy groups in SU-8. The specific reaction process is shown in Scheme 1. The unlinked regions are washed away

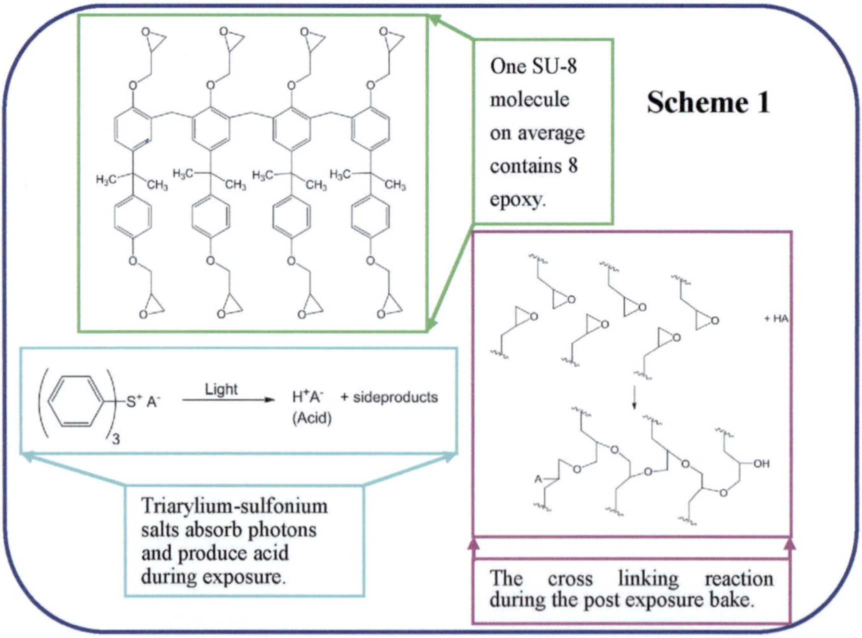

Scheme 1 A scheme describing the SU-8 photoresist and the photosensitive cationic polymerization process

first by propyleneglycolmetheylether acetate and then cleaned with acetone, leaving behind a copy of interference pattern permanently embedded in the polymer film.

As the response of the resist is nonlinear to the light intensity, the feature of the unit cell can be tuned by the exposure time and light intensity. For example, the templates for typical plasmonic bowtie structures [69] have been obtained by controlling the exposure dose [70]. Furthermore, dot-pattern and hole-pattern could both be achieved from the same interference pattern by switching the resist between negative-tone and positive-tone [37].

5 Applications of Templates

Fabrication of large-area periodic micro/nano-structures by HL method has found applications in a lot of research fields. For example, enhancing light extraction from light-emitting diodes (LEDs) [71] or organic light-emitting diodes (OLEDs) [72, 73] is very important to improve the performance of the related devices. Similarly, it can find applications in reducing reflection in solar cells [74]. HL method can also be utilized to incorporate gain medium into corrugated structures and afford distributed feedback (DFB) to produce lasing [75].

Until now, the most documented application of HL has been in the formation of templates for PhCs, especially for 3D structures. For 3D PhCs, full confinement of photons can only be achieved with complete PBGs. Photoresist-based patterns generated by HL obviously do not have enough refractive index contrast to realize this goal. Therefore, other higher refractive index materials are necessary to infiltrate into the templates.

There are several ways of affording templates of PhCs including the self-assemble method and EBL, etc. As we have mentioned in the above context, EBL is time-consuming and expensive. The self-assemble method is cheap and fast, but the crystal structure of the assembled opal is limited to *fcc* or *hcp*. Furthermore, as the opal templates form a close-packed system with a solid filling fraction of 74 %, the maximum achievable filling fraction of frame materials after the inversion process is only 26 %, which would limit the application of the templates such as generating complete PBG. For the templates fabricated by HL, we can control the filling fraction of the sacrificial polymer template by simply tuning the exposure dose. In the following, we will illustrate some typical application examples for templates generated by the HL method.

Templates to be backfilled normally are those structures with high symmetry since they are easy to obtain complete bandgap. *Fcc*-type structures including diamond and woodpile are first candidates. Among them, woodpile structure seems the best candidate because it can exhibit a complete PBG with a relatively low minimum refractive index contrast of about 1.9. Furthermore, it can be easily fabricated compared to diamond structures [17]. For the materials infiltrated into templates, the common requirements are high refractive index, no absorption at the application wavelength range and easy to be infiltrated. Infiltration could be accomplished either by a dry process, including the dry chemical vapor deposition (CVD), the atomic layer deposition (ALD) and melting, or wet chemical methods, such as the liquid phase, the sol–gel reaction, the electrochemical reaction, etc.

Silicon is one of the most attractive semiconductors because of its particular properties. Its refractive index is as high as 3.4 in the infrared wavelength range. Therefore, it is the first candidate people would like to use in the application of PhCs for obtaining complete PBG. Unfortunately, silicon deposition requires a high processing temperature which will decompose the polymer template, thus we cannot inverse Si directly. Normally, two steps of infiltration are necessary. The first step is to back-fill SiO_2 [76] or Al_2O_3 [77, 78] into the polymer templates by the CVD or ALD method. Then Si is infiltrated by a CVD process.

Titania (TiO_2) is one of the most popular materials to be infiltrated into various kinds of templates created by the HL [79–81] or self-assembly [82–84] method. It combines a high refractive index, which is about 2.6 at 500 nm wavelength for anatase [85] with a large electronic band gap of 3 eV and it can be made with high purity. King et al. [80] utilized the HL method to create PhCs with space group *R32*. Then amorphous TiO_2 was deposited into the SU-8 template, which uses alternating pulses of $TiCl_4$ and water vapor. Both of these two stages are exothermic and proceed rapidly below 100 °C, a temperature that is compatible with the polymer template. At last, the top layer of TiO_2 was removed by argon-ion milling to expose a section through the polymer template and followed by oxygen plasma etching the SU-8 template. Figure 13a, b shows the SEM images of the resulting structures. The peak reflectivity was increased from 20 % for the template (Fig. 13c) to 65 % for the TiO_2/air structure (Fig. 13d), which provides evidence that the inverted PhCs owns high optical quality.

Compared to dry processes, wet chemical methods, such as electrodepositions and sol–gel reactions are more attractive for their simplicity, cost-effectiveness, and ease of fabrication. Based on the single-prism HL method, Park et al. [79] created 3D SU-8 polymer templates with the *fcc* symmetry and then applied sol–gel chemistry to obtain 3D TiO_2 inverse structures, as shown in Fig. 14a, b. As calcination is required to form TiO_2 in the sol–gel process, the degree of film shrinkage was as large as ~34 % and the long-range structural order is lost. The advantageous of this method is that the maximum number of TiO_2 layers depends on the number of SU-8 templates, not the TiO_2 sol–gel reaction. Therefore, the sol–gel process can produce thicker inverse structures.

As the deposition takes place preferentially at the bottom of the structure, materials can be completely filled into the polymer template by the electrodeposition method. Furthermore, the electrodeposition occurs at room temperature from water-based electrolytes. Therefore, the electrodeposition method has been utilized for infiltration widely. For example, anatase TiO_2 was backfilled into a diamond-like SU-8 template generated by four umbrella-like visible laser beams

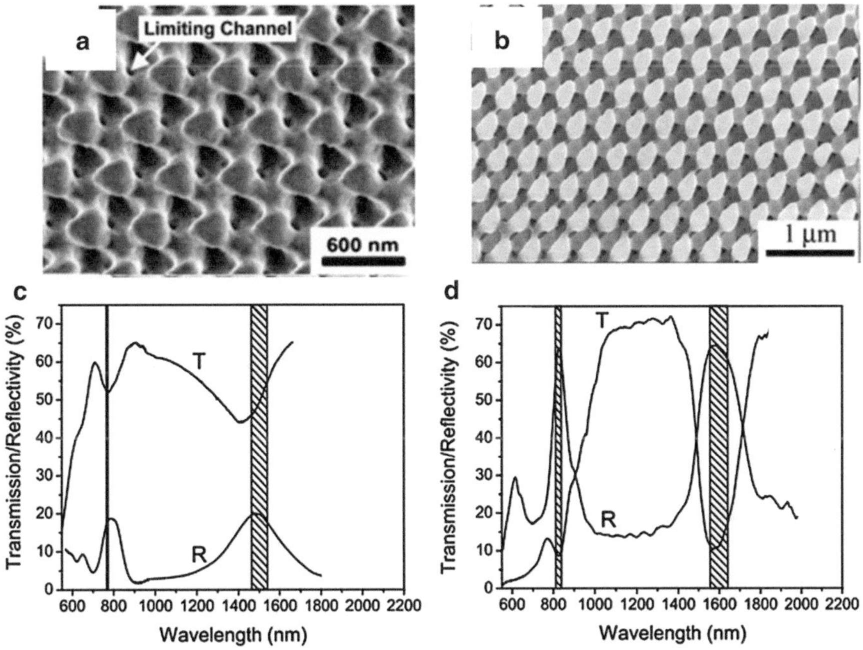

Fig. 13 (111) surface of (**a**) holographically generated SU-8 template and (**b**) TiO_2 inverse structures. Microreflectivity (*R*) and microtransmission (*T*) for (**c**) the polymer template and (**d**) a TiO_2 photonic crystal after the template had been removed. The *shaded bands* indicate the calculated position and width of the fundamental and second order (111) photonic bandgaps. Reproduced with permission from [80]

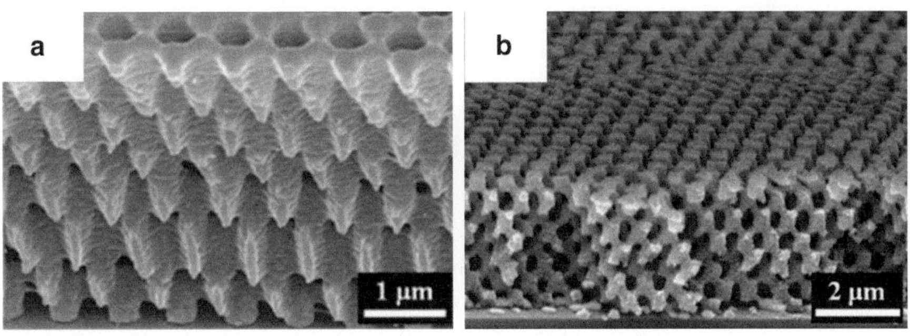

Fig. 14 SEM images of the (**a**) SU-8 template and (**b**) TiO_2 structures inversed by the sol–gel method. Reproduced with permission from [79]

[81]. They conquered the shrinkage problem during calcination by combining a preannealing step and a slow heating rate. Figure 15a, b demonstrates the template and inversed TiO_2 structures. The corresponding measured peak reflectivity shown in Fig. 15c, d increases from 24 % for the SU-8 template to 76 % for the inversed TiO_2 (anatase) PhCs, indicating increased optical quality of the film.

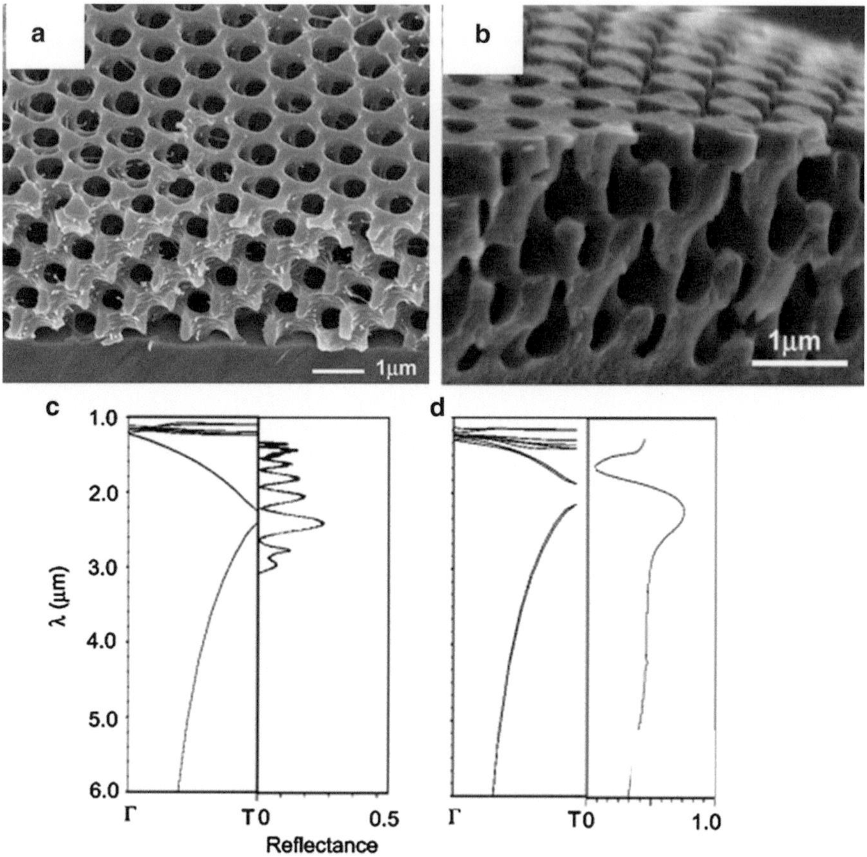

Fig. 15 Cross-sectional view of SEM images of the (**a**) SU-8 template and (**b**) inverse 3D titania crystals after preannealing and calcination. (**c**) and (**d**) FTIR reflectance spectra of 3D diamond-like structures in the (111) direction for SU-8/inverse TiO_2. In each panel, the *left figure* is a simulated spectrum, and the *right one* is the experimental data. Reproduced with permission from [81]

Electrodeposition of ZnO has also been used to fill into the SU-8 template made by the HL method [86]. In this case, strong enhancement in the spontaneous emission rate and reduction in lasing threshold was observed in the ZnO reversed PhC.

Cu_2O is another appropriate material to be infiltrated because it also has high refractive index of ~2.6 and is transparent at wavelengths longer than ~600 nm. Furthermore, dense crystalline Cu_2O can be directly electrodeposited from aqueous solution without the high temperature process. Miyake et al. [87] and Park et al. [16] fabricated 3D patterns with *fcc*-like symmetry and woodpile structures in SU-8 via HL, respectively. Then they used the electrodeposition method to back-fill Cu_2O into the templates. Cu_2O grew through the cavity of the template from the ITO substrate and completely covers the external surface of the template in 70 min. After the electrodeposition, the overlayer of Cu_2O was polished off and finally the

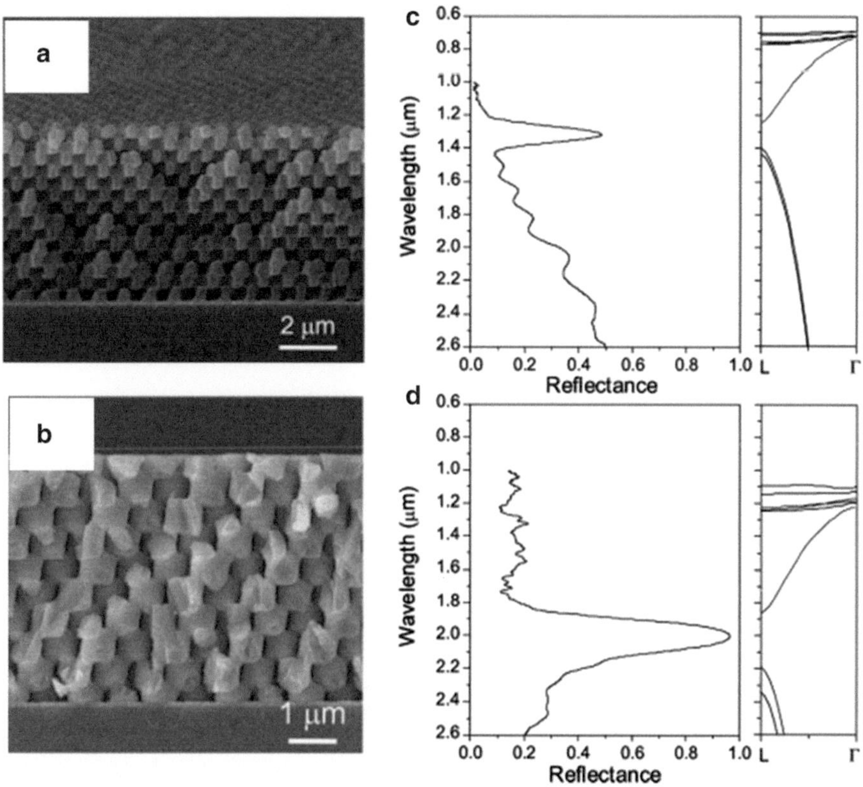

Fig. 16 SEM images of fractured cross-section of (**a**) the SU-8 template and (**b**) inverse 3D Cu_2O photonic crystals after removal of the polymer template by RIE. (**c**) and (**d**) Reflection spectra and calculated photonic-band structure in (111) direction for polymer template/inverse Cu_2O photonic crystal. Reproduced with permission from [87]

polymer template was removed by reactive-ion etching, resulting in a Cu_2O inverse structure. The template and inverse structure is shown in Fig. 16a, b. Due to the low temperature inversion process, the resultant Cu_2O structure matches closely with polymer templates. Furthermore, as shown in Fig. 16c, d, the peak reflection increased from 50 % to 100 % after inversion, providing evidence for complete inversion with high quality of the resultant PhCs. In the case of woodpile PhCs, a complete PBG of 5.3 % in the visible range has been achieved [16].

Besides the normally used SU-8 and other polymer-based photoresist, there are other medium periodic structures can be recorded into. For example, metallodielectric quasiperiodic microstructures have been recorded into silver halide gelatin emulsion holographic plates with a high resolution [24]. Hydrogel with hydrophilicity and biocompatibility has been applied to recording thermoresponsive 3D PhCs [88]. Ma et al. [89] obtained wide band gap photonic planar structures in dichromate gelatin emulsions because of their self-supporting and swelling properties.

Although the HL method can generate perfect periodic and quasiperiodic structures, it can also find other applications which are irrelevant with PBG effects. In those cases, periodicity is not necessary. For example, fully open 3D hydrogel networks with high surface area have been generated by HL on neural prosthetic devices. It can enhance the performance of the device by serving as a mechanical buffer, improving the impedance and delivering neurotrophins [90]. Small pores generated in SU-8 by HL have also been utilized to store Ag nanoparticles to form antibacterial nanocomposition [91].

6 Conclusions

HL is one of the most important techniques for fabricating periodic and quasiperiodic micro/nano-structures over large area. Among the methods producing multiple beams for interference, a single-prism based HL has attracted much interests because the refracting/reflecting prism can split and recombine multiple beams simultaneously, which can compact and simplify the optical experimental setup. Functional defects with accurate position embedded in the periodic and quasiperiodic structures can extend their applications in PhCs. In this chapter, we introduced the principle, advantages, experimental process, and applications of the single-prism HL. Furthermore, we demonstrated a HL-based method to create functional defects and periodic structures in a single step. Lastly, we illustrated some applications of the templates created by HL. The tremendous progress in single-prism/element HL technique producing high quality PhCs and concomitant development of creating functional defects at precise location within the PhCs we introduced here would bring the field a step closer to realizing practical PhC devices for commercial applications.

References

1. E. Yablonovitch, Phys. Rev. Lett. **58**, 2059 (1987)
2. S. John, Phys. Rev. Lett. **58**, 2486 (1987)
3. P. Lodahl, A.F. Driel, I.S. Nikolaev, A. Irman, K. Overgaag, D. Vanmaekelbergh, W.L. Vos, Nature **430**, 654 (2004)
4. S. Noda, M. Fujita, T. Asano, Nat. Photonics **1**, 449 (2007)
5. A. Tandaechanurat, S. Ishida, D. Guimard, M. Nomura, S. Iwamoto, Y. Arakawa, Nat. Photonics **5**, 91 (2011)
6. P. Michler, Science **290**, 2282 (2000)
7. M.H. Qi, E. Lidorikis, P.T. Rakich, S.G. Johnson, J.D. Joannopoulos, E.P. Ippen, H.I. Smith, Nature **429**, 538 (2004)
8. A. Blanco, E. Chomski, S. Grabtchak, M. Ibisate, S. John, S.W. Leonard, C. Lopez, F. Meseguer, H. Miguez, J.P. Mondia, G.A. Ozin, O. Toader, H.M. Driel, Nature **405**, 437 (2000)

9. M. Deubel, G. Freymann, M. Wegener, S. Pereira, K. Busch, C.M. Soukoulis, Nat. Mater. **3**, 444 (2004)
10. N. Yamamoto, S. Noda, A. Chutinan, Jpn. J. Appl. Phys. **37**, L1052 (1998). Part (2)
11. M. Campbell, D.N. Sharp, M.T. Harrison, R.G. Denning, A.J. Turberfield, Nature **404**, 53 (2000)
12. V. Berger, O. Gauthier-Lafaye, E. Costard, J. Appl. Phys. **82**, 60 (1997)
13. S. Shoji, S. Kawata, Appl. Phys. Lett. **76**, 2668 (2000)
14. L.Z. Cai, X.L. Yang, Y.R. Wang, Opt. Lett. **27**, 900 (2002)
15. T.Y.M. Chan, O. Toader, S. John, Phys. Rev. E **73**, 046610 (2006)
16. D. Xu, K.P. Chen, K. Ohlinger, Y. Lin, Appl. Phys. Lett. **93**, 031101 (2008)
17. S.G. Park, Adv. Mater. **23**, 2749 (2011)
18. K.M. Ho, C.T. Chan, C.M. Sokoulis, Phys. Rev. Lett. **65**, 3152 (1990)
19. K.M. Ho, C.T. Chan, C.M. Sokoulis, R. Biswas, M. Sigalas, Solid State Commun. **89**, 413 (1994)
20. W. Mao, G. Liang, Y.Y. Pu, H. Wang, Z. Zeng, Appl. Phys. Lett. **91**, 261911 (2007)
21. W. Mao, G. Liang, H. Zou, R. Zhang, H. Wang, J. Opt. Soc. Am. B **23**, 2046 (2006)
22. L.J. Wu, Y.C. Zhong, K.S. Wong, G.P. Wang, L. Yuan, Appl. Phys. Lett. **88**, 091115 (2006)
23. Y. Yang, Q.Z. Li, G.P. Wang, Appl. Phys. Lett. **93**, 061112 (2008)
24. X. Wang, C.Y. Ng, W.Y. Tam, C.T. Chan, P. Sheng, Adv. Mater. **15**, 1526 (2003)
25. Y. Yang, S.H. Zhang, G.P. Wang, Appl. Phys. Lett. **88**, 251104 (2006)
26. K. Shen, G.M. Jiang, W.D. Mao, S. Baig, R.M. Wang, Appl. Opt. **52**, 6474 (2013)
27. Y.K. Pang, J.C.W. Lee, H.F. Lee, W.Y. Tam, C.T. Chan, P. Sheng, Opt. Express **13**, 7615 (2005)
28. Y.V. Miklyaev, D.C. Meisel, A. Blanco, G. Freymanna, K. Busch, W. Koch, C. Enkrich, M. Deubel, M. Wegener, Appl. Phys. Lett. **82**, 1284 (2003)
29. J.B. Yeo, S.D. Yun, H.Y. Lee, J. Appl. Phys. **102**, 084502 (2007)
30. N.D. Lai, W.P. Liang, J.H. Lin, C.C. Hsu, C.H. Lin, Opt. Express **13**, 9605 (2005)
31. K. Ohlinger, H. Zhang, Y. Lin, D. Xu, K.P. Chen, Opt. Mater. Express **1**, 1034 (2011)
32. J. Lutkenhaus, F.A. Farro, D. George, K. Ohlinger, H. Zhang, Z. Poole, K.P. Chen, Y. Lin, Opt. Mater. Express **2**, 1236 (2012)
33. G. Wang, C. Tan, Y. Yi, H. Shan, J. Mod. Opt. **50**, 2155 (2003)
34. L.J. Wu, Y.C. Zhong, C.T. Chan, K.S. Wong, G.P. Wang, Appl. Phys. Lett. **86**, 241102 (2005)
35. J.D. Joannopoulos, P.R. Villeneuve, S. Fan, Nature **386**, 143 (1997)
36. J. Scrimgeour, D.N. Sharp, C.F. Blanford, O.M. Roche, R.G. Denning, A.J. Turberfield, Adv. Mater. **18**, 1557 (2006)
37. X. Zhang, S. Strauf, Appl. Phys. Lett. **102**, 093110 (2013)
38. H.M. Su, Y.C. Zhong, X. Wang, X.G. Zheng, J.F. Xu, H.Z. Wang, Phys. Rev. E **67**, 056619 (2003)
39. X. Mai, R. Moshrefzadeh, U.J. Gibson, G.I. Stegeman, C.T. Seaton, Appl. Opt. **24**, 3155 (1985)
40. J. Boor, N. Geyer, U. Gosele, V. Schmidt, Opt. Lett. **34**, 1783 (2009)
41. J.Q. Chen, W. Jiang, X.N. Chen, L. Wang, S.S. Zhang, R.T. Chen, Appl. Phys. Lett. **90**, 093102(1–3) (2007)
42. S.G. Park, S.M. Yang, Nanoscale **5**, 4110 (2013)
43. W.J. Xiong, Y. Xu, Y.J. Xiao, X.X. Lv, L.J. Wu, Photonics Nanostruct. (2014). doi:10.1016/j.photonics.2014.10.008
44. J. Xu, R. Ma, X. Wang, W.Y. Tam, Opt. Express **15**, 4287 (2007)
45. M. Born, E. Wolf, *Principles of Optics*, 7th edn. (Cambridge University Press, Cambridge, 1999)
46. X.H. Sun, X.M. Tao, Y.Y. Wang, Appl. Phys. A **98**, 255 (2010)
47. Y.K. Pang, J.C. Lee, C.T. Ho, W.Y. Tam, Opt. Express **14**, 9113 (2006)
48. W.Y. Tam, Appl. Phys. Lett. **89**, 251111 (2006)
49. Y.C. Zhong, L.J. Wu, H.M. Su, K.S. Wong, H.Z. Wang, Opt. Express **14**, 6837 (2006)
50. P.V. Braun, S.A. Rhinne, F. Garcia-Santamaria, Adv. Mater. **18**, 2665 (2006)

51. M. Fujita, S. Takahashi, Y. Tanaka, T. Asano, S. Noda, Science **308**, 1296 (2005)
52. O. Painter, A. Husain, A. Scherer, P.T. Lee, I. Kim, J.D. O'Brien, P.D. Dapkus, IEEE Photon Technol. Lett. **12**, 1126 (2000)
53. P. Ferrand, M. Egen, R. Zentel, J. Seekamp, S.G. Romanov, C.M.S. Torres, Appl. Phys. Lett. **83**, 5289 (2003)
54. E. Vekris, V. Kitaev, G. Freymann, D.D. Perovic, J.S. Aitchison, G.A. Ozin, Adv. Mater. **17**, 1269 (2005)
55. Q. Yan, Z. Zhou, X.S. Zhao, S.J. Chua, Adv. Mater. **17**, 1917 (2005)
56. S. Jeon, J. Park, R. Cirelli, S. Yang, C.E. Heitzman, P.V. Braun, P.J.A. Kenis, J.A. Rogers, Proc. Natl. Acad. Sci. **101**, 12428 (2004)
57. G. Lee, S.H. Song, C. Oh, P. Kim, Opt. Lett. **29**, 2539 (2004)
58. T. Kondo, S. Juodkazis, V. Mizeikis, S. Matsuo, H. Misawa, New J. Phys. **8**, 250 (2006)
59. Y. Lin, A. Harb, K. Lozano, D. Xu, K.P. Chen, Opt. Express **17**, 16625 (2009)
60. X.S. Xie, M. Li, J. Guo, B. Liang, Z.X. Wang, A. Sinistkii, Y. Xiang, J.Y. Zhou, Opt. Express **15**, 7032 (2007)
61. P.Q. Zhang, X.S. Xie, Y.F. Guan, J.Y. Zhou, K.S. Wong, L. Yan, Appl. Phys. B Lasers Opt. **104**, 113 (2011)
62. J. Li, Y. Liu, X. Xie, P. Zhang, B. Liang, L. Yan, J. Zhou, G. Kurizki, D. Jacobs, K.S. Wong, Y. Zhong, Opt. Express **16**, 12899 (2008)
63. M. Mitchell, *An Introduction to Genetic Algorithms* (MIT Press, Cambridge, 1996)
64. J. Luthenhaus, D. George, M. Moazzezi, U. Philipose, Y. Lin, Opt. Express **21**, 26227 (2013)
65. J. Xavier, J. Joseph, Appl. Phys. Lett. **104**, 081104 (2014)
66. Y. Yang, Q.Z. Li, H.P. Wang, Opt. Express **16**, 11275 (2008)
67. J.H. Moon, J. Ford, S. Yang, Polym. Adv. Technol. **17**, 83 (2006)
68. A. Campo, C. Greiner, J. Micromech. Microeng. **17**, R81 (2007)
69. A. Kinkhabwala, Z. Yu, S. Fan, Y. Avlasevich, K. Müllen, W.E. Moerner, Nat. Photonics **3**, 654 (2009)
70. X. Zhang, M. Theuring, Q. Song, W.D. Mao, M. Begliarbekov, S. Strauf, Nano Lett. **11**, 2715 (2011)
71. R.H. Horng, Y.L. Tsai, T.M. Wu, D.S. Wuu, C.H. Chao, IEEE J. Sel. Top. Quantum Eletron. **15**, 1327 (2009)
72. Y.R. Do, Y.C. Kim, Y.W. Song, C.O. Cho, H. Jeon, Y.J. Lee, S.H. Kim, Y.H. Lee, Adv. Mater. **15**, 1214 (2003)
73. K.B. Choi, S.J. Shin, T.H. Park, H.J. Lee, J.H. Hwang, J.H. Park, B.Y. Hwang, Y.W. Park, B.K. Ju, Org. Electron **15**, 111 (2014)
74. L. Müller-Meskamp, Y.H. Kim, T. Roch, S. Hofmann, R. Scholz, S. Eckardt, K. Leo, A.F. Lasagni, Adv. Mater. **24**, 906 (2012)
75. M.M. Mróz, G. Sforazzini, Y. Zhong, K.S. Wong, H.L. Anderson, G. Lanzani, J. Cabanillas-Gonzalez, Adv. Mater. **25**, 4347 (2013)
76. N. Tétreault, G. Freymann, M. Deubel, M. Hermatschweiler, F. Pérez-Willard, S. John, M. Wegener, G.A. Ozin, Adv. Mater. **18**, 457 (2006)
77. V. Ramanan, E. Nelson, A. Brzezinski, P.V. Braun, P. Wiltzius, Appl. Phys. Lett. **92**, 173304 (2008)
78. D.J. Shir, E.C. Nelson, D. Chanda, A. Brzezinski, P.V. Braun, J.A. Rogers, J Vac. Sci. Technol. B **28**, 783 (2010)
79. S.G. Park, T.Y. Jeon, S.M. Yang, Langmuir **29**, 9620 (2013)
80. J.S. King, E. Graugnard, O.M. Roche, D.N. Sharp, J. Scrimgeour, R.G. Denning, A.J. Turberfield, C.J. Summers, Adv. Mater. **18**, 1561 (2006)
81. Y. Xu, X. Zhu, Y. Dan, J.H. Moon, V.W. Chen, A.T. Johnson, J.W. Perry, S. Yang, Chem. Mater. **20**, 1816 (2008)
82. J.E.G.J. Wijnhoven, W.L. Vos, Science **218**, 802 (1998)
83. J.E.G.J. Wijnhoven, L. Bechger, W.L. Vos, Chem. Mater. **13**, 4486 (2001)
84. S.L. Kuai, V.V. Truong, J. Appl. Phys. **96**, 5982 (2004)

85. J. Bartels (ed.), *Landolt-BoÈrnstein, Zahlenwerte, Funktionen, Optische Konstanten* (Springer, Berlin, 1962)
86. Y.C. Zhong, Z.N. Yue, G.K.L. Wong, Y.Y. Xi, Y.F. Hsu, A.B. Djurišić, J.W. Dong, W.J. Chen, K.S. Wong, Appl. Phys. Lett. **97**, 191102 (2010)
87. M. Miyake, Y.C. Chen, P.V. Braun, P. Wiltzius, Adv. Mater. **21**, 3012 (2009)
88. J.H. Kang, J.H. Moon et al., Adv. Mater. **20**, 3061 (2008)
89. R. Ma, J. Xun, W.Y. Tam, Appl. Phys. Lett. **89**, 081116 (2006)
90. J.H. Jang, S.J. Jhaveri, B. Rasin, C.Y. Koh, C.K. Ober, E.L. Thomas, Nano. Lett. **8**, 1456 (2008)
91. O. Akhavan, M. Abdolahad, R. Asadi, J. Phys. D Appl. Phys. **42**, 135416 (2009)

Part III
Chemistry and Physics at Work

Strong Coupling in Organic and Hybrid-Semiconductor Microcavity Structures

David G. Lidzey and David M. Coles

Abstract In this chapter we aim to introduce the reader to the phenomenon of strong exciton–photon coupling in photonic structures with an emphasis on planar microcavities. We will consider both organic and hybrid organic–inorganic systems. A simple description of classical strong coupling will be introduced through the use of a coupled oscillator model. The popular experimental techniques used to characterize strongly coupled microcavities will be described and we explain how to interpret experimental data. We give a brief history of the field of organic polaritonics in microcavities beginning at the first reported observation and covering the early experimental studies involving commonly used materials including J-aggregates, porphyrins, small organic molecules and molecular crystals. We then discuss more recent investigations aimed at determining the dynamics of polariton populations. A combination of steady state and ultrafast pump-probe experiments have allowed us to resolve the processes that lead from excited molecular states to observable polariton populations. An account of hybrid organic–inorganic polariton states (exciton hybridization) in microcavities is given, followed by the recent observation of polariton-mediated energy transfer between hybridized organic excitons. Recent milestones in the field of organic polaritons are described, specifically room-temperature organic polariton lasing and polariton condensation. Finally, there is a short review of optical systems other than planar microcavities that have been shown to support strong coupling of organic and hybrid exciton states

Keywords Bose–Einstein condensation • Distributed Bragg reflectors • Lasing • Microcavity • Polaritons • Strong coupling

---

D.G. Lidzey (✉)
Department of Physics and Astronomy, University of Sheffield, Sheffield S3 7RH, UK
e-mail: d.g.lidzey@sheffield.ac.uk

D.M. Coles
Department of Materials, University of Oxford, Oxford OX1 3PH, UK
e-mail: david.coles@materials.ox.ac.uk

D. Comoretto (ed.), *Organic and Hybrid Photonic Crystals*,
DOI 10.1007/978-3-319-16580-6_11

1 Microcavity Geometry

A planar microcavity consists simply of two mirrors separated by a length L which is generally on in the order of 100s of nanometers. An ideal cavity structure is depicted in Fig. 1.

The wavevector ($\boldsymbol{k}$) of a cavity photon mode depends upon the angle the cavity mode propagates with respect to the surface normal (θ_{int}) as well the cavity length (L) as given by

$$\boldsymbol{k}=\frac{m\pi}{n_{\text{cav}}L}\frac{1}{\cos\theta_{\text{int}}}, \tag{1}$$

where n_{cav} is the refractive index of the intracavity region. The cavity mode has wavevector components that are parallel ($k_{\|}$) and perpendicular ($k_{\perp}$) to the mirror surface. The perpendicular component is determined by the cavity length, while the parallel component depends upon the angle of propagation. At normal incidence the parallel component of the wavevector is zero.

$$k_{\perp}=\frac{m\pi}{n_{\text{cav}}L}=k_0. \tag{2}$$

$$k_{\|}=k_0\tan\theta_{\text{int}}. \tag{3}$$

The energy of a cavity mode is then

$$E_{\gamma}(\theta)=E_0\left(1-\frac{\sin^2\theta_{\text{ext}}}{n_{\text{cav}}}\right)^{-\frac{1}{2}}, \tag{4}$$

where $E_0=\frac{hck_{\perp}}{2\pi}$ is the cavity cutoff energy and $\theta_{\text{ext}}=\sin^{-1}(n_{\text{cav}}\sin\theta_{\text{int}})$.

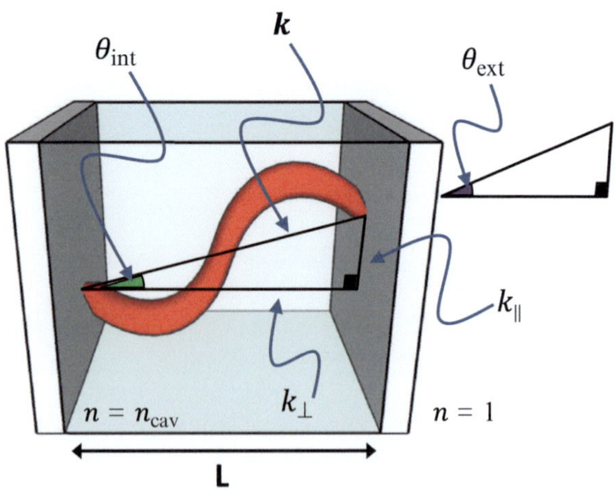

Fig. 1 A planar microcavity with a confined 1λ ($m=2$) optical mode at an angle θ_{int}

The mirrors of a microcavity most often comprise of either a metallic film or dielectric distributed Bragg reflectors (DBRs). While metallic mirrors have the advantage of being simple to deposit, they can introduce optical absorption losses and hence reduce the Q-factor of the cavities. DBRs are one-dimensional photonic crystals created from alternating layers of high and low refractive index transparent materials as discussed in other chapters. In cavities designed for strong coupling however, inorganic dielectrics are generally used due to the higher attainable reflectivities. The reflection at each layer boundary interferes constructively to create a highly reflective structure (often exceeding 99.99 %) over a range of wavelengths termed the reflectivity stopband. DBRs do not exhibit the absorption loses intrinsic to metallic mirrors. This makes DBRs a common choice for microcavity mirrors. The thickness of the constituent layers should be equal to one quarter of required stopband central wavelength divided by the refractive index of the material.

The reflectivity, R at centre of the stopband of a DBR consisting of N pairs of high-low index is approximately

$$R = 1 - 4\frac{n_{\text{air}}}{n_{\text{sub}}}\left(\frac{n_1}{n_2}\right)^{2N}, \tag{5}$$

where n_{sub} is the refractive index of the substrate, n_{air} is the refractive index of air and $n_1(n_2)$ is the refractive index of the low (high) index material. Therefore, increasing N, and increasing the refractive index contrast of the layers results in a higher reflectivity stopband. The stopband width $\Delta\lambda_{\text{sb}}$ increases with increasing refractive index contrast. This can be approximated by [1]

$$\Delta\lambda_{\text{sb}} = \frac{\lambda_{\text{DBR}}}{4\pi}\sin^{-1}\left(\frac{n_2 - n_1}{n_2 + n_1}\right). \tag{6}$$

The DBR stopband position is dependent on angle through Eq. (7). So long as θ remains small, the stopband centre is given by [2]

$$\lambda_{\text{DBR}}(\theta) = \lambda_{\text{DBR}}(0)\cos\left(\frac{n_{\text{air}}}{n_{\text{eff}}}\theta\right), \tag{7}$$

where n_{eff} is the effective refractive index of the mirror (equal to $\sqrt{n_1 n_2}$ for layers of thickness $\lambda/4n$). The operation of a DBR is based on the constructive interference of multiple reflections at consecutive material boundaries. For TE polarized light there is always a finite reflectivity at a boundary, while for TM polarized light the reflectivity at a boundary drops to zero at the Brewster angle. Hence we see that the reflectivity of the stopband is maintained at none zero angles only for TE polarized light, while the TM stopband reduces in reflectivity for increasing angle. For this reason consideration must be given to the polarization of light used in a microcavity experiment.

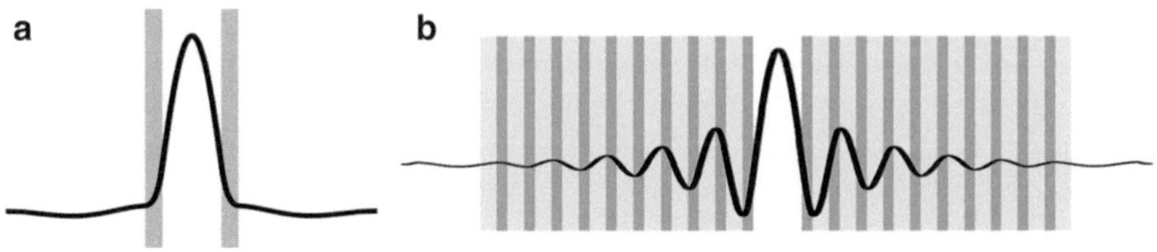

Fig. 2 Electric field distribution of a $\lambda/2$ cavity mode in a cavity consisting of (**a**) two 75 nm thick silver mirrors (**b**) two DBRs of ten bilayers of $n_1 = 1.5$, $n_2 = 2.2$

When a cavity is constructed using DBR mirrors, the penetration of light into the mirrors should be taken into account as it effectively extends the cavity length beyond its physical bounds. Figure 2 shows the electric field distribution of the $\lambda/2$ cavity mode confined in (a) a metallic cavity and (b) a dielectric cavity. In the metallic cavity the optical mode is well confined in the cavity region with the field decaying exponentially in the mirrors. This extension of the cavity mode is generally much smaller than the cavity length and can be neglected. This is in contrast to the dielectric cavity case in which the electric field extends into the mirrors. At normal incidence we can define the optical field penetration depth into the DBRs, L_{DBR}, as

$$L_{\mathrm{DBR}} = \frac{\lambda_{\mathrm{DBR}}}{2n_{\mathrm{cav}}} \frac{n_1 n_2}{(n_2 - n_1)}. \tag{8}$$

The energy of the cavity mode in a cavity with two DBRs is given by

$$E_{\gamma,\ \mathrm{DBR\ cav}} = \frac{E_\gamma L + E_{\mathrm{DBR}} L_{\mathrm{DBR}}}{L + L_{\mathrm{DBR}}}, \tag{9}$$

where $E_{\mathrm{DBR}} = hc/\lambda_{\mathrm{DBR}}$. It can be seen that the cavity energy is the weighted average of the 'ideal' cavity energy and the energy of the stopband centre. A comprehensive treatment of the cavity mode in DBR cavities including angular and polarization dependence can be found in [3].

2 Semi-Classical Description of Strong Coupling

A strongly coupled system of the type described in this chapter (light-matter) requires two things: an optical resonator and a material system capable of supporting excitons. We introduce some parameters to describe the properties of the components.

An important parameter used to describe an optical resonator (or cavity) is the Q-factor (Q). This is defined by the ratio of energy stored within the cavity to the energy loss per optical time period multiplied by a factor of 2π. Experimentally, we can easily measure the Q-factor of a cavity by observing the transmission spectrum.

The Q-factor is then given by the ratio of the resonance energy (E_γ) of the transmission peak to its energetic full-width-half-maximum (ΔE). Alternatively, Q can be expressed in terms of the resonance wavelength (Eq. 10).

$$Q = \frac{E_\gamma}{\Delta E} = \frac{\lambda_\gamma}{\Delta \lambda}. \tag{10}$$

We can then relate the cavity Q-factor to the length of time that a photon is trapped within the cavity (τ_{cav}) through Eq. (11). Here we have just incorporated Heisenberg's Uncertainty Principle ($\tau \Delta E \leq \hbar/2$) into Eq. (10).

$$\tau_\gamma \approx \frac{Q\hbar}{2E_\gamma} = \frac{Q\lambda_\gamma}{2\pi c} = \frac{1}{\kappa_\gamma}, \tag{11}$$

where $\hbar$ and c are the reduced Planck constant and the speed of light, respectively, and κ_γ is the cavity photon decay rate. When a two-level system (whether it be an atom, quantum well or organic molecule) is placed inside the cavity, an interaction occurs due to the modified local density of photonic states with respect to free space. The strength of this interaction is described by the parameter g_0, often called the atom-photon coupling constant [4].

$$g_0 = \sqrt{N}\left(\frac{\pi \mu^2 c}{n_{cav}^2 \lambda_\gamma \varepsilon_0 \hbar V}\right)^{\frac{1}{2}}, \tag{12}$$

where N is the number of excitons in the cavity, μ is the dipole moment of the exciton (assumed to be parallel to the cavity E-field), n_{cav} is the refractive index of the cavity region, ε_0 is the vacuum permittivity and V is the mode volume of the cavity (the volume in which the field is confined, see [5]). We also define the total decay rate of the excitonic system into states not supported by the cavity (e.g. emitting a photon that is non-resonant with the cavity, non-radiative decay, emission in a direction not confined by the cavity, etc.) as $\kappa_{exc,nr}$. Two distinct coupling regimes can thus be identified depending on the relative values of κ_γ, $\kappa_{exc,nr}$ and g_0. If $g_0 < (\kappa_\gamma, \kappa_{exc,nr})$, the system is said to be in the so-called weak coupling regime. Here, the cavity acts to modify the free space decay dynamics of the exciton and the system can be treated as a perturbation of the uncoupled cavity and exciton using Fermi's golden rule. The density of states in a microcavity is described by a Lorentzian function that peaks a wavelength λ that is resonant with the central wavelength of the cavity mode. At this point the density of states is enhanced with respect to free space and hence if the exciton wavelength is resonant with the cavity mode the radiative decay rate is enhanced. Away from the cavity mode, the density of states and hence the exciton radiative decay rate is suppressed. The Purcell factor F_P describes the maximum enhancement in the exciton radiative decay rate that is obtainable in a given cavity.

If $g_0 > (\kappa_\gamma, \kappa_{\mathrm{exc,nr}})$, the system is in the strong coupling regime and the cavity can no longer be considered to cause a perturbation in the radiative decay dynamics of the exciton, and instead the exciton and cavity mode exchange energy with a frequency of $\Omega = 2g_0$, where Ω is the Rabi frequency. This can be visualized as the exciton absorbing a cavity photon, re-emitting back into the cavity mode and absorbing the photon again. The exciton and photon can be considered coupled oscillators and described by a two-level coupled oscillator model (Eq. 13). The cavity and exciton modes become mixed, forming a new quasi-particle called a cavity polariton.

$$\begin{pmatrix} E_\gamma(\theta) & \hbar\Omega/2 \\ \hbar\Omega/2 & E_{\mathrm{x}} \end{pmatrix} \begin{pmatrix} \alpha_\gamma(\theta) \\ \alpha_{\mathrm{x}}(\theta) \end{pmatrix} = E_{\mathrm{P}} \begin{pmatrix} \alpha_\gamma(\theta) \\ \alpha_{\mathrm{x}}(\theta) \end{pmatrix}. \tag{13}$$

E_γ, E_{x} and E_{P} are the cavity energy, exciton energy and polariton energy, respectively. The eigenvector contains α_γ and α_{x} which are the polariton mixing coefficients and describe the 'amount' of cavity mode and exciton mode mixed into the polariton, respectively. Equation (13) can be solved for E_{P} to give an expression with two unique solutions.

$$E_{\mathrm{P}} = \frac{E_\gamma^2 + E_{\mathrm{x}}^2}{2} \pm \frac{1}{2}\sqrt{\left(E_\gamma - E_{\mathrm{x}}\right)^2 + (\hbar\Omega)^2}. \tag{14}$$

$$\alpha_\gamma^2 = \frac{E_\gamma - E_{\mathrm{P}}}{E_\gamma + E_{\mathrm{x}} - 2E_{\mathrm{P}}}, \qquad \alpha_{\mathrm{x}}^2 = 1 - \alpha_\gamma^2. \tag{15}$$

We plot the solutions to Eq. (14), after substituting E_γ for the angular dependent energy of a microcavity mode as described by Eq. (4), as blue lines in Fig. 3a. The exciton and photon energies are shown as red and black dashed lines, respectively. We have shown a case where the photon energy lies below the exciton energy at normal incidence, i.e. the cavity is negatively detuned ($\Delta = E_\gamma - E_{\mathrm{x}} < 0$). This is generally the desired case for strongly coupled microcavities. The two polariton branches undergo an anticrossing at the point where the exciton and photon are energetically resonant (simply referred to as 'resonance'), with one branch being higher in energy than the exciton and one being lower. We term these branches the upper polariton branch (UPB) and lower polariton branch (LPB), respectively. Figure 3b, c shows the mixing coefficients of the UPB and LPB, respectively. It can be seen that at small angles the UPB has a large excitonic mixing fraction and hence is exciton-like. The dispersion at this point is also shallow like the flat exciton dispersion. At large angles the polaritons contain a large photonic mixing fraction and the dispersion becomes steeper and approximates the cavity mode dispersion. The situation is reversed for the LPB, with it being photon-like at small angles and exciton-like at large angles. At resonance both the UPB and LPB are 50 % photon and 50 % exciton.

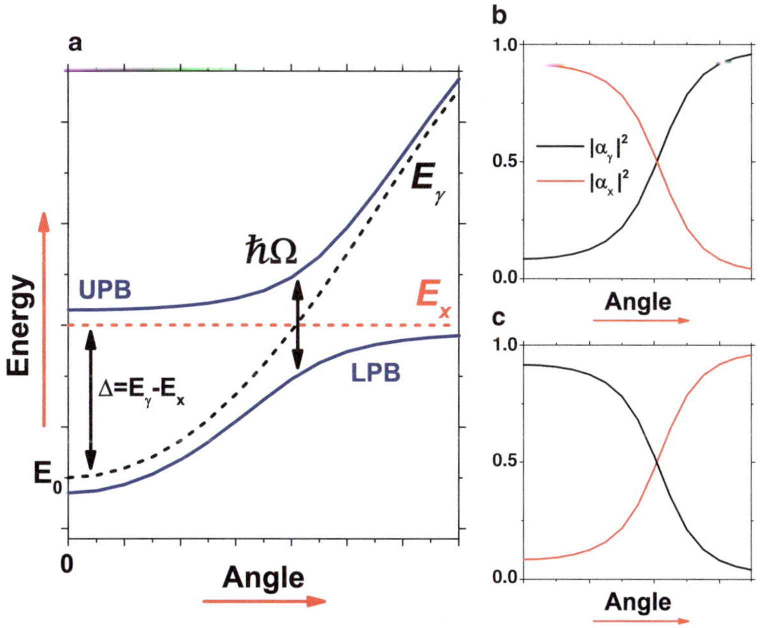

Fig. 3 (**a**) Dispersion relation for a strongly coupled, negatively detuned cavity. *Black* and *red dashed lines* are the cavity photon mode (E_γ) and exciton (E_x) energy dispersions, while the polariton branch dispersions are *solid blue lines*. The Rabi splitting energy $\hbar\Omega$ is shown at exciton–photon resonance. Polariton mixing fractions are shown for (**b**) the upper polariton branch and (**c**) the lower polariton branch

3 Commonly Used Experimental Techniques

The most commonly used experimental techniques to observe strong coupling in microcavities involve angular dependent measurements to tune the cavity mode energy and hence probe polariton states along the dispersion curve, and in particular to observe the anticrossing of polariton branches that is a characteristic feature of strong coupling. It should be noted that the cavity mode energy can also be tuned by changing the cavity length (through the L dependence of E_0), and so by incorporating a wedge into the cavity region the cavity mode energy can be scanned with position on the sample. This is a common technique used in the characterization of inorganic microcavity structures where the cavity growth process can naturally result in a wedged structure.

In Fig. 4 we show schematic setups for the most common angular dependent measurements used to characterize strongly coupled planar microcavity structures. Part (a) shows a transmission experiment. In the example shown, a fibre coupled white light source is collimated before passing through a polarizer to TE polarize the light. The light is focused on the cavity surface. Consideration should be taken as to the numerical aperture of the focusing lens. A high NA lens will focus light

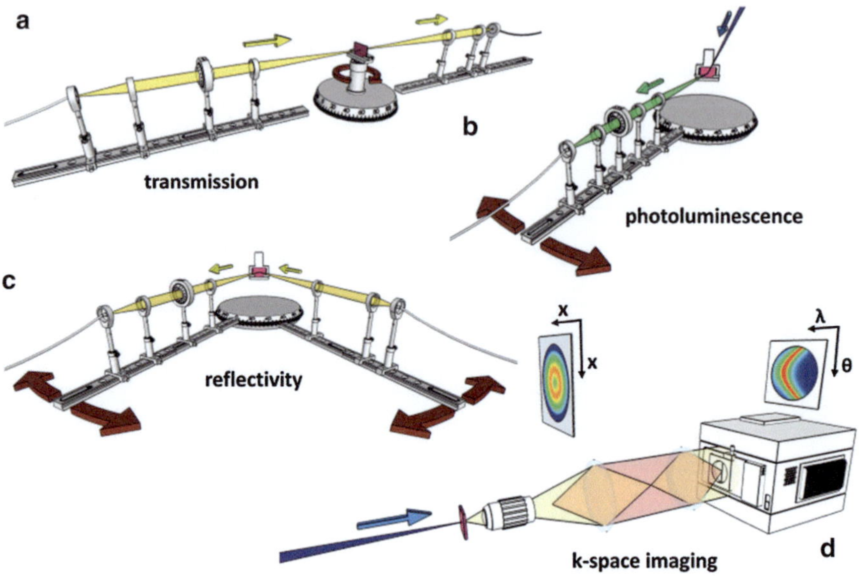

Fig. 4 Commonly used angular dependent experiments for characterizing strongly coupled planar microcavity structures. (**a**) Transmission. (**b**) Photoluminescence. (**c**) Reflectivity. (**d**) k-space imaging

over a larger angular range which has the effect of broadening the observed polariton modes. However, a very low NA lens will focus the light to a large spot, which may also broaden the modes due to thickness variations in the active layer. The light transmitted through the cavity is collected before being sent to a spectrometer. The excitation and collection optics can remain fixed while the sample is rotated on a rotation stage. Care must be taken to ensure that the cavity is located at the rotation axis of the stage so that the exciton spot remains on the same point on the sample while it is rotated. The measured transmission spectra should be divided by a reference spectrum taken without the sample present (or a blank substrate). For transmission measurements, the microcavity structure should be fabricated on a transparent substrate, and both mirrors should have a finite transmission.

Part (b) shows an angular dependent photoluminescence setup. Here, for non-resonant excitation, a microcavity is excited with a laser of shorter wavelength than the exciton energy. The angle of excitation is not critical, however if DBR mirrors are used, the efficiency of excitation can be maximized by choosing an excitation wavelength and angle that correspond to a reflectivity minimum of the DBR. PL emission is collected from the sample with a lens mounted on a goniometer. The collected light is collimated and TE polarized before being fibre coupled to a spectrometer. A long pass filter is used to block any scattered excitation light. Angular dependent PL measurements are usually Lambertian corrected to account for the change in solid angle collection as the angle is scanned. This is simply done by dividing the spectrum collected at a given angle by the cosine of that angle.

A typical reflectivity setup is shown in part (c). Here, the microcavity is mounted on the optical axis of a two armed goniometer. A fibre coupled white light source is focused on the cavity surface. The reflected light is collected, TE polarized and fibre coupled to a spectrometer. The excitation spot should remain at the same position on the sample as the angles are scanned.

K-space (or Fourier space) imaging is a method whereby PL emission from the sample is collected with a high NA lens (usually a microscope objective) positioned one focal length away from the cavity surface. One focal length behind the objective is the Fourier image where the emission angle is mapped onto image position. Two further lenses image this Fourier plane onto the slit of a spectrometer. The NA of the lens limits the angle of emission that can be collected, hence this imaging method is commonly employed in the study of inorganic cavities where the strong coupling region occurs over a relatively small angular range.

4 Early Experiments on Organic Microcavities

The Frenkel-type excitons supported by organic materials have fundamental differences to the Wannier–Mott excitons of inorganic materials. Frenkel excitons have relatively large exciton biding energies (100s of meV) due to the small dielectric constants of organic materials. This results in organic excitons and hence organic polaritons being stable at room temperature, while inorganic excitons will generally disassociate if not held at cryogenic temperatures. A second difference is the higher oscillator strengths of organic materials that results in Rabi splittings that can be an order of magnitude larger than seen in inorganic microcavities. Furthermore, organic Frenkel excitons tend to be more localized than inorganic Wannier–Mott excitons; a property that may well define the efficiency by which polaritons interact with each other.

Cavity exciton-polaritons were first observed by Weisbuch et al. in 1992 through reflectivity measurements on an inorganic III–V semiconductor microcavity structure grown monolithically with GaAs quantum wells positioned between two GaAlAs-AlAs DBRs [6]. It was 6 years later that Lidzey [7] and co-workers demonstrated that an organic semiconductor, in this case a porphyrin, could also strongly couple to a cavity.

The cavity consisted of a single DBR mirror (SiO_2–SiN) and a silver mirror forming a $\lambda/2$ cavity filled with a 100 nm film of tetra-(2,6-*t*-butyl)phenol-porphyrin zinc (4TBPPZn) dispersed in a polystyrene matrix. Fig. 5, left, shows white light reflectivity spectra taken over a series of angles. Clear dips are seen that undergo an anticrossing about the porphyrin absorption energy. If the dip positions are plotted as a function of observation angle, as is shown in Fig. 5a, this anticrossing is clearly seen in the dispersion and a Rabi splitting energy of 160 meV can be calculated by

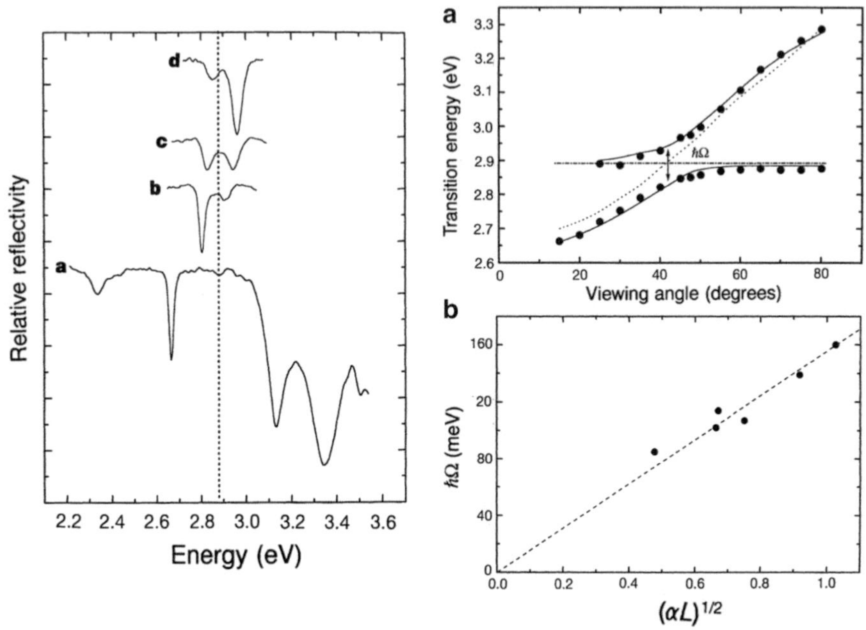

Fig. 5 *Left*: Series of white-light reflectivity spectra at different angles for a strongly coupled system of a layer of porphyrin molecules placed between a DBR and metallic mirror. (**a**) Dispersion relation of a strongly coupled microcavity containing an organic porphyrin dye (**b**) Rabi splitting energy as a function of the square root of the film absorption. From [7]

fitting the data with Eq. (14). From Fig. 15b, it can be seen that the Rabi splitting energy is proportional to the square root of the organic layer absorption, a result consistent with a macroscopic classical approach to strong coupling [8–10]. Here, the porphyrin Soret-band transition that is coupled to the cavity mode has a very low fluorescence quantum yield, and hence the polaritons in this structure were non-radiative.

The first observation of organic polariton emission came shortly afterwards [11]. Photoluminescence experiments are of vital importance in understanding polariton dynamics since the polariton population of a state is proportional to the emission intensity divided by the photonic fraction of that state (this accounts for photon-like polaritons being more emissive than exciton-like polaritons). Therefore, by recording PL intensity as a function of angle, and dividing the polariton branch emission intensity by the photonic fraction at that angle, the polariton population distribution can be calculated. Here the active layer in the microcavity consisted of a layer of cyanine dye (2, 2′-dimethyl-8-phenyl-5, 6, 5′, 6-dibenzothiacarbocyanine chloride) in a transparent polymer matrix. This dye has a relatively broad absorption and emission band, but undergoes electrostatically driven self-assembly in certain solvents (in this case a water/methanol mix) to form one- or two-dimensional nanostructures called J-aggregates [12, 13]. The excited state wavefunction

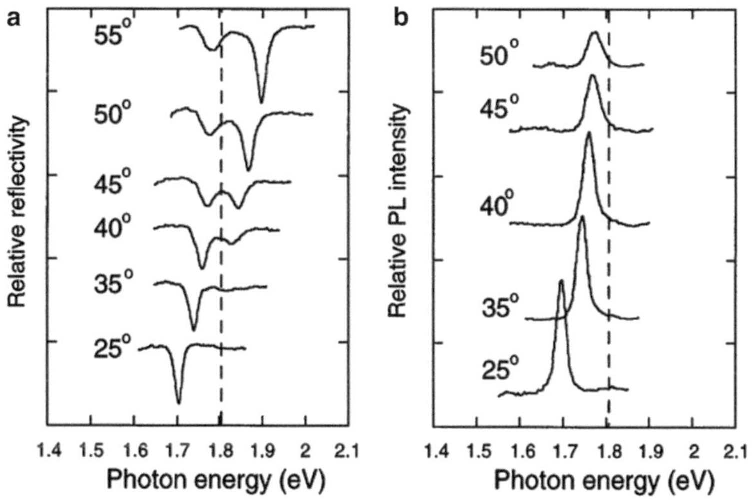

Fig. 6 (**a**) Reflectivity and (**b**) photoluminescence spectra for a series of angles recorded from a microcavity containing a J-aggregate dye. From [11]

becomes delocalized over several molecules causing a redshift in the exciton energy, and motional narrowing leads to a much reduced exciton linewidth that is well suited to strong coupling as it approximates a two-level system. By mixing the J-aggregate solution with an inert polymer, in this case poly(vinyl alcohol), the aggregates can be fixed and spincast into a thin solid film.

The characteristic anticrossing of two dips in the transmission spectra can be seen in (Fig. 6a), corresponding to the LPB and UPB. In the PL spectra (Fig. 6b), only one peak is seen that has an energy that is coincident with that of the lower branch. There is no sign of emission from the UPB. We discuss this asymmetry in Sect. 6. A model was developed to account for the observed asymmetry in LPB/UPB emission intensity [14] that described the creation of exciton states following non-resonant high-energy excitation (into the 'exciton reservoir' states). The reservoir excitons could then scatter into the UPB and LPB at a rate that is proportional to the exciton fraction of the final polariton state. It was assumed that polaritons were also able to transfer from the UPB to LPB at a rate proportional to the product of the exciton fractions of the initial and final polariton states. This transfer was thought to be facilitated by interaction with the vibrational landscape of the dye and polymer matrix. Comparison with experimental results revealed this inter-branch scattering to occur on a timescale of ~30 fs. This model was a foundation for more complex models (not involving direct inter-branch transfer) that now provide a more detailed understanding of processes that determine polariton populations.

5 Steady State Studies of Polariton Populations

The importance of the vibrational landscape on populating polariton states was demonstrated using a microcavity consisting of a J-aggregate layer between two high reflectivity DBRs [15]. The J-aggregate used was 5,6-dichloro-2-[[5,6-dichloro-1-ethyl-3-(4-sulfobutyl)-benzimidazol-2-ylidene]-propenyl]-1-ethyl-3-(4-sulfobutyl)-benzimidazolium hydroxide, sodium salt, inner salt (TDBC) dispersed in a polyvinyl alcohol (PVA) matrix. The TDBC J-aggregate, which forms in aqueous solutions, has a narrow absorption band centred at 586 nm with a linewidth of <10 nm, and PL emission at 593 nm with a linewidth of <16 nm. The normalized absorption and photoluminescence spectra of the TDBC monomer and J-aggregate are shown in Fig. 7. Here, the monomeric form of the dye is formed by dissolving the dye in methanol which acts as a good solvent and prohibits aggregate formation.

Angular dependent structure is observed in the LPB emission intensity in the form of a number of broad peaks. The LPB emission intensity as a function of the energy separation between the exciton reservoir and the LPB ($E_{\mathrm{x}} - E_{\mathrm{LPB}}$) for several temperatures is plotted in Fig. 8a. Part (b) shows the Raman intensity spectrum of TDBC on the same energy scale. It can be seen that several sharp Raman modes are present, with the modes at energies 40, 84 and 150 meV (indicated with grey dashed vertical lines) being coincident with the energy separation of the reservoir and resonances 1, 2 and 3. A higher energy Raman mode at 184 meV appears to be coincident with resonance 4. The presence of localized regions of PL intensity on the LPB (and hence localized increased polariton

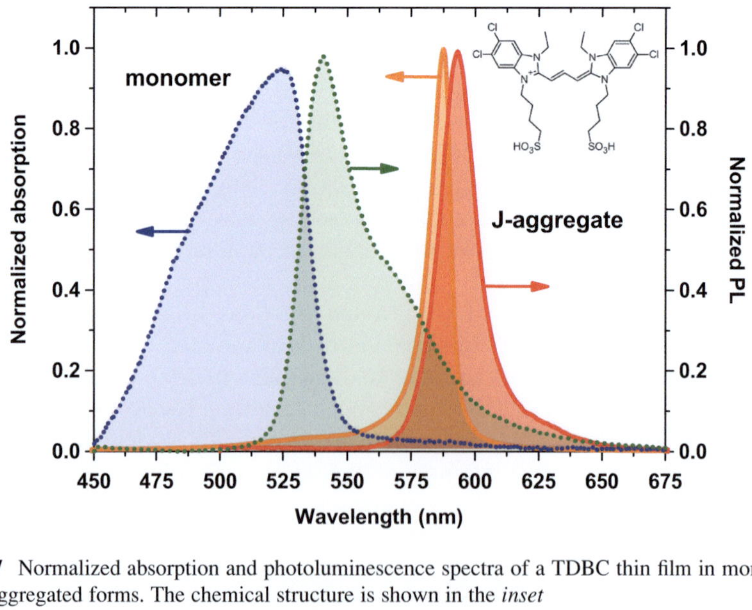

Fig. 7 Normalized absorption and photoluminescence spectra of a TDBC thin film in monomer and aggregated forms. The chemical structure is shown in the *inset*

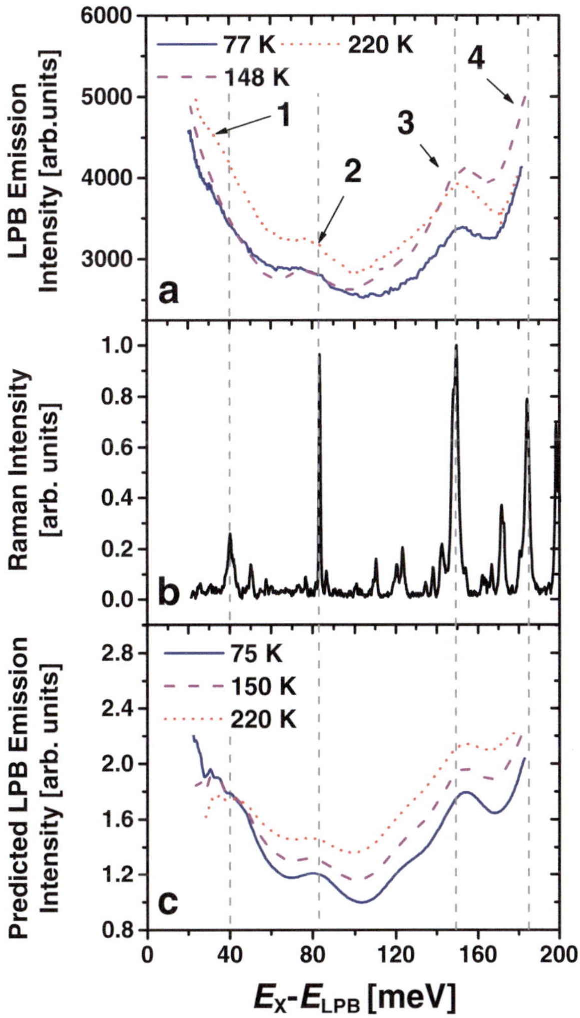

Fig. 8 (**a**) Lambertian-corrected LPB emission intensity as a function of energy separation between the exciton reservoir and the LPB for cavity A. (**b**) Raman spectrum of TDBC. (**c**) Modelled LPB intensity. From [15]

population) at points where the energetic separation of the LPB and exciton reservoir is resonant with a Raman active molecular vibration indicates that these vibrations play a role in transferring excitons from the reservoir into polariton states along the LPB. Indeed, these modes act as a fast and direct pathway between the exciton and polariton states by allowing excitons to 'dump' energy into the vibrational modes. The fast buildup of polaritons in states that meet the Raman resonance condition was also observed by pump-probe time domain measurements by Somaschi et al. [16]. Numerical simulations have been performed to better understand the relaxation pathways by which excitons relax to polariton states. Complete details can be found in [17–21].

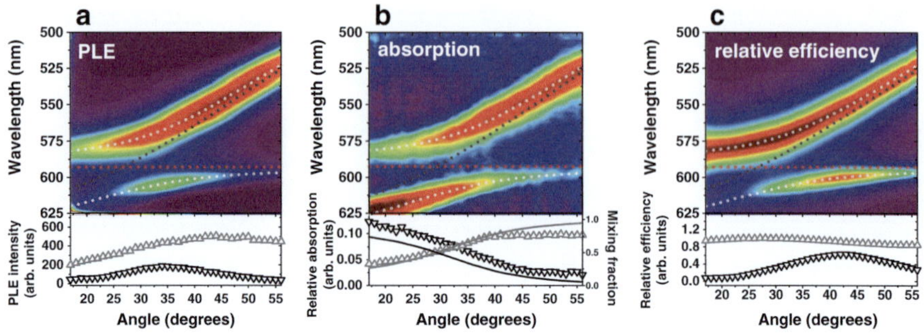

Fig. 9 (**a**) Angular and wavelength dependent PLE signal recorded at the LPB wavelength at normal incidence for a metallic cavity containing TDBC. (**b**) Microcavity absorption recorded simultaneously with the PLE signal. (**c**) Scattering efficiency into the $k_{\|} = 0$ on the LPB. The *lower parts of each panel* show the magnitude of the signal for the LPB (*black down triangles*) and UPB (*grey up triangles*). From [22]

The scattering of populations between polariton states has been studied via photoluminescence excitation (PLE) spectroscopy [22]. Here, a metallic microcavity containing TDBC was excited as a function of wavelength and angle while emission was collected from the bottom of the LPB at normal incidence. The absorption of polariton states was measured simultaneously by recording the light reflected from the cavity. The angular and wavelength dependent PLE signal is shown in Fig. 9a. The lower panel shows the magnitude of the PLE signal for each branch. The UPB PLE signal is greater than that of the LPB for all angles and generally increases with increasing angle, while a peak can be seen in the LPB PLE signal at angles greater than the resonance angle. The absorption of the cavity is shown in Fig. 9b. Also shown in the lower panels are the photonic mixing coefficients of the LPB and UPB (black and grey solid lines, respectively). It can be seen that the profile of the cavity absorption closely follows the photonic component. This is not surprising since light couples into the cavity and reaches the absorptive active region through this component.

The relaxation efficiency into the LPB ground state can be mapped by dividing the PLE signal by the absorption. This efficiency map is shown in Fig. 9c. The scattering efficiency of the UPB is relatively constant with angle. Polaritons in UPB states non-radiatively scatter to the exciton reservoir on timescales of the order of 10s of fs [23] (see Sect. 6). Excitons can then scatter into LPB polariton states through interaction with molecular vibrations or via optical pumping of the photonics component of the polariton.

The LPB scattering efficiency decreases as the photon fraction of the polariton state increases (i.e. towards reduced external viewing angles). This increase in photon fraction suppresses their ability to scatter into lower energy states, hence the reducing PLE signal. A similar effect is seen in inorganic microcavities, and has been termed the relaxation 'bottleneck' [24–26]. The relaxation efficiency of the UPB polaritons towards the LPB ground state is at all times greater than that of polariton states on LPB. Since the majority of UPB polaritons return to the exciton reservoir (an assumption supported by the fact that the PL emission from the UPB is very weak), it indicates that direct scattering to LPB states around $k_{\|} = 0$ is more efficient from the exciton reservoir than from polariton states along the LPB.

6 Ultrafast Studies of Polariton Dynamics

From steady state measurements and modelling, many aspects of polariton dynamics can be inferred. However, in order to directly observe the redistribution of states in a strongly coupled system time-resolved measurements are required. The timescales involved in polariton dynamics are dominated by the photon lifetime of the cavity which is generally on the order of 10's of femtoseconds in organic cavity systems. Ultrafast pump-probe spectroscopy is one such technique that can probe these timescales. The transmission of the sample (T) is measured with a weak probe laser pulse which measures the ground state absorption, which is followed by a short intense pump pulse. This puts the system into the excited state. The sample is then probed again with transmission of the probe pulse being modified by an amount ΔT due to the excited state population. There are three possible effects that may be observed. Firstly, the transmission of the probe pulse may be increased due to the reduced population in the ground state of the system (bleaching). Secondly, the transmission of the probe pulse may be decreased due to absorption from the first excited state into a higher energy excited state (photoinduced absorption (PIA)). Finally, the transmission of the probe pulse may appear to increase due to stimulated emission from the excited state. Changing the delay in pump and probe pulse arrival times allows the excited state dynamics to be observed with high temporal resolution.

The first investigation into the ultrafast dynamics of organic polaritons was performed by Song et al. [27] on a strongly coupled J-aggregate microcavity. The pump-probe technique was used to excite polariton states in the microcavity which were then probed as a function of angle. It was found that the magnitude of the optical non-linearity ($\Delta T/T$) of the strongly coupled system depended superlinearly on the exciton–photon mixing of the polariton states, with a maximum signal being observed at resonance. The maximum non-linearity was an order of magnitude greater than in a bare J-aggregate film.

It had previously been demonstrated that in inorganic microcavities, by resonantly pumping the LPB at a 'magic angle', polaritons could be made to scatter from each other, with one scattering to a higher energy polariton state (the idler) and one to the bottom of LPB (the signal) [28, 29]. This scattering mechanism could be stimulated by a weak probe pulse at $k_{\|} = 0$ on the LPB leading to a large amplification of the transmitted light. It could also lead to a large buildup of polaritons at the bottom of the LPB leading to non-linear behaviour. The system must be excited at a specific angle such that signal and idler states exist on the polariton dispersion such that energy and angular momentum can be conserved in the scattering process. Savvidis et al. [30] investigated whether such an effect could be reproduced in organic microcavities containing zinc porphyrin as the coupling media. It was observed that following a pump pulse, the transmission of the probe pulse *decreased*, in contrast to the amplification seen in inorganic cavities. This was attributed to excitons scattering into long-lived triplet states thereby reducing the oscillator strength of the strongly coupled singlet excitons.

A study of the temporal processes in organic microcavities was performed by Virgili et al. [23]. Ultrafast pump-probe spectroscopy was performed on both the bare TDBC J-aggregate dye and a strongly coupled microcavity consisting of a layer of TDBC in PVA between two DBR mirrors. The pump and probe pulses had a temporal duration of 15 fs and permitted two-dimensional differential transmission ($\Delta T/T$) maps to be recorded as a function of probe frequency and delay. Here, it was shown that the polariton population in the UPB, LPB and exciton reservoir could be followed through the PIA associated with a transition of the coupled exciton to a higher lying two-exciton state.

The time-dependent intensity of the PIA signals in the strongly coupled cavity is shown in Fig. 10 as open circles for the (a) UPB, (b) exciton and (c) LPB. The UPB PIA consists of a 'fast' decay, reaching $\Delta T/T = 0$ at 250 fs after the pump pulse. The exciton population shows a slow rise time followed by a slow decay of time constant approximately 5 ps. It was concluded that the slow rise time was a result of upper branch polaritons scattering back to the exciton reservoir since the decay time of UPB states and the rise time of the exciton signal were approximately equal. The reservoir decay was slower than was seen in the control film, likely due to the effectively reduced excitation density due to the finite reflectivity of the DBR mirrors at the exciton energy. The LPB PIA signal had a fast rise time with a bi-exponential decay with time constants on the order of 150 fs and ~10 ps.

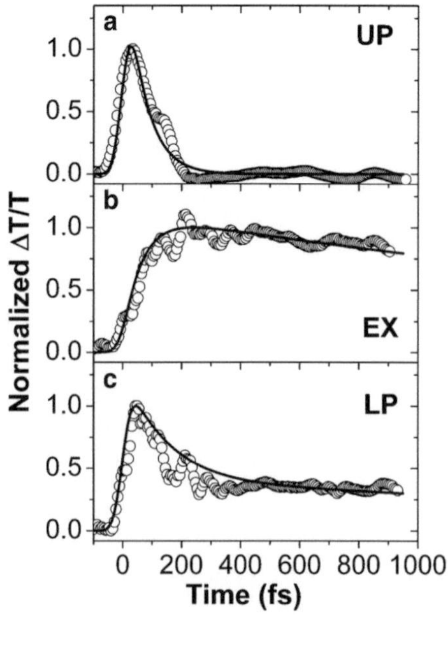

Fig. 10 Time-dependent PIA signals (*open circles*) for (**a**) UPB (2.03 eV), (**b**) exciton reservoir (2.13 eV) and (**c**) LPB (2.17 eV). *Solid black lines* are fits from the model. From [23]

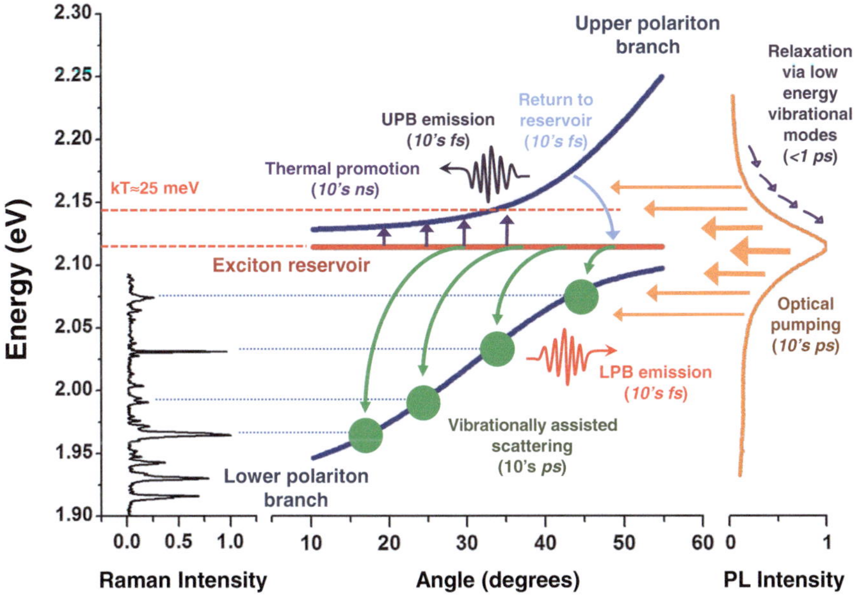

Fig. 11 Schematic of the important processes involved in populating the polariton branch states

To quantify the transients presented in Fig. 10, a simple model of population transfer between a three energy level system was used. This is an extension of the model proposed in [14] which only considered population transfer between a two-level system: LPB and UPB. The results of fitting the model to the PIA data are shown as solid lines in Fig. 10. It was proposed that the UPB population scattered to the exciton reservoir on a timescale of 150 fs. This agrees well with theoretical predictions [10, 31, 18] and, importantly, is on the same timescale as the radiative lifetime of the UPB which results in a fast non-radiative depopulation of the UPB and explains the weak UPB emission intensity. It was also found that reservoir excitons scattered into lower-branch polariton states with a time constant of 3.2 ps. A range of times for this process have been theoretically predicted, ranging from 9 [31, 32] to 350 ps [20]. The processes are summarized in Fig. 11 with approximate time constants for each process.

7 Hybrid-Semiconductor Polaritons

When two different exciton resonances are located within a microcavity, they may both simultaneously couple to the same cavity mode. This can lead to the formation of the so-called hybrid polaritons that correspond to an optically driven mixing between exciton states. Such systems are of significant interest, as mixing between different types of exciton states (e.g. Frenkel and Mott-Wannier) could be used to create optoelectronic devices having entirely new functionalities (e.g. all-optical switching, low-threshold polariton lasing, etc.). This field is in its relative infancy and theoretical understanding is less well developed than for systems based on a single exciton-state coupled to a single photon-mode. Nevertheless, a series of interesting effects have recently been demonstrated.

Hybridized polariton states were first observed in inorganic microcavities [33], with quantum wells of different energies grown at adjacent antinodes of a microcavity. Hybrid organic polariton states were observed soon after where two different organic dyes were located within the same microcavity [34]. Here, as expected, three polariton branches were clearly observed in reflectivity spectra. The existence of three polariton branches is expected on the basis of a three-level coupled oscillator model as shown by Eq. (16).

$$\begin{pmatrix} E_\gamma(\theta) & \hbar\Omega_1/2 & \hbar\Omega_2/2 \\ \hbar\Omega_1/2 & E_{\mathrm{x}1} & 0 \\ \hbar\Omega_2/2 & 0 & E_{\mathrm{x}2} \end{pmatrix} \begin{pmatrix} \alpha_\gamma(\theta) \\ \alpha_{\mathrm{x}1}(\theta) \\ \alpha_{\mathrm{x}2}(\theta) \end{pmatrix} = E_{\mathrm{P}} \begin{pmatrix} \alpha_\gamma(\theta) \\ \alpha_{\mathrm{x}1}(\theta) \\ \alpha_{\mathrm{x}2}(\theta) \end{pmatrix}, \tag{16}$$

where $E_{\mathrm{x}1}(E_{\mathrm{x}2})$ is the energy of exciton species 1(2), $\hbar\Omega_1(\hbar\Omega_2)$ is the Rabi splitting energy due to the interaction of the cavity mode and exciton 1 (exciton 2) and $|\alpha_\gamma|^2$, $|\alpha_{\mathrm{x}1}|^2$ and $|\alpha_{\mathrm{x}2}|^2$ are the mixing coefficients of the photon, exciton 1 and exciton 2. By solving the characteristic equation three polariton branches are obtained.

The polariton dispersion for a strong-coupled microcavity containing two different J-aggregated molecular dyes is shown in Fig. 12a, with upper (UPB), middle (MPB) and lower (LPB) branches obtained. The mixing components of the MPB are shown in part (b). At ~32° maximum mixing occurs and there are approximately equal parts of both exciton species and photon in the MPB. The reflectivity spectrum at the approximate point of maximum mixing is shown in part (c) where all three branches are visible.

Optically driven hybridization of vibronic progressions of an exciton in a single material was then demonstrated by Holmes and Forrest [35]. Here, a thermally deposited thin film of the organic molecule 3,4,7,8 napthalenete-tracarboxylic dianhydride (NTCDA) was used as the strong coupling material. A large intra-molecular phonon energy results in a clearly pronounced vibronic replica ($0 \rightarrow 1$) peak in the NTCDA absorption spectrum. The $0 \rightarrow 0$ and $0 \rightarrow 1$ transitions can simultaneously couple to the cavity mode. Kéna-Cohen et al. later demonstrated hybrid polaritons formed by the vibronic replicas in an organic single crystal [36]. This would later lead to the first observation of organic polariton lasing (see Sect. 8).

A different approach to hybridization of organic excitons was taken by Lodden and Holmes [37]. Here, a photonic structure was created (similar to a DBR), in which the strong coupling materials (the small organic molecules tetraphenyl-porphyrin (TPP) and octaethylporphyrin (OEP)) were incorporated directly into the structure. The optically active layers were embedded in a matrix of bathocuproine (BCP), which is a wide bandgap material and whose blue absorption

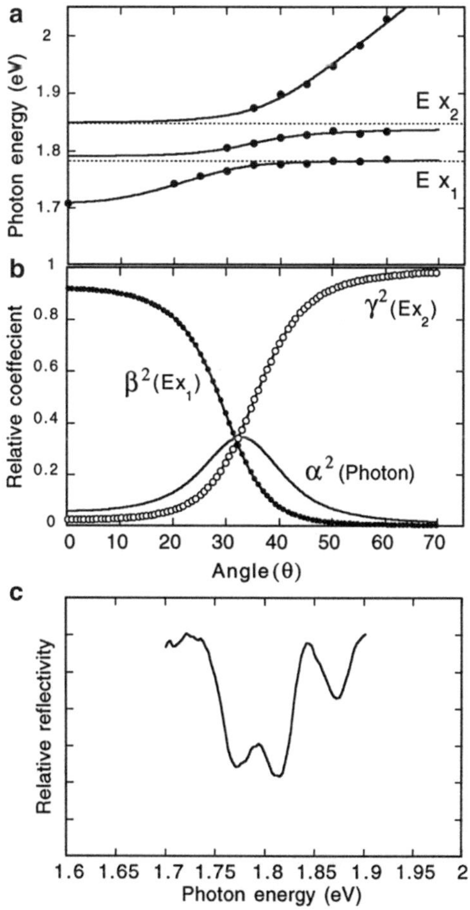

Fig. 12 (**a**) Polariton dispersion from reflectivity measurements from a microcavity structure containing two organic dyes simultaneously coupled to the cavity mode. (**b**) Mixing coefficients of the middle polariton branch. (**c**) Reflectivity spectrum measured close to the point of maximum mixing showing three dips corresponding to the three polariton branches. From [34]

band does not overlap with those of the strongly coupled molecules. It was shown through transfer matrix modelling that a single layer of the active materials would not have a high enough oscillator strength to result in strong coupling, hence it was concluded that the collective response of all of the layers in the photonic structure permitted the system to enter the strong coupling regime. This coupling is therefore the result of an interaction that extends across the entire structure that is over a micron thick.

Agranovich et al. had earlier predicted that by having an organic material and inorganic quantum wells in the same microcavity, the organic Frenkel excitons and inorganic Wannier–Mott excitons could be mixed to form hybrid polariton states that contain the properties of both exciton types [38]. This would result in polariton states that possess a high oscillator strength with a large interaction cross-section—properties desirable for non-linear switching applications.

Experimental observation of hybrid organic–inorganic polaritons was demonstrated simultaneously by two independent groups. Holmes et al. [39] coupled InGaP quantum wells to the small organic molecule TPP. Wenus et al. [40] instead coupled inorganic perovskites (PEATBP) and the molecular dye zinc-tetraphenylporphyrin (ZnTPP). Here it was found that the MPB polariton states were emissive, despite the fact that the ZnTPP Soret-band transition that was mixed into the polariton state was non-emissive. It was concluded therefore that the MPB polaritons acquired their emissive nature from the luminescent perovskites. It was also speculated that the MPB polaritons acted as energy relaxation pathway that coupled excitons from the PEATBP to the Q-band of the ZnTPP.

A more detailed study of the energy relaxation pathway created between two exciton species in a microcavity was recently reported [41]. A film containing two different J-aggregates (TDBC and NK-2707, which is redshifted from TDBC by ~40 nm) was placed in a metallic cavity. Angular resolved PL measurements following non-resonant excitation revealed that the LPB contained the largest polariton population at all external viewing angles, with the MPB and UPB having progressively smaller polariton populations. This result suggested that the MPB was rapidly and non-radiatively depopulated by polariton scattering to the lower-lying exciton level, again acting as an energy transfer pathway.

A simple phenomenological model was developed that was able to predict the luminescence intensity from the UPB, MPB and LPB as a function of viewing angle as shown in Fig. 13. The model described the following picture of the relaxation dynamics in this hybrid polariton system. The UPB is populated by both radiative pumping and thermal promotion of excitons from the TDBC reservoir. The relaxation of UPB polaritons to the upper-exciton reservoir is much faster than its radiative emission rate. The model suggests that the radiative decay channel is only responsible for depopulating 20 % of the hybrid polaritons in this branch. The MPB is almost entirely populated by the scattering from the higher energy exciton reservoir while it is depopulated through two competing processes: non-radiative relaxation into the lower-lying exciton reservoir and direct radiative emission. It was calculated that non-radiative relaxation accounts for the depopulation of over 90 % of the polaritons in this branch, explaining the strong suppression of luminescence from most of the MPB. Therefore once a TDBC exciton scatters to an MPB state, it is highly likely to quickly relax to the lower-lying exciton reservoir and hence the MPB has the crucial function of exchanging population between the two exciton species. It was argued that such hybrid polariton states act as a model for the energy transfer processes that can occur in certain photosynthetic complexes.

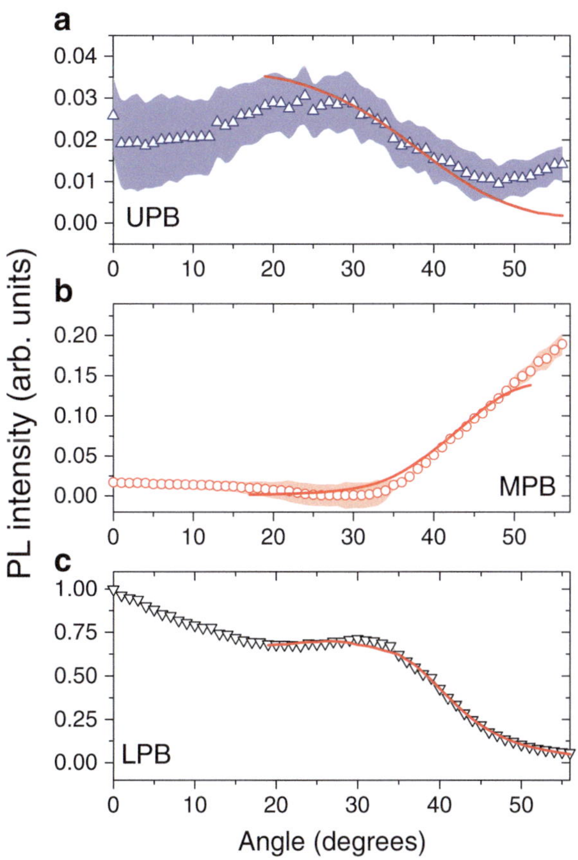

Fig. 13 Observed polariton emission intensity (*open symbols*) following non-resonant excitation at 473 nm, and modelled intensity (*solid lines*) for the (**a**) UPB, (**b**) MPB and (**c**) LPB. The *shaded regions* represent the experimental uncertainty. From [41]

8 Room-Temperature Polariton Lasing and Polariton Condensation

In 1999, Senellart and Bloch [42] showed that stimulated emission could occur from the $k_{\parallel} = 0$ states on the LPB. Since the emission occurs from the lowest energy polariton state, no population inversion is required, making polariton lasing distinct from photon lasing discussed in 7th, 16th and 20th chapters. Polariton lasing is therefore often called 'thresholdless', however in practice a threshold excitation energy is required to overcome the LPB bottleneck.

The first observation of lasing from a strongly coupled organic microcavity was made by Akselrod et al. [43]. There, a weakly coupled fluorescent dye was incorporated into the cavity along with a strongly coupled J-aggregate. Above threshold, the weakly coupled dye would lase at a wavelength that overlapped with the bottom of the LPB. This overcame the problem of the poor photostability and strong exciton–exciton annihilation present in J-aggregates that act to greatly increase the lasing threshold [44].

Polariton lasing from organic microcavities was first observed by Kéna-Cohen and Forrest in 2010 [45]. The novel microcavity structure used was fabricated by cold welding two DBRs together using gold stripes evaporated onto the DBR surfaces and pressing them together at high pressure to create channels 2 mm wide. Liquid anthracene melt was drawn into the channels through capillary action and slowly allowed to cool to form large single crystals. A schematic of the cavity structure is shown in Fig. 14a.

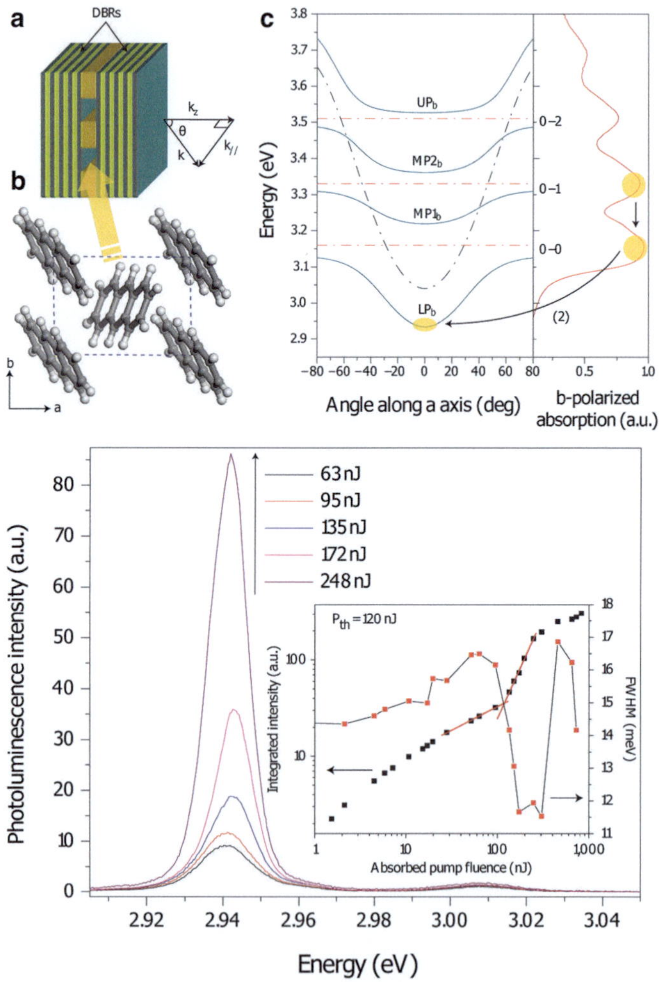

Fig. 14 (**a**) Cold-welded microcavity structure. (**b**) Anthracene molecular packing in the single crystal and the crystal axes **a** and **b**. (**c**) Polariton dispersion of the *a*-polarized polariton branches (*solid blue lines*). The uncoupled exciton and photon energies are shown as *dot-dashed red* and *black lines*, respectively. The **b**-polarized anthracene absorption is shown in the right panel (*solid red line*). *Bottom*: PL emission spectra of the LPB collected at normal incidence for a series of pulse excitation energies. The *inset* shows the integrated emission intensity and linewidth as a function of absorbed pump fluence. Reprinted by permission from [45]. Copyright 2010 by Macmillan Publishers Ltd

Anthracene crystals display vibronic replicas in the absorption spectrum, and hence the strongly coupled system displayed multiple polariton branches. There is also strong polarization anisotropy due to the molecular packing in the crystal. This leads to the splitting of each of the vibrational replicas into two bands, one for each of the crystal axes, depicted in Fig. 14b and labelled **a** and **b**. This in turn results in a polarization splitting of the polariton branches.

Figure 14c shows the polariton dispersion of the cavity for the **a**-polarized polariton branches (which is observed to be TM-polarized). The anthracene absorption spectrum (left) displays three peaks that strongly couple to the cavity mode, resulting in four polariton branches (an upper, lower and two middle).

The microcavity was designed such that the photoluminescence peak from the $0 \rightarrow 1$ vibronic replica overlapped with the $k_{\|} = 0$ state of the LPB. This resulted in efficient optical pumping of the polariton ground state, the main process responsible for populating these states [46]. The LPB ground state emission spectra for a series of pumping intensities is shown in Fig. 14 (bottom). The inset shows the integrated PL intensity and emission linewidth as a function of absorbed pump fluence. A simultaneous transition to a superlinear power dependence in the PL intensity and a narrowing of the emission linewidth is indicative of lasing. It was later shown that the lasing threshold could be reduced by almost an order of magnitude by cooling the sample to 12 K [47].

Since the constituent components of polaritons (excitons and photons) are bosons, the polariton itself can be described as a boson. Direct experimental evidence of bosonic behaviour was first observed in inorganic cavities [48] by Savvidis et al. in 2000, where final state stimulation of polaritons lead to an amplification of a light pulse that was probing the LPB ground state [28]. The most striking example of the bosonic nature of particles is the Bose–Einstein condensate. At first sight, polaritons appear to be well suited to forming a condensate. The photonic component of the polariton results in an effective mass on the order of 10^{-4} that of the free electron mass [49]. The critical temperature for Bose—Einstein condensation is inversely proportional to the effective mass [50], hence the small mass could allow critical condensation temperatures of 100 K for GaAs quantum well-based microcavities, and even higher in wide bandgap semiconductors such as ZnO [51]. However, the bosons comprising a Bose–Einstein condensate need to be in thermal equilibrium, a state that is hindered by the short polariton lifetime. Condensation of polaritons was demonstrated in 2006 by Kasprzak et al. when polaritons at $k_{\|} = 0$ on the LPB in a CdTe microcavity were shown to occupy a single quantum state [49]. The condensate formed at 19 K, well above the temperatures required to observe condensation in the 'traditional' atomic systems (generally on the order of μK). It is worth noting that since the condensate is formed of polaritons in the same state ($k_{\|} = 0$), when they radiatively decay the emission is coherent, therefore by creating a polariton condensate you also create a polariton laser (although one that does not operate on the principle of stimulated emission). This type of polariton condensate is termed 'non-equilibrium' since the population of polaritons in the condensate is not stable—it is continuously decaying and being replenished. This is in contrast to the atomic condensates which are formed by a stable population of atoms.

The observation of polariton condensates in inorganic systems of course raised the question about whether it would be possible to observe the phenomena in organic systems. Since the organic Frenkel excitons remain bound at room temperature, it may be expected that a room-temperature polariton condensate could be achieved. Indeed, this was recently demonstrated by two independent groups—Daskalakis et al. [52] at Imperial College London and Plumhof et al. [53] based at IBM Zurich. A thermally evaporated thin film of the oligofluorene 2,7-bis[9,9-di(4-methylphenyl)-fluoren-2-yl]-9,9-di(4-methylphenyl)fluorene (TDAF) was used by the Imperial group while the IBM group used the ladder polymer methyl-substituted ladder-type poly(*p*-phenylene) (MeLPPP).

Figure 15a–c show the angular resolved PL from the MeLPPP microcavities for increasing pump pulse intensities. At small pulse intensity, the near-parabolic emission from the LPB is clearly seen. As the pulse intensity is increased a second emission feature becomes visible that is blue-shifted from the LPB by up to 10 meV (although still lies well below the cavity mode energy, white solid line). At high

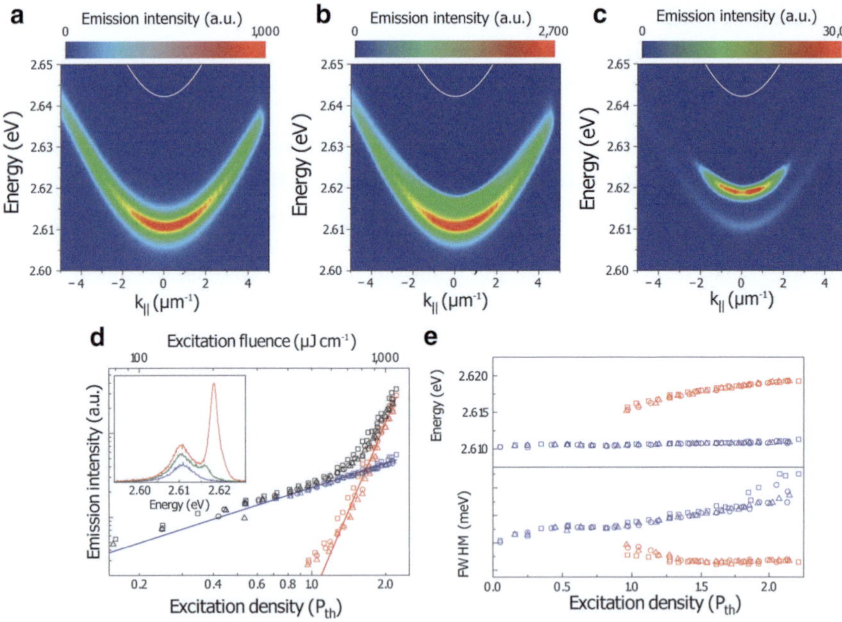

Fig. 15 Angle-resolved microcavity emission from an MeLPPP-based microcavity at pump intensities of (**a**) 0.22 P_{th}, (**b**) P_{th} and (**c**) 1.7 P_{th} where $P_{th} \sim 500\ \mu J\ cm^2$. (**d**) Excitation intensity dependence of the emission intensity of the LPB (*blue symbols*), condensate (*red symbols*) and total intensity (*black symbols*). The *inset* shows the normal incidence PL spectra taken at pump intensities of 0.5 P_{th} (*blue line*), P_{th} (*green line*) and 1.5 P_{th} (*red line*). (**e**) Emission peak position and linewidth dependence of the LPB (*blue symbols*) and condensate (*red symbols*) on the pump intensity. Reprinted by permission from [53]. Copyright 2013 by Macmillan Publishers Ltd

pulse intensities, the blue-shifted peak is clearly visible and is more intense than the bare LPB peak. The blue-shifted peak had a linewidth of approximately half that of the LPB peak and became visible at a pulse intensity threshold of $P_{th} \sim 500\ \mu J\ cm^2$. The inset of Fig. 15d shows the normal incidence PL spectra taken at pump intensities of 0.5 P_{th} (blue line), P_{th} (green line) and 1.5 P_{th} (red line), where the appearance of the blue peak is clearly seen. The excitation intensity dependence of the LPB peak (blue symbols) and blue peak (red symbols) is shown in Fig. 15d, while the total integrated intensity is shown as black symbols which shows a non-linear increase in emission intensity. The emission peak position and linewidth dependence on the pump intensity is shown in Fig. 15e where the blue symbols are the LPB and red symbols are the blue-shifted peak. It was also shown that above threshold, the emission lifetime reduces by a factor of more than 4.

While the non-linear response, narrow linewidth and lifetime collapse of the blue-shifted peak are the first signs that the emission is coherent in nature, it is also necessary to demonstrate spatial and temporal coherence of the emitting states to fully confirm the observation of polariton condensation. This can be done by creating interferograms by imaging the sample emission at normal incidence and splitting it along two perpendicular paths. At the end of one path the light is reflected, while at the end of the other the image is inverted and reflected back. The two images are then recombined. Spatial and temporal coherence manifests as interference fringes in the combined image (the basis of the Michelson interferometer). By introducing a small delay in one of the paths (by increasing the path length), the coherence time can be mapped. Figure 16a, b shows time-averaged interferograms, again for the MeLPPP-based cavities, for excitation intensities of 0.8 P_{th} and 5 P_{th}, respectively, with zero delay between the paths. Below threshold, no interference fringes are observed, while above threshold fringes are seen that cover the entire emission region. This is evidence that the emission originates from the same quantum state and can therefore be described as a polariton condensate. Parts (c) and (d) show single-shot (imaging the emission following a single excitation pulse) interferograms at zero delay and a 3 ps delay. It was observed that after 3 ps the interference fringes vanish and hence this is the limit of lifetime of the condensate. A threshold dependent large scale coherence was also observed in the TDAF microcavities.

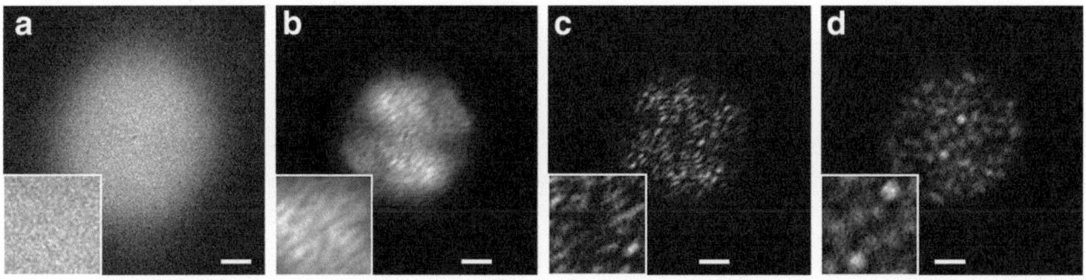

Fig. 16 Time-averaged interferograms for excitation intensities of (**a**) 0.8 P_{th} (**b**) 5 P_{th} with zero delay between the paths. Single-shot interferograms at (**c**) zero delay and (**d**) 3 ps delay. Reprinted by permission from [53]. Copyright 2013 by Macmillan Publishers Ltd

It is important to show that the above threshold emission occurs from polariton states, and that the system has not reverted to the weak coupling regime and is undergoing conventional photon lasing. Evidence of this is provided by temperature dependent measurements. The condensation threshold increases with reducing temperature. This implies that the formation of the condensate depends upon some thermal process which becomes more efficient with increasing temperature, such as scattering with molecular vibrations. A conventional laser displays the opposite threshold temperature dependence since the quantum yield of organic materials increases with decreasing temperature. Furthermore, the blueshift observed in the condensate emission is indicative of exciton–polariton and polariton–polariton repulsive interactions and has been well documented in inorganic structures [54, 49]. A theoretical treatment of the blueshift based on exciton density was performed by the Imperial group and a reasonable comparison with the experimental data obtained.

9 Organic Polariton Light-Emitting Diodes

In order for polariton based devices to become of practical use, it is highly desirable that such devices can be operated electrically. An organic polariton light-emitting diode (LED) was first demonstrated by Tischler et al. in 2005 [55].[1] This LED used the TDBC J-aggregate as the strong-coupling material which was deposited by a layer-by-layer technique. The anode consisted of a semi-transparent silver layer on indium-tin-oxide (ITO). The silver also acted as one of the cavity mirrors. Poly-TPD was used as a hole transport layer to inject charges into the J-aggregate. The cathode used was a silver mirror on top of a Mg:Ag electron injection layer. BCP was used an electron transport/hole blocking layer.

Angular dependent reflectivity spectra (Fig. 17a) showed an anticrossing of the polariton branches. Applying a voltage across the device results in an emission peak that follows the LPB dispersion. There is also a weak emission feature at the energy of the UPB, but it is far weaker than the LPB emission, as was observed in optically pumped microcavities. Electroluminescence (EL) spectra recorded at a series of angles are shown in Fig. 17c where the LPB emission is clearly visible. An expanded view of the UPB emission is shown in part (b). The polariton dispersions constructed from the reflectivity and EL spectra are shown in parts (d) and (e) for two LEDs of different cavity thicknesses, achieved by varying the thickness of poly-TPD layer.

An alternative method for electrical creation of polaritons was demonstrated by Lodden et al. in 2010 [59, 60]. By embedding an efficient electroluminescent material in a microcavity having an emission that overlaps with the absorption of

[1] It is worth noting that organic polariton electroluminescence preceded the observation of the inorganic analogue by 3 years [56–58].

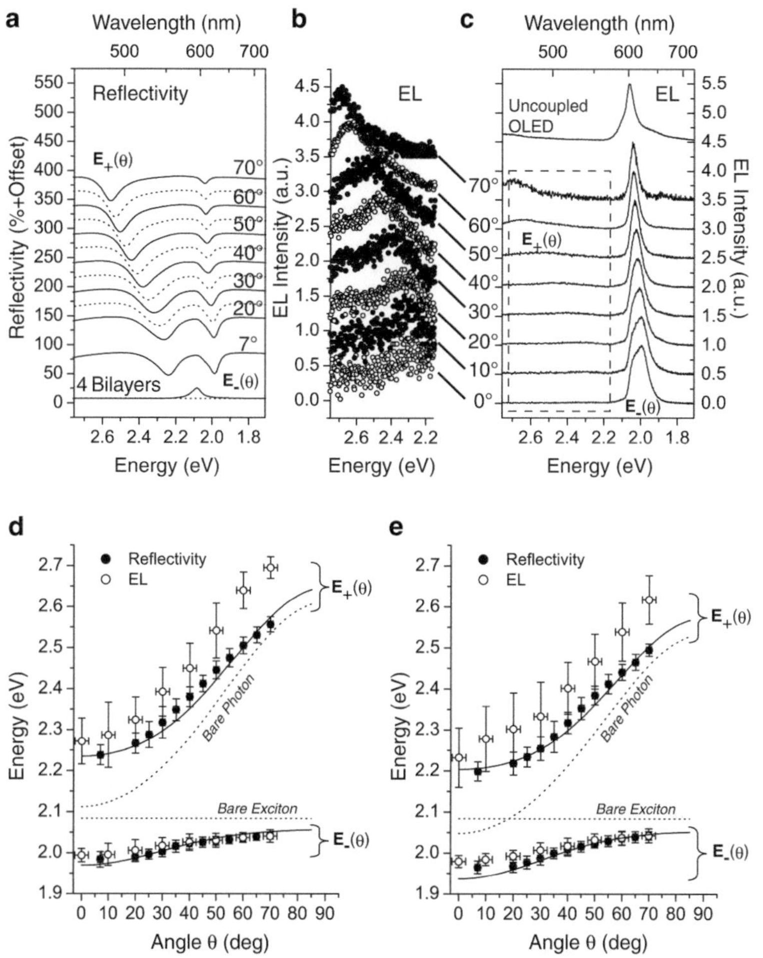

Fig. 17 (**a**) Angle-resolved reflectivity spectra of a strongly coupled organic polariton LED. (**b**) Expanded view of the angle-resolved UPB emission region taken from the full spectra shown in (**c**). (**d**) and (**e**) show polariton dispersions for cavities of different thicknesses. Reprinted with permission from [55]. Copyright 2005 by the American Physical Society

a material that underwent strong coupling, polariton states were optically pumped by the emission of the electroluminescent material.

The external efficiency of a strongly coupled polariton LED based on TDBC was compared with that of a weakly coupled device by Christogiannis et al. [61]. It was found that the polariton LED was six times less efficient than a weakly coupled control LED. This was attributed to the slow scattering of excitons into polariton states, hence most excitons remain in the reservoir rather than scattering into the emissive polariton branches. It was also concluded that to create a polariton condensate in an organic LED would require a significant enhancement in scattering rates between the exciton reservoir and the LPB to build up a large polariton density. Here, the use of a hybrid-polariton diode structure may become important; by creating an organic–inorganic polariton hybrid, a high injection density could be sustained by the inorganic component with efficient room-temperature emission from the organic-semiconductor polariton component.

10 Strong Coupling in Other Systems

Whilst microcavities provide a simple and fascinating system with which to study strong coupling in organic and hybrid materials, they are by no means the only systems where matter and electromagnetic radiation can interact strongly. Surface plasmons (SPs) [62] are electromagnetic fields confined at a metal-air interface and provide an equally rich playground for the study of polaritons. Due to surface plasmons having a momentum larger than light in free space, they cannot be excited directly by shining light onto a metal surface. Methods of exciting surface plasmons include prism, grating or defect coupling which all act to increase the momentum of light relative to free space. The first demonstration of strong coupling between surface plasmons and organic excitons was by Bellessa et al. [63].

Here, a thin film of TDBC in a PVA matrix was deposited onto a 50 nm thick silver film, and was placed on a hemispherical prism, as shown in Fig. 18a. Angular dependent reflectivity spectra (Fig. 18b) and the dispersion relation (Fig. 18c) clearly show the split-peak structure and anticrossing characteristic of strong coupling. The Rabi splitting energy is of the same order of magnitude as observed in microcavity structures. Further experiments involved extraction of polariton luminescence through adding a corrugation to the organic-metal layer [64] and investigation of the effect of plasmon polarization [65]. Patterning of the metal surface with features such as sub-wavelength sized holes [66] or slits [67] also allows for direct coupling to SP modes.

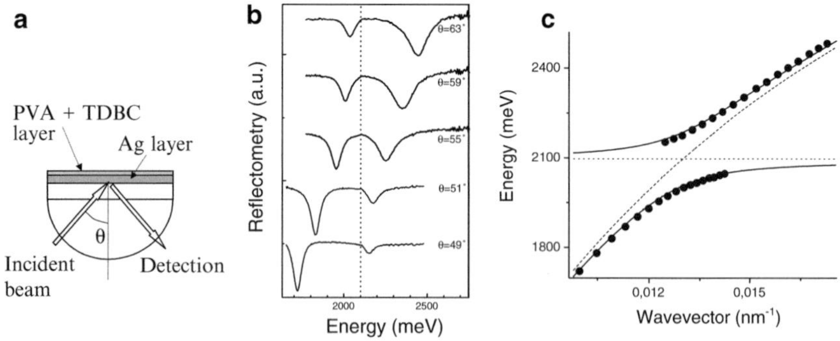

Fig. 18 (**a**) Experimental setup for the observation of strong coupling between surface plasmons and organic exciton. (**b**) Reflectivity spectra at a series of angles. (**c**) Polariton dispersion from reflectivity spectra. Reprinted with permission from [63]. Copyright 2004 by the American Physical Society

Perovskites (organic–inorganic hybrid materials) have also been shown to strongly couple with SPs [68], and it was later shown that two energetically detuned excitons could simultaneously couple to an SP creating hybrid SP-polaritons [69]. Strong coupling between J-aggregates and SPs on a silver film has been shown to enable coherent emission of the aggregates over micron length scales [70].

It is not only metal films that support surface plasmons. Other metallic structures that have been shown to support strong coupling between SPs and organic materials included gold nanoparticles [71], nanovoids [72], nanorods [73] and silver nanodisks [74]. Other structures that have supported strong coupling in organic or hybrid materials include distributed feedback gratings made by etching channels into a quartz substrate and filling the channels with the strong coupling material [75–78] and the 2D photonic crystal analogue [79], as well as a 2D photonic crystal made from silica microspheres and infiltrated with organic–inorganic hybrid perovskite, bis-(phenethylammonium) tetraiodoplumbate (PAPI) [80]. Bloch surface waves, which occur at the interface of a photonic crystal (e.g. a DBR) and a dielectric medium and propagate in the plane of the DBR, have also recently been shown to strongly couple to J-aggregates [81].

11 Summary and Outlook

Organic polaritons have proved to be a rich area of research for 15 years and have continued to deliver fascinating and surprising results right up to the present. Their properties can be distinguished from polaritons formed in cavities based on GaAs-based QW systems, through room-temperature operation, increased light–matter interaction strength and ease of sample fabrication. In this chapter we have reviewed the basics of classical strong coupling, described the common experimental techniques used to characterize microcavities and provided a condensed literature review of some of the most important experimental results in the field. With the recent observation of polariton lasing, polariton condensation and polariton-mediated energy transfer, new and exciting avenues of study have been opened up with possible applications in ultrafast optical switches, energy harvesting systems, low-threshold lasers, optical amplifiers and systems for polariton logic. Organic–inorganic hybrid polariton LEDs meanwhile offer a potential means to generate and utilize polariton states in a practical device. We hope that the reader will be encouraged to join the growing number of researchers in the area of organic microcavities and contribute towards taking organic microcavities into the realm of practical applications.

References

1. D.I. Babic, S. Corzine, IEEE J. Quantum Electron. **28**, 514 (1992)

2. M. Cardona, R. Merlin, *Light Scattering in Solids IX: Novel Materials and Techniques* (Springer, New York, 2007)
3. G. Panzarini, L. Andreani, A. Armitage et al., Phys. Solid State **41**, 1223 (1999)
4. M. Fox, *Quantum Optics: An Introduction*. Oxford Master Series (Oxford University Press, Oxford, 2006)
5. K. Ujihara, Jpn. J. Appl. Phys. **30**, L901 (1991)
6. C. Weisbuch, M. Nishioka, A. Ishikawa, Y. Arakawa, Phys. Rev. Lett. **69**, 3314 (1992)
7. D.G. Lidzey, D.D.C. Bradley, M.S. Skolnick et al., Nature **395**, 53 (1998)
8. L.C. Andreani, V. Savona, P. Schwendimann, A. Quattropani, Supperlattices Microstruct. **15**, 453 (1994)
9. M.S. Skolnick, T.A. Fisher, D.M. Whittaker, Semicond. Sci. Technol. **13**, 645 (1998)
10. V.M. Agranovich, M. Litinskaia, D.G. Lidzey, Phys. Rev. B **67**, 085311 (2003)
11. D.G. Lidzey, D.D.C. Bradley, T. Virgili et al., Phys. Rev. Lett. **82**, 3316 (1999)
12. G. Scheibe, Angew. Chem. **50**, 212 (1937)
13. E.E. Jelley, Nature **138**, 1009 (1936)
14. D.G. Lidzey, A.M. Fox, M.D. Rahn et al., Phys. Rev. B **65**, 195312 (2002)
15. D.M. Coles, P. Michetti, C. Clark et al., Adv. Funct. Mater. **21**, 3691 (2011)
16. N. Somaschi, L. Mouchliadis, D. Coles et al., Appl. Phys. Lett. **99**, 143303 (2011)
17. P. Michetti, G.L. Rocca, Physica E **40**, 1926 (2008)
18. P. Michetti, G.C. La Rocca, Phys. Rev. B **71**, 115320 (2005)
19. P. Michetti, G.C. La Rocca, Phys. Rev. B **79**, 035325 (2009)
20. P. Michetti, G.C. La Rocca, Phys. Rev. B **77**, 195301 (2008)
21. P. Michetti, G.C. La Rocca, Phys. Rev. B **82**, 115327 (2010)
22. D.M. Coles, R.T. Grant, D.G. Lidzey et al., Phys. Rev. B **88**, 121303 (2013)
23. T. Virgili, D. Coles, A.M. Adawi et al., Phys. Rev. B **83**, 245309 (2011)
24. M. Müller, J. Bleuse, A. André, H. Ulmer-Tuffigo, Physica B **272**, 476 (1999)
25. M. Müller, J. Bleuse, R. André, Phys. Rev. B **62**, 16886 (2000)
26. A.I. Tartakovskii, M. Emam-Ismail, R.M. Stevenson et al., Phys. Rev. B **62**, R2283 (2000)
27. J.-H. Song, Y. He, A.V. Nurmikko et al., Phys. Rev. B **69**, 235330 (2004)
28. P.G. Savvidis, J.J. Baumberg, R.M. Stevenson et al., Phys. Rev. Lett. **84**, 1547 (2000)
29. M. Saba, C. Ciuti, J. Bloch et al., Nature **414**, 731 (2001)
30. P.G. Savvidis, L.G. Connolly, M.S. Skolnick et al., Phys. Rev. B **74**, 113312 (2006)
31. M. Litinskaya, P. Reineker, V. Agranovich, J. Lumin. **119–120**, 277 (2006)
32. M. Litinskaya, P. Reineker, V. Agranovich, J. Lumin. **110**, 364 (2004)
33. J. Wainstain, C. Delalande, D. Gendt et al., Phys. Rev. B **58**, 7269 (1998)
34. D.G. Lidzey, D.D.C. Bradley, A. Armitage et al., Science **288**, 1620 (2000)
35. R.J. Holmes, S.R. Forrest, Phys. Rev. Lett. **93**, 186404 (2004)
36. S. Kéna-Cohen, M. Davanço, S.R. Forrest, Phys. Rev. Lett. **101**, 116401 (2008)
37. G.H. Lodden, R.J. Holmes, Phys. Rev. Lett. **109**, 096401 (2012)
38. V. Agranovich, H. Benisty, C. Weisbuch, Solid State Commun. **102**, 631 (1997)
39. R.J. Holmes, S. Kéna-Cohen, V.M. Menon, S.R. Forrest, Phys. Rev. B **74**, 235211 (2006)
40. J. Wenus, R. Parashkov, S. Ceccarelli et al., Phys. Rev. B **74**, 235212 (2006)
41. D.M. Coles, N. Somaschi, P. Michetti et al., Nat. Mater. **13**, 712 (2014)
42. P. Senellart, J. Bloch, Phys. Rev. Lett. **82**, 1233–1236 (1999)
43. G.M. Akselrod, E.R. Young, M.S. Bradley, V. Bulović, Opt. Express **21**, 12122 (2013)
44. G.M. Akselrod, Y.R. Tischler, E.R. Young et al., Phys. Rev. B **82**, 113106 (2010)
45. S. Kéna-Cohen, S. Forrest, Nat. Photonics **4**, 371 (2010)
46. L. Mazza, S. Kéna-Cohen, P. Michetti, G.C. La Rocca, Phys. Rev. B **88**, 075321 (2013)
47. M. Slootsky, Y. Zhang, S.R. Forrest, Phys. Rev. B **86**, 045312 (2012)
48. P.G. Lagoudakis, Nat. Mater. **13**, 227 (2014)
49. J. Kasprzak, M. Richard, S. Kundermann et al., Nature **443**, 409 (2006)
50. B. Deveaud (ed.), *The Physics of Semiconductor Microcavities* (Wiley, Chichester, 1999)
51. A. Das, J. Heo, A. Bayraktaroglu et al., Opt. Express **20**, 11830 (2012)

52. K.S. Daskalakis, S.A. Maier, R. Murray, S. Kéna-Cohen, Nat. Mater. **13**, 271 (2014)
53. J.D. Plumhof, T. Stöferle, L. Mai et al., Nat. Mater. **13**, 247 (2014)
54. H. Deng, G. Weihs, D. Snoke et al., Proc. Natl. Acad. Sci. **100**, 15318 (2003)
55. J.R. Tischler, M.S. Bradley, V. Bulović et al., Phys. Rev. Lett. **95**, 036401 (2005)
56. S.I. Tsintzos, N.T. Pelekanos, G. Konstantinidis et al., Nature **453**, 372 (2008)
57. D. Bajoni, E. Semenova, A. Lemaître et al., Phys. Rev. B **77**, 113303 (2008)
58. A.A. Khalifa, A.P.D. Love, D.N. Krizhanovskii et al., Appl. Phys. Lett. **92**, 061107 (2008)
59. G.H. Lodden, R.J. Holmes, Phys. Rev. B **82**, 125317 (2010)
60. G.H. Lodden, R.J. Holmes, Appl. Phys. Lett. **98**, 233301 (2011)
61. N. Christogiannis, N. Somaschi, P. Michetti et al., Adv. Opt. Mater. **1**, 503 (2013)
62. W.L. Barnes, A. Dereux, T.W. Ebbesen, Nature **424**, 824 (2003)
63. J. Bellessa, C. Bonnand, J.C. Plenet, J. Mugnier, Phys. Rev. Lett. **93**, 036404 (2004)
64. C. Bonnand, J. Bellessa, C. Symonds, J.C. Plenet, Appl. Phys. Lett. **89**, 231119 (2006)
65. C. Bonnand, J. Bellessa, J.C. Plenet, Phys. Rev. B **73**, 245330 (2006)
66. J. Dintinger, S. Klein, F. Bustos et al., Phys. Rev. B **71**, 035424 (2005)
67. A. Salomon, R.J. Gordon, Y. Prior et al., Phys. Rev. Lett. **109**, 073002 (2012)
68. C. Symonds, J. Bellessa, J.C. Plenet et al., Appl. Phys. Lett. **90**, 091107 (2007)
69. T.K. Hakala, J.J. Toppari, A. Kuzyk et al., Phys. Rev. Lett. **103**, 053602 (2009)
70. S. Aberra Guebrou, C. Symonds, E. Homeyer et al., Phys. Rev. Lett. **108**, 066401 (2012)
71. D. Djoumessi Lekeufack, A. Brioude, A.W. Coleman et al., Appl. Phys. Lett. **96**, 253107 (2010)
72. Y. Sugawara, T.A. Kelf, J.J. Baumberg et al., Phys. Rev. Lett. **97**, 266808 (2006)
73. G.A. Wurtz, P.R. Evans, W. Hendren et al., Nano Lett. **7**, 1297 (2007)
74. J. Bellessa, C. Symonds, K. Vynck et al., Phys. Rev. B **80**, 033303 (2009)
75. T. Fujita, Y. Sato, T. Kuitani, T. Ishihara, Phys. Rev. B **57**, 12428 (1998)
76. F. Sasaki, S. Kobayashi, S. Haraichi, Appl. Phys. Lett. **81**, 391 (2002)
77. F. Sasaki, S. Haraichi, S. Kobayashi, IEEE J. Quantum Electron. **38**, 943 (2002)
78. M. Shimizu, T. Ishihara, Appl. Phys. Lett. **80**, 2836 (2002)
79. R. Shimada, A. Yablonskii, S. Tikhodeev, T. Ishihara, IEEE J. Quantum Electron. **38**, 872 (2002)
80. K. Sumioka, H. Nagahama, T. Tsutsui, Appl. Phys. Lett. **78**, 1328 (2001)
81. S. Pirotta, M. Patrini, M. Liscidini et al., Appl. Phys. Lett. **104**, 051111 (2014)

Plasmonic and Photonic Crystals

Robert Brückner, Vadim G Lyssenko, and Karl Leo

Abstract In the last years, photonic crystals gained much attention and various new designs have been discussed. To create useful optoelectronic devices, sophisticated structures and hybrid devices consisting of dielectric and organic media combined with highly conductive materials such as metals have to be engineered. In the first part of this chapter, we discuss various designs of photonic crystals including distributed Bragg reflectors (DBRs) and organic microcavities, exhibiting high-quality photonic cavity modes. In a next step, the challenge of combining highly absorptive metallic layers with DBRs or organic microcavities is discussed. As a consequence of the interaction of the photonic cavity mode with the metallic structures, new hybridized modes—so-called Tamm plasmon-polaritons (TPPs)—can be observed. Higher-dimensional functionality of hybrid photonic crystals can be realized by periodically structuring the metal layer. The structuring leads to a strong confinement of the fields which is evident from a clear discretization of the modes. Additionally, surface plasmon-polaritons (SPPs) are excited, propagating at the interface between the silver and the adjacent dielectric layers. Modes up to very high order numbers are detectable as quasi-linear periodic lines in the dispersion pattern. Finally, the coherence properties of the devices are discussed, including periodic arrays of localized cavity modes and metal-based Tamm plasmons.

Keywords Photonic crystals • Microcavities • Vertical cavity surface emitting laser • VCSEL • oVCSEL • Distributed Bragg Reflector • DBR • Laser • Lasing • Tamm state • Tamm Plasmon-Polariton • TPP • Surface Plasmon-Polariton • SPP • Superlattice • Bragg-scattering • Metal contacts

R. Brückner (✉)
Center for Advancing Electronics Dresden Technische Universität Dresden, Dresden, Germany

Institut für Angewandte Photophysik, Technische Universität Dresden, 01062 Dresden, Germany
e-mail: Robert.Brueckner@iapp.de

V.G. Lyssenko • K. Leo
Institut für Angewandte Photophysik, Technische Universität Dresden, 01062 Dresden, Germany
e-mail: Vadim.Lyssenko@iapp.de; Karl.Leo@iapp.de

D. Comoretto (ed.), *Organic and Hybrid Photonic Crystals*,
DOI 10.1007/978-3-319-16580-6_12

1 Introduction to Photonic and Plasmonic Crystals

In this chapter, we concentrate on microcavities which are 1D and with additional structuring, 2D photonic crystals. We show ways to implement metal into these structures to realize hybrid metal-photonic crystals [1]. Due to the lossy nature of metals, special designs have to be developed which support plasmonic excitations and in some cases even the coupling to photonic modes. As many examples for photonic crystals are given in this book, we concentrate on photonic crystals with special functionality. Inspired by nature, photonic crystals are referred to as periodic structures which interact with impinging light. This way, the reflection and/or transmission properties are strongly influenced and many applications are obvious. Further functionality is added by modifying the 1D photonic crystal structures, for example, by breaking the perfect periodicity of the crystal or by adding active materials, i.e. media exhibiting optical net gain such as organic materials, or, the opposite, passive materials with high absorption such as metals. As a result, interesting devices which exhibit potential application features can be obtained.

In the first part, we will introduce into the theory of 1D photonic crystals. Next, defect states in such crystals are discussed and are exemplarily represented by simple microcavity structures. To enhance the functionality and complexity, higher-dimensional systems are realized by laterally structuring only one single layer within the 1D photonic structure. This approach allows to study 2D crystal properties, although experimentally realized much easier than fully periodic systems in two or even three dimensions. With this basis, initial spectroscopic experiments inspired by solid-state physics are shown, and first metal layers in optically active microcavities are discussed, including the coherence properties. The optically passive metal leads to plasmonic excitations contributing to the emission pattern of the sample. The strong dependence on the metal not only allows the control of optical modes, but also makes new applications for sensing or lasers possible.

1.1 *Distributed Bragg Reflectors*

Reflections at metallic surfaces originate from the high density of free electrons in metals (see Sect. 2.2). However, electrons may start oscillating collectively, increasing the probability of collisions resulting in a conversion of energy into heat. Thus, the reflectivity of metals in the visible range is limited.

In general, electromagnetic fields separate into a transmitted and a reflected part in the presence of an optical boundary. For instance, in the visible spectrum, the amount of reflected light at a single dielectric, i.e., transparent surface is only a few percent. To increase the reflectivity at a certain design wavelength λ_d, a sequence of dielectric layers with alternating refractive indices (n_j) is considered, forming a simple 1D photonic crystal. In this case, multiple reflections occur which can interfere constructively if the thickness of each layer is $d_j = \lambda_d/4n_j$, where $j = \text{H}$, L represent high and low refractive medium, respectively. Consequently, the phase

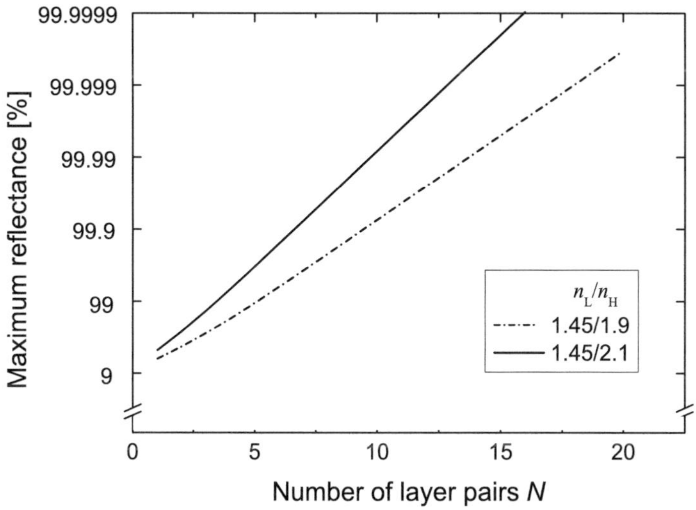

Fig. 1 Calculated maximum reflectance versus number of layer pairs for the typical material combination of $n_H = 2.1$ (TiO_2) or 1.9 (ZrO_2) and $n_L = 1.45$ (SiO_2), with $n_s = 1.5$ and $n_0 = 1.0$. Taken from [1]

shift of the reflected parts equals $2 \cdot \pi/2$. Additionally, the wave experiences a phase shift of π when being reflected at an optically dense interface. The resulting overall phase shift is then 2π, meaning that successive reflections from neighboring interfaces will be in-phase which is prerequisite for constructive interference. This situation is analogous to Bragg reflections of X-rays at crystal planes. Hence, a device with $2N + 1$ layers, starting and ending with the high refractive medium is referred to as distributed Bragg reflector (DBR).

Figure 1 illustrates the maximum reflectivity of a DBR depending on the number of layer pairs. Two sets of refractive indices are evaluated which are frequently used to design mirrors in the visible spectrum which is TiO_2/SiO_2 ($n_H/n_L = 2.1/1.45$) and ZrO_2/SiO_2 ($n_H/n_L = 1.9/1.45$), respectively. The calculated curves do neither account for lossy media nor the respective material dispersion. Hence, for realistic media, the losses prevent the reflectance from approaching 100 %, so that additional layer pairs would not improve the reflectance further.

In chapter "Polymer Multilayer by Spin Coating" by Scotognella et al., DBRs made of polymers are under study for interesting application. In Fig. 2, an approach to calculate the optical properties called transfer matrix algorithm [2] is applied to calculate the reflectance spectra of DBRs with varying number of layer pairs ranging from $N = 3, 7, 10$ and again for both sets of refractive indices. With increasing number of layers, the reflectance within a certain spectral range increases. This range is called the stop band and is analogous to the band gap occurring in semiconductor crystals. In both cases, the origin of the band gap is the periodicity of either the electronic crystal or the DBR as a crystal for photons. At this point, the expression "photonic crystal" with a pronounced "forbidden band" (at least for

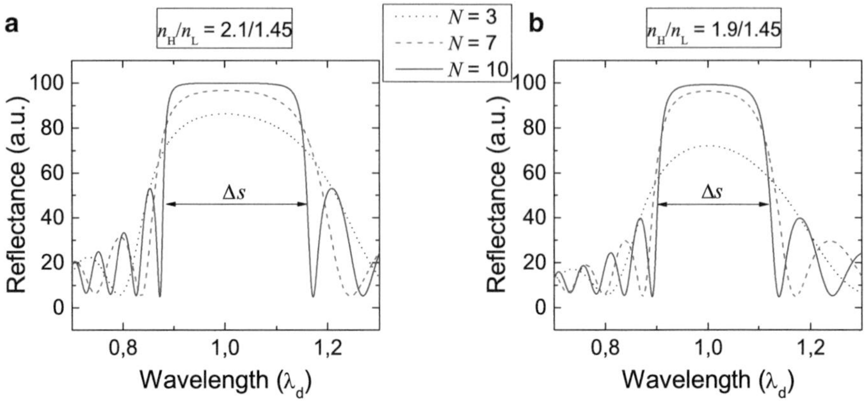

Fig. 2 Calculated reflectance of loss-less DBR mirrors composed of up to 21 alternating layers of n_H and n_L at normal incidence. The design is described as (H L)*N* H. With increasing number of layer pairs *N*, the stop band width $\Delta\ s$ extends. (**a**) $n_H/n_L = 2.1/1.45$ represent the material combination of TiO_2/SiO_2 and (**b**) $n_H/n_L = 1.9/1.45$ represent the material combination of ZrO_2/SiO_2. Taken from [1]

TE-polarized light) is fully motivated [3, 4]. In order to design optical elements based on DBRs it is important to estimate the extension of the stop band as [5]:

$$\Delta s = \frac{2\lambda_d \Delta n}{\pi n_{eff}}. \tag{1}$$

The maximum reflectivity is found in the center of the stop band exactly at the design wavelength λ_d. Δn denotes the difference in refractive indices of the composing materials, and n_{eff} is the effective refractive index of the DBR. If the index contrast is small (as it is here), n_{eff} can be approximated by the average value of n_H and n_L.

To the left and right of the stop band, side bands are found that exhibit oscillatory behavior with large amplitudes down to zero reflectance. If plotted in units of photon energy, the period of these oscillations is constant.

1.2 Microcavity Structures

DBRs can be utilized to arrange high-quality interferometric setups. One possible geometry is the Fabry–Pérot interferometer, where light is reflected multiple times between two parallel plates. If the distance between the two mirrors is of the order of the wavelength of light, the system is called micro-resonator or microcavity. For normal incidence $\theta = 0$, standing waves can occur if the optical thickness of the cavity L_c determined by the distance d_c between the mirrors is a multiple of the optical half-wavelength:

$$L_c = n_c d_c = \frac{m\lambda_c}{2}, \tag{2}$$

where $m = 1, 2, 3, \ldots$ is the mode number, n_c and d_c are the refractive index and the physical thickness of the cavity layer, respectively. Note that phase delays in the mirrors are neglected.

To describe the simplest microcavity in which the cavity thickness corresponds to an half-wavelength layer, the qualitative description of the phase of a DBR can easily be applied. The situation corresponds to a DBR with a large number of layers, with the special case of one layer in the center exhibiting twice the optical thickness. This layer can be regarded as a defect which will cause a characteristic defect state. The phase shift of a wave traveling through a $\lambda/2$ layer equals $\phi_c = k_c d_c = \pi$, instead of $\pi/2$ for all other single quarter-wavelength layers of a DBR. As a consequence, the former condition for constructive interference on reflection is canceled and destructive interference at the design wavelength will occur.

In Fig. 3, the altered reflection spectra for two different typical material combinations of a half-wavelength microcavity structure are depicted. Exactly at the design wavelength, where in the case of a DBR the maximum reflectance was achieved, a resonant state of maximal transmittance occurs due to the canceled interference. This phenomenon is again analogous to defect states in solid state physics, which may occur within the electronic band gap of a crystal.

In Fig. 4b a scanning electron micrograph of a half-wavelength microcavity is depicted. The different layers and materials are clearly visible.

In contrast to two-beam interferometers, in a Fabry–Pérot-type etalon, the number of beams which can interfere with each other only depends on the optical losses. By using highly reflective mirrors, such as DBRs, and non-absorbing optical

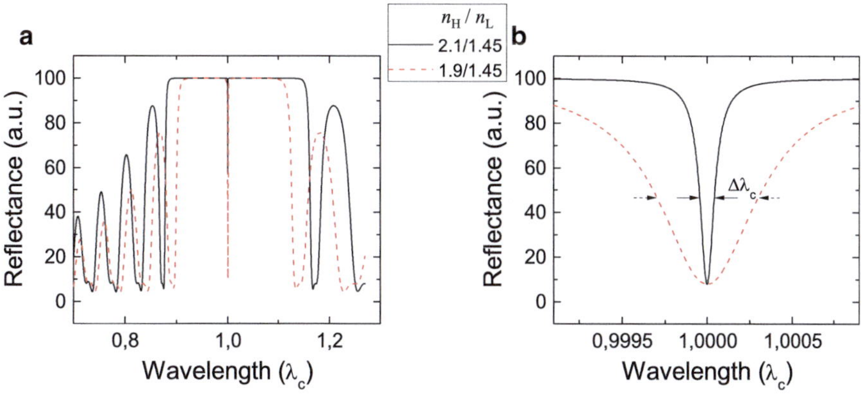

Fig. 3 Calculated reflectance spectra of loss-less microcavities composed of two DBRs and a cavity layer at normal incidence. The design is (H L)10 H 2L H (L H)10. (**a**) The localized cavity resonance appears at the cavity wavelength λ_c, in the center of the stop band. (**b**) The resonances exhibit a Lorentzian shape. The width $\Delta\lambda_c$ of the lines is in accordance with the estimation of the Q factor from Eq. (3) for loss-less media. The residual reflectance does not approach zero, since the minimum reflectance is determined by the boundary between ambient and substrate. Taken from [1]

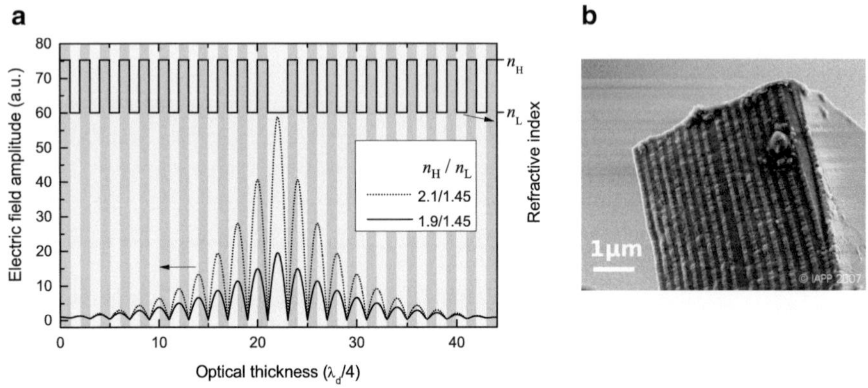

Fig. 4 (**a**) Calculated electric field amplitude for a (H L)10 H 2L H (L H)10 microcavity for different combinations of refractive indices. For these stacks, an increase of index contrast of 0. 2 causes a threefold higher field enhancement. Light is incident from the right with unit amplitude. The surrounding media are air $n_0 = 1$ and a glass substrate $n_S = 1.5$. Taken from [1]. (**b**) SEM micrograph of a dielectric microcavity, resembling a 1D photonic crystal with a defect layer. © iapp

media, the number of round-trips inside the Fabry–Pérot resonator can exceed 10^9 [6] leading to an extremely high spectral resolution.

To quantify the resolution, one evaluates the quality factor (Q factor) of the microcavity as the ratio of the resonant cavity wavelength to its spectral width [7]:

$$Q = \frac{\lambda}{\Delta\lambda} = \frac{4m}{n_c} \frac{\pi\sqrt{R}}{1 - R}. \tag{3}$$

Here, the mirror reflectances are assumed to be equal (and close to unity) and the cavity medium as loss-less. The Q factor scales linearly with the mode number and therefore with the cavity thickness [8, 9]. Note that the Q factor depends on all losses, i.e. not only on the limited reflectivity of the mirrors, but also on the absorption of the resonant light inside the structure.

In general, the Q factor is a measure of the rate at which optical energy is dissipated from within the cavity (due to absorption, scattering, or mirror leakage), and Q^{-1} corresponds to the fraction of energy which is lost in a single round-trip.

Even though the microcavity is actually transparent at its resonant wavelengths, photons of the corresponding energy cannot simply pass the device, but are trapped inside the structure. After a certain time, the cavity lifetime,

$$\tau_c = \frac{Q\lambda_c}{2\pi c}, \tag{4}$$

photons can escape from the cavity, resulting in a characteristic transmission resonance as demonstrated in Fig. 3b.

The accumulation of photons inside the cavity leads to the generation of an intense electric field there, see Fig. 4a. This field enhancement is determined by the Q factor.

The spectral distance between non-overlapping neighboring resonances $\lambda_{c,m}$ and $\lambda_{c,m+1}$ is called free spectral range (FSR) and scales inversely proportional to the cavity length. Especially for microcavities with $m = 1$, the FSR typically exceeds the stop band width Δs, allowing only a single resonance to exist. Hence, microcavities are capable of low-threshold single mode lasing, which will be discussed later.

1.3 Cavity Photon Dispersion in Quantized Systems

In this section, we want to briefly discuss fundamentals of photons in quantized systems. A short view to the simple formalism helps to interpret angular-resolved experiments revealing the corresponding dispersion relation. If the dispersion relation of a photonic structure lies inside the light cone $\omega = ck$ (with ω the angular frequency, c the speed of light, and k the wave number), photons can couple to it according to energy and momentum conservation. The characteristic energy dispersion of a Fabry–Pérot resonator is determined by the design of the resonator (schematic depicted in Fig. 7) and the resulting allowed components of the **k**-vector:

$$\mathcal{E} = \hbar\omega = \hbar c|\mathbf{k}| = \hbar c\sqrt{k_z^2 + k_y^2 + k_x^2}. \tag{5}$$

Here, we define the sample substrate as lying in the $x - y$-plane, with its normal being parallel to the z-direction. Thus, for planar microcavities, photons are confined by the mirrors, while they are free to propagate within the cavity plane ($x - y$). The resulting confinement in z-direction yields $k_z = \frac{2\pi}{\lambda_c}$. Taking the condition for λ_c according to Eq. (2) and the relation $|\mathbf{k}| = \frac{k_z}{\cos\theta}$, where θ is the external angle taken with respect to the cavity normal, one can write the photon dispersion for one-dimensionally confined microcavities [10–13] as:

$$\mathcal{E} = \frac{\hbar c\pi}{d_c n_c \cos\theta}. \tag{6}$$

In case of 2D lateral confinement, the refractive index is uniaxially modulated in the resonator plane by preparing, for example, wire-shaped structures. Typically, the wire width L_y is of the order of the wavelength. Hence, photons are still free to propagate along the x-axis, whereas the boundary conditions at the discontinuity of the cavity refractive index in y-direction require the electric field to vanish there. This condition leads to $k_y = m_y\pi/(n_c L_y)$, where m_y denotes the number of nodes in the electric field distribution across the wire (in y-direction). The discretization of the dispersion for a wire-shaped, two-dimensionally confined cavity reads:

$$\mathcal{E} = \frac{\hbar c \pi}{n_c} \sqrt{\frac{1}{L_c^2 \cos^2\theta} + \frac{(m_y + 1)^2}{L_y^2}}. \tag{7}$$

Finally, a further discontinuity of the refractive index in x-direction leads to a complete 3D confinement yielding no momentum-dependent term:

$$\mathcal{E} = \frac{\hbar c \pi}{n_c} \sqrt{\frac{1}{L_c^2 \cos^2\theta} + \frac{(m_y + 1)^2}{L_y^2} + \frac{(m_x + 1)^2}{L_x^2}}, \tag{8}$$

where m_x and L_x are defined similarly to the previous case.

The interpretation of these formulae reveals interesting aspects. For 1D confinement, according to Eq. (6), the dispersion relation can be approximated as parabolically-shaped, see Fig. 5a. The apex of the parabola corresponds to the cavity design wavelength. With increasing angle, the mode shifts towards shorter wavelengths or higher energies as it is typical for Fabry–Pérot resonators.

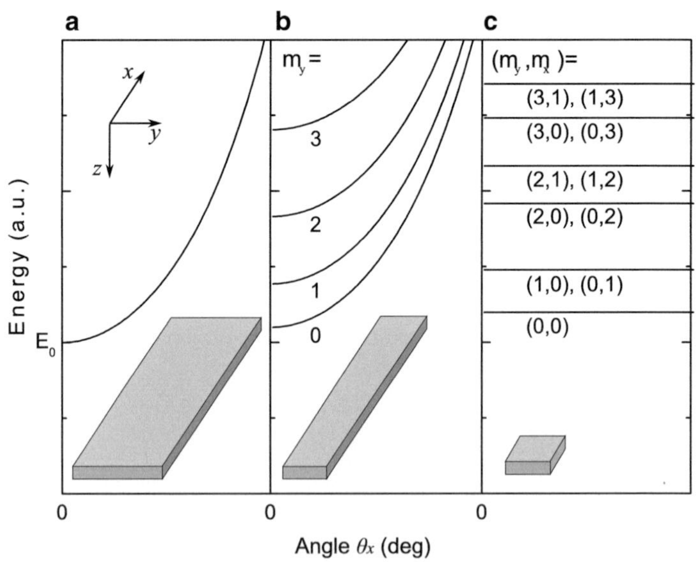

Fig. 5 Photon dispersion energy vs. angle θ in x-direction for structures providing different degrees of confinement. (**a**) The basic dispersion for a planar structure is approx. parabolic in shape. The apex of the curve corresponds to the design wavelength of the microcavity. (**b**) 2D confinement ($L_y = 5 \cdot L_z$) leads to a discretization in energy. The eigenenergy for the lowest quantum number is shifted to higher values. The difference in energy scales quadratically with increasing mode number. (**c**) In case of 3D confinement, with $L_x = L_y = 5 \cdot L_z$, photons are not allowed to propagate in any direction. The split modes show no dispersion and are degenerate for non-zero mode numbers due to the symmetric structure. After [12]

A reduction of the dimensionality (leading to a higher dimensional mode-confinement) by introducing a discontinuity of the refractive index of length L_y (in y-direction) into the structure leads to the formation of split parabolic modes replacing the formerly single parabolic mode (see Fig. 5b). The splitting in energy scales quadratically with the number of nodes m_y along this direction according to Eq. (7). The ground state energy is slightly shifted to higher values with respect to the unstructured case. Additional confinement of the device in x-direction by L_x leads to a further modification of the dispersion. The split modes exhibit no dispersion and are degenerate for non-zero mode numbers m_y and m_x, if $L_y = L_x$, see Fig. 5c.

Such structures exist in nature (even more complex ones—cf. chapter "Biological Photonic Crystals" by Grimann and Fuhrmann–Lieker for biological photonic crystals) or can be realized experimentally by means of photo- and electron beam lithography techniques [14, 15], shadow mask deposition [16], or by applying acousto-optical lattices [17]. The corresponding photon dispersions can be mapped by angle-resolved photoluminescence spectroscopy. In this case, the experimental quantity corresponding to the in-plane momentum is the angle with respect to the surface normal, linked via $k_{\parallel} = \sqrt{k_x^2 + k_y^2} \propto \sin\theta$. A detailed description of the band structure of higher dimensionality is found in chapter "Photonic Crystals: An Introductory Survey" by Liscidini and Andreani.

2 Hybrid Metal-Organic Photonic Crystals

2.1 Motivation

In the previous section, we have addressed the basics of microcavities as photonic crystals with a distinct defect state (resonator layer). In order to realize novel *optoelectronic devices* with an organic gain medium implemented into a microcavity, electrical contacts are needed. In case of organic light emitting diodes (OLEDs) [18] and organic solar cells (OSCs) [19], two types of contacts are widely established. First, transparent conducting oxides (TCOs) are commonly used for bottom contacts. Generally, they provide reasonable conductivity and comparably high transmittance of over 80 % in the visible range (400–780 nm). The TCO most commonly used is In_2O_3:Sn (ITO), with typical layer thicknesses of 80–150 nm, resistivities as low as 10^{-3} to $10^{-4}\,\Omega$ cm, and sheet resistances of 10–40 Ω/sq.

Despite the various advantages, there are some drawbacks of using ITO. Typically, ITO is produced by means of sputtering from a target under high vacuum or by pulsed laser ablation deposition. These techniques limit the use of ITO as top contact since, generally, the topmost organic layers may get damaged.

The second way to efficiently contact an organic active layer is the utilization of thin metallic layers, which can easily be deposited by means of thermal evaporation under high vacuum [19] (for details see Sect. 3).

In inorganic structures, photon lasers have been realized [14, 15, 20] and potential polariton lasers [21–23] have been investigated. Up to now, there are no electrically driven *organic* solid-state devices featuring coherent emission.

In organic-based structures, researchers investigated different coherent phenomena like polariton lasing [24] (cf. chapter "Strong Coupling in Organic and Hybrid-Semiconductor Microcavity Structures" by Lidzey and Coles), Bose–Einstein condensation of *non-interacting* photons [25–28], and photon lasing [29–31]. A common feature of these structures is their all-dielectric nature with non-doped mirrors. However, for injecting charge carriers or applying electric fields to the active medium [32, 33], either highly doped and therefore conductive mirrors or suitable electrode structures are needed. To dope mirrors in order to achieve reasonable conductivities, sophisticated techniques have to be applied and difficulties like lattice-matching needs to be managed, e.g. by means of metal-organic vapor phase epitaxy (MOVPE) or molecular beam epitaxy (MBE).

An estimation of the equivalent charge carrier density to overcome the lasing threshold yields several $\mathrm{kA/cm^2}$, completely neglecting the additional losses due to interaction with charge carriers [34] or excited state absorption [35]. Recently, we applied extremely high current densities of up to $6.2\,\mathrm{kA/cm^2}$ into an OLED (at nanosecond pulses) [36]. These promising results show that OLEDs can sustain such high densities and are therefore suitable candidates to become implemented into an optical resonator. However, these high current densities rule out the use of TCOs as contacts. Considerable ohmic losses would lead to substantial heating, regarded to be detrimental to organic materials.

This fact motivates the study of *metallic structures* inside organic microcavities which may enable electrical contacting. In addition, metal also may serve as broadband high-reflectivity mirror and act as heat sink due to the high thermal conductivity [37]. Most challenging are the apparently unavoidable optical losses as a common feature of all metals. In the following, we recall the optical properties of metals and show possible metal-based excitations in metal-dielectric photonic crystals. Later we will discuss the technical implementation of metals into optical structures and present also reasonable and simple designs.

2.2 Optics of Metals

This section considers a situation where strong optical losses in metals alter or even prevent the propagation of electromagnetic waves. Metals are highly reflective for frequencies up to the visible range of the electromagnetic spectrum. Therefore, they are widely used as mirrors. The high density of free electrons causing this effect also allows for current flows in metals. With this in mind, one can describe the propagation of electromagnetic waves in highly absorbing metals and derive the wave equation. These equations are identical with the corresponding equations for dielectric media, but with a generalized complex-valued dielectric function $\widehat{\varepsilon} = \varepsilon^{'} + i\varepsilon^{''}$ and the complex refractive index $\widehat{n} = n + i\kappa$. All details including the following excerpt can be found in the Springer book by Stefan Maier [38].

In the optical frequency range, $\widehat{\varepsilon}$ can be obtained experimentally by determining the complex refractive index $\widehat{n}$ as:

$$\begin{aligned} \varepsilon^{'} &= n^2 - \kappa^2 \\ \varepsilon^{''} &= 2n\kappa \\ n^2 &= \frac{\varepsilon^{'}}{2} + \frac{1}{2}\sqrt{\varepsilon^{'2} + \varepsilon^{''2}} \\ \kappa &= \frac{\varepsilon^{''}}{2n}. \end{aligned} \tag{9}$$

κ is called the attenuation index or extinction coefficient and is easily experimentally accessible. It describes the attenuation of the amplitude of the intensity $I(x)$ of a beam propagating through the medium with absorption coefficient α along the x-direction, according to the law of Lambert–Beer:

$$I(x) = I_0 e^{-\alpha x} \qquad \text{with} \qquad \alpha(\omega) = \frac{2\kappa(\omega)\omega}{c}. \tag{10}$$

This damping leads to a limited penetration depth into conducting media, which is typically in the order of a small fraction of the wavelength λ of the light.

However, the high electron densities in metals cause strong attenuation of the light but there are ways to exploit the optical properties of metals by choosing the right metals and right designs. This opens the way to the *exciting* field of metal *excitations*.

2.3 Plasmonic Excitations

In this section, we want to briefly discuss possible electromagnetic excitations which occur in combination with metals, referred to as plasmonic excitations. In metals, various plasmons can be created i.e. volume plasmons, localized particle plasmon-polaritons, and surface plasmon-polaritons (SPPs). However, due to their dispersion relation, SPPs cannot simply be excited by light. Complex geometries such as the Kretschmann–Raether configuration [39] or gratings are necessary to provide the corresponding momentum given by the resulting reciprocal lattice vector. The latter is especially interesting in combination with photonic crystals. We will present a simple approach to access SPPs in a two-dimensionally structured microcavity later in Sect. 4.

A rather new type of plasmonic excitation is the so-called Tamm plasmon-polariton (TPP, named after the physicist Igor Tamm), which can occur at the interface of a metal layer attached to a photonic crystal exhibiting a band gap. These excitations prove extraordinarily important, since at certain resonant wavelengths (at the eigenenergies of TPPs), the absorption can be drastically lowered, so that even lasing action of TPPs can be obtained.

2.4 Surface Plasmon-Polaritons

Besides volume plasmons which are collective longitudinal oscillations of the electrons in metal, SPPs are surface states propagating at the interface between a dielectric medium and a conductor. In this case, the electromagnetic field couples strongly to the electron plasma of the metal and is evanescently confined in directions normal to the interface.

Confinement of the wave to the surface requires $\mathrm{Re}\{\varepsilon_1\} < 0$ and $\varepsilon_2 > 0$, i.e. SPPs only exist at the interface between a conductor (with ε_1) and an insulator (with ε_2). The necessary continuity of field components at the interface, combined with the requirement to fulfill the wave equation results in the dispersion relation of SPPs [38]:

$$\beta = k_0 \sqrt{\frac{\varepsilon_1 \varepsilon_2}{\varepsilon_1 + \varepsilon_2}}, \tag{11}$$

where k_0 denotes the vacuum wave number of the propagating wave. The characteristic dispersion branch of SPPs normalized to the respective plasma frequency is plotted in Fig. 6, assuming a dielectric constant of metal with negligible damping. The black solid line lying to the right of the respective light line (cf. Sect. 1.3) corresponds to the propagating SPP. Due to momentum conservation, SPPs cannot be excited optically without applying special phase-matching techniques, e.g. by utilizing gratings or prisms. In Sect. 4, a technique is described to efficiently Bragg-scatter the SPP dispersion branches exceeding the wave vector of light by embedding a thin silver grating into an organic microcavity.

Assuming a generalized complex dielectric function (cf. Eq. (9)) accounting for the absorption of metals in the wave equation yields non-zero solutions only for transverse magnetic (TM or p) modes. On the other hand, transverse electric (TE or s) modes are not supported and do therefore not exist. Another detailed description can be found in [40].

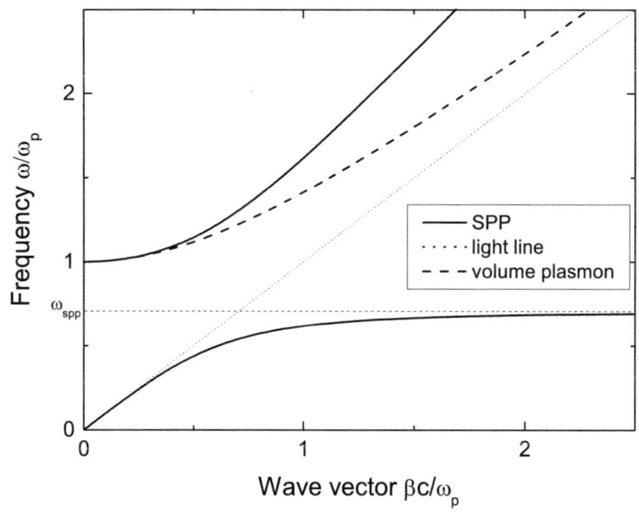

Fig. 6 Dispersion relation of the SPPs occurring at the interface between silver and air (*black solid line*, lower branch). This branch approaches the characteristic SPP frequency given by $\omega_{spp} = \omega_p/\sqrt{1+\varepsilon_2}$. Due to strong coupling of photons ($\omega = ck$) to the electron sea, the original volume plasmon resonance (*red dashed line*) is altered exhibiting a larger curvature (*black solid line*, upper branch). After [38]

2.5 Tamm Plasmon-Polaritons

In solid-state physics, electronic states can form at the surface of a crystal, if their eigenenergies are within the band gap and the environment is of higher potential energy [41]. Similar to these so-called Tamm electronic surface states, *optical Tamm states* can form at the interface between two photonic structures having overlapping band gaps [42]. The corresponding wave-function decays exponentially into both structures. In the following, the DBRs represent the photonic structures, forming a microcavity. In such metal-free microcavities where two DBRs are separated by a certain distance, the thickness of the cavity layer, the occurring localized cavity resonance can also be termed *optical Tamm state*.

Recently, the complex interplay between a metal layer and a periodic dielectric structure—such as a DBR—has been studied by Kaliteevski et al. [43]. According to their considerations, this configuration results in the formation of localized surface states at the interface. The confinement of the electromagnetic field is due to the negative dielectric constant of metals below the plasma frequency ω_p and the photonic band gap (stop band) of the DBR [42]. Since this phenomenon is again in analogy with the Tamm electronic surface states, these states are called *Tamm Plasmon-Polaritons* or *Tamm plasmons* [43–45] with interesting features:

- they are localized inside the DBR stop bands;
- the dispersion $\omega(\mathbf{k}_{||})$ is parabolic in shape;

- in contrast to SPPs, Tamm states can be excited in both polarizations, in TE and TM;
- the splitting between the TE and TM polarized branches is proposed to scale quadratically with the in-plane *k* vector;
- their in-plane wave vector is close to zero and they can therefore be excited via photons;
- the eigenenergy depends on the thicknesses of both the metal layer itself and the adjacent dielectric layers.

Studying a gold covered DBR composed of GaAs/AlAs, Sasin et al. showed experimentally the existence of Tamm plasmons [46]. In a further step, strong coupling between Tamm states, formed in a silver-covered AlAs/GaAlAs DBR, and InGaAs quantum-well excitons was demonstrated experimentally by Symonds et al. [47]. Recently, a more complex geometry was studied theoretically by Kaliteevski et al. [48]. They demonstrated that a metal layer *on top* of a microcavity with embedded quantum-well generates hybrid states consisting of exciton polaritons coupled to Tamm plasmons. As a potential application, Liew et al. proposed exciton-polariton integrated circuits by utilizing Tamm states [49]. The authors discussed the electrical sensitivity of Tamm-plasmon-exciton-polariton modes to generate and control bistable states in semiconductor microcavities.

Both Tamm plasmons and surface plasmons forming at colloidal photonic crystals are also discussed in the chapter "Plasmonic Nanostructures" by Romanov. Further surface states, e.g. Bloch surface waves (BSWs) which can occur at the surface of photonic crystals are described in chapter "New Sensing Strategies Based on Surface Modes in Photonic Crystals".

3 Experiments on Planar Metal-Organic Photonic Crystals

The above-mentioned works emphasize the importance of hybrid structures composed of (partly) active dielectric multilayers and metallic components in a planar configuration. The occurring low-loss photonic states are ideal candidates to study non-linear phenomena such as lasing in electrically contacted devices. In the following, we go a step ahead and demonstrate in a more complex structure the coexistence and coupling of various optical and plasmonic modes in a planar half-wavelength organic microcavity.

In particular, we discuss microcavity structures which include a thin silver layer attached to the topmost high-refractive index quarter-wavelength layer of the bottom DBR, and expect, therefore, the formation of a superimposed metal-cavity state, as described above.

In Fig. 7a, a schematic of the sample featuring different regions is shown. First, a metal-free area is shown on the left. In the center, a wedge-shaped silver layer is embedded with a gradient in thickness up to 40 nm. In the right part, the metal layer exhibits a constant thickness. The cavity layer is formed by an organic gain

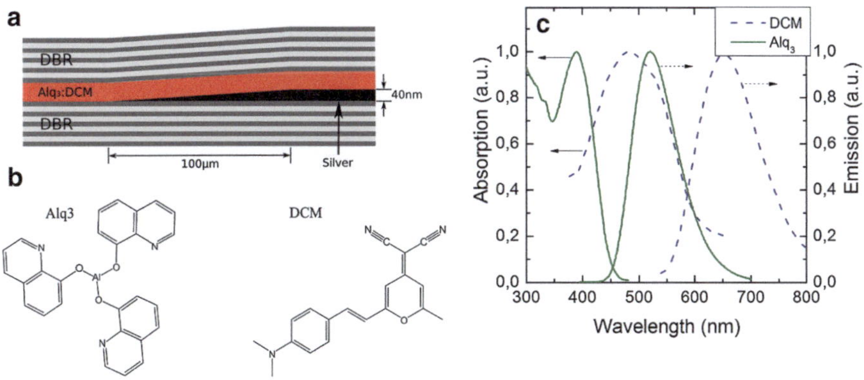

Fig. 7 (**a**) Schematic cross-section of the organic microcavity with an embedded metal layer. The wedge-shaped silver layer (*black*) exhibits a thickness gradient from 0 to 40 nm within a distance of 100 µm. The DBRs are composed of 21 alternating layers of TiO_2 and SiO_2. (**b**) An example of efficient dye molecule blend, Alq_3 (*left*) and DCM (*right*). (**c**) Emission and absorption spectra of Alq_3 and DCM

medium, here the host–guest system of tris-8-hydroxyquinoline aluminum (Alq_3) doped with 2wt.% of 4-dicyanomethylene-2-methyl-6-p-dimethylaminostyryl-4H-pyran (DCM) [50]. The corresponding molecular structures are depicted in Fig. 7b. Besides the broad emission and absorption characteristics, further advantages of using dyes are the high purities which can be achieved by means of sublimation, the processability in high vacuum with very precise layer control providing an accuracy in the order of a few nanometers, and the possibility of easy doping by thermal co-evaporation.

According to [51–53], the doping ratio of 2 % by weight of DCM molecules into the Alq_3 matrix is regarded as optimum with respect to the lasing threshold. Quantum confinement arguments demand that with increasing DCM concentration, the peak of the emission band shifts to longer wavelengths due to the aggregation of adjacent DCM molecules.

The samples are optically pumped in a confocal micro-photoluminescence setup using a 405 or 532 nm laser to excite into the absorption bands of either Alq_3 or DCM. The measured emission properties of the metal-organic hybrid cavity at normal detection angle ($0^\circ \pm 1.5^\circ$) are depicted in Fig. 8a. A sharp cavity mode in the center of the photonic stop band at 632 nm is visible when no metal is present inside the cavity. The corresponding Q factor at this position is approximately 800. When the excitation spot approaches and crosses the metal border, a new line emerges in addition to the cavity mode and shifts to larger wavelengths with increasing metal thickness. The Q factor obtained from this resonance (labeled as TS1) is still in the order of 400. For silver layers thicker than ~ 20 nm, a second new line evolves from the long-wavelength sideband of the DBRs, moving to higher energies with increasing silver thickness (labeled as TS2). The spectral shift of the new modes converge until reaching a certain constant separation (of ~ 26 nm) at silver thicknesses above 35–40 nm. At the same time, the emission into the cavity

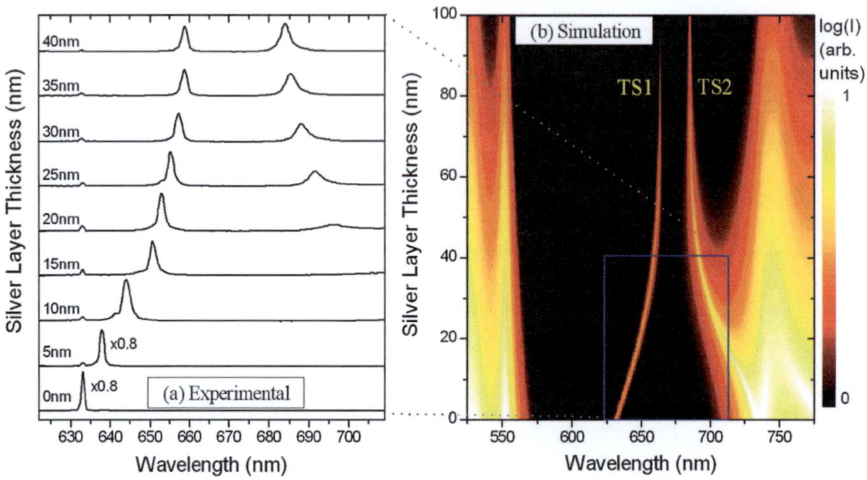

Fig. 8 (**a**) Emission spectra of the microcavity recorded at different silver thicknesses from 0 to 40 nm (approx. values). Both peaks approach each other down to a minimal separation. The intensity of the original cavity resonance at 632 nm decreases drastically, but remains visible even for thick silver layers. (**b**) Numerically simulated transmission spectra for a microcavity with an incorporated silver layer of continuously variable thickness. The *blue rectangle* corresponds to the region where experimental data are obtained. After [54], Copyright 2011 by The American Physical Society

resonance substantially decreases while the intensities of the two new lines increase.

The observed low-energy state is similar to the previously discussed TPP at the interface between metal and a DBR. Additionally, the sample features a second DBR and a half-wavelength organic cavity layer on top of the silver layer. Therefore, we observe an additional hybridized mode in the spectra. According to the considerations in [48]—describing the case of a cavity with a metal layer on top of the DBR—we see two coexisting hybrid cavity-metal resonances in such structures.

Here, the observed and discussed Tamm plasmons are called Tamm state 1 (TS1) and Tamm state 2 (TS2), according to Fig. 8a, b.

In order to verify this interpretation, a transfer matrix formalism [42, 43] is utilized to numerically determine the transmission properties of a half-wavelength organic microcavity with included silver layer of increasing thickness (Fig. 8b).

At zero silver thickness, only one cavity resonance (TS1) exists at the design wavelength $\lambda_c = 632$ nm. Upon increasing the silver thickness up to $\sim$40 nm, we observe a continuous shift of the cavity resonance towards lower energies and the emergence of a new mode (TS2) from the DBR side band shifting to higher energies. Both spectral movements of the modes saturate at a silver thickness of $\sim$40–50 nm, at a minimal spectral distance of 22 nm. Any further increase of the metal thickness merely leads to a decrease of the overall electric field amplitudes due to a higher absorption in the silver. The reason for these spectral shifts is the coupling of both modes which is discussed in the following. Later we will also discuss the impact of laterally structured metallic layers implemented into organic microcavities and how their optical properties can be influenced allowing to fully control the output emission.

3.1 *Hybridization of Photonic and Plasmonic Modes*

The eigenenergies of both modes do not only depend on the thickness of the metal layer and the adjacent high-refractive index layer of the DBR [43], but also on the thickness of the cavity layer, as shown in Fig. 9. Here, the numerically calculated transmission spectra of the device are plotted, where a 40 nm silver layer is incorporated and the thickness of the cavity layer is varied. It is visible that at cavity layer thicknesses of $1\lambda_c/4$ and $3\lambda_c/4$ only one state (TS2) is located in the stop band (however, for thicker cavity layers, the FSR decreases and higher order cavity modes move into the stop band). Note that due to the coupling of both modes, they are mixed to some extent, which becomes apparent at the transition region from TS1 into TS2. At any other thickness of the cavity layer, two states exist simultaneously: Resonance TS1 shows a significant red-shift across the stop band as the cavity thickness increases.

However, resonance TS2 depends only weakly on the cavity thickness, emphasizing that the origin of the field localization is different from that of an ordinary cavity mode. At cavity thicknesses of approx. $\lambda_c/2$ and λ_c, anticrossing occurs, demonstrating that two different fields of the same wavelength (i.e. ~ 686 nm) are not allowed to propagate through such a structure. Finally, both modes hybridize, as with increasing cavity thickness the one state is continuously converted into the

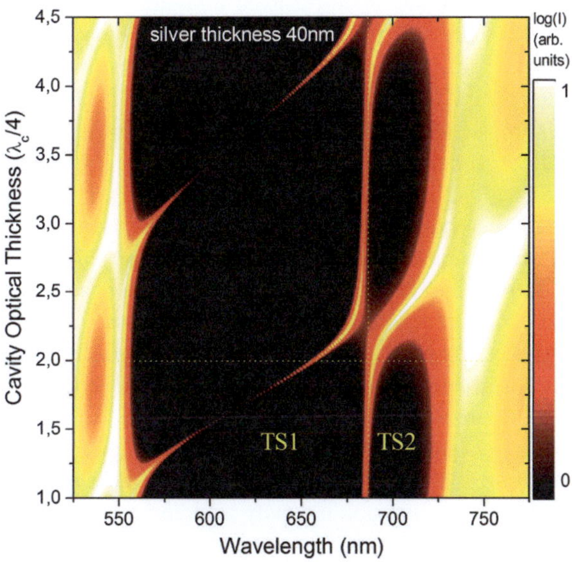

Fig. 9 The simulation shows the dependence of the eigenenergies of the Tamm states on the thickness of the dielectric cavity layer. Both resonances cannot occupy the same wavelength in the stop band, which is identified by an anticrossing behavior (*dotted vertical line*). The *dotted horizontal line* indicates the position where experimental data are recorded. Adapted from [54], Copyright 2011 by The American Physical Society

other and vice versa. A clear distinction between the modes is difficult. The underlying origin of this behavior becomes evident when discussing the electric field distribution inside the metal-organic microcavity for both resonances [54].

3.2 *Coherent Emission from Metal-Organic Photonic Crystals*

Organic optoelectronic devices are particularly sensitive to heating due to their low thermal conductivity. The strong optical absorption of metallic components in such devices may trigger chemical degradation and crystallization of the active layers. In fact, any absorption of photons limits optical coherence, which may render metallic components incompatible with lasing [55]. However, recently, Symonds et al. [56] reported lasing from a GaAs/silver micro-structure, albeit restricted to liquid nitrogen temperature. In micro-structures involving organic semiconductors and metals, such observations are expected at room temperature. Here, two experimental configurations are typically considered: Traditional structures based on surface plasmons (cf. Sect. 2.4), which enhance the interactions of light to organic materials [57, 58] or structures based on Tamm plasmons (cf. Sect. 2.5) [43, 46].

Next, we study the optical response of the hybrid cavity under intense optical pumping [59]. For this purpose, nanosecond pulses of 532 nm wavelength are utilized to pump the optically active DCM molecules in either of the regions. The rapid molecular relaxation [60] ensures that pumping generates an incoherent exciton population. In this experiment, regions with homogeneous metal 15 μm away from metal-free regions are pumped via the focused beam (diameter of 3 μm).

In Fig. 10, the angle-resolved emission patterns of an optimized metal-organic cavity in dependence on the excitation energy are presented. This representation is most suitable to study the optical characteristics of photonic crystals, cf. Sect. 1.3. The spectral and angular resolution resembles the dispersion relation, i.e. the fixed relation between energy and momentum both of which are conserved quantities. Further, the intensity distribution along the dispersion relation gives insights into the coupling of modes, the distribution of the density of states and reveals even non-linear phenomena. In a laterally unstructured device, microcavity photons exhibit a parabolic dispersion curve, similar to the free-electron dispersion relation.

In panel (a) of Fig. 10, both the cavity state and the TS1 appear. Surprisingly, in this configuration, the original cavity resonance is supported by scattering at defects and wave-guiding along the sample plane. Examining the Q factors (which determine a potential laser threshold) of the modes leads to values of ~ 1400 for the cavity state and about ~ 650 for TS1.

Our Q factors are comparable [56, 61] or higher [46] than values reported in literature for epitaxially grown samples measured at cryogenic temperatures. In panel (b), the emission intensity is increased twice to 0.11 nJ. An asymmetry of the

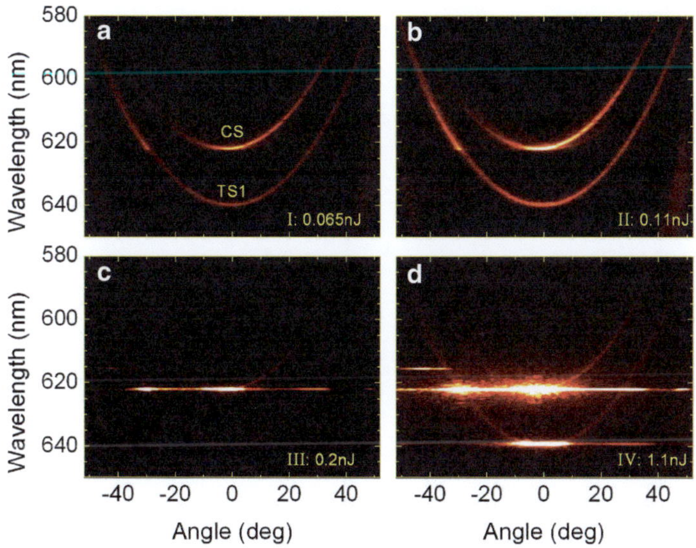

Fig. 10 Angular dispersion of the CS and TS1 in dependence on the excitation energy. The different excitation energies are marked with roman numerals and correspond to the marks and labels in the input–output curve in Fig. 11. (**a**) At low excitation energy, both modes are apparent with different intensities. (**b**) Increasing the pump energy leads to linearly increased spontaneous emission until (**c**), the threshold of the cavity mode is reached. The coherent emission couples also into the TS1 parabola (at negative angles). (**d**) By further increasing the excitation energy, the emission of the TS1 mode at $k = 0$ increases super-linearly indicating lasing. The dispersion-less feature, which is visible at 615 nm and at angles smaller than $-35°$, appears with reaching the CS coherence threshold and is ascribed to a reflection of the output signal in the detection path of the spectrograph. False colors are arbitrarily scaled for each panel. From [59]—Reproduced by permission of The Royal Society of Chemistry

cavity parabola becomes more pronounced. The branch covering negative angles is less intense compared to positive angles.

Instead, starting from this photon energy, the intensity of the TS1 is increased. This might be explained by scattering from cavity mode photons into the TS1 dispersion with conserved energy. The asymmetric scattering towards negative angles is due to the fact that in this direction the metallic layer is continuous, whereas the metal edge is at the opposite site. The extent of the modes in lateral direction can exceed 10 μm [62–64] causing an overlapping of a particular mode into non-supported areas. Hence, in different directions different modes are supported.

But why is this behavior important? The extent of modes into non-supported regions can be exploited in samples with laterally structured metallic layers and are explained in Sect. 4.

By further increasing the excitation energy, the coherence threshold of the cavity state is reached at 620 nm (panel (c), excitation energy 0.2 nJ). At $k = 0$ of the corresponding parabola, a super-linear increase in emission intensity is observed

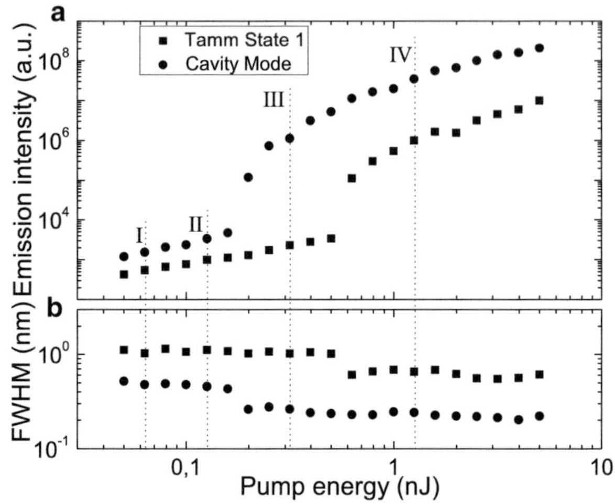

Fig. 11 (**a**) Input–output characteristics of the cavity mode (*red dots*) and the Tamm state 1 (*black squares*), respectively. The integrated emission of the TS1 is several times weaker compared to the cavity mode. Both modes exhibit a linear increase in intensity both below and above the coherence threshold. (**b**) The linewidth of the CS (TS1) reduces from ~ 0.45 nm (~ 1 nm) below threshold to ~ 0.25 nm (~ 0.6 nm) above it. After [1]

accompanied by a reduction in linewidth (see also Fig. 11). The coherent light emission is scattered into the TS1 dispersion at an angle of $\sim 32°$. This scattering process into a second mode is in full analogy to Kastler-ring emission, theoretically described by Alfred Kastler in 1962 [65] and experimentally demonstrated in a coupled organic microcavity by Schneeweiss et al. [66].

A further increase in pump energy leads to the observation of a second non-linear threshold, this time of the TS1 at ~ 640 nm, see panel (d) in Fig. 10.

The input–output behavior of the lasing characteristics is presented in Fig. 11. According to the two times lower Q factor of TS1, its respective mean coherence threshold is found to be two to three times higher than for the CS. Both the difference of the eigenenergies and of the thresholds which scales inversely proportional with the Q factor indicates that both coherent modes oscillate completely independent from each other. While the linewidth of the cavity mode reduces from ~ 0.45 nm below threshold to ~ 0.25 nm above it, the linewidth of TS1 decreases from ~ 1 nm to ~ 0.6 nm.

Remarkably, in this experiment no degradation or destruction of the Alq_3:DCM sample is noticed, even for excitation energies 1.5 orders of magnitude above the coherence threshold of the CS (limited by the maximum output power of the pump laser). The roman numerals (I–IV) in Fig. 11a denote different excitation energies corresponding to the cases presented in Fig. 10a–d.

In summary, we could show that a simple photonic crystal featuring defect states can show interesting optical properties and even the implementation of metallic layers with the prospect as being used as electrodes open new opportunities to design novel optoelectronic devices. It is possible to show coherence from metal-organic photonic crystals with very low thresholds. Interestingly, the lasing characteristics can be maintained even though absorptive elements are present. These are very important findings and demonstrate that simplest structural designs can lead to interesting features.

In the following, it can be shown that reducing the amount of metal by structuring the layer into stripes of constant width and thickness leads to more flexibility, lower thresholds, and a greater variety of modes including their control.

4 Experiments on Structured Microcavities and Multi-dimensional Mode Confinement

Up to this point, we have considered simple geometries of photonic crystals realized as microcavities with a certain periodicity in one dimension including potential defects introduced by a cavity layer, metal layer or both. The optical properties are determined by the materials chosen and the exact design parameters. As a result, the microcavity photons show a parabolic dispersion relation similar to the free electron dispersion in a one-dimensional electronic crystal. In the following, more sophisticated structures of higher dimensionality are under study. Established ways to realize efficient photon confinement in microcavity structures are, e.g., by means of optical mesas or dots [16], pillars [67], or photonic molecules [10]. However, another elegant way of bringing well-known phenomena for electrons in solid-state physics into optical laboratories is to place such confining potentials periodically into the microcavity in order to create a super-lattice.

Here we concentrate on a grating of silver photonic wires embedded into microcavities and see various excitations of cavity photons, Tamm plasmons, and SPPs. The use of the organic laser dye DCM diluted into the organic matrix Alq_3 not only enables the investigation of the resulting complex dispersion relations but also allows for the observation of coherent phenomena. Therefore, this system exemplarily represents several classes of photonic crystals of different dimensionality, materials and degree of mode coupling.

In Fig. 12, the sample as a 2D metal-organic photonic crystal is shown. The metal layer is periodically structured into patterns of deep metallic photonic wires. In the following, we describe the impact of both the 2D confinement due to single deep photonic wires and the impact of the periodicity on the mode structure.

At first, the angle-resolved emission spectra of the structure is discussed. In Fig. 13a, the structuring of the metal layer leads not only to a superimposing of the dispersion characteristics of the cavity state in a metal-free MC and the Tamm

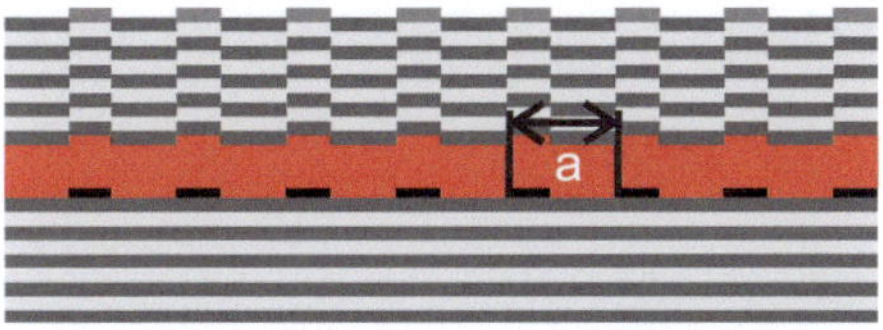

Fig. 12 Schematic of the structured microcavity sample. The silver layer is structured by means of photo-lithography into a periodic pattern (stripe period $a = 7.4$ or $11.1\,\mu m$), yielding a 2D metal-organic photonic crystal

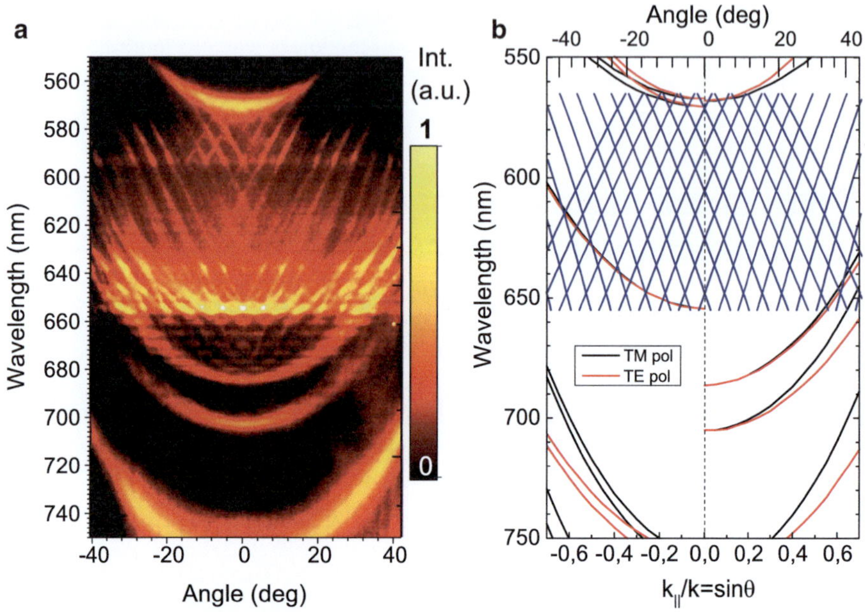

Fig. 13 (**a**) The angle-resolved emission spectrum of the metal organic microcavity. Here, the two Tamm states and the cavity mode coexist, replicated in *k*-space due to Bragg scattering according to the period of $a = 7.4\,\mu$m. The upper Tamm state discretizes due to optical confinement. Bragg-scattered surface plasmon-polaritons appear as quasi-linear dispersions for wavelengths below 650 nm. (**b**) An analytical approximation of the occurring modes. In the *left part* is the cavity parabola shown. In the *right part* are the eigenenergies of the Tamm states shown (without confining lateral potential). The *solid blue lines* are Bragg-scattered copies of surface plasmon polaritons, apparent as almost linear lines. Adapted with permission from [68], Nature Publishing Group 2012

states in MCs with metal but to a much more complicated pattern [68]. In order to understand most phenomena apparent, many aspects and especially the nature of light coupling to metal has to be studied.

As expected, the emission spectrum of the structured region reveals all three resonances, that is, the cavity mode, TS1, and TS2, at their appropriate energies for small angles. The dispersion branches discretize in *k*-space and the interior of the Tamm parabola is filled with discrete sub-states. These effects are caused by the refractive index modulation along the metal grating, which generates a periodic photonic potential along the cavity plane. In the direction perpendicular to the grating, Tamm plasmons are optically confined to the 3.0-μm-wide metal stripes and ordinary cavity modes to the 4.4 μm gaps between them. The latter is several times smaller than the mode radius in a comparable planar microcavity [9, 64]. The resulting discretization is similar to that seen in three-dimensional photonic dots [16].

Additional to the optical confinement, the periodicity of our structure normal to the growth direction also plays a role. As a result of Bragg scattering, the original parabolic cavity dispersion is replicated in k-space according to the reciprocal lattice vector $G = 2\pi/a$. The parabolae cross at integer multiples of the lateral component of k, $k_{||} = \pm m \cdot \pi/a$. At these points, the density of states is enhanced and bright spots appear in the experimentally measured spectra in k-space. Later on, these will act as condensation points for stimulated emission.

As we included a grating structure and efficient Bragg scattering can occur, we are also able to observe almost linear dispersion curves below 650 nm. The slope of these linear dispersion branches passing through 655 nm at $\theta = 0$ and through 580 nm at $\theta = \pm 23.5°$ can be used to derive an effective refractive index:

$$n_{eff} = \frac{\Delta k_{||}}{\Delta k_0} \tag{12}$$

resulting in $n_{\text{eff}} = 3.58$ at 655 nm [69]. As this value exceeds the refractive indices of all materials involved, SPPs at the interface between the organic cavity layer and a silver stripe provide the only possible explanation. The wave vector of a SPP is $k_{SPP} = k_{Alq3}[\varepsilon_{Ag}/(\varepsilon_{Ag} + \varepsilon_{Alq3})]^{1/2}$ [38] (cf. also with Sect. 2.4) along the cavity. However, due to momentum conservation, this mode cannot escape from the DBR resonator into air, cf. Fig. 6.

Here, the dispersion relation of this surface plasmon (visible as almost linear lines) is Bragg-scattered by integer multiples of the length of the reciprocal lattice vector, G, provided by the period of the embedded grating. The resulting replica dispersion branches at $k = k_{SPP} \pm nG$ may penetrate into the air cone for sufficiently large n and be observed outside the cavity. This is the basic principle of plasmonic crystals and may be extended in three dimensions accordingly. In our structure, the optical feedback in longitudinal direction given by the high reflectance of the dielectric mirrors drastically enhances the scattering effect. In literature periodic 2D and 3D plasmonic structures are common. More unconventional approaches utilize gratings with an additional chirp [70] or even quasi-periodic structures such as Moiré patterns [71].

For clarity, Fig. 13b shows the resonances of the cavity mode, TS1, and TS2 calculated applying the transfer matrix algorithm. These modes overlap with the quasi-linear surface plasmon dispersions, $k = k_{SPP} \pm nG$, shown for very high orders $n = 13$–30. The consequences of this overlap appear most clearly in the spectrum of Fig. 13a, where high-order Bragg-scattered surface plasmons, a replica of the cavity dispersion, and the Tamm resonance all interfere.

4.1 Coherence in 2D Metal-Organic Photonic Crystals

Owing to its non-linear nature, stimulated emission amplifies the roughness of the energy landscape both in real and in k-space. In turn, the development of coherence

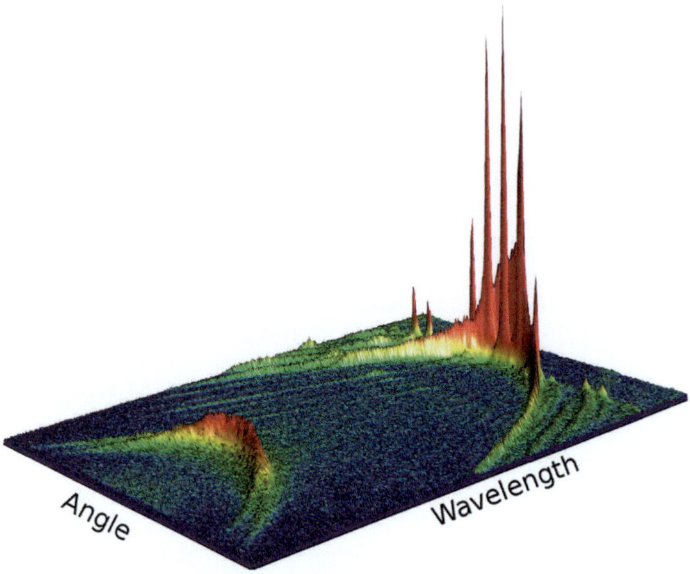

Fig. 14 Lasing of 2π-states in a hybrid photonic-plasmonic crystal. Coherent emission occurs *red-shifted* at positions where Bragg-scattered cavity parabolas cross. The *straight lines* within the stop band are replica of SPPs traveling at the interface between the silver stripes and adjacent dielectric (or organic, respectively) layers

is strongly influenced by the disrupted cavity dispersion. In the experiment shown in Fig. 14, we pump a patterned metal-organic microcavity with an energy above 15 nJ per nanosecond pulse, off-resonantly in a very small spot with a diameter of $< 2\,\mu$m. At this threshold, a 2.5-fold increase of the excitation energy results in a 300-fold increase of the cavity mode emission, with considerable spectral narrowing. Here, besides emission from $k_{||} = 0$, the spectrum reveals coherent features at even multiples of $k_{||} = \pi/a$, with energies ~ 3 meV higher than the dispersion minimum. This corresponds to the energy of the cavity parabola at $k_{||} = 2\pi/a$. In analogy to results published for an inorganic semiconductor microcavity by Lai et al. [72], we call this state 2π-state.

This coherent emission occurs in an emission pattern linearly polarized perpendicular to the orientation of the metal grating. As pumping occurs off-resonantly, it is safe to conclude that we observe a self-organized coherent process. At the same time, the incoherent spontaneous emission at oblique angles $k_{||} > 0$ increases slightly sub-linearly, conserving the shape and spectral distribution of the dispersion relation.

Interestingly, the thresholds of coherent emission obtained from the metal-stripe sample compared to thresholds obtained from metal-free regions are identical (not shown), within the experimental error. This implies that an optimized structure (lasing modes matching the gain maximum, the ratio of metal stripe width to metal-free stripes changed to minimize losses, and so on) will lead to very low lasing thresholds due to this additional lateral confinement, allowing it to be competitive with the best metal-free devices.

All details of the coherence and phase-locking including an analysis can be found in [68].

5 Conclusion

In summary, we have discussed novel effects in modified microcavity structures containing organic active materials. After introducing into the physics of 1D photonic crystals, we discussed the occurrence of sharp cavity resonances for filtering or lasing applications. Higher dimensional mode confinement allows to fully control the dispersion relation of photonic crystals and with that the output emission.

Further, we discussed the impact of metal implemented into photonic crystals and showed ways to overcome or exploit the metal's losses by introducing into the concepts of SPPs and TPPs. To not only prove these basics but also to gain a full control, we showed various of those states in simple experiments. The special design and the use of highly purified organic laser dye molecules enables both the investigation of complex dispersion patterns in the emission geometry and the coherence properties due to providing a broad gain spectrum.

In the last section, metallic layers are structured periodically and the impact on the optics is discussed based on the theoretical excursion at the beginning of the chapter.

Acknowledgements This work has been supported by Deutsche Forschungsgemeinschaft (DFG projects LE 747/37-1 and LE 747/41-1) and via the excellence cluster CFAED. Stimulating discussions with Reinhard Scholz and Andreas Mischok contributed to the chapter.

References

1. R. Brückner, Ph.D. Thesis, TU Dresden, 2012
2. P. Yeh, A. Yariv, C.S. Hong, J. Opt. Soc. Am. **67**, 423 (1977)
3. S. John, Phys. Rev. Lett. **58**, 2486 (1987)
4. E. Yablonovitch, Phys. Rev. Lett. **58**, 2059 (1987)
5. A. Yariv, 3rd edn. (Wiley, 1989)
6. S. Kuhr, S. Gleyzes, C. Guerlin, J. Bernu, U.B. Hoff, S. Deléglise, S. Osnaghi, M. Brune, J.M. Raimond, S. Haroche, E. Jacques, P. Bosland, B. Visentin, Appl. Phys. Lett. **90**, 164101 (2007)
7. A.V. Kavokin, J.J. Baumberg, G. Malpuech, F.P. Laussy, 1st edn. (Oxford University Press, 2007)
8. K. Ujihara, Jpn. J. Appl. Phys. **30**, L901 (1991)

9. R. Brückner, M. Sudzius, H. Fröb, V.G. Lyssenko, K. Leo, J. Appl. Phys. **109**, 103116 (2011)
10. M. Bayer, T. Gutbrod, J. Reithmaier, A. Forchel, T. Reinecke, P. Knipp, A. Dremin, V. Kulakovskii, Phys. Rev. Lett. **81**, 2582 (1998)
11. M. Bayer, T. Reinecke, F. Weidner, A. Larionov, A. McDonald, A. Forchel, Phys. Rev. Lett. **86**, 3168 (2001)
12. G. Dasbach, M. Bayer, M. Schwab, A. Forchel, Semicond. Sci. Technol. **18**, S339 (2003)
13. T. Gutbrod, M. Bayer, A. Forchel, P. Knipp, T. Reinecke, A. Tartakovskii, V. Kulakovskii, N. Gippius, S. Tikhodeev, Phys. Rev. B **59**, 2223 (1999)
14. S. Reitzenstein, T. Heindel, C. Kistner, A. Rahimi-Iman, C. Schneider, S. Höfling, A. Forchel, Appl. Phys. Lett. **93**, 061104 (2008)
15. S. Reitzenstein, A. Bazhenov, A. Gorbunov, C. Hofmann, S. Munch, A. Löffler, M. Kamp, J.P. Reithmaier, V.D. Kulakovskii, A. Forchel, Appl. Phys. Lett. **89**, 51107 (2006)
16. M. Langner, R. Gehlhaar, C. Schriever, H. Fröb, V.G. Lyssenko, K. Leo, Appl. Phys. Lett. **91**, 181119 (2007)
17. E. Cerda-Méndez, D. Krizhanovskii, M. Wouters, R. Bradley, K. Biermann, K. Guda, R. Hey, P. Santos, D. Sarkar, M. Skolnick, Phys. Rev. Lett. **105**, 116402 (2010)
18. S. Reineke, F. Lindner, G. Schwartz, N. Seidler, K. Walzer, B. Lüssem, K. Leo, Nature **459**, 234 (2009)
19. J. Meiss, M.K. Riede, K. Leo, Appl. Phys. Lett. **94**, 013303 (2009)
20. C. Böckler, S. Reitzenstein, C. Kistner, R. Debusmann, A. Löffler, T. Kida, S. Höfling, A. Forchel, L. Grenouillet, J. Claudon, J.M. Gérard, Appl. Phys. Lett. **92**, 091107 (2008)
21. D. Bajoni, E. Semenova, A. Lemaître, S. Bouchoule, E. Wertz, P. Senellart, J. Bloch, Phys. Rev. B **77**, 113303 (2008)
22. A.A. Khalifa, A.P.D. Love, D.N. Krizhanovskii, M.S. Skolnick, J.S. Roberts, Appl. Phys. Lett. **92**, 61107 (2008)
23. S.I. Tsintzos, N.T. Pelekanos, G. Konstantinidis, Z. Hatzopoulos, P.G. Savvidis, Nature **453**, 372 (2008)
24. S. Kéna-Cohen, S.R. Forrest, Nat. Photonics **4**, 371 (2010)
25. J. Klaers, F. Vewinger, M. Weitz, Nat. Phys. **6**, 512 (2010)
26. J. Klaers, J. Schmitt, F. Vewinger, M. Weitz, Nature **468**, 545 (2010)
27. J. Klaers, J. Schmitt, F. Vewinger, M. Weitz, Phys. Unserer Zeit **42**, 58 (2011)
28. J. Klaers, J. Schmitt, T. Damm, F. Vewinger, M. Weitz, Phys. Rev. Lett. **108**, 160403 (2012)
29. V.G. Kozlov, S.R. Forrest, Curr. Opin. Solid State Mater. Sci. **4**, 203 (1999)
30. V. Bulovic, V.G. Kozlov, V.B. Khalfin, S.R. Forrest, Science **279**, 553 (1998)
31. M. Koschorreck, R. Gehlhaar, V.G. Lyssenko, M. Swoboda, M. Hoffmann, K. Leo, Appl. Phys. Lett. **87**, 181108 (2005)
32. A. Perumal, B. Lüssem, K. Leo, Org. Electron. **13**, 1589 (2012)
33. A. Perumal, M. Fröbel, S. Gorantla, T. Gemming, B. Lüssem, J. Eckert, K. Leo, Adv. Funct. Mater. **22**, 210 (2012)
34. M.D. McGehee, A.J. Heeger, Adv. Mater. **12**, 1655 (2000)
35. V. Kozlov, G. Parthasarathy, P. Burrows, V. Khalfin, J. Wang, S. Chou, S. Forrest, IEEE J. Quantum Electron. **36**, 18 (2000)
36. D. Kasemann, R. Brückner, H. Fröb, K. Leo, Phys. Rev. B **84**, 115208 (2011)
37. C.Y. Lu, S.W. Chang, S.L. Chuang, T.D. Germann, D. Bimberg, Appl. Phys. Lett. **96**, 251101 (2010)
38. S.A. Maier (Springer, 2007)
39. E. Kretschmann, Z. Phys. **241**, 313 (1971)
40. A. Yariv, 6th edn. (Oxford University Press, 2007)
41. I. Tamm, Z. Phys. **76**, 849 (1932)
42. A.V. Kavokin, I.A. Shelykh, G. Malpuech, Phys. Rev. B **72**, 233102 (2005)
43. M. Kaliteevski, I. Iorsh, S. Brand, R. Abram, J. Chamberlain, A. Kavokin, I. Shelykh, Phys. Rev. B **76**, 165415 (2007)

44. I.A. Shelykh, M. Kaliteevskii, A.V. Kavokin, S. Brand, R.A. Abram, J.M. Chamberlain, G. Malpuech, Phys. Status Solidi A **204**, 522 (2007)
45. S. Brand, M.A. Kaliteevski, R.A. Abram, Phys. Rev. B **79**, 85416 (2009)
46. M.E. Sasin, R.P. Seisyan, M.A. Kalitteevski, S. Brand, R.A. Abram, J.M. Chamberlain, A.Y. Egorov, A.P. Vasilev, V.S. Mikhrin, A.V. Kavokin, Appl. Phys. Lett. **92**, 251112 (2008)
47. C. Symonds, A. Lemaître, E. Homeyer, J.C. Plenet, J. Bellessa, Appl. Phys. Lett. **95**, 151114 (2009)
48. M. Kaliteevski, S. Brand, R.A. Abram, I. Iorsh, A.V. Kavokin, I.A. Shelykh, Appl. Phys. Lett. **95**, 251108 (2009)
49. T. Liew, A. Kavokin, T. Ostatnický, M. Kaliteevski, I. Shelykh, R. Abram, Phys. Rev. B **82**, 033302 (2010)
50. P. Hammond, Opt. Commun. **29**, 331 (1979)
51. V.G. Kozlov, V. Bulovic, S.R. Forrest, Appl. Phys. Lett. **71**, 2575 (1997)
52. V.G. Kozlov, V. Bulovic, P.E. Burrows, M. Baldo, V.B. Khalfin, G. Parthasarathy, S.R. Forrest, Y. You, M.E. Thompson, J. Appl. Phys. **84**, 4096 (1998)
53. S.R. Forrest, V.G. Kozlov, V. Bulovic, Appl. Phys. Lett. **71**, 2575 (2001)
54. R. Brückner, M. Sudzius, S. Hintschich, H. Fröb, V. Lyssenko, K. Leo, Phys. Rev. B **83**, 033405 (2011)
55. M. Wouters, Nat. Photonics **6**, 271 (2012)
56. C. Symonds, A. Lemaître, P. Senellart, M.H. Jomaa, S. Aberra Guebrou, E. Homeyer, G. Brucoli, J. Bellessa, Appl. Phys. Lett. **100**, 121122 (2012)
57. J. Bellessa, C. Bonnand, J. Plenet, J. Mugnier, Phys. Rev. Lett. **93**, 036404 (2004)
58. S. Aberra Guebrou, C. Symonds, E. Homeyer, J.C. Plenet, Y.N. Gartstein, V.M. Agranovich, J. Bellessa, Phys. Rev. Lett. **108**, 66401 (2012)
59. R. Brückner, V.G. Lyssenko, S. Hofmann, K. Leo, Faraday Discuss. **174**, 183 (2014)
60. R. Kersting, U. Lemmer, R. Mahrt, K. Leo, H. Kurz, H. Bässler, E. Göbel, Phys. Rev. Lett. **70**, 3820 (1993)
61. O. Gazzano, S. Vasconcellos, K. Gauthron, C. Symonds, J. Bloch, P. Voisin, J. Bellessa, A. Lemaître, P. Senellart, Phys. Rev. Lett. **107**, 247402 (2011)
62. K. Ujihara, Jpn. J. Appl. Phys. **33**, 1059 (1994)
63. G. Björk, H. Heitmann, Y. Yamamoto, Phys. Rev. A **47**, 4451 (1993)
64. F. Becker, R. Brückner, M. Langner, S. Hintschich, M. Sudzius, H. Fröb, V. Lyssenko, K. Leo, Superlattice. Microst. **49**, 193 (2011)
65. A. Kastler, Appl. Opt. **1**, 17 (1962)
66. P. Schneeweiss, M. Sudzius, R. Gehlhaar, M. Hoffmann, V.G. Lyssenko, H. Fröb, K. Leo, Appl. Phys. Lett. **91**, 51118 (2007)
67. J. Reithmaier, M. Röhner, H. Zull, F. Schäfer, A. Forchel, P. Knipp, T. Reinecke, Phys. Rev. Lett. **78**, 378 (1997)
68. R. Brückner, A.A. Zakhidov, R. Scholz, M. Sudzius, S.I. Hintschich, H. Fröb, V.G. Lyssenko, K. Leo, Nat. Photonics **6**, 322 (2012)
69. R. Scholz, R. Brückner, K. Leo, in preparation (2014)
70. J.S. Bouillard, S. Vilain, W. Dickson, G.A. Wurtz, A.V. Zayats, Sci. Rep. **2**, 829 (2012)
71. S.M. Lubin, A.J. Hryn, M.D. Huntington, C.J. Engel, T.W. Odom, ACS Nano **7**, 11035 (2013)
72. C.W. Lai, N.Y. Kim, S. Utsunomiya, G. Roumpos, H. Deng, M.D. Fraser, T. Byrnes, P. Recher, N. Kumada, T. Fujisawa, Y. Yamamoto, Nature **450**, 529 (2007)

One-Dimensional Photonic Crystals for Light Management in Organic Solar Cells

Marina Mariano, Paola Mantilla-Pérez, Pablo Romero-Gómez, Alberto Martínez-Otero, Xavier Elias, Rafael Betancur, Silvia Colodrero, and Jordi Martorell

Abstract An effective charge collection in the majority of organic solar cells is achieved when the photo-converting layer is just a few tens of nanometer in thickness. In such conditions light management is an essential ingredient to reach the best performance from such type of organic devices. This chapter introduces an inverse integration procedure to optically optimize the architecture of organic solar cells. The relevant role played by the electron and hole transporting layers in such optical optimization of the architecture is discussed. Next, inverse integration is considered to design disordered one-dimension photonic crystal type structures to enhance photocurrent generation in semi-transparent solar cells. It is shown that the light management provided by the photonic structure is essential to achieve a performance which optimizes the balance among device transparency and efficiency.

Keywords Light trapping • Organic solar cells • Thin films • Photonic crystals • Dye sensitized solar cells • Power conversion efficiency • Transparency

---

M. Mariano • P. Mantilla-Pérez • P. Romero-Gómez • A. Martínez-Otero • X. Elias
R. Betancur • S. Colodrero
ICFO-Institut de Ciencies Fotoniques, Mediterranean Technology Park,
08860, Castelldefels Barcelona, Spain
e-mail: m.marianojuste@gmail.com; paola.mantilla@icfo.es; pablo.romero@icfo.es; alb.martinxc@gmail.com; xavier.elias@icfo.es; rabetanc@gmail.com; silvia.colodrero@icfo.es

J. Martorell (✉)
ICFO-Institut de Ciencies Fotoniques, Mediterranean Technology Park,
08860, Castelldefels Barcelona, Spain

Departament de Fisica i Enginyeria Nuclear, Universitat Politècnica de Catalunya,
Terrassa, Spain
e-mail: jordi.martorell@icfo.es

D. Comoretto (ed.), *Organic and Hybrid Photonic Crystals*,
DOI 10.1007/978-3-319-16580-6_13

1 Introduction

Provided no other renewable energy source is as abundant as sunlight, photovoltaic (PV) technology has emerged as one of the most promising candidates to meet current and future energy demands of clean and sustainable electricity. Since the early silicon solar devices developed around 1950, there has been a tremendous advance in the field bringing power conversion efficiencies (PCEs) from an initial 6 % to higher than 44 % [1]. Although most of the commercially available PV systems are made from crystalline silicon wafers, other semiconductors have been used to develop thin film cells with the aim of cutting the material usage and eventually reduce the manufacturing costs.

Indeed, several of the thin film PV technologies being developed may offer relatively high PCEs with submicron thick photon absorber layers. This is particularly true for the organic photovoltaic (OPV) solar cell technology that is based on an architecture where the thickness of absorber photo-converting layers is in the order of 100 nm. Additionally, when compared to their inorganic counterparts, organic electronic materials have lower cost and can be processed at room temperature. The organic PV technology has other advantages as, for instance, an intrinsic flexibility of the organic material layers, which may be very useful in a roll-to-roll production. This may increase the manufacturing speed and, eventually, further reduce fabrication costs. Besides a potentially lower cost, organic PV materials can be engineered to tune their electronic and optical properties to fulfill specific applications [2], which may be difficult to meet with the standard PV modules.

The low charged carrier mobility in donor or acceptor organic materials prevents the use of the thicker active layers that would be needed for a more efficient photon harvesting. On the other hand, organic semiconductors have relatively strong absorption coefficients (typically more than 10^5 cm^{-1}), which partly balance such low charge-carrier mobility, leading to relatively high photon absorption even in very thin devices [3]. In any case, in the majority of OPV cells the low charge mobility results in devices where the total active layer thickness is close to 100 nm or even less. The semi-transparency of the absorber layer associated with such small thickness may turn out to be the strongest asset for OPVs to compete in the photovoltaic production of electrical energy.

In recent years, many authors recognized that such intrinsic semi-transparency of the absorber layer opens the door for an effective integration in transparent elements, such as windows in buildings and automobiles, or screens in electronic equipment [4, 5]. In a common OPV device, the active layer semi-transparency cannot be used because such device is capped with a non-transparent metal layer that serves as one of the electrodes of the cell and as mirror to reflect back into the device photons that have not been absorbed during the first path. To re-harvest, invisible to the human eye, IR or UV photons lost in semi-transparent devices require either changes in the cell architecture or the use of materials with an enhanced absorption in the UV and especially in the near IR.

The aim of this chapter is to demonstrate that simple photonic structures as 1-D photonic crystals can be used to effectively re-harvest the IR and UV lost photons. In Sect. 2 of the chapter the concept and operational principles of organic solar cells and semi-transparent cells are reviewed. Special attention is given to the materials and deposition procedure required to make an OPV cell semi-transparent. In Sect. 3 it is shown that when a 1-D photonic crystal is incorporated in the structure of the cell and its intrinsic periodicity is broken adding an additional degree of freedom in the structural design, the light management becomes very effective and semi-transparent cells with an optimal performance can be obtained. Details on the performance of such cells are also given in Sect. 3. The use of different light trapping strategies proposed to increase light absorption in dye sensitized solar cells is also reviewed along Sect. 4. Finally some concluding remarks are given in Sect. 5.

2 Operational Principles of OPV Cells

2.1 Photon Absorption to Charge Collection in OPV Cells

The process of converting light into electrical current in an OPV cell is accomplished by four consecutive steps: (1) Absorption of a photon leading to the formation of an excited state, known as the electron-hole pair (exciton), (2) exciton diffusion to a region where (3) charge separation occurs and, (4) the charge transport and collection by the anode (holes) and cathode (electrons) to supply a direct current for the external electrical load. The electric current that an organic solar cell delivers is proportional to the number of created charges that are collected at the electrodes. This number will depend on the product of the fraction of photons absorbed ($\eta_{\mathrm{A}}(\lambda)$, being λ the wavelength), the fraction of excitons that diffuse to the donor–acceptor interface (η_{ED}), the fraction of excitons that are dissociated (η_{CS}), and finally the fraction of charges collected by the electrodes (η_{CC}) [6]. A schematic drawing of the energy levels forming an OPV with their four fundamental steps involved in the photon to charge conversion mechanism is shown in Fig. 1.

From the efficiencies above and the photon flux (cf. Fig. 2) one may obtain the short circuit current density

$$J_{\mathrm{sc}} = q\eta_{\mathrm{ED}}\eta_{\mathrm{CS}}\eta_{\mathrm{CC}}\int \eta_A(\lambda)\Phi(\lambda)\frac{\lambda}{hc}\mathrm{d}\lambda, \tag{1}$$

where q is the charge of the electron, $\Phi(\lambda)$ corresponds to the standard AM1.5G spectral irradiance of the sun, h is the Plank's constant, and c is the velocity of light.

The energy difference between the lowest unoccupied molecular orbital (LUMO) of the donor and highest occupied molecular orbital (HOMO) of the acceptor provides the driving force for the dissociation of excitons (cf. Fig. 1). This dissociation physically occurs at the interface between the donor and the acceptor materials. To maximize this area and compensate for a short exciton diffusion length, the concept of bulk heterojunction (BHJ) was introduced.

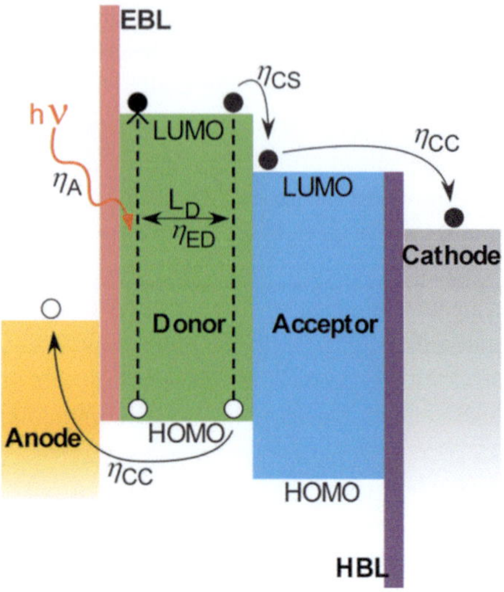

Fig. 1 Energy levels of an OPV cell where the different steps and the corresponding efficiencies in the photon absorption to charge collection process are shown. The electron and hole blocking layers (EBL and HBL, respectively) are shown too

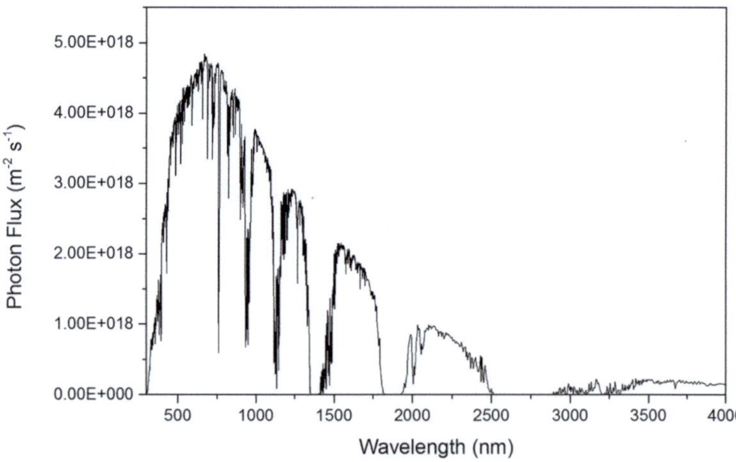

Fig. 2 Sun photon flux at the surface of the earth corresponding to the standard AM1.5G spectral irradiance of the sun

The BHJ in an OPV cell involves mixing donor–acceptor materials in the active layer of the device. This concept was first demonstrated by Hiramoto et al. through the co-evaporation of donor and acceptor molecules under high-vacuum conditions [7]. The first efficient BHJ polymer solar cells were independently realized in 1995 by the groups of Heeger and Friend in polymer–fullerene and polymer–polymer blends [8, 9]. Polymer–fullerene BHJ systems currently provide some of the highest efficiency OPV devices. Single junction polymer solar cell efficiencies are currently close to 10 %. The most common and promising materials used nowadays in the active BHJ layer in OPV cells are the PTB7 family donor polymer mixed with the $PC_{71}BM$ fullerene derivative [10–13].

2.2 *OPV Device Architecture*

The standard OPV device architecture consists of a transparent bottom electrode for the hole collection, a thin layer of absorber material and a top metal electrode for the electron collection (cf. Fig. 3a) [11, 14]. In the inverted structure, the role of the electrodes is swapped; thus, the electrons are collected by the bottom transparent electrode and the holes by the top metal electrode (cf. Fig. 3b) [15, 16].

In order to improve charge collection, transporting layers are introduced in the device structure between the electrodes and the BHJ layer (cf. Figs. 1 and 3). Hole transporting layers (HTL) are deposited between the anode and the active layer and, similarly, electron-transporting layers (ETL) between the active layer and the cathode for the standard device geometry. A good ETL, as well as a good HTL, must meet a list of requirements like transparency, good electrical properties, and chemical stability [17, 18]. Those transporting layers have multiple functions: to reduce the energy barrier between the active layer and the electrode, thus forming ohmic contacts for effective charge extraction; to form selective contacts for a single type of carrier, such as an electron-transporting layer that also serves as a hole-blocking layer; and to help position the absorber layer towards the maximum of the electric-field intensity distribution [19, 20].

The most commonly used device structure for a BHJ organic solar cell comprises a conductive transparent electrode, typically metal oxides like indium tin oxide (ITO) or fluorine-doped tin oxide (FTO), on glass or plastic substrates covered by a thin hole conducting layer, such as the polymer poly(3,4-ethylene-dioxythiophene)-poly(styrenesulfonate) (PEDOT:PSS), over which the absorber layer is deposited from a solution by spin coating, doctor blading, ink-jet printing, spray coating, etc. [21, 22]. Finally, a thin metal layer is deposited (Al, Ca/Al,

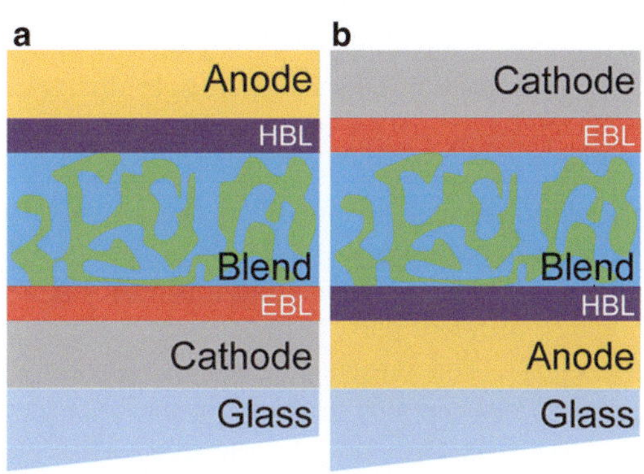

Fig. 3 BHJ architecture depicted for OPV cells in (**a**) direct and (**b**) inverted configurations. The donor and the acceptor are intermixed forming the BHJ or blend

LiF/Al, for instance), usually by thermal evaporation. This structure has been used in the realization of devices with very good performance, reaching a PCE above 9 % [23]. The collection process is reversed in the inverted configuration, i.e., collecting the holes by the top electrode and the electrons by the bottom electrode. In such a structure, the top metal electrode would be a high work function electrode, like silver or gold (Ag or Au), thus eliminating the oxidation problem, while the bottom electrode should be a transparent electron conducting layer (inorganic oxides like zinc oxide (ZnO) or titanium oxide (TiOx) are good examples), eliminating the problem of the acidic PEDOT:PSS on ITO or FTO.

2.3 Light Harvesting in OPV Cells

As indicated the overall thickness of the absorber layer in an OPV cell is about 100 nm. Consequently, it is very important to optimize the thicknesses and position of the absorbing materials to ensure that the electromagnetic field distribution is such that most of the light intensity is concentrated in the active layer. One of the simplest approaches to enhance the absorption in the active layer is replacing the buffer layer in contact with the back reflector for a material with a reduced extinction coefficient. This enhances the reflectivity of the back electrode resulting in an increase of the J_{sc} [11]. Light absorption enhancements can be further improved by tuning the thicknesses of all the cell layers to obtain an optimal optical interference.

Light absorption enhancements by texturing in two or three dimensions have been considered in several forms and instances. When hexagonal arrays of nanocolumns [24, 25] or nanoholes [26] were embossed in the blend layer to increase light trapping by scattering, enhanced photocurrents were observed. Such enhancement in the absorption by light scattering is also seen when the nanostructure is integrated in one of the buffer layers [27]. Other thin film technologies, such as quantum dot solar cells, applied a light-trapping scheme with a nano-imprinted electrode capable of both diffracting light and collecting the photo-generated carriers [28, 29]. Texturing the electrodes was also applied in OPV cells using a periodic [30] or a random [31, 32] grating. The increase in photon absorption resulted from light diffracted after the incident light had been reflected back by the rear electrode.

Other approaches to increase the effective light path inside OPV cells considered the use of plasmonic structures [33]. The presence of metallic nanoparticles in the buffer layer not only increased the degree of light absorption but also enhanced the degree of exciton dissociation [34, 35]. Different shapes have been proposed to enhance different parts of the spectra, such as nanocubes [36] or plasmonic oligomers [37]. The presence of nanostructures is not restricted to the cell buffer layers. They can also be found in the active layer [38] or as a grating in the rear electrode [39]. Recent studies showed a plasmonic OPV with a PCE above 8 % [40].

This latter configuration used the light scattered by plasmonic nanoparticles to achieve a 20 % enhancement in the light absorption.

There are other types of structures with the ability to redirect or even trap light inside the devices. One of the first exploited configurations was the V-shape tandem device with two individual cells combined in a folded geometry [41]. The unabsorbed light from the first cell is re-directed towards the adjacent cell and absorbed there. At small entrance angles almost all the light is trapped within the structure and eventually absorbed by the active layers. The extrapolation of a light trapping by multiple reflections but in single cells was applied to a microprism array substrate [42, 43]. Other light trapping configurations have been explored with the optical elements separated from the thin film devices, such as a pyramidal rear reflector [44] or an array of microlenses [45]. The former one induced four passes of light through the active layer due to the total internal reflection at the original light incidence surface. In the latter one, the enhancement was attributed to a combination of two processes: a longer optical path length inside the absorber layer because the curved microlenses surface refracted the incoming light and, secondly, light reflecting off one of the microlenses striking a neighboring one where the lens curvature results in an incident angle favorable for transmission.

Diminishing the size of such optical elements laid over planar devices down to submicron scale has been explored, too. A configuration of nanospheres on top of a high-index photovoltaic absorber layer from a thin film cell was proposed to couple light into such high-index material and obtain an enhanced light absorption. The more effective coupling originated from whispering gallery modes (WGM) inside the spheres [46, 47]. An alternative approach to improve the absorption using WGM considered a low-quality factor spherical nanoshell structure that facilitated the coupling of light into such resonant modes and substantially enhanced the light path inside the material [48]. Such structures to assess the WGM have been experimentally shown to trap light and enhance the absorption in silicon thin films [49, 50]. WGM coupling in a configuration based on implementing a thin film cell on the backside of an array of parallel fibers has been proposed. Numerical calculations performed using the parameters of a thin film organic cell demonstrated that light coupling becomes more effective as the angle for the incident light relative to the fiber array normal increases up to an optimal angle close to 55°. It was proposed that such WGM coupling may perform as an effective partial tracking of the sun movement for over 100° without any mechanical help [51].

2.4 Semi-Transparent OPV Cells

Despite an extremely thin absorber layer, the majority of the fabricated OPV devices are opaque because the rear contact is, typically, a non-transparent metal layer. This layer performs as one of the electrodes of the cell and as mirror to reflect back into the device photons not absorbed after the first path. It has not been until recently that semi-transparent top electrodes have been fabricated with similar electrical properties as the ones which are deposited directly on the substrate [52–59]. As opposed to a 100 nm ITO layer which is commonly used as bottom transparent electrode in the many different kinds of OPV cells, there is no standard material or fabrication procedure for the top electrode in semi-transparent OPV cells. In many instances the physical or chemical properties of the absorber layer used may determine the nature of the top semi-transparent electrode. In opaque OPV cells, typically, a 100 nm thick layer of Ag or Al is deposited by thermal evaporation, a procedure and materials that are relatively innocuous to the previously deposited layers of the cell. When this last layer is thinned down to less than 20 nm to make it semi-transparent, obtaining a homogenous layer with low resistivity is no longer a straightforward task. An alternative to the ultra-thin films of evaporated Al or Ag is a solution-processed nanowire electrode which has been used in several occasions leading to optically and electrically good performing devices [59–63]. Another solution-processed alternative is a top electrode of PEDOT:PSS [64, 65]. A non-annealed 120 nm thick layer of ITO was also tested as top electrode in [66]. A thin layer of Au as top electrode was used recently in semi-transparent cells exhibiting efficiencies above 5 % [67].

However, among all these different kinds of semi-transparent top electrodes, the most widely used is the dielectric/thin metal/dielectric structure [14, 52, 53, 57, 68, 69]. As noted fabrication of such thin film structure on top of the polymeric materials may damage the active material causing a decrease in the quality of the electrical properties of the cells. In order to prevent such cell degradation, the three thin layer structure may be deposited by thermal evaporation due to the low evaporation temperature of the silver and some dielectric layers. In order to reduce the percolation threshold of Ag layers deposited by thermal evaporation, the film must be deposited at a high deposition rate of 5.5–6 Å s^{-1} while the substrate is simultaneously cooled down to −5 °C [14, 52, 57]. Such elevated evaporation rates and low substrate temperatures generally reduce surface diffusion of Ag atoms and thus alter the standard nucleation process leading to rather homogenous layers of Ag with thicknesses below 10 nm.

3 1-D Photonic Crystals to Manage Light in Semi-Transparent OPV Cells

When the top electrode of an OPV cell is made semi-transparent, the capacity to trap the electromagnetic field in the active layer diminishes. Irrespective of the type of semi-transparent top electrode used, this occurs at all wavelengths leading to devices exhibiting efficiencies which are about 60 % the one of the corresponding opaque cell. In the majority of the potential applications of semi-transparent cells, transparency for the IR or UV photons would not be needed. Several approaches that modify the cell architecture have been proposed to re-harvest such photons.

A combination of a Bragg reflector and an anti-reflective coating have been used to increase near IR photon harvesting demonstrating that the efficiency of small molecule OPV cells could be increased from 1.3 % to 1.7 % [66]. An alternative option to re-harvest red light considered the use of cholesteric liquid crystals as wavelength dependent reflectors [70]. Enhancing near IR photons harvesting is also possible by considering donor polymers with an absorption band shifted close to 800 nm. Following such strategy semi-transparent OPV cells were reported to exhibit a power conversion efficiency (PCE) above 4 % [60, 70]. The use of high performance polymers or tandem cells has also been considered in semi-transparent cells [71, 72]. In that case, either efficiencies above 5 % can be achieved at the expense of a rather limited luminosity, or high transmission in the visible is obtained at the expense of a reduced near IR photon harvesting which eventually limits the overall efficiency of the device.

An optimal approach is to combine high performance polymer PV materials with a photonic configuration to re-harvest UV and near IR photons when the top electrode is thinned down. This may bring the PCE of visibly transparent cells to the high limit efficiencies that were established on a model based on the Schockley–Queisser theory [73].

As indicated in Sect. 2, the single junction opaque OPV cells that exhibit the highest efficiency are cells made with BHJs of the PTB7:$PC_{71}BM$ family [12, 74, 75]. Although lowering the band gap of such kind of materials is a desirable objective to increase the efficiency for semi-transparent as well as non-transparent cells, attempts to do so have not yet yielded efficiencies higher than the ones measured with PTB7:$PC_{71}BM$ for standard or inverted architectures [13]. The extinction coefficient of this blend spanning from 300 to 800 nm averages 0.2 is shown in Fig. 4. With such extinction coefficient and a 100 nm thick layer, approximately half the photons at any given wavelength do not get absorbed during the first light path.

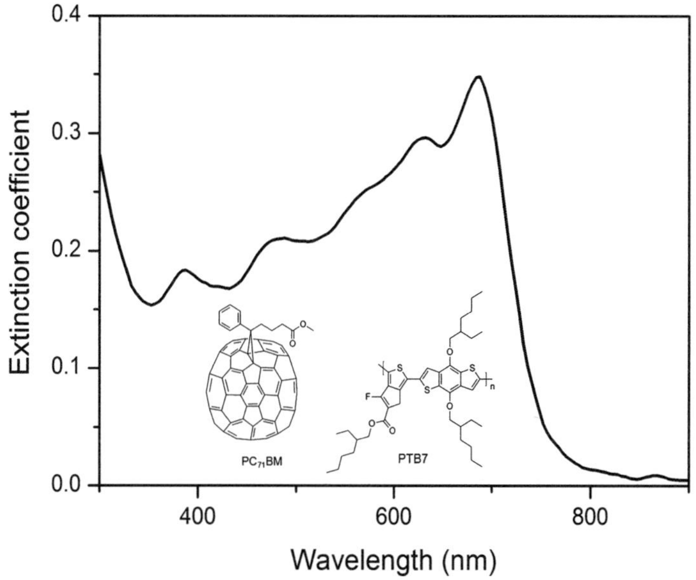

Fig. 4 The extinction coefficient of the PTB7:$PC_{71}BM$ blend

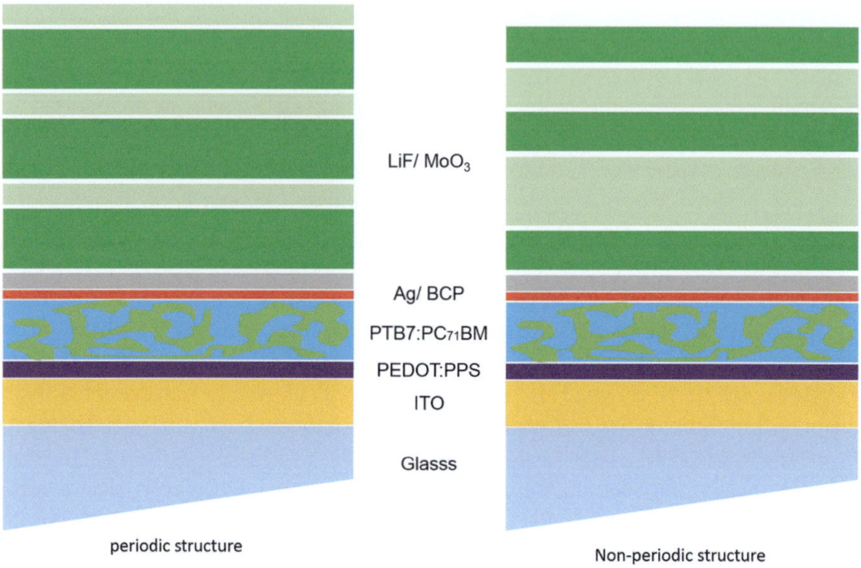

Fig. 5 OPV PTB7:PC_{71}BM cell incorporating a periodic 1-D photonic crystal of six layers (*left*), and a non-periodic multilayer of five layers (*right*)

To enhance trapping of the electromagnetic field at such wavelengths, one may consider 1-D photonic structures incorporated above the top semi-transparent electrode to reflect the IR and UV and transmit the visible as shown on the left drawing in Fig. 5. The reflectivity spectra of devices incorporating such periodic multilayers are shown in Fig. 6. The multilayers are designed to satisfy the Bragg condition to get maximum reflectivity at a near IR wavelength range. However, in a photovoltaic device, interference must be the optimal one at all wavelengths of the portion of the solar spectrum being absorbed by the active material. One way to better reach the goal of a broadband photonic control using simple one-dimensional structures is to increase the degrees of freedom and use a numerical inverse problem solving method. For a semi-transparent OPV cell there are essentially two parameters that will determine its level of performance: the efficiency in converting light to electricity and the device visible transmission or luminosity. The numerical inverse problem solving method must be implemented to design a photonic multilayer, where the constraint of a periodic structure is removed, that maximizes the contribution to the J_{sc} for wavelengths below near UV and above near IR while keeping the device visible transparency above the desired lower limit value.

As shown in Fig. 7, the J_{sc} obtained following such procedure increases rapidly when layers are added in the photonic crystal but saturates beyond five layers. For the five layer structure, the calculated J_{sc} is 76.3 % that of the corresponding opaque cell. On the contrary, for an optimal six layer periodic structure, the best efficiency that can be reached is 72 % that of the opaque cell. The better performance of the non-periodic structure is attributed to a reflectivity, shown in Fig. 6, that adapts

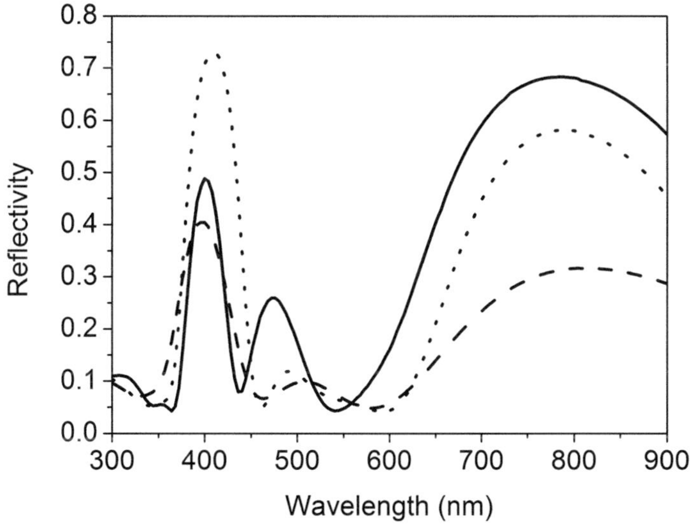

Fig. 6 Reflectivity of the entire OPV device incorporating a periodic 1-D photonic crystal of four layers (*dashed line*), of six layers (*dotted line*) and a non-periodic multilayer of five layers (*solid line*)

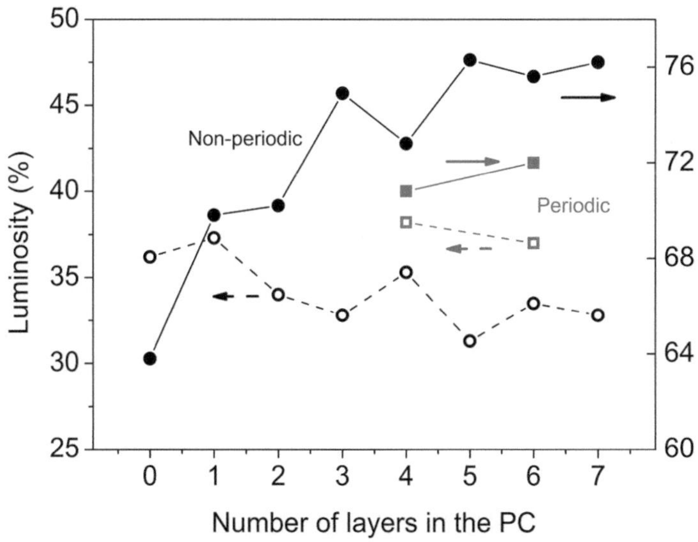

Fig. 7 As a function of the number of layers, numerically determined relative short circuit current (*solid black circles*) and luminosity (*open black circles*) for devices incorporating the non-periodic multilayer, and short circuit current (*solid grey squares*) and luminosity (*open grey squares*) for devices incorporating optimal periodic 1-D photonic crystals of 2 and 3 periods. The short circuit currents are given relative to the corresponding one from an equivalent opaque cell

optimally not only to the absorption spectrum of the absorber blend (cf. Fig. 4) but also to the sun photon flux shown in Fig. 2. As seen in Fig. 6, the reflectivity of the non-periodic structure is enhanced for the near IR photons at the expense of a reduction for the near UV photons when compared to the reflectivity of the six layer periodic structure. This is in correspondence to a larger photon flux in near IR band relative to the UV (cf. Fig. 2). The reflectivity in the visible is maintained low in both cases, ensuring a visible device transparency or luminosity close to 30 %.

The non-periodic configurations predicted above have been tested in an experimental realization that considered a PTB7:$PC_{71}BM$ cell with a standard inverted architecture. Semi-transparent PTB7:$PC_{71}BM$ cells, exhibiting a 30 % visible light transmission and a 5.6 % efficiency were reported [14]. Such performance was achieved by using the inverse integration described above to design and fabricate a non-periodic multilayer to optimally trap near IR and near UV photons. The light harvesting recovery led to a short circuit current for the semi-transparent cell equivalent to 77 % the one from its opaque counterpart. In addition, modifying the layer structure it was shown that the device color could be broadly tuned without significantly altering its performance [14].

4 Light Trapping Approaches in DSSCs

Increase of photon trapping capacity into the absorbing layer has also been explored for other semi-transparent cells such as the dye sensitized solar cells (DSSCs) [76–78]. In this case, dye molecules adsorbed on the surface of a wide band gap semiconductor are responsible for light absorption, whereas charge transport processes take place at the porous inorganic network and a liquid component, respectively. A significant difference with respect to the OPV cells is that any architecture potentially integrable within the device should be porous enough to allow proper liquid electrolyte diffusion and, therefore, correct device functioning. In fact, this particular property of DSSCs was for a long time the main limiting factor for the application of optical elements that, on the contrary, had been commonly employed in other thin film photovoltaic technologies [79–81]. Instead, apart from the development of newly designed and efficient sensitizers [82–84], approaches based on diffuse scattering phenomena were for a long time the only ones widely employed to increase DSSC efficiency [85–87]. Those gave rise to a large enhancement of the light harvesting efficiency, but also implied the loss of transparency of the devices.

Periodic structures, such as 3-D and 1-D photonic crystals, were also implemented in DSSCs to control the optical absorption properties of such devices. In the former case, inverse opals composed of air voids periodically arranged in a semiconductor oxide matrix were coupled to a standard TiO_2 nanocrystalline electrode. The effect of integrating such kind of 3-D photonic crystal on the device performance was studied both theoretically and experimentally [88, 89]. Besides a photocurrent improvement, the use of these periodic architectures allowed controlling the DSSC transparency, contrarily to what happened when diffuse scattering strategies were employed. Although certain contribution coming from optical resonances was demonstrated [90], the mechanism behind the observed improvement was mainly the back reflection of wavelengths lying within the photonic band gap of the periodic structure.

The use of novel porous 1-D photonic crystals was proposed as a viable pathway to enhance the photovoltaic performance of DSSCs due to the possibility of achieving higher reflectivity when compared to their 3-D counterparts [91]. Such structures were made through the alternate deposition of different types of metal oxide nanoparticles by means of solution-processed techniques [92]. After a careful optimization of their structural and optical properties, efficiency enhancements up to 70 % were obtained [93, 94]. A theoretical analysis also demonstrated the relevance of the interplay between the effect of the electrode thickness, the dye absorption spectrum and the characteristics of the Bragg reflections for such kind of devices [95], an excellent agreement with the experimental results being attained for the case of using thin absorbing layers of around 2–3 μm in thickness.

Another light trapping strategy implied the use of one-dimensional gratings moulded on the surface of the nanocrystalline TiO_2 active layer [96, 97]. To do so, soft-lithography technique allowed the fabrication of polymeric replicas, which were prepared from masters with the desired periodic surface relief pattern, that were afterwards imprinted on the surface of the semiconductor oxide layer. This approach led to J_{sc} improvements up to 15 % for a broad range of angles of incidence, which are attributed to an increase in the light optical path of diffracted beams within the absorbing layer. Figure 8 illustrates the different configurations mentioned to increase the photon to current conversion efficiency in semi-transparent DSSCs.

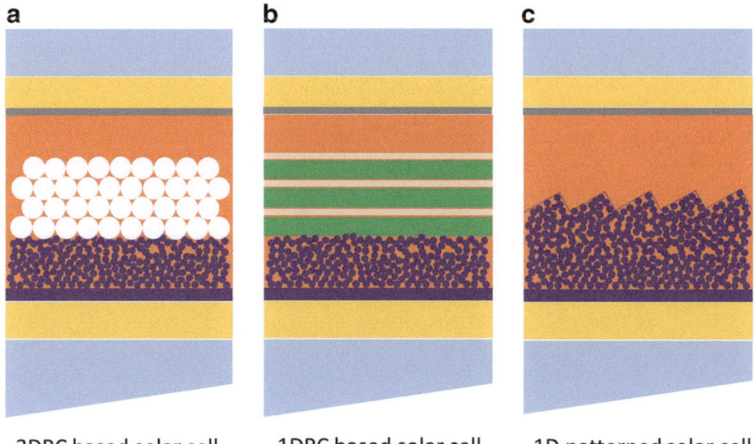

Fig. 8 Scheme of light trapping approaches proposed to enhance the optical absorption of semi-transparent DSSCs through the coupling of: (**a**) 3-D photonic crystals, (**b**) 1-D photonic crystals, and (**c**) 1-D patterning of the active layer, respectively

5 Conclusions

In semi-transparent OPV cells, light harvesting is reduced when high transmissivity top electrodes are used and the device loses its capacity for light trapping in the active layer. Although the capacity for photon trapping in any thin film solar device would always be limited, in this chapter it is demonstrated that the photon management provided by a one-dimensional structure of only five nano layers is sufficient to bring back the light harvesting capacity of the device to 77 % that of the opaque cell. It is shown that when the intrinsic periodicity of 1-D photonic crystals is eliminated, the PCE of the semi-transparent cell is larger when compared to the PCE of the cell incorporating a standard 1-D periodic structure. A study of the theoretical efficiency limits of transparent cells determined that cells with a 30 % luminosity have the potential to exhibit a PCE close to 88 % that of the Shockley–Queisser limit [73]. Combining the photonic control proposed here with the use of red shifted absorption low band gap polymers it is expected that the performance of semi-transparent cells may get very close to such theoretical prediction. Finally, the laboratory scale results discussed in this chapter may, open the door to design and fabrication of new types of PV modules with a large capacity to be incorporated in buildings as windows leading to a very good integration of electrical power generation sources in highly populated urban areas.

References

1. NREL, http://www.nrel.gov/ncpv/, Content Last Updated: June 05, 2015
2. S.R. Forrest, Nature **428**, 911 (2004)
3. C. Winder, N.S. Sariciftci, J. Mater. Chem. **14**, 1077 (2004)
4. R.F. Bailey-Salzman, B.P. Rand, S.R. Forrest, Appl. Phys. Lett. **88**, 233502 (2006)
5. H.P. Kim, H.J. Lee, A.R.B. Yusoff, J. Jang, Sol. Energy Mater. Sol. Cells **108**, 38 (2013)
6. H. Hoppe, N.S. Sariciftci, J. Mater. Res. **19**, 1924 (2004)
7. M. Hiramoto, H. Fujiwara, M. Yokoyama, J. Appl. Phys. **72**, 3781 (1992)
8. G. Yu, J. Gao, J.C. Hummelen, F. Wudl, A.J. Heeger, Science **270**, 1789 (1995)
9. J.J.M. Halls, C.A. Walsh, N.C. Greenham, E.A. Marseglia, R.H. Friend, S.C. Moratti, A.B. Holmes, Nature **376**, 498 (1995)
10. Y.Y. Liang, Z. Xu, J.B. Xia, S.T. Tsai, Y. Wu, G. Li, C. Ray, L.P. Yu, Adv. Mater. **22**, E135 (2010)
11. A. Martinez-Otero, X. Elias, R. Betancur, J. Martorell, Adv. Opt. Mater. **1**, 37 (2013)
12. Z.C. He, C.M. Zhong, S.J. Su, M. Xu, H.B. Wu, Y. Cao, Nat. Photonics **6**, 591 (2012)
13. L.Y. Lu, L.P. Yu, Adv. Mater. **26**, 4413 (2014)
14. R. Betancur, P. Romero-Gomez, A. Martinez-Otero, X. Elias, M. Maymo, J. Martorell, Nat. Photonics **7**, 995 (2013)
15. R. Betancur, A. Martinez-Otero, X. Elias, P. Romero-Gomez, S. Colodrero, H. Míguez, J. Martorell, Sol. Energy Mater. Sol. Cells **104**, 87 (2012)
16. N. Formica, D.S. Ghosh, A. Martinez-Otero, T.L. Chen, J. Martorell, V. Pruneri, Appl. Phys. Lett. **103**, 183304 (2013)
17. I. Litzov, C.J. Brabec, Materials **6**, 5796 (2013)
18. S. Murase, Y. Yang, Adv. Mater. **24**, 2459 (2012)

19. P. Peumans, A. Yakimov, S.R. Forrest, J. Appl. Phys. **93**, 3693 (2003)
20. J. Drechsel, B. Mannig, F. Kozlowski, M. Pfeiffer, K. Leo, H. Hoppe, Appl. Phys. Lett. **86**, 244102 (2005)
21. B.T. de Villers, C.J. Tassone, S.H. Tolbert, B.J. Schwartz, J. Phys. Chem. C **113**, 18978 (2009)
22. W.Y. Nie, R. Coffin, J.W. Liu, C. MacNeill, Y. Li, R.E. Noftle, D.L. Carroll, Int. J. Photoenergy **2012**, 175610 (2012) doi:10.1155/2012/175610
23. M. Scharber, N.S. Sariciftci, Prog. Polym. Sci. **38**, 1929 (2013)
24. D.-H. Ko, J.R. Tumbleston, L. Zhang, S. Williams, J.M. DeSimone, R. Lopez, E.T. Samulski, Nano Lett. **9**, 2742 (2019)
25. Y.C. Liu, C. Kirsch, A. Gadisa, M. Aryal, S. Mitran, E.T. Samulski, R. Lopez, J. Phys. D. Appl. Phys. **46**, 024008 (2013)
26. Y.S. Hsiao, F.C. Chien, J.H. Huang, C.P. Chen, C.W. Kuo, C.W. Chu, P.L. Chen, J. Phys. Chem. C **115**, 11864 (2011)
27. F.S. Zu, X.B. Shi, J. Liang, M.F. Xu, C.S. Lee, Z.K. Wang, L.S. Liao, Appl. Phys. Lett. **104**, 243904 (2014)
28. M.M. Adachi, A.J. Labelle, S.M. Thon, X. Lan, S. Hoogland, E.H. Sargent, Sci. Rep. **3**, 2928 (2013)
29. A. Mihi, F.J. Beck, T. Lasanta, A.K. Rath, G. Konstantatos, Adv. Mater. **26**, 443 (2014)
30. K.S. Nalwa, J.M. Park, K.M. Ho, S. Chaudhary, Adv. Mater. **23**, 112 (2011)
31. S.I. Na, S.S. Kim, J. Jo, S.H. Oh, J. Kim, D.Y. Kim, Adv. Funct. Mater. **18**, 3956 (2008)
32. X. Yu, X.M. Yu, J.J. Zhang, Z.Y. Hu, G.S. Zhao, Y. Zhao, Sol. Energy Mater. Sol. Cells **121**, 28 (2014)
33. H. Atwater, A. Polman, Nat. Mater. **9**, 205 (2010)
34. J.L. Wu, F.C. Chen, Y.S. Hsiao, F.C. Chien, P.L. Chen, C.H. Kuo, M.H. Huang, ACS Nano **5**, 959 (2011)
35. L.Y. Lu, Z.Q. Luo, T. Xu, L.P. Yu, Nano Lett. **13**, 59 (2013)
36. S.W. Baek, G. Park, J. Noh, C. Cho, C.H. Lee, M.K. Seo, H. Song, J.Y. Lee, ACS Nano **4**, 3302 (2014)
37. F. Pastorelli, S. Bidault, J. Martorell, N. Bonod, Adv. Opt. Mater. **2**, 171 (2014)
38. C.C.D. Wang, W.C.H. Choy, C.H. Duan, D.D.S. Fung, W.E.I. Sha, F.X. Xie, F. Huang, Y. Cao, J. Mater. Chem. **22**, 1206 (2011)
39. J.B. You, X.H. Li, F.X. Xie, W.E.I. Sha, J.H.W. Kwong, G. Li, W.C.H. Choy, Y. Yang, Adv. Energy Mater. **2**, 1203 (2012)
40. Q.Q. Gan, F.J. Bartoli, Z.H. Kafafi, Adv. Mater. **25**, 2385 (2013)
41. Y.H. Zhou, F.L. Zhang, K. Tvingstedt, W.J. Tian, O. Inganas, Appl. Phys. Lett. **93**, 033302 (2008)
42. M. Niggemann, M. Glatthaar, P. Lewer, C. Muller, J. Wagner, A. Gombert, Thin Solid Films **511**, 628 (2006)
43. M. Niggemann, M. Riede, A. Gombert, K. Leo, Phys. Status Solidi A **205**, 2862 (2008)
44. W.R. Cao, J.D. Myers, Y. Zheng, W.T. Hammond, E. Wrzesniewski, J.G. Xue, Appl. Phys. Lett. **99**, 023306 (2011)
45. J.D. Myers, W.R. Cao, V. Cassidy, S.H. Eom, R.J. Zhou, L.Q. Yang, W. You, J.G. Xue, Energy Environ. Sci. **5**, 6900 (2012)
46. J. Grandidier, D.M. Callahan, J.N. Munday, H.A. Atwater, Adv. Mater. **23**, 1272 (2011)
47. J. Grandidier, R.A. Weitekamp, M.G. Deceglie, D.M. Callahan, C. Battaglia, C.R. Bukowsky, R. Colton, C. Ballif, R.H. Grubbs, H.A. Atwater, Phys. Status Solidi A **210**, 255 (2013)
48. Y. Yao, J. Yao, V.K. Narasimhan, Z.C. Ruan, C. Xie, S.H. Fan, Y. Cui, Nat. Commun. **3**, 664 (2012)
49. G. Kang, H. Park, D. Shin, S. Baek, M. Choi, D.H. Yu, K. Kim, W.J. Padilla, Adv. Mater. **25**, 2617 (2013)

50. J. Yin, Y.S. Zang, C. Yue, X. He, J. Li, Z.H. Wu, Y.Y. Fang, Phys. Chem. Chem. Phys. **15**, 16874 (2013)
51. M. Mariano, F.J. Rodríguez, P. Romero-Gomez, G. Kozyreff, J. Martorell, Sci. Rep. **4**, 4959 (2014)
52. C. Tao, G.H. Xie, C.X. Liu, X.D. Zhang, W. Dong, F.X. Meng, X.Z. Kong, L. Shen, S.P. Ruan, W.Y. Chen, Appl. Phys. Lett. **95**, 053303 (2009)
53. L. Shen, Y. Xu, F.X. Meng, F.M. Li, S.P. Ruan, W.Y. Chen, Org. Electron. **12**, 1223 (2011)
54. C. Tao, G.H. Xie, F.X. Meng, S.P. Ruan, W.Y. Chen, J. Phys. Chem. C **115**, 12611 (2011)
55. T. Winkler, H. Schmidt, H. Flugge, F. Nikolayzik, I. Baumann, S. Schmale, T. Weimann, P. Hinze, H.H. Johannes, T. Rabe, S. Hamwi, T. Riedl, W. Kowalsky, Org. Electron. **12**, 1612 (2011)
56. H. Jin, C. Tao, M. Velusamy, M. Aljada, Y.L. Zhang, M. Hambsch, P.L. Burn, P. Meredith, Adv. Mater. **24**, 2572 (2012)
57. N.P. Sergeant, A. Hadipour, B. Niesen, D. Cheyns, P. Heremans, P. Peumans, B.P. Rand, Adv. Mater. **24**, 728 (2012)
58. Z.K. Liu, J.H. Li, Z.H. Sun, G.A. Tai, S.P. Lau, F. Yan, ACS Nano **6**, 810 (2012)
59. J. Krantz, T. Stubhan, M. Richter, S. Spallek, I. Litzov, G.J. Matt, E. Spiecker, C.J. Brabec, Adv. Funct. Mater. **23**, 1711 (2012)
60. C.C. Chen, L.T. Dou, R. Zhu, C.H. Chung, T.B. Song, Y.B. Zheng, S. Hawks, G. Li, P.S. Weiss, Y. Yang, ACS Nano **6**, 7185 (2012)
61. W. Gaynor, J.Y. Lee, P. Peumans, ACS Nano **4**, 30 (2010)
62. F. Guo, X.D. Zhu, K. Forberich, J. Krantz, T. Stubhan, M. Salinas, M. Halik, S. Spallek, B. Butz, E. Spiecker, T. Ameri, N. Li, P. Kubis, D.M. Guldi, G.J. Matt, C.J. Brabec, Adv. Energy Mater. **3**, 1062 (2013)
63. J.H. Yim, S.-Y. Joe, C. Pang, K.M. Lee, H. Jeong, J.Y. Park, Y.H. Ahn, J.C. de Mello, S. Lee, ACS Nano **8**, 2857 (2014)
64. J.H. Yim, S.Y. Joe, C. Pang, K.M. Lee, H. Jeong, J.Y. Park, Y.H. Ahn, J.C. de Mello, S. Lee, Sol. Energy Mater. Sol. Cells **108**, 38 (2013)
65. Z. Tang, Z. George, Z.F. Ma, J. Bergqvist, K. Tvingstedt, K. Vandewal, E. Wang, L.M. Andersson, M.R. Andersson, F.L. Zhang, O. Inganas, Adv. Energy Mater. **2**, 1467 (2013)
66. R.R. Lunt, V. Bulovic, Appl. Phys. Lett. **98**, 113305 (2011)
67. F. Pastorelli, P. Romero-Gomez, R. Betancur, A. Martinez-Otero, P. Mantilla-Perez, N. Bonod, J. Martorell. doi:10.1002/aenm.201400614
68. C.Y. Chang, L.J. Zuo, H.L. Yip, C.Z. Li, Y.X. Li, C.S. Hsu, Y.J. Cheng, H.Z. Chen, A.K.Y. Jen, Adv. Energy Mater. **4**, 1301645 (2014). doi:10.1002/aenm.201301645
69. S. Schubert, J. Meiss, L. Muller-Meskamp, K. Leo, Adv. Energy Mater. **3**, 438 (2013)
70. Y. Galagan, M.G. Debije, P.W.M. Blom, Appl. Phys. Lett. **98**, 043302 (2011)
71. L. Dou, W.-H. Chang, J. Gao, C.-C. Chen, J. You, Y. Yang, Adv. Energy Mater. **25**, 825 (2013)
72. K.S. Chen, J.F. Salinas, H.L. Yip, L.J. Huo, J.H. Hou, A.K.Y. Jen, Energy Environ. Sci. **5**, 9551 (2012)
73. R.R. Lunt, Appl. Phys. Lett. **101**, 043902 (2012)
74. C.C. Chueh, S.C. Chien, H.L. Yip, J.F. Salinas, C.Z. Li, K.S. Chen, F.C. Chen, W.C. Chen, A.K.Y. Jen, Adv. Energy Mater. **3**, 417 (2013)
75. A. Martínez-Otero, X. Elias, R. Betancur, J. Martorell, Adv. Opt. Mater. **1**, 37 (2012)
76. B. O'Regan, M. Grätzel, Nature **353**, 737 (1991)
77. M.K. Nazeeruddin, A. Kay, I. Rodicio, R. Humphry-Baker, E. Müller, P. Liska, N. Vlachopoulos, M. Grätzel, J. Am. Chem. Soc. **115**, 6382 (1993)
78. M. Grätzel, Acc. Chem. Res. **42**, 1788 (2009)
79. D. Zhou, R. Biswas, J. Appl. Phys. **103**, 093102 (2008)
80. P. Bermel, C. Luo, L. Zeng, L.C. Kimerling, J.D. Joannopoulos, Opt. Express **15**, 16986 (2007)

81. K.R. Catchpole, A. Polman, Opt. Express **16**, 21793 (2008)
82. F. Gao, Y. Wang, D. Shi, J. Zhang, M. Wang, X. Jing, R. Humphry-Baker, P. Wang, S.M. Zakeeruddin, M. Grätzel, J. Am. Chem. Soc. **130**, 10720 (2008)
83. H. Imahori, T. Umeyama, S. Ito, Acc. Chem. Res. **42**, 1809 (2009)
84. T. Horiuchi, H. Miura, S. Uchida, Chem. Commun. 3036 (2003)
85. A. Usami, Chem. Phys. Lett. **277**, 105 (1997)
86. S. Hore, C. Vetter, R. Kern, H. Smit, A. Hinsch, Sol. Energy Mater. Sol. Cells **90**, 1176 (2006)
87. F.E. Galvez, E. Kemppainen, H. Miguez, J. Halme, J. Phys. Chem. C **116**, 11426 (2012)
88. A. Mihi, H. Míguez, J. Phys. Chem. B **109**, 15968 (2005)
89. A. Mihi, M.E. Calvo, J.A. Anta, H. Míguez, J. Phys. Chem. C **112**, 13 (2008)
90. A. Mihi, H. Míguez, I. Rodríguez, S. Rubio, F. Meseguer, Phys. Rev. B **71**, 125131 (2005)
91. S. Colodrero, A. Mihi, L. Häggman, M. Ocaña, G. Boschloo, A. Hagfeldt, H. Míguez, Adv. Mater. **21**, 764 (2009)
92. S. Colodrero, M. Ocaña, H. Míguez, Langmuir **24**, 4430 (2008)
93. S. Colodrero, A. Forneli, C. López-López, L. Pellejà, H. Míguez, E. Palomares, Adv. Funct. Mater. **22**, 1303 (2012)
94. D. Colonna, S. Colodrero, H. Lindstrom, A. Di Carlo, H. Míguez, Energy Environ. Sci. **5**, 8238 (2012)
95. G. Lozano, S. Colodrero, O. Caulier, M.E. Calvo, H. Míguez, J. Phys. Chem. C **114**, 3681 (2010)
96. J. Kim, J.K. Koh, B. Kim, J.H. Kim, E. Kim, Angew. Chem. Int. Ed. **51**, 6864 (2012)
97. C. López-López, S. Colodrero, A. Jiménez-Solano, G. Lozano, R. Ortiz, M.E. Calvo, H. Míguez, Adv. Energy Mater. **2**, 879 (2014)
98. G.-M. Ng, E.L. Kietzke, T. Kietzke, T. Kietzke, L.W. Tan, P.K. Liew, F.R. Zhu, Appl. Phys. Lett. **90**, 103505 (2007)
99. F.C. Chen, J.L. Wu, K.H. Hsieh, W.C. Chen, S.W. Lee, Org. Electron. **9**, 1132 (2008)
100. J.S. Huang, G. Li, Y. Yang, Adv. Mater. **20**, 415 (2008)
101. R. Koeppe, D. Hoeglinger, P.A. Troshin, R.N. Lyubovskaya, V.F. Razumov, N.S. Sariciftci, ChemSusChem **2**, 309 (2009)
102. J. Meiss, K. Leo, M.K. Riede, C. Uhrich, W.M. Gnehr, S. Sonntag, M. Pfeiffer, Appl. Phys. Lett. **95**, 213306 (2009)
103. H. Schmidt, H. Flugge, T. Winkler, T. Bulow, T. Riedl, W. Kowalsky, Appl. Phys. Lett. **94**, 243302 (2009)
104. S. Tanaka, A.A. Zakhidov, R. Ovalle-Robles, Y. Yoshida, I. Hiromitsu, Y. Fujita, K. Yoshino, Synth. Met. **159**, 2326 (2009)
105. A. Bedeloglu, A. Demir, Y. Bozkurt, N.S. Sariciftci, Renew. Energy **35**, 2301 (2010)
106. T. Ameri, G. Dennler, C. Waldauf, H. Azimi, A. Seemann, K. Forberich, J. Hauch, M. Scharber, K. Hingerl, C.J. Brabec, Adv. Funct. Mater. **20**, 1592 (2010)
107. Q.F. Dong, Y.H. Zhou, J.N. Pei, Z.Y. Liu, Y.W. Li, S.Y. Yao, J.B. Zhang, W.J. Tian, Org. Electron. **11**, 1327 (2010)
108. J.Y. Lee, S.T. Connor, Y. Cui, P. Peumans, Nano Lett. **10**, 1276 (2010)
109. F. Nickel, A. Puetz, M. Reinhard, H. Do, C. Kayser, A. Colsmann, U. Lemmer, Org. Electron. **11**, 535 (2010)
110. A.K. Pandey, I.D.W. Samuel, IEEE J. Sel. Top. Quantum Electron. **16**, 1560 (2010)
111. Y.H. Zhou, H. Cheun, S. Choi, W.J. Potscavage, C. Fuentes-Hernandez, B. Kippelen, Appl. Phys. Lett. **97**, 153304 (2010)
112. X.Z. Wang, G.-M. Ng, J.-W. Ho, H.-L. Tam, F.R. Zhu, IEEE J. Sel. Top. Quantum Electron. **16**, 1685 (2010)
113. A. Colsmann, A. Puetz, A. Bauer, J. Hanisch, E. Ahlswede, U. Lemmer, Adv. Energy Mater. **1**, 599 (2011)
114. Y.Y. Lee, K.H. Tu, C.C. Yu, S.S. Li, J.Y. Hwang, C.C. Lin, K.H. Chen, L.C. Chen, H.L. Chen, C.W. Chen, ACS Nano **5**, 6564 (2011)

115. J.E. Lewis, E. Lafalce, P. Toglia, X.M. Jiang, Sol. Energy Mater. Sol. Cells **95**, 2816 (2011)
116. J. Meiss, F. Holzmueller, R. Gresser, K. Leo, M. Riede, Appl. Phys. Lett. **99**, 193307 (2011)
117. R.J. Peh, Y.R. Lu, F.L. Zhao, C.-L.K. Lee, W.L. Kwan, Sol. Energy Mater. Sol. Cells **95**, 3579 (2011)
118. A. Bauer, T. Wahl, J. Hanisch, E. Ahlswede, Appl. Phys. Lett. **100**, 073307 (2012)
119. A. Colsmann, M. Reinhard, T.-H. Kwon, C. Kayser, F. Nickel, J. Czolk, U. Lemmer, N. Clark, J. Jasieniak, A.B. Holmes, D. Jones, Sol. Energy Mater. Sol. Cells **98**, 118 (2012)
120. J.-W. Kang, Y.-J. Kang, S. Jung, D.S. You, M. Song, C.S. Kim, D.-G. Kim, J.-K. Kim, S.H. Kim, Org. Electron. **13**, 2940 (2012)
121. Y.H. Kim, L. Muller-Meskamp, A.A. Zakhidov, C. Sachse, J. Meiss, J. Bikova, A. Cook, A.A. Zakhidov, K. Leo, Sol. Energy Mater. Sol. Cells **96**, 244 (2012)
122. Z. Tang, Z. George, Z.-F. Ma, J. Bergqvist, K. Tvingstedt, K. Vandewal, E. Wang, L.M. Andersson, M.R. Andersson, F.-L. Zhang, O. Inganas, Adv. Energy Mater. **2**, 1467 (2012)
123. M.J. Tan, S. Zhong, R. Wang, Z.X. Zhang, V. Chellappan, W. Chen, Appl. Phys. Lett. **103**, 063303 (2013)
124. S.A. Carter, M. Angelopoulos, S. Karg, P.J. Brock, J.C. Scott, Appl. Phys. Lett. **70**, 2067 (1997)
125. B. Maennig, J. Drechsel, D. Gebeyehu, P. Simon, F. Kozlowski, A. Werner, F. Li, S. Grundmann, S. Sonntag, M. Koch, K. Leo, M. Pfeiffer, H. Hoppe, D. Meissner, N.S. Sariciftci, I. Riedel, V. Dyakonov, J. Parisi, Appl. Phys. Mater. Sci. Process. **79**, 1 (2004)
126. M. Sanyal, B. Schmidt-Hansberg, M.F.G. Klein, A. Colsmann, C. Munuera, A. Vorobiev, U. Lemmer, W. Schabel, H. Dosch, E. Barrena, Adv. Energy Mater. **1**, 363 (2011)
127. C.N. Hoth, S.A. Choulis, P. Schilinsky, C.J. Brabec, J. Mater. Chem. **19**, 5398 (2009)
128. Z.Q. Liu, P. Zhan, J. Chen, C.J. Tang, Z.D. Yan, Z. Chen, Z.L. Wang, Opt. Express **21**, 3021 (2013)
129. J. Meiss, T. Menke, K. Leo, C. Uhrich, W.M. Gnehr, S. Sonntag, M. Pfeiffer, M. Riede, Appl. Phys. Lett. **99**, 043301 (2011)

New Sensing Strategies Based on Surface Modes in Photonic Crystals

Emiliano Descrovi, Paola Rivolo, Luca Boarino, Natascia De Leo, and Fabrizio Giorgis

Abstract During the past decade the investigations on the use of optical nanostructures for sensing applications have become an active area of research. The near-field interactions of sensed analytes with optical modes sustained by photonic crystals have resulted in a general increase of the detection performances, both in labelled and label-free configurations. As a particular case, one-dimensional photonic crystals (1DPC) represents an interesting opportunity for enhancing the light–matter interaction mediated by Bloch Surface Waves (BSWs). In this chapter we introduce the concept of surface modes on planar and corrugated 1DPC based on dielectric stratified structures providing several illustrative examples in sensing applications. The use of 1DPC surface modes as optical transducers presents some advantages, such as spectral and polarization tunability and low losses. Moreover, functionalizing layers tailored for specific applications can be included within the definition of the photonic crystal itself, thus leading to hybrid organic/inorganic structures. Some recent advances on the use of a photonic crystal platform for biosensing applications are presented and discussed, including label-free and improved fluorescence-based approaches wherein the intimate coupling of organic emitters to BSWs can be fruitfully exploited.

Keywords Bloch Surface Waves • Dielectric multilayers • Enhanced fluorescence • Surface functionalizations • Optical sensors • Molecular recognition • Label-free detection

---

E. Descrovi • P. Rivolo • F. Giorgis (✉)
Department of Applied Science and Technology, Politecnico di Torino, Corso Duca degli Abruzzi 24, 10129 Torino, Italy
e-mail: emiliano.descrovi@polito.it; paola.rivolo@polito.it; fabrizio.giorgis@polito.it

L. Boarino • N. De Leo
Nanofacility Piemonte, Istituto Nazionale di Ricerca Metrologica, Strada delle Cacce 91, 10135 Torino, Italy
e-mail: l.boarino@inrim.it; n.deleo@inrim.it

D. Comoretto (ed.), *Organic and Hybrid Photonic Crystals*,
DOI 10.1007/978-3-319-16580-6_14

1 Introduction

It is nowadays well known that in photonic crystals (PCs), electromagnetic waves are prohibited to propagate in all directions within their photonic band gap (PBG) [1, 2]. However, when appropriately terminated, PCs can support Bloch Surface Waves (BSWs) with frequencies lying inside the PBG [3].

For a two-dimensional (2D) PC the existence of BSWs has been shown theoretically [4] and experimentally [5], but no particular applications have been exploited. On the other side, BSWs can be easily coupled to one-dimensional PCs (1DPCs, stacks of periodic alternating dielectric layers). In such a case, BSWs can propagate at the interface between a homogeneous medium and the truncated periodic structure [6]. In particular, BSW dispersion curves are located within forbidden bands of the 1DPC, beyond the "light line" of the homogenous medium with refractive index n (corresponding to the linear dispersion relation $\omega = kc/n$, ω is the light frequency, k the wavevector, and c the vacuum light velocity). This results in an exponential decay of the field envelope inside the periodic structure and an exponential decay of the field in the homogeneous medium because of total internal reflection. BSW on dielectric 1DPCs are equivalent to surface plasmon polaritons (SPP) on thin metal films and share some common characteristics. However, since dielectrics are characterized by much lower extinction than metals, BSW resonances appear much narrower than those observed for SPP. Moreover, differently from surface plasmon resonances (SPRs) on metallic thin films, the BSW spectral position can be tuned by a proper design of the 1DPC from the near-Infrared [7] to the visible wavelength range [8], for both transverse-electric (TE) and transverse-magnetic (TM) polarizations [9]. Thus, suitably designed 1DPCs can open the access to regions where SPR, due to limitations introduced by the metals dispersion, cannot operate.

In the last decade, the use of BSWs coupled to 1DPCs has been widely developed in the optical sensing area [10, 11], where the capability to confine and guide light in micro- and nano-structures can be exploited to enhance sensitivity by increasing the effective light–matter interaction. This approach has been developed besides other sensing strategies based on pure plasmon resonances (see the chapter by F. Marabelli et al. in this book).

Bloch surface wave sensors, operating in conventional SPR-like detection scheme [12, 13] show limit of detection comparable to SPP sensors (i.e., ultimate resolution approaching 10^{-7} refractive index units variations), but yield increased figures of merit [14, 15].

Beside the conventional SPR-like scheme, based on reflectivity intensity and phase detections [16, 17], other improved detection configurations based on enhancement phenomena mediated by BSWs such as light diffraction [18], Raman scattering [19], and fluorescence [20, 21] have been demonstrated. The results are really promising in the context of high-sensitivity sensors.

In several cases concerning with sensing platforms, the top-surface of the 1DPC is loaded by organic layers. These layers can be aimed to couple surface modes without any discrete optical element [22] or to control the emission pattern of light sources [23, 24]; in many studies, such layers behave as functional coatings giving a chemical surface selectivity of the 1DPCs with respect to specific analytes [11].

In this chapter different detection schemes based on hybrid organic/inorganic structures coupling BSWs devoted to biosensing applications are presented and critically discussed. In detail, the following sections concern with SPR-like biodetection on Antibody/Anti-antibody systems and BSW-enhanced fluorescence spectroscopy applied in the framework of labeling and label-free approaches.

2 Photonic Crystals Growth and Optical Properties

1DPCs (periodic multilayers, presented and discussed also in other contributions such as 4th, 8th, 11th, and 13th chapters) are here investigated and applied in sensing experiments. The structures were fabricated by Plasma Ion Assisted Deposition (PIAD) and Plasma Enhanced Chemical Vapor Deposition (PECVD). The first technique is able to grow thin films and stratified structures based on stoichiometric amorphous materials like a-SiO_2, a-Ta_2O_5, a-TiO_2, a-Al_2O_3 with extremely low absorption losses without substrate heating. The second one allows the deposition of homogeneous hydrogenated amorphous dielectric binary/ternary alloys such as a-SiOx:H, a-$Si_{1\text{-}x}N_x$:H, a-$Si_{1\text{-}x}C_x$:H, a-SiO_xN_y:H, a-SiO_xC_y:H at a substrate temperature ranging within 200–250 °C. In particular, thin films grown by PECVD present the peculiarity to have optical gaps and refractive indices tunable in a wide range, depending on the elemental composition, which can be easily controlled by the gas feedstocks [25]. Figure 1 shows the refractive index of PECVD a-$Si_{1\text{-}x}N_x$:H thin films with several atomic nitrogen content ($x = \mathrm{N}/(\mathrm{N} + \mathrm{Si})$).

The multilayers, grown on glass substrates (i.e., Corning 7059, BK7) are composed of a periodic stack of high and low refractive index layers, terminated with an eventual ultrathin film used to tailor the Bloch surface modes. Figure 2 shows a SEM cross-sectional viewgraph of typical a-$Si_{1\text{-}x}N_x$:H/a-Si_3N_4:H based 1DPC, where it is possible to appreciate the high quality of the interfaces related to layers with different composition (e.g., optical refractive index).

The experimental determination of the BSW dispersion curve was performed in the Kretschmann–Raether spectroscopic configuration [26], as shown in Fig. 3. A collimated and TE (or TM)-polarized beam is impinged on the 1DPC through the input facet of a glass coupling prism. Reflectance maps $R(\theta, \lambda)$ are obtained by means of a detection scheme in which the sample is rotated with respect to the incident beam at wavelength λ. θ is the angle between the normal to the 1DPC layers and the direction of the incident beam at the prism/multilayer interface. The reflected light is impinged onto a photodiode. The collection lens and the photodiode are rotated according to the angular position of the sample.

In addition to the experimental measurements, the optical response of our structures with the help of a well-established modal method (C-method) is calculated [27]. Typical calculated and measured TE-polarized reflectance maps of

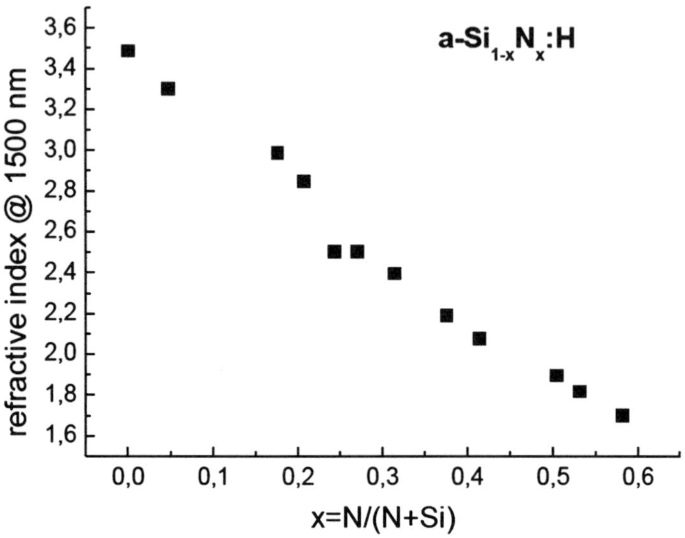

Fig. 1 Refractive index of a-$Si_{1\text{-}x}N_x$:H thin films grown by PECVD measured at 1500 nm as a function of atomic N content

Fig. 2 FESEM cross-sectional viewgraph of a representative a-$Si_{1\text{-}x}N_x$:H/a-Si_3N_4:H based 1DPC

ten-period multilayered structures based on a-$Si_{1\text{-}x}N_x$:H/a-Si_3N_4:H are both shown in Fig. 4. Bright and dark tones indicate high and low reflectance regions, respectively. The narrow dark band running well beyond the light line identifies the BSW dispersion curve. It is worth to underline the remarkably narrow dips associated with the BSW resonances (see inset of Fig. 4b).

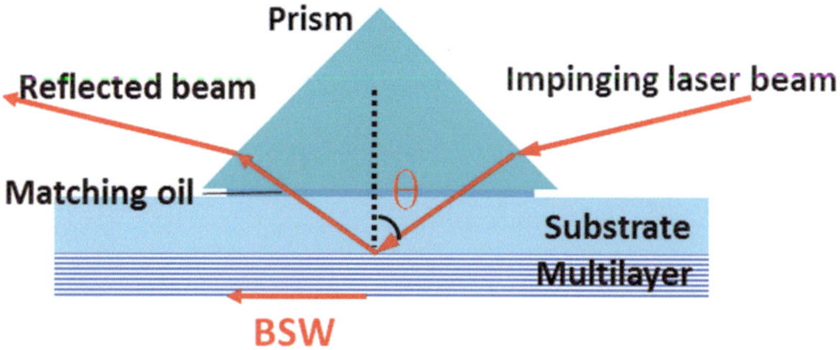

Fig. 3 Sketch of the typical detection scheme for BSW analysis: Kretschmann–Raether configuration

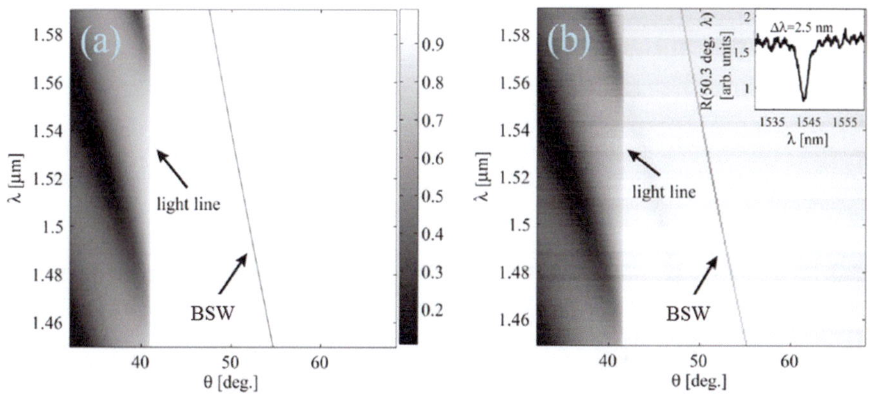

Fig. 4 (**a**) Calculated and (**b**) measured reflectance maps of a ten-period multilayered structure based on a-$Si_{1-x}N_x$:H/a-Si_3N_4:H in the (θ,λ) plane. The BSW dispersion curve is associated with the narrow dark band lying beyond the *light line* (adapted from [26])

3 Detection Scheme 1: SPR-Like for Label-Free Detection

Among all techniques, surface plasmon resonance (SPR) optical biosensors have been recognized to be extremely attractive for simple and fast biodetection and constitute nowadays a mature technology [28, 29]. In this technique a surface plasmon polariton (SPP) [30], propagating either along a gold or silver thin layer, is used to sense the refractive index changes in a very thin layer at the sensor surface, mainly in real-time conditions.

As stated in the Introduction section, in alternative to SPPs, electromagnetic BSW modes propagating at the interface between a homogeneous medium and a truncated periodic structure [31], such as a one-dimensional photonic crystal (1DPC) have been recently proposed [10, 32]. In this section, we report on time resolved sensing experiments demonstrating the direct, label-free, and specific binding of the widely used antibody/anti-antibody reference system constituted by target molecules (Goat Anti-Rabbit IgG tagged with Horse Radish Peroxidase label (HRP), PAbII in the following) and their corresponding receptor molecules (Rabbit IgG, AbI in the following) [11].

A uniform receptors monolayer was transferred onto the surface of a silicon based 1DPC (four-period stack based on a-$Si_{1-x}N_x$:H/a-Si_3N_4:H layers) by soft printing with a PolyDiMethylSiloxane (PDMS) stamp [33]. The recognition was then carried out by measuring the spectral shift of the BSW resonance upon exposition to the target molecules in aqueous solution. The response to the exposition to a non-specific target, Goat Anti Mouse IgG also tagged with HRP (NAbII in the following) is reported too and demonstrates the sensor's selectivity.

In detail, 1DPC biosensors prepared with the superficial AbI layer were mounted on the BK7 coupling prism (np = 1.501) of an ATR experimental setup in the Kretschmann configuration, which is commonly used to excite either SPP on conductive layers [34] or BSW on 1DPCs [26]. The sensor back face is contacted to the prism facet by means of an index matching oil. The probe laser beam can therefore tunnel through the 1DPC and efficiently excite the BSW at the free 1DPC/ analyte truncation interface.

In the experimental setup, already discussed in the previous section, a collimated and TE-polarized beam is expanded from a fibered, tunable diode laser source and used to illuminate the sample through the prism. The wavelength can be tuned in the $\lambda \in$ [1,450 nm, 1,590 nm].

The sample is illuminated at an angle θ with respect to the normal to the prism base, as shown in the inset of Fig. 5. A first lens weakly focuses the beam on the 1DPC surface and a second lens collects the reflected light and focuses it onto the detector (PD). Reflectance profiles at fixed wavelength were obtained by rotating

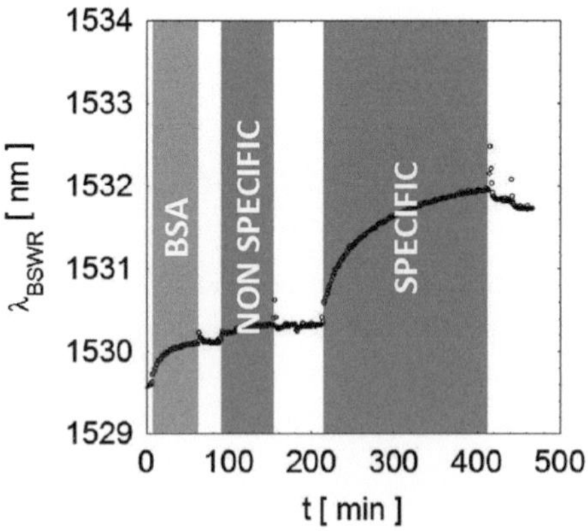

Fig. 5 Sensorgram obtained during a sensing experiment carried out by ATR upon injection of BSA, NAbII and PAbII. The *white regions* correspond to rinsing steps with ddH_2O; during such rinsing steps, it may happen to record steep and permanent shifts due to the formation of air microbubbles in the sensed zone. *Inset*: scheme of the biosensor (adapted from [11])

the prism and the photodiode with respect to the incident beam in a θ–2θ configuration. The 1DPC is topped by a PolyVinylChloride (PVC) flow cell.

In a typical measurement, at a fixed θ_0, the reflectance spectrum in a convenient wavelength range around the BSW resonance for such θ_0 is measured (wavelength interrogation) [29]. Starting with the cell filled with double distilled water (ddH_2O), several analytes are injected in ddH_2O solution and detect the variation of the reflectance dip at fixed time intervals. The reflectance curve is fitted in real time, extracting the time dependent BSW resonance wavelength λ_{BSWR}.

After the surface of the sensors was prepared by micro-contact printing the AbI receptors, their performance was checked by sequentially exposing them to Bovin Serum Albumin (BSA), non-specific and specific target analytes in water solution. A typical sensorgram is shown in Fig. 5. In this specific case, the reflectance spectrum is sampled every 54 s and the fitted λ_{BSWR} is plotted as a function of the exposition time starting from ddH_2O environment. The sensorgram was obtained during experiments carried out at $\theta_0 = 64.47°$, corresponding to an initial BSW resonance $\theta_0 = 1{,}529.6$ nm in ddH_2O.

As a first step a BSA solution with concentration 290 μg/mL (35 μg/cm^2) is injected in the cell, in order to saturate all the non-specific free sites where the primary antibody was not printed. The surface density was calculated taking into account the cell volume and the 1DPC exposed surface. In Fig. 5, the time interval $t \in [6.7 \text{ min}, 62.5 \text{ min}]$ corresponds to BSA incubation. The λ_{BSWR} first shows a sudden shift by $\Delta\lambda_{RI} = 0.10$ nm due to the change of the volume refractive index of the solution and then slowly shifts to larger values due to the formation of a BSA adlayer. The kinetics of the slow shift can be fitted by a single exponential $\lambda_{BSWR}(\mathrm{t}) = \lambda_0 + \Delta\lambda_{RI} + \Delta\lambda_{BSA} \cdot [1 - \exp(-\mathrm{t}/\tau_{BSA})]$, with $\tau_{BSA} = 13.7$ min and $\Delta\lambda_{BSA} = 0.38$ nm. At the end of incubation the cell is rinsed with ddH_2O and, after some transient effects due to pressure changes in the flow cell, the resonance remains in the same position, indicating that BSA is firmly adsorbed to the surface.

As a second step an assay with a non-specific antibody, as negative reference, was performed by injecting a non-specific 125 μg/mL (15 μg/cm^2) NAbII solution. In Fig. 5, the time interval $t \in [88.5 \text{ min}, 154.6 \text{ min}]$ corresponds to NAbII incubation. The λ_{BSWR} shows again a sudden shift by $\Delta\lambda_{RI} = 0.10$ nm due to the change of the solution and then a very slight increase, due to the presence of residual non-specific binding sites. Starting from the initial position of the resonance, the kinetics of the slow shift can be fitted by a single exponential $\lambda_{BSWR}(\mathrm{t}) = \lambda_0 + \Delta\lambda_{RI} + \Delta\lambda_{NEG} \cdot [1 - \exp(-\mathrm{t}/\tau_{NEG})]$, with $\tau_{NEG} = 17.5$ min and $\Delta\lambda_{NEG} = 0.10$ nm. At the end of incubation the cell is rinsed with ddH_2O and, after transient pressure changes, the resonance remains in the position reached at the end of the incubation time. The residual shift of the resonance $\Delta\lambda_{NEG}$ is due to NAbII binding to non-specific sites and sets the resolution of the sensor.

Following the same procedure used for the precedent steps, after washing the flow cell, the PAbII solution is injected in ddH_2O with concentration 125 μg/mL (15 μg/cm^2). In Fig. 5, the time interval $t \in [212.5 \text{ min}, 414.5 \text{ min}]$ corresponds to PAbII incubation. As usual, the λ_{BSWR} shows a sudden shift by $\Delta\lambda_{RI} = 0.27$ nm due to the change of the solution. Then a large shift of the λ_{BSWR} is observed, that can be

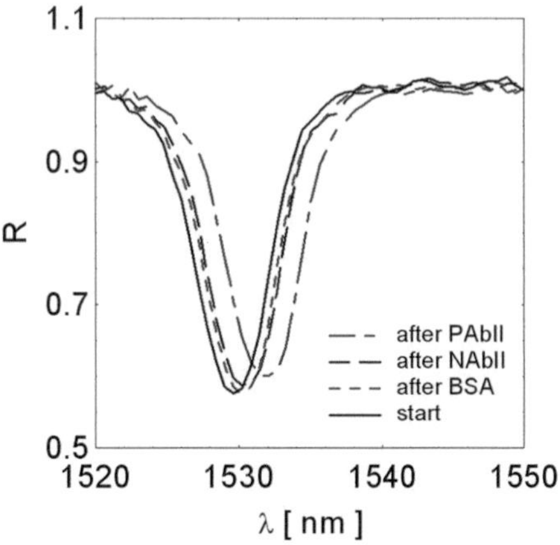

Fig. 6 Absolute reflectance spectra measured at $\theta_0 = 64.47°$ during the sensing experiment with sensorgram given in Fig. 1. The spectra were obtained after BSA, NAbII and PAbII incubation, after rinsing twice in ddH_2O (adapted from [11])

fitted by means of a double exponential law $\lambda_{BSWR}(t) = \lambda_0 + \Delta\lambda_{RI} + \Delta\lambda_{POS1} \cdot [1 - \exp(-t/\tau_{POS1})] + \Delta\lambda_{POS2} \cdot [1 - \exp(-t/\tau_{POS2})]$, with $\tau_{POS1} = 90$ min, $\tau_{POS2} = 12.8$ min, $\Delta\lambda_{POS1} + \Delta\lambda_{POS1} = 1.50$ nm. At the end of incubation the cell was rinsed twice with ddH_2O and, after transient pressure changes, the resonance shows a large residual shift $\Delta\lambda_{POS} = \Delta\lambda_{POS1} + \Delta\lambda_{POS2} = 1.40$ nm. Since the time dependence of the BSW resonance shift diverges from a single exponential behavior, we can foresee that the binding kinetics is strongly influenced by mass transport/diffusion dynamics.

Figure 6 shows the absolute reflectance spectra sampled during the recording of the sensorgram reported in Fig. 5 at some specific times. They were taken at the beginning of the experiment and after exposing the sensor to BSA, NAbII and PAbII. All spectra were obtained in ddH_2O after rinsing.

The comparison between the response to the negative and positive assay, carried out on the basis of data shown in Figs. 5 and 6, clearly demonstrates that the sensor is responding selectively to the exposition to non-specific and specific species, given its suitable functionalization.

In order to investigate the limit of resolution, several positive assays with different PAbII concentrations $[C_{POS}]$ are performed, each time using a fresh sensing substrate. Comparison with the level reached during the negative assay indicates that the minimum concentration that can be efficiently used to detect the specific antibody PAbII, in presence of non-specific species with concentration 125 μg/mL, is about 2 μg/mL, corresponding to about 10 pM. Analysis of the detection noise indicates that, in the absence of non-specific response, the minimum wavelength shift could be pushed down to $\Delta\lambda_{BSWR,MIN} = 10^{-2}$ nm, leading to a potential minimum detectable concentration $[C_{POS,MIN}] = 0.2$ μg/mL, i.e. 1 pM. The same minimum concentration limit has been reported by Liedberg et al. [28] for the same anti-IgG by an SPR biosensor.

4 Detection Scheme 2: Enhanced Fluorescence for Labelled Detection

One of the most widespread strategies aimed at improving the fluorescence emission in biosensing applications is based on field-enhancement effects occurring close to nanostructured surfaces [35]. Whilst field-enhancement effects on a solid surface are typically provided by plasmonic means, an alternative approach is based on the exploitation of BSWs. As a main advantage, the use of dielectric materials avoids undesired Forster resonance energy transfer, which strongly quenches the fluorescence signal at close proximity to metals in plasmon enhanced fluorescence schemes. Similarly to surface plasmon enhanced fluorescence, the amplification of fluorescence signal originating from fluorophores attached to 1DPC structures has been proposed based on the enhanced field intensity associated with BSWs on one-dimensional photonic crystals [20, 21]. However, since the evanescent tail of BSW is extending several hundreds of nanometers from the crystal surface, fluorescence could be potentially enhanced also in a small volume included within the outer medium (typically water or air). Hybrid approaches combining flat or structured multilayers with an intrinsically 3D arrangement of specific bioreceptors can take advantage of the above-mentioned BSW evanescent tail extension. Functionalization based on hydrogel technology can be fruitfully employed to this aim [36].

As an example, a new type of planar 1DPC structure that improves the sensitivity of fluorescence assays is presented here. Such an improvement can be ascribed to several aspects. Firstly, the field-enhancement effect is provided by a resonant coupling of laser excitation to BSW. In this way, that the excitation rate of fluorophore labels is increased. Secondly, a three-dimensional hydrogel-binding matrix is attached on the 1DPC surface in order to capture higher amounts of target analyte molecules in aqueous buffer. Thirdly, the profile of the evanescent field of BSW is tuned in order to probe quite large distances from the surface at which the hydrogel-binding matrix is attached (Fig. 7a). Finally, the fluorescence light associated with the affinity binding of target analyte is collected with high efficiency by minimizing its leakage into the substrate and by enhancing the intensity emitted toward a detector located in front of the sensor chip (Fig. 7b). The advantage of this approach is demonstrated in a model immunoassay experiment in comparison with a standard fluorescence (TIRF) readout.

The 1DPC structure is designed to enhance the field intensity at an excitation wavelength $\lambda = 633$ nm, wherein BSW exists. At the same time, the 1DPC blocks the propagation of emitted fluorescence into the glass substrate at the emission wavelength $\lambda = 670$ nm. To do so, the 1DPC is constituted by two portions S1 and S2, based on a-$Si_{1-x}N_x$:H/a-Si_3N_4:H stacks. The design of the segment S1 was carried out to enable ATR-based coupling of a light beam to BSW at a wavelength of 633 nm but with no NSW available at the emission wavelength of 670 nm. This configuration would avoid any fluorescence leakage into the BSW. Furthermore, an additional Bragg mirror segment S2 is employed to reflect the fluorescence directly emitted into the multilayer. The resulting effect is illustrated in Fig. 8a, wherein the

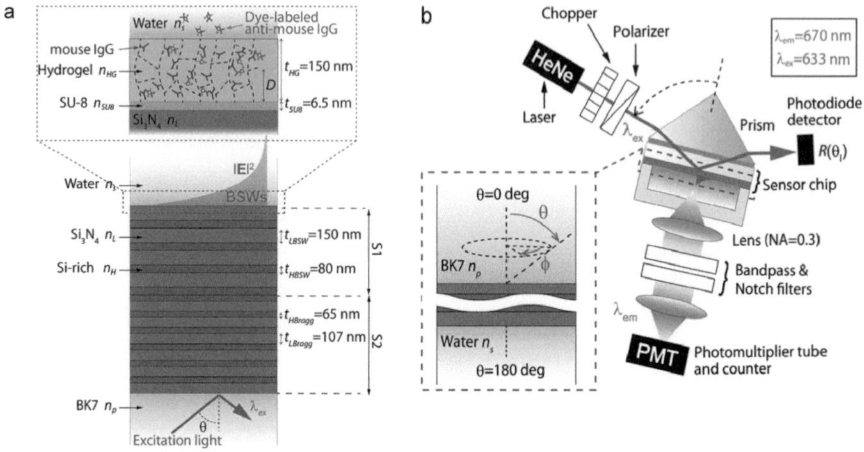

Fig. 7 (**a**) Multilayer made of two portions of periodic low and high refractive index layers designed to support BSWs (S1) and serve as Bragg mirror (S2) with attached hydrogel matrix for the capture of target analyte; (**b**) Experimental setup for Bloch surface wave-enhanced fluorescence spectroscopy (adapted from [37])

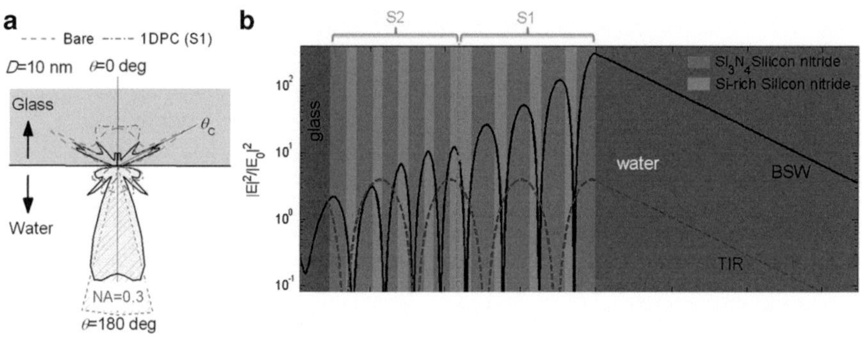

Fig. 8 (**a**) Angular emission of fluorescence emitted by a randomly oriented dipole at 10 nm from the surface of the 1DPC; (**b**) Electric field intensity for resonant coupled BSW on the 1DPC (*solid*) and for totally internally reflection at a glass–water interface (*dashed*). Angle of incidence is 64.45° for both cases (adapted from [37])

calculated forward emission from molecules at a distance $D = 10$ nm from the surface can be appreciated.

As shown by simulations in Fig. 8b, the resonant coupling to BSW on the top of the 1DPC structure at the excitation wavelength is associated to a strong electric field intensity at the 1DPC interface. The BSW evanescent field probes the aqueous medium (i.e., the hydrogel matrix) on the top of the 1DPC with the penetration depth comparable to the thickness of the attached hydrogel-binding matrix, i.e. 150 nm. For comparison, the electric field intensity distribution for a total internal reflection (TIR) condition at the interface between a bare glass substrate

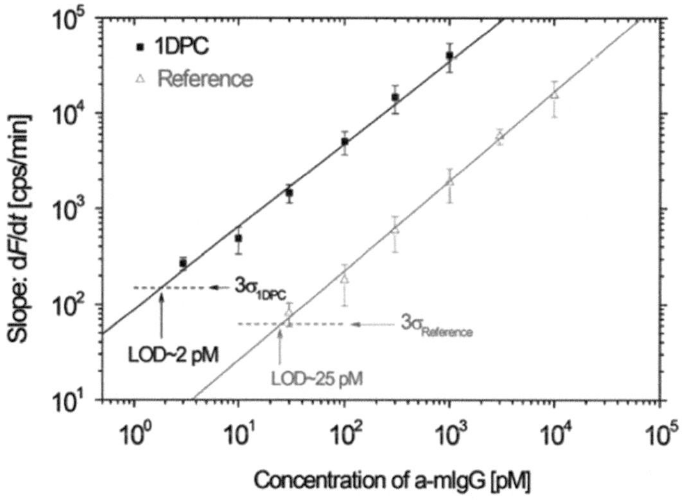

Fig. 9 Titration curves for a-mIgG sensing using a fluorescence-based assay on the 1DPC and on bare glass slide. Experimental data are fitted with a linear function and the baseline noise, error bars and LOD are indicated (adapted from [37])

and an aqueous medium is also shown. At least two orders of magnitude in near-field intensity can be reached with the use of the 1DPC considered here.

In building a titration curve, a comparison between the 1DPC and a bare glass slide can be performed. The angle of incident excitation laser beam (wavelength 633 nm) is fixed at the BSW resonance and the fluorescence signal F is measured in time upon the sequential flow of samples with a-mIgG molecules. For the 1DPC chip, the a-mIgG concentration between 0.3 pM and 1 nM is used. For bare glass, the analyte concentration was varied between 10 pM and 10 nM. Each sample is flowed over the surface with mIgG-functionalized hydrogel-binding matrix for 10 min followed by 10 min rinsing with PBS-T. Although the higher background due to a higher autofluorescence from the 1DPC, the limit of detection of 2 pM was obtained for the sensor chip with the 1DPC, as compared to 25 pM for the regular bare glass (Fig. 9).

As an alternative, it is possible to employ diffraction gratings directly fabricated on the 1DPC by electron beam lithography and reactive ion etching processes. In this way, any use of prism or oil-contact optics for BSW excitation and/or fluorescence collection can be avoided. The photonic structure is still based on a dielectric multilayer that is surface-patterned with linear subwavelength corrugations and then functionalized by 3-aminopropyltriethoxylsilane (APTES) and glutaraldehyde [38]. The overall system can be optically interrogated by means of an extended excitation laser beam. The resulting fluorescence is collected through a collection optics arranged in a simple wide-field configuration. In Fig. 10, an exemplary comparison between the improvement in fluorescence detection on the diffractive structures and on the flat 1DPC regions is presented.

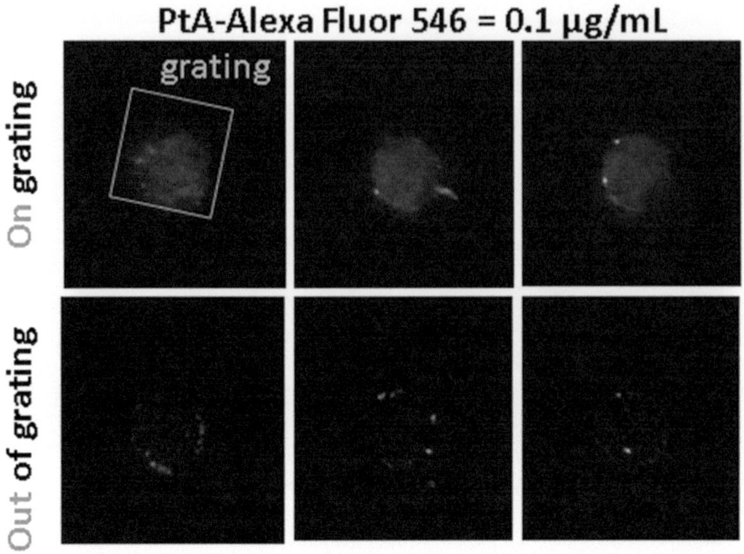

Fig. 10 Fluorescence images of AlexaFluor564 ptA spotted onto the corrugated and flat 1DPC surface via a PDMS contact printing membrane. Spots are approximately 150 μm in size. An overall increase of fluorescence intensity of about ten times can be detected corresponding to the corrugated regions

5 Detection Scheme 3: Enhanced Fluorescence for Label-Free Detection

Similarly to SPR, as discussed in Sects. 1–3, the coupling to a BSW in the Kretschmann–Raether configuration, is detected as a dip in the reflectance spectrum. However, due to the reduced losses in the dielectric 1DPC dielectric, the width of the BSW resonance is much smaller than that observed for SPR, by a factor ranging between 20 and 50, depending on the 1DPC design [39] and the depth of the dip is very small. Accordingly, the limit of detection (LOD) will be strongly affected by the noise level of the reflectance measurement. BSW-based refractometric biosensors show a sensitivity, defined as the ratio of the resonance spectral shift over a corresponding perturbation of the external medium refractive index, that is generally smaller as compared to the SPR case (e.g., 1,100 nm/RIU as reported in [32]). Despite such sensitivity limitation, BSW can show a better performance for sensing applications owing to the reduced width of their resonance, provided the signal-to-noise ratio is improved. To address this issue, the mechanism of BSW-coupled emission can be exploited.

A refractometric scheme has been proposed, based on the monitoring of the shift of a BSW-coupled luminescence peak radiated from a proper (fluorescent) 1DPC in which BSWs are resonantly laser-excited [39]. BSWs are coupled by prism-illuminating the 1DPC with a laser light ($\lambda = 532$ nm) incident at a given angle in accordance with the BSW dispersion curve. A strong near-field enhancement is produced at the last 1DPC interface, which can be usefully employed for boosting the fluorescence emission of organic dyes on the 1DPC top.

Similarly to the well-known Surface Plasmon Coupled Emission (SPCE), the emitted fluorescence couples into the BSW and leaks in the glass substrate with a specific angle according to the BSW dispersion curve. When a refractive index perturbation occurs close to the 1DPC surface, the BSW resonance shifts. In our particular implementation, such a shift is spectrally tracked by detecting the BSW-coupled fluorescence leaking out of the prism with a fixed angle. Although a fluorescence emission is exploited here, this technique can be advantageously defined as label-free.

The 1DPC is based on six layers of alternating low (L) and high (H) index materials (SiO_2 and Ta_2O_5, respectively) deposited by PIAD [40]. On top of this stack, a 100 nm polyacrylic Acid (PPAA) is plasma-deposited. The structure is able to sustain TE-polarized BSW in the visible range, when the outer medium contacting the last (PPAA) layer is an aqueous solution. Plasma-Polymerized Acrylic Acid (PPAA) thin functional films are obtained in a 13.56 MHz Radio Frequency-Plasma Enhanced Chemical Vapour Deposition (RF-PECVD) reactor by means of a pulsed plasma discharge. This coating exposes at the surface a high number of carboxylic groups (–COOH) able to react with the amino groups (–NH2) of biomolecules. The PPAA layer is then impregnated with an ethanolic solution of Cy3 dye at a concentration of 10^{-5} M. Fluorescence is resonantly excited through BSW ($\lambda = 532$ nm) in the Kretschmann configuration and then collected at a specific leakage angle by means of a fibered spectrometer (Fig. 11a). The top 1DPC surface is contacted to a polymeric microfluidics for liquid analyte handling.

The BSW-coupled radiation mechanism can be experimentally observed by performing an angularly resolved spectral analysis of the fluorescence emitted from the Cy3-doped PPAA layer. As a result, narrow spectral peaks (3 nm wide) are angularly dispersed as shown in Fig. 11b, according to the BSW dispersion curve.

The refractometric capabilities of such a detection scheme can be tested by measuring a physical perturbation occurring on a water solution contacting the 1DPC surface. Measurements are performed by real-time detecting the BSW-coupled fluorescence spectrum emitted at a given detection angle θ_{fluo}. A typical sensorgram showing the resonance position shift as a function of time is illustrated in Fig. 12, wherein a sequence of several glucose concentrations corresponding to different Δn are injected into the flow cell.

As an example of biosensing experiment, the covalent binding of protein G onto the PPAA film is real-time monitored (Fig. 13). First, the BSW peak position is stabilizing in a PBS buffer. Subsequently, a 300 μL volume of the protein G solution is injected and recirculated. A monotonic redshift of the BSW resonance is observed, indicating the approach of an asymptotic condition. The progressive binding of protein G to the –COOH groups of the PPAA surface can be appreciated. The plot shows the BSW peak shift $\Delta\lambda_{fluo}$ as a function of time. After washing and recirculating with PBS, the BSW peak experiences an initial quick blueshift that stabilizes after few minutes at a $\Delta\lambda_{fluo} = 0.6$ nm with respect to the initial position,

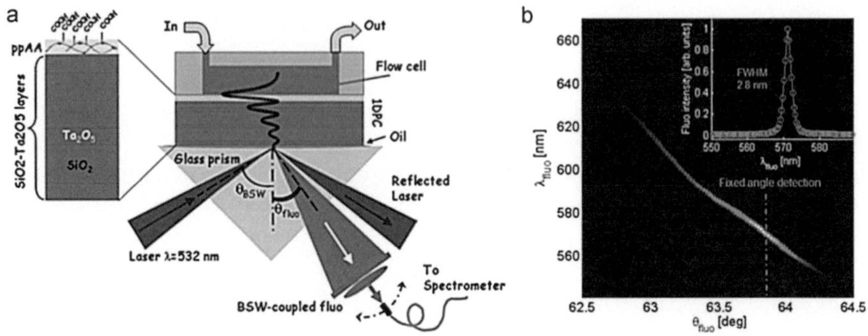

Fig. 11 (**a**) Sketch of the photonic structure and the experimental configuration for the refractometric sensing exploiting BSW-coupled emission; (**b**) angularly resolved spectral emission of Cy3-impregnated PPAA coupled into BSW. At specific detection angle, the fluorescence is characterized by narrow spectral peaks 3 nm wide (adapted from [39])

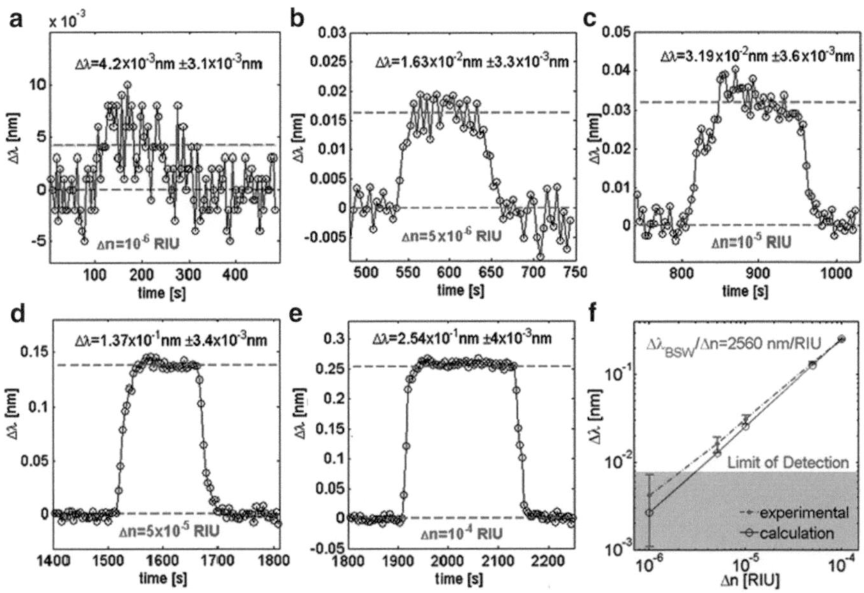

Fig. 12 (**a**)–(**e**) Real-time tracking of the fluorescence peak position as the refractive index of the aqueous medium is perturbed as indicated, (**f**) calculated (*circles*) and measured (*points*) sensitivity $\Delta\lambda_{BSW}/\Delta n$ for refractive index perturbations Δn of the medium in contact to the 1DPC surface (adapted from [39])

taking into account the reduced bulk refractive index of the solution. Thus, the overall redshift can be essentially ascribed to the surface coverage of protein G. In order to ensure that no protein G overlayers (i.e., excess of protein G molecules eventually stitched on the surface due to electrostatic forces and not by virtue of the amidic bond), the flow rate has been increased at $t = 60$ min. This effect is

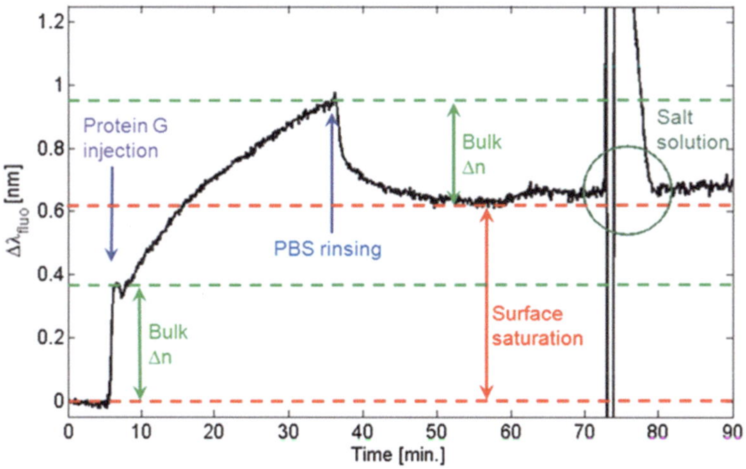

Fig. 13 Real-time sensorgram related to the covalent binding of protein G onto the functional PPAA film during incubation in PBS (adapted from [40])

perturbing the measurement of the BSW peak. However, no substantial blueshift due to the removal of an eventual excess of protein G is produced.

6 Conclusions

In summary, the biosensing potential of Bloch Surface Waves coupled to 1D photonic crystals in several detection schemes was demonstrated. The schemes concern with polarized reflectance and enhanced fluorescence measurements, which allow labeled and label-free recognition.

The real-time monitoring of species adsorption on specific receptors printed on the free surface of a 1DPC descriptively demonstrated the reliability of the selective recognition of specific secondary antibody targets.

A Photonic Crystal supporting BSWs with an integrated Bragg mirror was designed to amplify the fluorescence signal emitted from fluorophore labels by combining BSW-enhanced excitation rate and directional emission. The sensing experiments, taking advantage of a three-dimensional hydrogel-binding matrix showed a sensitivity enhancement by more than one order of magnitude with respect to the standard TIRF configuration. As an alternative, diffraction gratings patterned on the 1DPC surface demonstrated enhanced fluorescence regimes avoiding any use of prism or oil-contact optics for BSW coupling.

Finally, BSW-controlled fluorescence is presented as a refractometric means for real-time tracking the covalent binding of protein G on a COOH-rich polymeric film. The proposed scheme overcomes the limitation in the signal-to-noise ratio occurring in standard angularly resolved ATR reflectometry when low-losses materials are used for the photonic crystal.

The presented results could be exploited in the fabrication of compact optofluidic devices in a biosensing micro-array approach.

Acknowledgements The research described above has been developed with noticeable contributions and fruitful discussions involving several researchers and students. In particular, Prof. Francesco Michelotti, Prof. Joseph R. Lakowicz, Prof. Hans Peter Herzig, Prof. Federico Bussolino, Prof. Fabrizio Pirri, Dr. Peter Munzert, Dr. Jakub Dostalek, Dr. Francesca Frascella, Dr. Serena Ricciardi, Dr. Mirko Ballarini, Dr. Pietro Mandracci, Dr. Alessandro Chiadò, Dr. Alessandro Virga, Dr. Lucia Napione, Dr. Angelica Chiodoni, Dr. Emanuele Enrico, Dr. Lorenzo Dominici, Mr. Angelo Angelini, and Ms. Valeria Moi are gratefully acknowledged.

We acknowledge the collaboration with NanoFacility Piemonte, INRiM, a laboratory supported by Compagnia di San Paolo. This research has received funding from the EU FP7 project BILOBA (Grant #318035), the Italian Flagship Project NANOMAX (Progetto Bandiera MIUR PNR 2011–2013), the Italian FIRB 2011 NEWTON (RBAP11BYNP), and the Piedmont Regional Project CIPE 2008 PHOENICS.

References

1. C.M. Soukoulis, *Photonic Band Gap Materials* (Kluwer, Dordrecht, 1996)
2. L.C. Andreani, M. Liscidini, Photonic crystals: an introductory survey, in *Organic and Hybrid Photonic Crystals*, ed. by D. Comoretto (New York, Springer, 2015)
3. J.D. Joannopoulos, R.D. Meade, J.N. Winn, *Photonic Crystals: Molding the Flow of Light* (Princeton University Press, Princeton, 1995)
4. J.M. Elson, P. Tran, Phys. Rev. B **54**, 1711 (1996)
5. W.M. Robertson, G. Arjavalingam, R.D. Meade, K.D. Brommer, A.M. Rappe, J.D. Joannopoulos, Opt. Lett. **18**, 528 (1993)
6. P. Yeh, A. Yariv, A.Y. Cho, Appl. Phys. Lett. **32**, 104 (1978)
7. E. Descrovi, T. Sfez, M. Quaglio, D. Brunazzo, L. Dominici, F. Michelotti, H.P. Herzig, O.J.F. Martin, F. Giorgis, Nano Lett. **10**, 2087 (2010)
8. A. Farmer, A.C. Friedli, S.M. Wright, W.M. Robertson, Sens. Actuators B **173**, 79 (2012)
9. J. Gao, A. Sarangan, Q. Zhan, Opt. Lett. **37**, 2640 (2012)
10. M. Shinn, W.M. Robertson, Sens. Actuators B **105**, 360 (2005)
11. P. Rivolo, F. Michelotti, F. Frascella, G. Digregorio, P. Mandracci, L. Dominici, F. Giorgis, E. Descrovi, Sens. Actuators B **161**, 1046 (2012)
12. E. Descrovi, F. Frascella, B. Sciacca, F. Geobaldo, L. Dominici, F. Michelotti, Appl. Phys. Lett **91**, 241109 (2007)
13. F. Michelotti, B. Sciacca, L. Dominici, M. Quaglio, E. Descrovi, F. Giorgis, F. Geobaldo, Phys. Chem. Chem. Phys **12**, 502 (2009)
14. A. Sinibaldi, N. Danz, E. Descrovi, P. Munzert, U. Schulz, F. Sonntag, L. Dominici, F. Michelotti, Sens. Actuators B **174**, 292 (2012)
15. A. Sinibaldi, E. Descrovi, F. Giorgis, L. Dominici, M. Ballarini, P. Mandracci, N. Danz, F. Michelotti, Biomed. Opt. Express **3**, 2405 (2012)
16. Y.H. Li, T.L. Yang, S.M. Song, Z.Y. Pang, G.Q. Du, S.H. Han, Appl. Phys. Lett. **103**, 041116 (2013)
17. Y. Li, T. Yang, Z. Pang, G. Du, S. Song, S. Han, Opt. Express **22**, 21403 (2014)
18. M. Liscidini, M. Galli, M. Patrini, R. Loo, C. Goh, C. Ricciardi, F. Giorgis, J.E. Sipe, Appl. Phys. Lett. **94**, 043117 (2009)

19. S. Pirotta, X.G. Xu, A. Delfan, S. Mysore, S. Maiti, G. Dacarro, M. Patrini, M. Galli, G. Guizzetti, D. Bajoni, J.E. Sipe, G.C. Walker, M. Liscidini, J. Phys. Chem. C **117**, 6821 (2013)
20. I.V. Soboleva, E. Descrovi, C. Summonte, A.A. Fedyanin, F. Giorgis, Appl. Phys. Lett. **94**, 231122 (2009)
21. M. Ballarini, F. Frascella, F. Michelotti, G. Digregorio, P. Rivolo, V. Paeder, V. Musi, F. Giorgis, E. Descrovi, Appl. Phys. Lett. **99**, 043302 (2011)
22. L. Yu, E. Barakat, T. Sfez, L. Hvozdara, J. Di Francesco, H.P. Herzig, Light Sci. Appl. **3**, 124 (2014)
23. M. Ballarini, F. Frascella, N. De Leo, S. Ricciardi, P. Rivolo, P. Mandracci, E. Enrico, F. Giorgis, F. Michelotti, E. Descrovi, Opt. Express **20**, 6703 (2012)
24. E. Descrovi, D. Morrone, A. Angelini, F. Frascella, S. Ricciardi, P. Rivolo, N. De Leo, L. Boarino, P. Munzert, F. Michelotti, F. Giorgis, Eur. Phys. J. D **68**, 53 (2014)
25. F. Giorgis, P. Mandracci, L. Dal Negro, C. Mazzoleni, L. Pavesi, J. Non-Crystalline Solids **266–269**, 588 (2000)
26. E. Descrovi, F. Giorgis, L. Dominici, F. Michelotti, Opt. Lett. **33**, 243 (2008)
27. L. Li, G. Granet, J.P. Plumey, J. Chandezon, Publ. Astron. Soc. Jpn. **5**, 141 (1996)
28. B. Liedberg, C. Nylander, I. Lundstroem, *Biosens. Bioelectron.* **10**, i–ix (1995)
29. M. Piliarik, J. Homola, Opt. Express **17**, 16505 (2009)
30. N. Danz, A. Kick, F. Sonntag, S. Schmieder, B. Höfer, U. Klotzbach, M. Mertig, Eng. Life Sci. **11**, 566 (2011)
31. P. Yeh, A. Yariv, C.-S. Hong, J. Opt. Soc. Am. **67**, 423–438 (1977)
32. F. Giorgis, E. Descrovi, C. Summonte, L. Dominici, F. Michelotti, Opt. Express **18**, 8087 (2010)
33. A. Bernard, E. Delamarche, H. Schmid, B. Michel, H.R. Bosshard, H. Biebuyck, Langmuir **14**, 2225 (1998)
34. F. Michelotti, L. Dominici, E. Descrovi, N. Danz, F. Menchini, Opt. Lett. **34**, 839–841 (2009)
35. J.R. Lakowicz, *Principles of Fluorescence Spectroscopy*, 3rd edn. (Springer, New York, 2006)
36. J. Dostalek, W. Knoll, *Biointerphases* **3**, Fd12–Fd22 (2008)
37. K. Toma, E. Descrovi, M. Toma, M. Ballarini, P. Mandracci, F. Giorgis, A. Mateescu, U. Jonas, W. Knoll, J. Dostálek, Biosens. Bioelectron. **43**, 108 (2013)
38. F. Frascella, C. Ricciardi, Functionalization protocols of silicon micro/nano-mechanical biosensors, in *Nanomaterial Interfaces in Biology* (Springer, Berlin, 2013), pp. 109–115
39. E. Descrovi, F. Frascella, M. Ballarini, V. Moi, A. Lamberti, F. Michelotti, F. Giorgis, C.F. Pirri, Appl. Phys. Lett. **101**, 131105 (2012)
40. F. Frascella, S. Ricciardi, P. Rivolo, V. Moi, F. Michelotti, P. Munzert, N. Danz, L. Napione, M. Alvaro, F. Giorgis, F. Bussolino, E. Descrovi, Sensors **13**, 2011 (2013)

Hybrid-Organic Photonic Structures for Light Emission Modification

Valentina Robbiano, Francesco Di Stasio, Salvatore Surdo, Shabbir Mian, Giuseppe Barillaro, and Franco Cacialli

Abstract We report the modification of the photoluminescence (PL) and decay rates of two different green-emitting conjugated polymers incorporated into photonic crystals with the stop-bands spectrally tuned on their emission. We observe both suppression (in the stop-band) and enhancement (at the high-energy band-edge) of the photoluminescence. Time-resolved measurements also reveal a concomitant modification of the emission lifetime that is enhanced at the band-edge and suppressed within the stop-band, thus confirming a variation of the radiative decay rate of the excitations in such photonic nanostructures. We propose two examples of fluorescent photonic composite systems. The first consists of a hybrid Si-organic system, obtained by infiltration of the polymer inside a rugate filter (a 1D photonic crystal). The second example is a fully organic system obtained by self-assembling of solvent-compatible microspheres-polymer system for obtaining a synthetic opal (a 3D photonic crystal) with a uniform distribution of the emitting material across the photonic structure.

Keywords 1D photonic crystals • 3D photonic crystals • Conjugated polymers • Conjugated polyrotaxanes • Functional materials @ photonic crystals • Hybrid photonic crystals • Light emission • Opal • Photoluminescence • Photophysics • Purcell effect • Rugate filters • Self-assembly • Silicon • Spontaneous emission

---

V. Robbiano • F. Cacialli (✉)
Department of Physics and Astronomy and London Centre for Nanotechnology, University College London, Gower Street, WC1E 6BT London, UK
e-mail: valentina.robbiano.12@ucl.ac.uk; f.cacialli@ucl.ac.uk

F. Di Stasio
Nanochemistry Department, Istituto Italiano di Tecnologia, Via Morego 30, 16163 Genoa, Italy
e-mail: francesco.distasio@iit.it

S. Surdo • G. Barillaro
Dipartimento di Ingegneria dell'Informazione, Università di Pisa, Via Diotisalvi 2, 56122 Pisa, Italy
e-mail: Salvatore.Surdo@iit.it; g.barillaro@iet.unipi.it

S. Mian
Department of Physics, McDaniel College, Westminster, MD 21157, USA
e-mail: smian@mcdaniel.edu

D. Comoretto (ed.), *Organic and Hybrid Photonic Crystals*,
DOI 10.1007/978-3-319-16580-6_15

1 Introduction

In the last 30 years, there has been an increasing interest in the development of novel photonic structures to control the light-emission properties of materials, as well as for the fabrication of semiconductor lasers. The concept of a photonic crystal (PhC) was introduced in 1987 with the studies of Yablonovitch [1] and John [2] on the inhibition of the spontaneous emission and light localization effects. These relate to structures whose periodicity in refractive index would enable to control, suppress, or enhance the propagation of photons, in analogy with the already well-established possibility to control electron propagation in "electronic" crystals, in which the periodicity of the electronic potential results in allowed and forbidden energy bands for the electrons. As well-described by Joannopolous and collaborators in their book [3], this is due to the possibility to re-write Maxwell equations in terms of a single master equation, in which the dielectric function plays a similar role to the potential in Schrödinger's equation. The field of photonic structures and crystals has then developed in a variety of directions, often dictated by the hurdles of fabrication of structures whose periodicity needs to be on the same lengthscale as that of the wavelength of the light one desires to control. This places some stringent requirements for PhCs that need to operate in the visible range, as one needs to generate photonically active patterns with typical dimensions/periodicities of the order of a few hundreds of nanometres. So far, the development and production of photonic crystals has been generally focused on top-down nanofabrication [4–7] making use of lithographic techniques such as electron-beam and focused ion-beam patterning (FIB). These are very expensive, and being essentially "serial" suffer from significant limitations in terms of either surface areas or sample throughput, in addition to being limited to the fabrication of mainly two-dimensional structures. Another major drawback is that the photonic effects are strongly inhibited by the residual surface roughness, thus spoiling the overall Q-factor of the cavities [8]. One technique that has shown more promise is the laser-assisted photopolymerization/activation of a cross-linker, to generate 3D nanostructures [9, 10] but for the as-prepared structures one needs some extremely careful control of relaxation effects of the structures after the polymerization to avoid distortion of the architectures and to ensure adherence to the intended pattern. By contrast, bottom-up approaches exploit the collective self-assembly of suitably sized building blocks (typically colloidal dielectric microspheres). This provides a much easier, faster and low cost alternative to top-down methods in preparing photonic crystals, enabling the preparation of both 2D (close-packed microspheres monolayers [11, 12]) and, more interestingly, 3D (artificial opals [13–16] and inverse opals [17, 18]) structures. Concerning opals and inverse opals, you could also read the chapter by C. López et al., in this volume. Since this is not a "serial" approach, it has the scope and potential to generate significantly larger samples. However, self-assembly methods lead to structures with a relatively high density of defects [19] that limit their application in optical devices. In addition, and somewhat in between the two approaches detailed above, we find a host of "non-conventional" top-down methods that are not "serial" (as e-beam and FIB) thereby allowing fabrication of structures over large (several square cm) areas. These methods consist, for example, in the spin-coating of polymers [20, 21] and the electrochemical micromachining of porous silicon [22–24]. Whereas spin-coating has been mainly used just to fabricate polymeric multilayers (e.g. full-plastic distributed Bragg reflectors), electrochemical micromachining, that consists of a controlled electrochemical etching of porous silicon, is a far more versatile technique and enables the preparation of inorganic PhCs with different dimensionalities (1D rugate filters and 2D arrays) with sub-micrometre accuracy [25, 26].

Photonic crystals exhibit a variety of optical properties that can be exploited in a variety of optoelectronics applications, such as optical fibres [27], sensors [28], solar cells [29] (see the chapter by Marina Mariano et al., in this volume), light-emitting diodes [30] and lasers [31]. In particular, PhCs are a fundamental component of laser sources, since they are necessary to provide optical feedback to the gain medium. For lasing application, see also the chapter by Seiichi Furumi, in this volume. A notable example of a laser based on a complex PhC structure is the work of Painter et al. where [32] controlled structural defects lead to the formation of optical microcavities with extremely high finesse, thus enabling laser emission from an embedded gain material with a very low stimulated emission threshold [21, 32]. An important property of PhCs is their ability to control the spontaneous emission of an embedded light-emitting material. The emission modification phenomena induced by PhCs are also reported in the chapters by L.C. Andreani and M Liscidini; F. Scotognella et al.; and Pieter-Jan Demeyer and Koen Clays, in this volume. In fact, the photonic band gap of a PhC can either inhibit or enhance spontaneous emission by modifying its radiative rate. When an active material is embedded into a PhC, its optical properties are strongly affected by the periodical dielectric environment, which alter the dispersion properties of photons (i.e. photonic density of states, p-DOS) [33]. Indeed, the p-DOS diminishes within the photonic band gap thus suppressing light-emission. On the other hand, at the edges of the photonic band gap, the p-DOS increases [34], i.e. a light-emission enhancement is observed. Both enhancement and suppression of the emitted light are accompanied by an increase and decrease of the radiative rate, respectively. Such an effect on the radiative rate can be used to modify, for example, the photoluminescence spectrum and intensity of the embedded material. In addition, the cross section of stimulated emission directly depends upon the radiative rate of the material. Therefore, increasing the radiative rate directly decreases the stimulated emission threshold.

To study the control of light, as well as to fabricate a variety of laser structures, conjugated polymers have been widely used as emitting materials. This is mainly due to their versatile processing, wide colour tunability (from NIR to near UV), high solid-state photoluminescence efficiency and strong optical absorption (up to $\sim 10^5$ cm^{-1}) [35, 36]. Furthermore, both conjugated molecules and polymers demonstrate low stimulated emission threshold [37–40] since they are an intrinsic four-level system [41], making them a promising gain material for the fabrication of solution processable semiconductor lasers in combination with hybrid and organic PhCs. The first optically pumped laser incorporating a neat conjugated polymer has been reported by Tessler et al. [31]. In this work, the authors studied the emission of poly(*p*-phenylene vinylene), PPV, prepared inside an optical cavity formed by a distributed Bragg reflector (DBR) and a silver mirror. Such photonic structures are relatively easily fabricated, but offer no lateral confinement and can support several in-plane modes causing optical losses. The most studied class of PhCs in the organic photonics field are distributed feedback (DFB) lasers [42, 43]. These devices consist of an active material film (in our case a conjugated polymer) deposited onto a corrugated substrate, which provide optical feedback by confining the light "in plane", similar to a waveguide. Many other configurations have been exploited for organic semiconductor lasers: flexible microcavities [21, 44], block-copolymer DBRs [45], opals [46], inverse opals [47] and opals microcavities [48, 49] infiltrated with dyes, random structures [50] and 2D photonic structures [7, 51].

Despite the numerous types of PhCs used, electrically pumped lasing has not been achieved yet, to the best of our knowledge, in conjugated polymer-based system. Restricting our discussion to these as gain materials, the main limitations that we can identify relate to (1) PL quenching by interchain interactions, [52] (2) polaron [53] and triplet absorption [54] (particularly severe for electrical pumping given that spin statistics strongly favour formation of triplets over singlets), (3) losses at the metallic electrodes (in the case of injection lasers) [55], (4) limited charge mobility [56] (preventing suitably large pumping current for injection lasers), and (5) material degradation under optical and electrical pumping [57]. Whereas several synthetic approaches have been used to reduce π–π stacking, increase distances between polymer chains [58] and reduce crystallinity [59]; for example by attaching side-groups to the conjugated backbone [60] or by threading the conjugated chain with molecular rings that act as spacers (i.e. as in polyrotaxanes) [61], significant progress is still needed to overcome the other limitations above. In general, both novel PhCs and conjugated polymers need to be developed and investigated to achieve CW optically pumped or electrically pumped lasing and in the following we report two examples from our experience in which we observed emission modification of two different conjugated polymers embedded into photonic crystals. The first system consists of 1D silicon PhCs, a rugate filter, where the luminescent polymer is infiltrated via dip-coating; the second one is a 3D PhC (synthetic opals) where the conjugated polymer is embedded into it. Such photonic structures enable emission modification of the conjugated polymers. In fact, we observed both suppression (in the stop-band) and enhancement (at the high-energy band-edge) of the photoluminescence. Remarkably, time-resolved measurements reveal a modification of the emission lifetime, which is enhanced at the band-edge and suppressed within the stop-band, and clearly point to a variation of the radiative decay rate of excitations in such hybrid-organic photonic nanostructures.

2 Hybrid Inorganic–Organic Photonic Structures: Rugate Filters Infiltrated with Polyfluorene Derivative

Porous silicon-based PhCs, such as rugate filters [62], have attracted significant interest since they provide a technologically robust route to fabrication of optical nanostructures with significant levels of complexity. Indeed, this class of photonic systems has been used widely for optoelectronic [63], photonic [64], sensing [28] and biological applications [65]. In particular, rugate filters are good candidates as vapour and liquid sensors [66–68], and as detectors for biological activities such as proteases [69, 70]. However, preparation of hybrid Si-organic structures, where an organic material is infiltrated into the pores of the rugate filter is not always straightforward, especially when pore sizes are comparable to the polymers gyration radii, or to the size of aggregated/phase-separated macromolecular mesostructures, thus causing pore clogging which often prevents complete infiltration [71]. One way to infiltrate macro-molecules inside the photonic structure is to grow the molecules "in situ" inside the pores starting from smaller monomers [72], via either electrochemical deposition [73, 74] or in situ chemical polymerisation [75]. The major drawback of this methodology is the intrinsic limitation to a relatively narrow set of monomers, and therefore not generally satisfactory. Interestingly, the particular ratio between the limited gyration radius (R_g, of the order of few nm) of poly[(9,9-di-*n*-octylfluorenyl-2,7-diyl)-alt-(benzo[2,1,3]thiadiazol-4,8-diyl)], (F8BT, Mw ~46 kDa) and the minimum pore sizes (approximately 10 times larger than R_g) allowed us to infiltrate the polymer inside the pores. In fact, we observe shifts of the reflectance peaks of the structures consistent with coating of the inner walls of the pores to the whole depth that can be probed in such optical experiments. Porous silicon rugate filters were prepared by electrochemical etching of (100)-oriented, highly boron doped p-type silicon wafers in a 3:1 (by vol.) solution of 48 % (by vol.) aqueous hydrofluoric acid and ethanol. The electrochemical etching was performed using a two-electrode configuration, with the silicon substrate acting as anode (working electrode) and a platinum ring immersed in the solution acting as cathode (and used as pseudo-reference electrode). A cosine-shaped current density waveform with peak-to-peak dynamics from 13.3 to 39.9 mA cm^{-2}, 50 repeats, and time period of 11 and 17.5 s was used to produce rugate filters with a stop-band in reflectance centred at 551 and 875 nm, respectively. The resulting porous silicon structures possess sinusoidally shaped periodic cavities with porosity values (defined as the ratio between the mass of etched silicon with respect to the total silicon subjected to etching) between 55.3 and 62.3 %; and a sinusoidally shaped refractive index as well, which varies between 1.95 and 2.21. To improve the infiltration process and prevent emission quenching from the heavily doped silicon layer, the rugate filters were partially oxidized into a tube furnace under pure-O_2 at 600 °C for 30 min. The oxidation leads to a general decrease of the effective refractive index of the rugate filter and, in turn, in a blue-shift of the stop-band to 515 nm (from 551 nm) and 771 nm (from 875 nm). Figure 1 shows Scanning Electron Microscope images at different magnification of one of

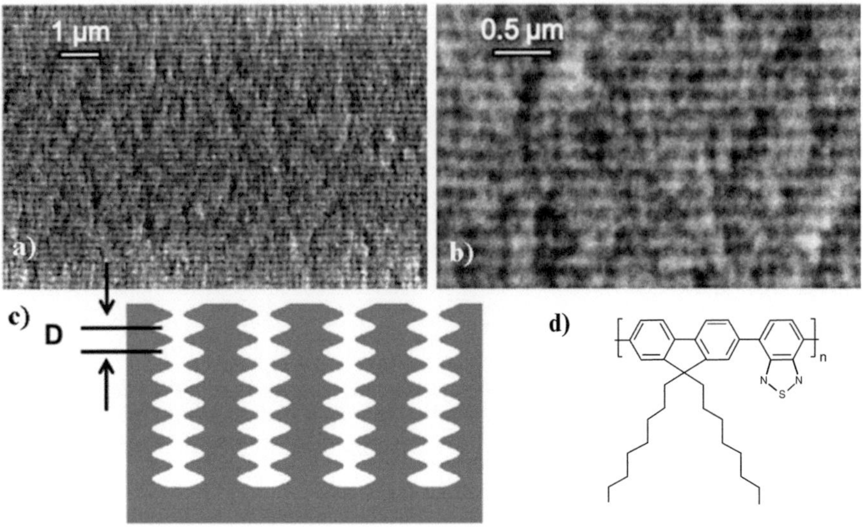

Fig. 1 (**a**, **b**) SEM cross sections at different magnifications of an as-prepared porous silicon rugate filter with resonant peak in reflectance centred at 875 nm. (**c**) Scheme of the cross section in (**a**) and (**b**) of the as-prepared rugate filter. *D* is the inter-planar spacing that is determined by the preparation conditions. (**d**) Chemical structure of F8BT

the as-prepared rugate filters and a scheme showing the same cross section. The size of pores is in the range of 80–100 nm for both structures, and the inter-planar spacing D is around 100 and 200 nm for the rugate filters with stop-band at 551 and 875 nm (before oxidation), respectively.

The F8BT was dissolved in a toluene solution 1 mg/ml, stirred overnight at room temperature and then filtered. The infiltration was obtained via dip-coating carried out at a controlled lifting rate (0.01 mm/min) and temperature (29 °C) inside an incubator. The excess polymer layer left onto the surface after dip-coating was removed using a cotton swab soaked with toluene. We consider that this procedure might have also helped the polymer infiltration. To assess the effect of the photonic structure on the F8BT spectrum, we compared the rugate whose stop-band is centred at 515, thus spectrally overlapping (i.e. tuned with the F8BT PL spectrum) with the rugate whose stop-band is centred at 771 nm and hence detuned with respect to the F8BT PL spectrum. The latter is used as reference to observe any PL modifications induced by the presence of the PhC since F8BT film has the same morphology as the one infiltrated into the tuned one.

Figure 2 shows the reflectance spectra for tuned and detuned samples before and after the infiltration of the F8BT and the typical PL spectrum of an F8BT film. Upon polymer infiltration, the reflectance peaks red-shifted by ~12 nm in both samples due to the increased effective refractive index of the F8BT with respect to air. In particular the reflectance peak for the tuned sample red-shifted from 515 to 528 nm

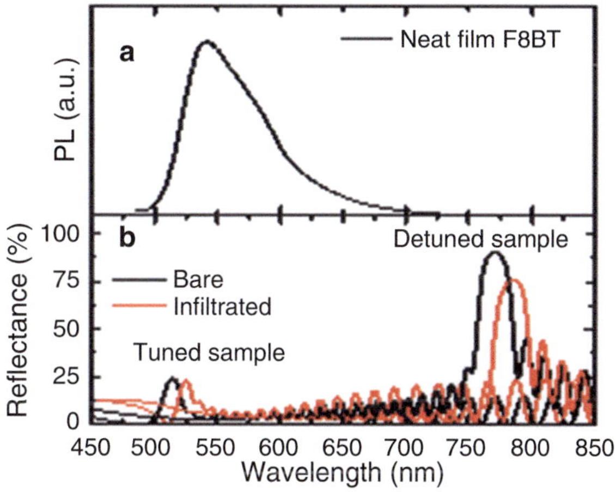

Fig. 2 (**a**) PL spectrum of a neat film of F8BT and (**b**) Reflectance spectra for the PhCs before (*black*) and after polymer infiltration (*grey*). The infiltration leads to a red-shift of the photonic stop-band (and thus of the reflectance peak)

after infiltration and for the detuned one it red-shifted from 771 to 784 nm. This is predicted by the Bragg–Snell law [76]:

$$m\lambda = 2D\sqrt{n_{\text{eff}}^2 - \sin^2\theta} \tag{1}$$

where m is the diffraction order, λ thc spectral position of the stop-band peak, n_{eff} the effective refractive index of the photonic crystal, θ the incidence angle and D the inter-planar spacing, that is determined by the preparation condition. The effective refractive index can be calculated by using the Lorentz–Lorenz equation and depends on the refractive indices of the media composing the PhC (in this case silicon, silicon dioxide and air before infiltration, and silicon, silicon dioxide, and F8BT after) weighted by their volume fraction.

$$\frac{n_{\text{eff}}^2 - 1}{n_{\text{eff}}^2 + 2} = \sum\nolimits_{\text{media}} f_{\text{medium}} \frac{n_{\text{medium}}^2 - 1}{n_{\text{medium}}^2 + 2} \tag{2}$$

where f is the volume fraction for each media.

Considering that the refractive indices of F8BT are 1.6 (extraordinary one) and 1.8 (ordinary one) [77] and the refractive index of air is 1, if the cavities have been completely filled by the polymer, according to Lorentz–Lorenz equation, a spectral shift of the reflectance peaks bigger than the one that observed from the measured spectra is expected. By analysing the reflectance peaks spectral shifts, it can be assumed that in this case the effective refractive index is not just a combination of the refractive indices of Si and F8BT and therefore the polymer does not entirely fill the nanostructured cavities. In fact, in this experiment, the infiltration consisted in the deposition of thin polymer films on the walls of the cavities. In this case, the final effective refractive index would be the combination of the ones of the silicon, silicon dioxide F8BT and air. It is reasonable to exclude a partial filling of

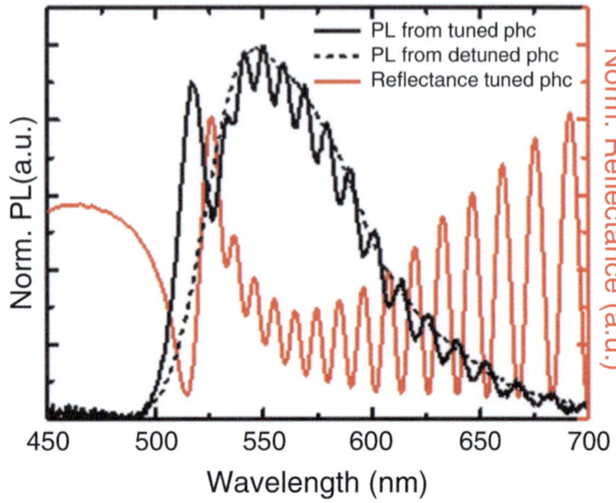

Fig. 3 PL spectra collected at normal incidence (0°) of the exciting beam for F8BT infiltrated into the tuned PhC (*solid line*) and for the F8BT infiltrated into the detuned PhC (*dashed line*). Reflectance spectrum (*grey line*) of the tuned PhC is shown

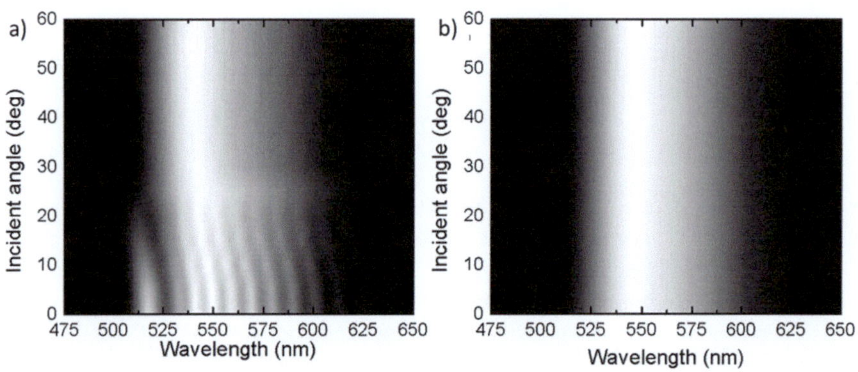

Fig. 4 Contour plots of the PL spectra as a function of the incidence angle for the F8BT infiltrated into the tuned (**a**) and detuned (**b**) PhC

the cavities due to the presence of just one reflectance peak, otherwise it would have been observed two reflectance peaks for the two PhCs stacking one over the other (i.e. the one made by Si and air and the one made by Si and polymer [78]).

To observe PL modification induced by the photonic crystal, PL measurements have been carried out both from the F8BT infiltrated into the tuned and detuned rugate filters; the second one has been used as reference. The PL spectra collected at normal incidence of the F8BT infiltrated into both rugate filters are shown in Fig. 3, together with the reflectance spectrum of the tuned rugate filter. Remarkably, the PL spectrum of F8BT is strongly modified by the optical features of the rugate filter: both the photonic stop-band and the interference fringes. In particular, the photonic band gap (PBG) partially suppresses light propagation at 527 nm and it also induces an increase in the PL intensity in the PBG high-energy edge. Interestingly, also the diffraction fringes modify the emission, leading to a sinusoidal modulation of the PL spectrum. In Fig. 4 we report the contour plot of the PL spectra as a function of

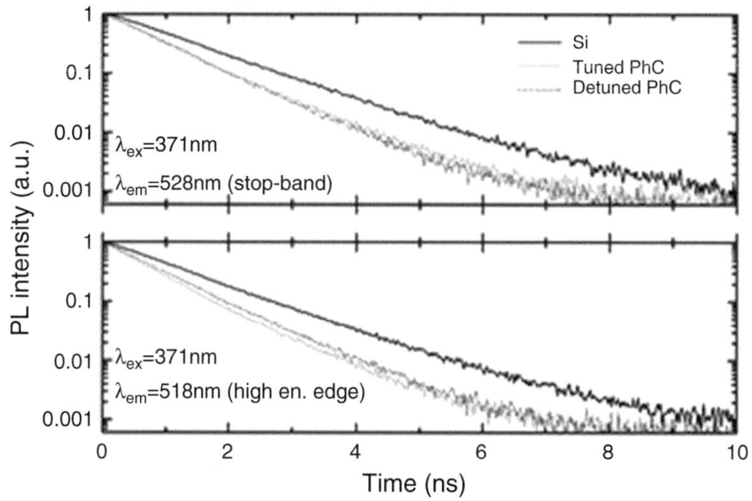

Fig. 5 Radiative decay of F8BT PL recorded at the stop-band (528 nm, *top panel*) and at the high-energy edge (518 nm, *bottom panel*). The measurement shown here was collected for the F8BT infiltrated into the tuned and detuned crystals (*grey solid line* and *black dashed lines*, respectively) and for a neat film of F8BT over a Silicon substrates (*black solid line*)

the angle of incidence of the laser beam for the F8BT infiltrated into the tuned PhC and detuned PhC. As expected, both the enhancement and the suppression are angle dependent because of the dispersion of the stop-band.

Final investigation consisted in the evaluation of the radiative decay of the F8BT PL (Fig. 5) both in the PBG (528 nm) and at the high-energy edge of the stop-band (518 nm). We focused on these two spectral lines because at such wavelengths there is a clear modification of the photonic density of state (p-DOS), as demonstrated by the reflectance spectra. In particular in the PBG there is a suppression of the photonic density of states, whereas at the high-energy edge of the stop-band the p-DOS is increased [33].

In previous studies the PL decay for neat film of F8BT deposited by spin-coating has been reported with a mono-exponential equation [79] with fluorescence decay lifetime varying from 0.89 to 1.4 nm depending on the wavelength, film thickness and substrate materials [80, 81]. In our case, for a neat film (thickness ~800 nm) on Si deposited by dip-coating, we obtained analogous values (1.15 ns at 518 nm and 1.18 ns at 528 nm, Table 1). Interestingly, we observed that when the polymer is infiltrated both in the tuned and detuned PhCs, the PL decay can be fitted with a double exponential expression:

$$I(t) = I_0 + I_1 e^{-\left(t/\tau_1\right)} + \mathrm{I}_2 e^{-\left(t/\tau_2\right)} \tag{3}$$

The lifetimes (τ_1 and τ_2) used in the fit-curves of the temporal evolution of the PL of the F8BT infiltrated into the crystals are reported in Table 1. The presence of

Table 1 Lifetime values (τ_1 and τ_2) extrapolated from the fit-curves of the PL decay

	$\tau_1(I_1)$	$\tau_2(I_2)$
High-energy edge (518 nm)		
PhC	690 ± 34 ps (0.8)	1,210 ± 60 ps (0.2)
Detuned	830 ± 41 ps (0.95)	1,590 ± 79 ps (0.05)
Film	1,150 ± 57 ps	
Stop-band (528 nm)		
PhC	860 ± 43 ps (0.94)	1,680 ± 84 ps (0.06)
Detuned	830 ± 41 ps (0.95)	1,600 ± 80 ps (0.05)
Film	1,180 ± 59 ps	

I_1 and I_2 indicate the relative contribute of each time constant. We estimated an error of 5 % on the measured value

a bi-exponential decay instead of a single exponential one can be attributed to a stronger interaction of the polymer with the Si substrate when inserted into the cavities. In fact, in this configuration the polymer layer is thinner compared to the neat film onto flat Si substrates. At the high-energy edge of the stop-band (518 nm), the decay for the F8BT infiltrated into the tuned rugate filter is faster than for the reference (i.e. the F8BT infiltrated into the detuned rugate filter) by ~17 % from the weighted mean of τ_1 and ~24 % τ_2 (Table 1), whereas within the stop-band (528 nm) the reverse situation occurs and the lifetimes are longer in the tuned sample compared to the detuned. Although these differences between the radiative decay rates [109] of the F8BT are relatively small, it should be considered that the photonic confinement for such samples happens only along the z-direction and that the rugate filter we have worked with had a narrow spectral width (516–533 nm) compared to the PL spectrum of F8BT (490–700 nm) [34].

3 Colloidal Photonic Crystals Self-assembled with Water-Soluble Conjugated Polyrotaxanes

Colloidal self-assembled PhCs represent another class of photonic structures that can be easily combined with conjugated polymers. Within self-assembled PhCs, synthetic opals [1, 2, 82, 83] have been especially popular thanks to their low cost of fabrication and ease of preparation. Furthermore, their functionalization with light-emitting materials can be easily achieved via infiltration by vapour phases [84], or solutions [82], allowing the investigation of a variety of photonic effects [85], including the modification of the emission spectra and radiative rates [34, 86]. Various materials such as metal nanoparticles [87], semiconductor nanocrystals [88] or conjugated molecules [89] have been incorporated into synthetic opals via infiltration. Nevertheless, achieving a homogenous distribution of the active materials in the opal via infiltration still represents a major challenge. To overcome this issue many fabrication techniques have been proposed, as, for example, infiltration of monomers followed by in situ polymerization [90], or layer-by-layer polymer

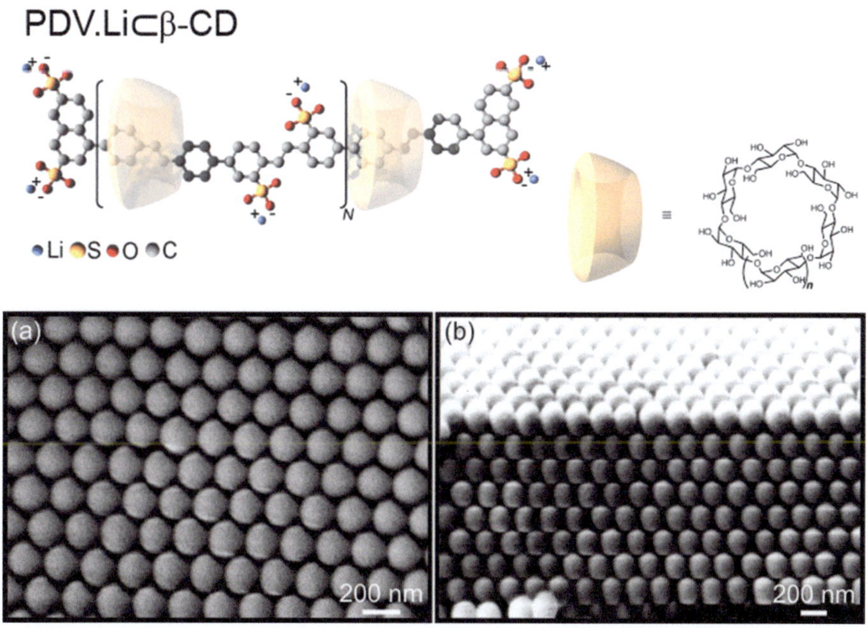

Fig. 6 *Top*: Chemical structure of PDV.Li⊂β-CD, threading ratio = 2, with an average number of repeat units $n = 10$. *Bottom*: SEM micrographs of an opal film incorporating PDV.Li⊂β-CD: (**a**) cryo-cleaved wall surface showing the internal structure of the opal, (**b**) film cross section. Strong contrast between the sphere and the interstices is observed, confirming that the latter are not filled up by the conjugated polyelectrolytes (the density of the conjugated moiety and of the spheres being comparable). Reproduced and adapted with permission from [98]

coating of the sphere surfaces [79]. Water-soluble conjugated polymers can be directly incorporated into the synthetic opal during growth, allowing for a "one-pot" chemical approach which does not require further sophisticated fabrication techniques. To this end, conjugated polyelectrolytic rotaxanes [58] (Fig. 6) are a well-suited class of materials for incorporation into synthetic opals since they demonstrate high photoluminescence quantum yield. Conjugated polyrotaxanes are supramolecular architecture in which a conjugated backbone, such as poly (4,4′-diphenylene vinylene) (PDV), is threaded through cyclodextrin rings (β-CD), that sterically impose increased inter-molecular distances. Such structures lead to higher photoluminescence quantum yield and reduced polaron formation compared to unthreaded materials [91], thus enabling optically pumped lasing [92], optical amplification [93] and fabrication of white-emitting organic light-emitting diodes [94, 95].

Opal films can be prepared using commercially available aqueous suspensions of monodisperse polystyrene nanospheres [96] (standard deviation < 5 %; refractive index, $n_{PS} = 1.59$), that can be diluted in de-ionized water as necessary to obtain the desired film thickness ($d \sim 5$ μm) upon complete water evaporation. β-CD-threaded PDV (PDV.Li⊂β-CD) was used as water-soluble luminescent semiconductor.

PDV.Li⊂β-CD is a polyelectrolytic derivative of poly-*para*-phenylene vinylene (PPV) where sulfonate side-groups balanced by Li^+ ions afford solubility in polar solvents (molecular structure in Fig. 6) [89, 97]. PDV.Li⊂β-CD incorporation was carried out by dissolving it in water, and then adding the solution to the nanospheres water-suspension before proceeding with the PhC growth process inside an incubator. To ensure that the presence of the polyrotaxane does not negatively affect the growth of the PhC, concentrations for the polyrotaxane water solutions ranging from 0.5 mg/ml to 1×10^{-3} mg/ml were investigated. The PhCs here discussed were prepared using a PDV.Li⊂β-CD concentration of 8×10^{-3} mg/ml which yielded undistorted, homogenous opals without deposition of an excess polymer layer on their surface, while preserving strong PL properties of the system.

The growth process was carried out at 45 ± 1 °C on soda-lime glass slides using the meniscus technique [89, 96, 99], thus obtaining a face-centred cubic lattice of nanospheres with the [110] direction perpendicular to the substrate. The obtained PhCs had an area of $\sim 7 \times 7$ mm^2. Scanning electron microscopy (SEM) micrographs (Fig. 6a, b) of the co-grown opal show no presence of lattice distortion induced by the incorporation of polyrotaxanes, nor an excess polymer layer on the opal surface for PhCs with 8×10^{-3} mg/ml PDV.Li⊂β-CD solution. Moreover, the interstices between nanospheres are empty, thereby preserving a desirable higher dielectric contrast within the structure than otherwise possible in case of substantial filling of the interstices. Preservation of the opal structure and optical properties upon polyrotaxane incorporation is demonstrated by the presence in reflectance spectra (Fig. 7a at wavelengths near and below 300 nm) of van Hove-like structures [100]. These features are due to diffraction along directions different from the incident one [33, 100] and are known to depend strongly on the order of

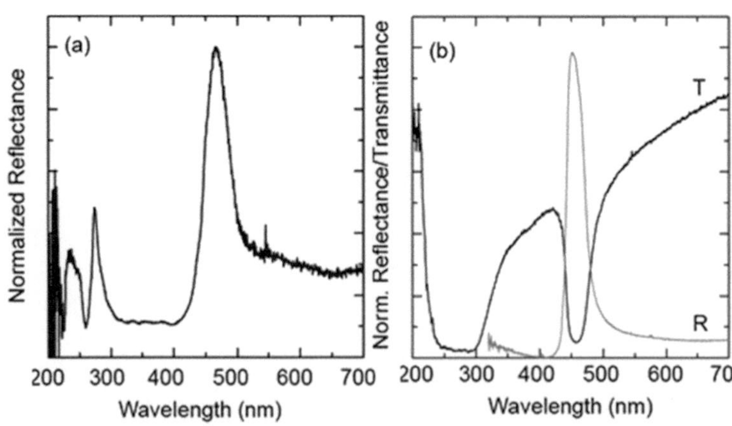

Fig. 7 (**a**) Reflectance spectrum of a typical co-grown opal with PDV.Li⊂β-CD fabricated with 200 nm polystyrene nanospheres showing van-Hove-like structures below 300 nm. (**b**) Reflectance (R) and transmittance (T) spectra of a polystyrene opal film co-grown with PDV.Li⊂β-CD (sphere diameter $a = 200$ nm, refractive index, $n_{PS} = 1.59$). The full-width-half-maximum of the PBG is not affected by the incorporation process. Reproduced and adapted with permission from [98]

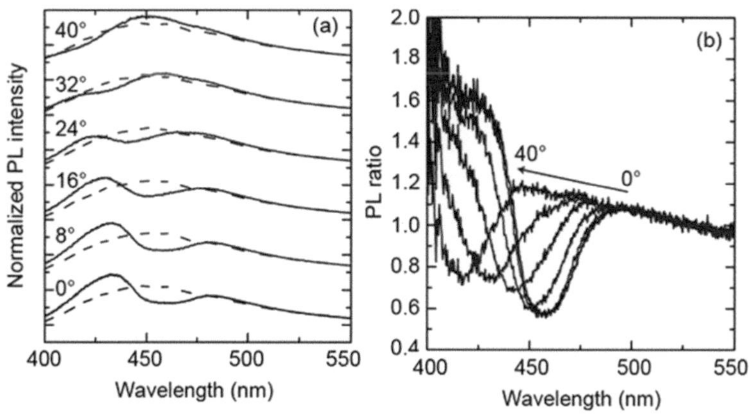

Fig. 8 (**a**) PL spectra of an opal film co-grown with PDV.Li⊂β-CD before (*solid line*) and after (*dashed line*) the thermal treatment at 75 °C at different incidence angle of the exciting beam [106], (**b**) ratio between the two PL spectra before and after the thermal process. In both panels the detection angle has been increased up to 40° with steps of 8°. Adapted and reproduced with permission from [98]

the system [33, 100–102], the microsphere quality [102], and the degree of incorporation of the luminescent material [84–87, 89, 103]. Indeed, they disappear when the structure is not enough regular, or when strongly absorbing materials are incorporated. In the case of co-grown opals, van Hove-like structures can still be observed both in the reflectance spectra [100] (Fig. 7a). All these features, combined with the additional effect of the PBG on the PL spectra (see Fig. 8a), can only be compatible with a minor incorporation of the rotaxinated polymer within the opals, such that the opal optical properties are not affected.

Figure 7b shows reflectance (R) and transmittance (T) spectra of the co-grown opal made with polystyrene beads with diameter $a = 200$ nm used in this study. As expected, the peak of reflectance at 455 nm ($E_B = 2.74$ eV) corresponds to a minimum in the transmittance spectrum. Significant changes in the optical properties of the PBG of the PhC upon incorporation of the rotaxanes are not observed. Both the spectral position of the stop-band and its full width at half-maximum ($\Delta E_B = 0.16$ eV) are unchanged compared to a bare opal [89, 97, 99], once again confirming the lack of lattice distortion and dielectric contrast modification, as expected from the SEM micrographs [76, 87, 103]. As a reference sample to study PL modifications induced by the PhC, the very same opal was thermally annealed at 75 °C for 5 min. Thermal annealing at 75 °C is enough to partially melt the polystyrene spheres, and thus destroy any photonic property of the structure [104] (as proven by the PL spectrum in Fig. 8a), without affecting the PL properties of the conjugated polyrotaxane, thanks to the higher thermal stability of this class of supramolecular systems [105].

To be able to compare the different properties, both measurements were carried out in the same area of the sample before and after the thermal process and the PL spectra were normalized at $\lambda = 550$ nm, far away from the PBG [34, 86, 89].

Figure 8 displays the PL spectra as a function of the incidence angle of the excitation beam for the co-grown PhCs and the reference sample. A 50 nm blue-shift of the PL peak and a small increase of the PL quantum yield (Φ_{PL}) to 38 ± 3 % (for the melted-opal "reference" samples) compared to a neat spin-coated film of PDV.Li⊂β-CD (Φ_{PL} ~ 18 ± 2 %) is observed [58, 107]. The PL blue-shift and the variation of Φ_{PL} can be assigned to electrostatic interactions of the Li cations with the negatively charged surface of the polystyrene nanospheres. This interaction changes the polymer structure, i.e. the electronic structure of semiconducting chains close to the surface. Furthermore, the opal growth is mainly driven by capillary forces [96] that would also act on the polyrotaxanes and impose additional structural constraints, with a concomitant modification of the PL dynamics of the polyrotaxane with respect to a neat-polymer film or solution [91]. This interpretation is further supported by what previously observed for poly(phenylene-ethynylene) polyelectrolytes neat and grafted to silica microspheres, for which a spectral blue-shift has also been observed. The PL spectrum of PDV.Li⊂β-CD (Fig. 8a) is strongly modified by the PBG, which partially suppresses (up to 33 %) light propagation between 448 and 482 nm (at 0° incidence). Instead, an increase of the PL intensity is observed at the short-wavelength edge of the PBG. As expected, both the enhancement and the suppression depend on the detection angle [34, 89] according to the dispersion of the PBG, thus inducing a directionality to the observed effect. The enhancement/suppression effects caused by the PhC with respect to the "melted reference" can be made more obvious by plotting the ratio between the two PL spectra, as in Fig. 8b: within the PBG the ratio is <1, but >1 at the short-wavelength edge of the PBG [34, 88, 89, 108].

Time-resolved PL measurements at the relevant wavelengths, i.e. at 437 nm (PL enhancement) and at 460 nm (PL suppression, Fig. 9) give further insight into the underlying photophysics of these photonic structures. In previous studies, the

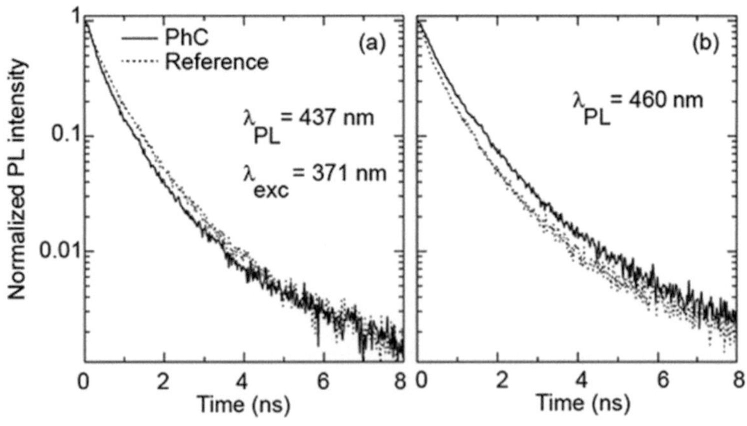

Fig. 9 Temporal evolution of the PDV.Li⊂β-CD PL measured at the short-wavelength edge (**a**, 437 nm) and inside (**b**, 460 nm) the PBG before (*solid line*) and after (*dashed line*) the thermal treatment at 75 °C. All measurements were carried out in air and at room temperature. Adapted and reproduced with permission from [98]

Table 2 Lifetime values (τ_1, τ_2, τ_3) used in the fit-curves the temporal evolution of the PL (the relative contribute of each time constant is included in parentheses)

	τ_1, ps (I_1)	τ_2, ps (I_2)	τ_3, ps (I_3)
437 nm (high-energy edge)			
PhC	300 ± 15 (0.33)	790 ± 40 (0.56)	2,720 ± 140 (0.11)
Reference	370 ± 20 (0.33)	860 ± 43 (0.57)	2,650 ± 130 (0.10)
460 nm (PBG)			
PhC	430 ± 20 (0.32)	950 ± 50 (0.55)	2,850 ± 140 (0.13)
Reference	365 ± 20 (0.34)	860 ± 40 (0.53)	2,750 ± 140 (0.13)

We estimated an error of 5 % on the measured value. Reproduced with permission from [98]

PL decay dynamics of polyrotaxane films have been fitted with a double exponential expression and assigned to the *intramolecular* singlet exciton and *inter*-molecular aggregate states [91]. Surprisingly, for the co-grown opal a triple-exponential expression: $I(t) = I_0 + I_1 e^{-\left(\frac{t}{\tau_1}\right)} + I_2 e^{-\left(\frac{t}{\tau_2}\right)} + I_3 e^{-\left(\frac{t}{\tau_3}\right)}$ must be used (Table 2). Whereas the longer decay times (τ_2 and τ_3), can be readily assigned to those previously identified ($\tau_{\text{exciton}} \sim 850$ ps, $\tau_{\text{aggregate states}} \sim 2{,}600$ ps) [91], the (fast) additional decay channel can be assigned to a new emissive species arising from the previously discussed interaction of the polyrotaxane with the nanospheres surfaces. This hypothesis is further supported by the observation of a three-exponential decay also in the annealed reference sample, suggesting that such a de-excitation pathway is observed only in the presence of polyrotaxane/colloid interaction. Furthermore, the weight of the different decay channels is similar at 437 and 460 nm for the PhC and the melted "reference". In addition, τ_3 is comparable in the opal, in the melted "reference" (2,650 ps at 437 nm and 2,750 ps at 460 nm) and in neat spin-coated films. At the short-wavelength edge of the stop-band (437 nm) the decay for the co-grown opal is shorter than for the reference (by 11 %, as estimated from the weighted mean of τ_1 and τ_2) whereas within the stop-band (460 nm) the reverse situation occurs (13 %, Table 2).

Indeed, a sizable effect on the radiative lifetime (Purcell effect) [109] can only be observed in photonic crystals for which the photonic density of states is strongly modified in the whole *k*-space. For synthetic polystyrene opals, the PBG is present only along the [110] (ΓL) crystallographic direction and it is spectrally narrow (447–482 nm) when compared to the width of the PL spectrum (405–650 nm), thus reducing the observable effect [34]. Furthermore, Φ_{PL} of PDV.Li$\subset\beta$-CD in the reference sample is ~38 ± 4 %, meaning that the radiative rate (K_R) is smaller than the non-radiative rate (K_{NR}), i.e. any modification caused by a redistribution of the photonic density of states will have a small effect on the PL lifetime. Interestingly, the observed PL lifetime modification is wavelength-dependent with a precise overlap with the PBG spectrum. According to this observation and previously reported experiments [34, 89] this effect can be assigned to the modification of the density of photonic states along the ΓL direction.

As a first order approximation, the modification of the PL lifetime can be considered as only induced by the redistribution of the photonic density of states, thus allowing an estimate of the variation of K_R. If we consider Φ_{PL} as the sum of contributions of the three emissive species, we obtain:

$$\Phi_{PL} = I_1\Phi_{PL1} + I_2\Phi_{PL2} + I_3\Phi_{PL3} \tag{4}$$

where I_x is the relative contribution to the total PL intensity (as obtained from the PL-decays) of each emissive species. As previously discussed, τ_2 and τ_3 are similar to previously measured values; we can assume that τ_2 corresponds to a Φ_{PL2} of ~35 %, since the relevant PL-lifetime is similar to the one measured in diluted PDV. Li$\subset\beta$-CD water solutions (for which emission is mainly due to the intrachain exciton) [91]. Instead, we can assume τ_3 to correspond to a Φ_{PL3} of ~3 % since the related PL-lifetime is similar to the one measured for PDV.Li films for which the emission is mainly due to aggregate states [110], hence we can calculate a Φ_{PL1} for the decay channel $1(\tau_1)$ of ~52 ± 5 %. From Φ_{PL} and τ we can calculate K_R and K_{NR} for all three decay channels, by using the following set of equations (for $x = 1, 2, 3$):

$$\Phi_{PLx} = \frac{K_{Rx}}{K_{Rx} + K_{NRx}}; \quad \frac{1}{K_{Rx}} = \frac{\tau_x}{\Phi_{PLx}} \tag{5}$$

Since we do not know Φ_{PL} for PDV.Li$\subset\beta$-CD after the photons redistribution takes place in the PhC, we can only calculate K_R and K_{NR} for the reference sample. Nevertheless, the variation of the photonic density of states only affects K_R, leaving K_{NR} unchanged (according to Fermi's golden rule, $K_R(\omega) \propto \frac{2\pi}{\hbar}\rho(\mathbf{r},\omega)$ [111], in which ρ is the photonic density of states). Therefore, knowing K_{NR} from the reference sample we can calculate K_R from $\tau = \frac{1}{K_R + K_{NR}}$.

Although this is a relatively simple approach, it permits to estimate an increase of K_R of 46 and 24 % at the high-energy PBG edge (437 nm), and a decrease of 31 and 27 % inside the PBG (460 nm) for τ_1 and τ_2, respectively (Table 3). τ_3 is unaffected by the redistribution of the photonic density of states; once again this can be explained by the large K_{NR} (for decay channel "3") that is not modified by the presence of the PhC leaving the modification of K_R too small to be measured.

Table 3 Radiative (K_R) and non-radiative (K_{NR}) rates for PDV.Li$\subset\beta$-CD incorporated inside the PhC and in the reference sample

	Decay channel 1 (τ_1)	Intrachain exciton (τ_2)	Aggregate states (τ_3)
437 nm (high-energy edge)			
$K_{R(PhC)}$ s^{-1} ($\times 10^8$)	19.8 ± 2	5.1 ± 0.5	0.10 ± 0.01
$K_{R(Reference)}$ s^{-1} ($\times 10^8$)	13.5 ± 2	4.1 ± 0.4	0.10 ± 0.01
460 nm (PBG)			
$K_{R(PhC}$ s^{-1} ($\times 10^8$)	9.5 ± 0.9	3.0 ± 0.3	0.10 ± 0.01
$K_{R(Reference)}$ s^{-1} ($\times 10^8$)	13.6 ± 2	4.1 ± 0.4	0.10 ± 0.01
K_{NR} s^{-1} ($\times 10^8$)	13.7 ± 2	7.5 ± 0.7	3.5 ± 0.3

All rates were calculated using the model proposed in the text. We include an error of 10 % on the calculated value. Reproduced with permission from [98]

4 Conclusion

In conclusion we have investigated two hybrid-organic photoactive systems. Wavelength and angle dependency of the PL suppression-enhancement has been observed due to the presence of the photonic stop-band. Furthermore, we observed a modification of the emission lifetime inside the stop-band and at its high-energy edge. Such variation of the radiative decay rate is assigned to the modulation of the photonic density of state induced by the photonic crystal that affects the infiltrated polymer.

References

1. Yablonovitch, E.: Phys. Rev. Lett. **58**, 2059 (1987)
2. John, S.: Phys. Rev. Lett. **58**, 2486 (1987)
3. Joannopulos, J.D., Meade, R.D., Win, J.N.: Photonic Crystals: Molding the Flow of the Light. Princeton University Press, Princeton (1995)
4. Noda, S., Chutian, A., Imada, M.: Nature **407**, 608 (2000)
5. Lin, S.Y., Fleming, J.G., Hetherington, D.L., Smith, B.K., Biswas, R., Ho, K.M., Sigalas, M. M., Zubrzycki, W., Kurtz, S.R., Bur, J.: Nature **394**, 251 (1998)
6. Mai, L., Ding, F., Stöferle, T., Knoll, A., Jan Offrein, B., Mahrt, R.F.: Appl. Phys. Lett. **103**, 243305 (2013)
7. Jebali, A., Mahrt, R.F., Moll, N., Erni, D., Bauer, C., Bona, G.-L., Bächtold, W.: J. Appl. Phys. **96**, 3043 (2004)
8. Moll, N., Stöferle, T., Schönenberger, S., Mahrt, R.F.: Opt. Express **17**, 20998 (2009)
9. Kawata, S., Sun, H., Tanaka, T., Takada, K.: Nature **412**, 697 (2001)
10. Maruo, S., Nakamura, O., Kawata, S.: Opt. Lett. **22**, 132 (1997)
11. Moon, G.D., Lee, T.I., Kim, B., Chae, G., Kim, J., Kim, S., Myoung, J.-M., Jeong, U.: ACS Nano **5**, 8600 (2011)
12. Belardini, A., Benedetti, A., Centini, M., Leahu, G., Mura, F., Sennato, S., Sibilia, C., Robbiano, V., Giordano, M.C., Martella, C., Comoretto, D., de Mongeot, F.B.: Adv. Opt. Mater. **2**, 208 (2014)
13. Koenderink, A.F., Johnson, P.M., Lopez, J.F.G., Vos, W.L.: CR Phys. **3**, 67 (2002)
14. Hartsuiker, A., Vos, W.L.: Langmuir **24**, 4670 (2008)
15. Comoretto, D., Grassi, R., Marabelli, F., Andreani, L.C.: Mat. Sci. Eng. C **23**, 61 (2003)
16. Bolognesi, A., Mercogliano, C., Yunus, S., Civardi, M., Comoretto, D., Turturro, A.: Langmuir **21**, 3480 (2005)
17. Johnson, N.P., McComb, D.W., Richel, A., Treble, B.M., Rue, R.M.D.L.: Synth. Met. **116**, 469 (2001)
18. Míguez, H., Meseguer, F., Lopez, C., López-Tejeira, F., Sunchez-Dehesa, J.: Adv. Mater. **13**, 393 (2001)
19. Galisteo-Lopez, J.F., Vos, W.L.: Phys. Rev. E **66**, 036616 (2002)
20. Frezza, L., Patrini, M., Liscidini, M., Comoretto, D.: J. Phys. Chem. C **115**, 19939 (2011)
21. Canazza, G., Scotognella, F., Lanzani, G., Silvestri, S.D., Zavelani-Rossi, M., Comoretto, D.: Laser Phys. Lett. **11**, 035804 (2014)
22. Lehmann, V., Föll, H.: J. Electrochem. Soc. **137**, 653 (1990)
23. Barillaro, G., Nannini, A., Piotto, M.: Sensors Actuators A Phys. **102**, 195 (2002)
24. Bassu, M., Surdo, S., Strambini, L.M., Barillaro, G.: Adv. Funct. Mater. **22**, 1222 (2012)

25. Barillaro, G., Strambini, L.M., Annovazzi-Lodi, V., Merlo, S.: IEEE J. Sel. Top. Quantum Electron. **15**, 1359 (2009)
26. Pacholski, C.: Sensors **13**, 4694 (2013)
27. Cregan, R.F., Mangan, B.J., Knight, J.C., Birks, T.A., Russell, P.S.J., Roberts, P.J., Allan, D.C.: Science **285**, 1537 (1999)
28. Liu, Y., Salemink, H.W.M.: Opt. Express **20**, 19912 (2012)
29. Guldin, S., Hüttner, S., Kolle, M., Welland, M.E., Müller-Buschbaum, P., Friend, R.H., Steiner, U., Tétreault, N.: Nano Lett. **10**, 2303 (2010)
30. Wiesmann, C., Bergenek, K., Linder, N., Schwarz, U.T.: Laser Photonics Rev. **3**, 262 (2009)
31. Tessler, N., Denton, G.J., Friend, R.H.: Nature **382**, 695 (1996)
32. Painter, O., Lee, R.K., Scherer, A., Yariv, A., O'Brien, J.D., Dapkus, P.D., Kim, I.: Science **284**, 1819 (1999)
33. Pavarini, E., Andreani, L.C., Soci, C., Galli, M., Marabelli, F., Comoretto, D.: Phys. Rev. B **72**, 045102 (2005)
34. Barth, M., Gruber, A., Cichos, F.: Phys. Rev. B **72**, 085129 (2005)
35. Samuel, I.D.W., Turnbull, G.A.: Mater. Today **28**, 28 (2004)
36. Samuel, I.D.W., Turnbull, G.A.: Chem. Rev. **107**, 1272 (2007)
37. Mróz, M.M., Sforazzini, G., Zhong, Y., Wong, K.S., Anderson, H.L., Lanzani, G., Cabanillas-Gonzalez, J.: Adv. Mater. **25**, 4347 (2013)
38. Gwinner, M.C., Khodabakhsh, S., Song, M.H., Schweizer, H., Giessen, H., Sirringhaus, H.: Adv. Funct. Mater. **19**, 1360 (2009)
39. Yang, Y., Turnbull, G.A., Samuel, I.D.W.: Appl. Phys. Lett. **92**, 163306 (2008)
40. Mróz, M.M., Perissinotto, S., Virgili, T., Gigli, G., Salerno, M., Frampton, M.J., Sforazzini, G., Anderson, H.L., Lanzani, G.: Appl. Phys. Lett. **95**, 31108 (2009)
41. Chénais, S., Forget, S.: Polym. Int. **61**, 390 (2012)
42. McGehee, M.D., Díaz-García, M.A., Hide, F., Gupta, R., Miller, E.K., Moses, D., Heeger, A. J.: Appl. Phys. Lett. **72**, 1536 (1998)
43. Kallinger, C., Hilmer, M., Haugeneder, A., Perner, M., Spirkl, W., Lemmer, U., Feldmann, J., Scherf, U., Müllen, K., Gombert, A., Wittwer, V.: Adv. Mater. **10**, 920 (1998)
44. Scotognella, F., Monguzzi, A., Cucini, M., Meinardi, F., Comoretto, D., Tubino, R.: Int. J. Photoenergy **2008**, 389034 (2008)
45. Yoon, J., Lee, W., Thomas, E.L.: Nano Lett. **6**, 2211 (2006)
46. Yoshino, K., Tatsuhara, S., Kawagishi, Y., Ozaki, M., Zakhidov, A.A., Vardeny, Z.V.: Appl. Phys. Lett. **74**, 2590 (1999)
47. Yoshino, K., Lee, S.B., Tatsuhara, S., Kawagishi, Y., Ozaki, M., Zakhidov, A.A.: Appl. Phys. Lett. **73**, 3506 (1998)
48. Jin, F., Li, C.-F., Dong, X.-Z., Chen, W.-Q., Duan, X.-M.: Appl. Phys. Lett. **89**, 241101 (2006)
49. Jin, F., Shi, L.-T., Zheng, M.-L., Dong, X.-Z., Chen, S., Zhao, Z.-S., Duan, X.-M.: J. Phys. Chem. C **117**, 9463 (2013)
50. Gottardo, S., Sapienza, R., Garcia, P.D., Blanco, A., Wiersma, D.S., Lopez, C.: Nat. Photonics **2**, 429 (2008)
51. Moll, N., Mahrt, R.F., Bauer, C., Giessen, H., Schnabel, B., Kley, E.B., Scherf, U.: Appl. Phys. Lett. **80**, 734 (2002)
52. Lemmer, U., Heun, S., Mahrt, R.F., Scherf, U., Hopmeier, M., Siegner, U., Go¨bel, E.O., Mu¨llen, K., Ba¨ssler, H.: Chem. Phys. Lett. **240**, 373 (1995)
53. Brédas, J.-L., Cornil, J., Heeger, A.J.: Adv. Mater. **8**, 447 (1996)
54. Segal, M., Baldo, M.A., Holmes, R.J., Forrest, S.R., Soos, Z.G.: Phys. Rev. B **68**, 075211 (2003)
55. Burin, A.L., Ratner, M.A.: J. Phys. Chem. A **104**, 4704 (2000)
56. Coropceanu, V., Cornil, J., da Silva Filho, D.A., Olivier, Y., Silbey, R., Brédas, J.-L.: Chem. Rev. **107**, 926 (2007)

57. Kozlov, V.G., Parthasarathy, G., Burrows, P.E., Khalfin, V.B., Wang, J., Chou, S.Y., Forrest, S.R.: IEEE J. Quantum Electron. **36**, 18 (2000)
58. Cacialli, F., Wilson, J.S., Michels, J.J., Daniel, C., Silva, C., Friend, R.H., Severin, N., Samori, P., Rabe, J.P., O'Connell, M.J., Taylor, P.N., Anderson, H.L.: Nat. Mater. **1**, 160 (2002)
59. Barta, P., Cacialli, F., Friend, R.H., Zagórska, M.: J. Appl. Phys. **84**, 6279 (1998)
60. Gigli, G., Barbarella, G., Favaretto, L., Cacialli, F., Cingolani, R.: Appl. Phys. Lett. **75**, 439 (1999)
61. Frampton, M.J., Anderson, H.L.: Angew. Chem. Int. Ed. **46**, 1028 (2007)
62. Ruminski, A.M., Barillaro, G., Secret, E., Huang, W.D., Potocny, A., Carion, U., Wertans, C., Sailor, M.J.: Adv. Opt. Mater. **1**, 510 (2013)
63. Presti, C.D., Irrera, A., Franzò, G., Crupi, I., Priolo, F., Iacona, F., Di Stefano, G., Piana, A., Sanfilippo, D., Fallica, P.G.: Appl. Phys. Lett. **88**, 033501 (2006)
64. Galli, M., Politi, A., Belotti, M., Gerace, D., Liscidini, M., Patrini, M., Andreani, L.C., Miritello, M., Irrera, A., Priolo, F., Chen, Y.: Appl. Phys. Lett. **88**, 251114 (2006)
65. Lin, V.S.-Y., Motesharei, K., Dancil, K.-P.S., Sailor, M.J., Ghadiri, M.R.: Science **278**, 840 (1997)
66. Ruminski, A.M., Moore, M.M., Sailor, M.J.: Adv. Funct. Mater. **18**, 3418 (2008)
67. Ruminski, A.M., Barillaro, G., Chaffin, C., Sailor, M.J.: Adv. Funct. Mater. **21**, 1511 (2011)
68. Surdo, S., Merlo, S., Carpignano, F., Strambini, L.M., Trono, C., Giannetti, A., Baldini, F., Barillaro, G.: Lab Chip **12**, 4403 (2012)
69. Kilian, K.A., Böcking, T., Gaus, K., Gal, M., Gooding, J.J.: ACS Nano **1**, 355 (2007)
70. Kilian, K.A., Lai, L.M.H., Magenau, A., Cartland, S., Böcking, T., Di Girolamo, N., Gal, M., Gaus, K., Gooding, J.J.: Nano Lett. **9**, 2021 (2009)
71. Tokranova, N.A., Novak, S.W., Castracane, J., Levitsky, I.A.: J. Phys. Chem. C **117**, 22667 (2013)
72. Ciampi, S., Böcking, T., Kilian, K.A., Harper, J.B., Gooding, J.J.: Langmuir **24**, 5888 (2008)
73. Hérino, R.: Mat. Sci. Eng. B. **69–70**, 70 (2000)
74. Harraz, F.A., Salem, M.S., Sakka, T., Ogata, Y.H.: Electrochim. Acta **53**, 3734 (2008)
75. Segal, E., Perelman, L.A., Cunin, F., Di Renzo, F., Devoisselle, J.M., Li, Y.Y., Sailor, M.J.: Adv. Funct. Mater. **17**, 1153 (2007)
76. Vos, W.L., Sprik, R., van Blaaderen, A., Imhof, A., Lagendijk, A., Wegdam, G.H.: Phys. Rev. B **53**, 16231 (1996)
77. Ramsdale, C.M., Greenham, N.C.: J. Phys. D. Appl. Phys. **36**, L29 (2003)
78. Jalkanen, T., Torres-Costa, V., Mäkilä, E., Kaasalainen, M., Koda, R., Sakka, T., Ogata, Y. H., Salonen, J.: ACS Appl. Mater. Interfaces **6**, 2884 (2014)
79. Kim, K., Webster, S., Levi, N., Carroll, D.L., Pinto, M.R., Schanze, K.S.: Langmuir **21**, 5207 (2005)
80. Lazzerini, G.M., Di Stasio, F., Fléchon, C., Caruana, D.J., Cacialli, F.: Appl. Phys. Lett. **99**, 243305 (2011)
81. Lazzerini, G.M., Mian, S., Di Stasio, F., Merari Masillamani, A., Crivillers, N., Reinders, F., Mayor, M., Samorì, P., Cacialli, F.: Appl. Phys. Lett. **101**, 153306 (2012)
82. Lopez, C.: Adv. Mater. **15**, 1679 (2003)
83. Rybin, M.V., Khanikaev, A.B., Inoue, M., Samusev, K.B., Steel, M.J., Yushin, G., Limonov, M.F.: Phys. Rev. Lett. **103**, 4 (2009)
84. Galisteo-Lopez, J.F., Ibisate, M., Sapienza, R., Froufe-Perez, L.S., Blanco, A., Lopez, C.: Adv. Mater. **23**, 30 (2011)
85. Pasquazi, A., Stivala, S., Assanto, G., Amendola, V., Meneghetti, M., Cucini, M., Comoretto, D.: Appl. Phys. Lett. **93**, 3 (2008)
86. Nikolaev, I.S., Lodahl, P., Vos, W.L.: Phys. Rev. A **71**, 10 (2005)
87. Morandi, V., Marabelli, F., Amendola, V., Meneghetti, M., Comoretto, D.: Adv. Funct. Mater. **17**, 2770 (2007)

88. Lodahl, P., van Driel, A.F., Nikolaev, I.S., Irman, A., Overgaag, K., Vanmaekelbergh, D., Vos, W.L.: Nature **430**, 654 (2004)
89. Berti, L., Cucini, M., Di Stasio, F., Comoretto, D., Galli, M., Marabelli, F., Manfredi, N., Marinzi, C., Abbotto, A.: J. Phys. Chem. C **114**, 2403 (2010)
90. McCaughey, B., Costello, C., Wang, D., Hampsey, J.E., Yang, Z., Li, C., Brinker, C.J., Lu, Y.: Adv. Mater. **15**, 1266 (2003)
91. Brovelli, S., Latini, G., Frampton, M.J., McDonnell, S.O., Oddy, F.E., Fenwick, O., Anderson, H.L., Cacialli, F.: Nano Lett. **8**, 4546 (2008)
92. Mroz, M.M., Perissinotto, S., Virgili, T., Gigli, G., Salerno, M., Frampton, M.J., Sforazzini, G., Anderson, H.L., Lanzani, G.: Appl. Phys. Lett. **95**, 3 (2009)
93. Brovelli, S., Virgili, T., Mroz, M.M., Sforazzini, G., Paleari, A., Anderson, H.L., Lanzani, G., Cacialli, F.: Adv. Mater. **22**, 3690 (2010)
94. Brovelli, S., Meinardi, F., Winroth, G., Fenwick, O., Sforazzini, G., Frampton, M.J., Zalewski, L., Levitt, J.A., Marinello, F., Schiavuta, P., Suhling, K., Anderson, H.L., Cacialli, F.: Adv. Funct. Mater. **20**, 272 (2010)
95. Brovelli, S., Cacialli, F.: Small **6**, 2796 (2010)
96. Dimitrov, A.S., Nagayama, K.: Langmuir **12**, 1303 (1996)
97. Di Stasio, F., Cucini, M., Berti, L., Comoretto, D., Abbotto, A., Bellotto, L., Manfredi, N., Marinzi, C.: J. Eur. Opt. Soc. **4**, 7 (2009)
98. Di Stasio, F., Berti, L., McDonnell, S.O., Robbiano, V., Anderson, H.L., Comoretto, D., Cacialli, F.: APL Mater. **1**, 042116 (2013)
99. Di Stasio, F., Berti, L., Burger, M., Marabelli, F., Gardin, S., Dainese, T., Signorini, R., Bozio, R., Comoretto, D.: Phys. Chem. Chem. Phys. **11**, 11515 (2009)
100. C. Karnutsch, C. Pflumm, G. Heliotis, J.C. deMello, D.D.C. Bradley, J. Wang, T. Weimann, V. Haug, C. Gartner, U. Lemmer, Appl. Phys. Lett. **90**, 131104 (2007)
101. F. Di Stasio, M. Cucini, D. Comoretto, unpublished
102. Comoretto, D., Robbiano, V., Canazza, G., Boarino, L., Panzarasa, G., Laus, M., Sparnacci, K.: Polym. Compos. **34**, 1443 (2013)
103. Morandi, V., Marabelli, F., Amendola, V., Meneghetti, M., Comoretto, D.: J. Phys. Chem. C **112**, 6293 (2008)
104. M. Cucini, M. Alloisio, A. Demartini, D. Comoretto, in *Biomimetic and Supramolecular System Research,* ed. by A.H. Lima (Nova Science, New York, 2008), p. 91
105. Kasiouli, S., Di Stasio, F., McDonnell, S.O., Constantinides, C.P., Anderson, H.L., Cacialli, F., Hayes, S.C.: J. Phys. Chem. B **117**, 5737 (2013)
106. M. Cucini, R. Narizzano, D. Comoretto, V. Morandi, F. Marabelli, in *Optical Properties of Artificial Opals Infiltrated with a Porphyrin*. International Conference on Optical Probes of p-conjugated polymers and functional self-assemblies, Turku, 11–15 June 2007
107. Latini, G., Parrott, L.J., Brovelli, S., Frampton, M.J., Anderson, H.L., Cacialli, F.: Adv. Funct. Mater. **19**, 3679 (2009)
108. Bechger, L., Lodahl, P., Vos, W.L.: J. Phys. Chem. B **109**, 9980 (2005)
109. Purcell, E.M.: Phys. Rev. **69**, 681 (1946)
110. Petrozza, A., Brovelli, S., Michels, J.J., Anderson, H.L., Friend, R.H., Silva, C., Cacialli, F.: Adv. Mater. **20**, 3218 (2008)
111. Kubo, S., Fujishima, A., Sato, O., Segawa, H.: J. Phys. Chem. C **113**, 11704 (2009)

Plasmonic Sensors on 2D Ordered Structures

Franco Marabelli, Andrea Valsesia, Silvia Giudicatti, Lucia Fornasari, Paola Pellacani, and Ana Frangolho

Abstract Colloidal lithography with polystyrene spheres allows for the fabrication of hybrid polymer/metal two-dimensional ordered surfaces. They consist of a hexagonal lattice of plasma-polymerized acrylic acid (ppAA) or poly(methyl methacrylate) (PMMA) pillars embedded in an optically thick gold film deposited on a glass substrate. Such a kind of nanostructured system has been shown to support either propagating Surface Plasmon Polaritons or "Mie-like" localized resonances, and appears to be particularly interesting for sensing applications. Tuning the structural parameters, a strong interaction among delocalized and localized plasmonic modes can be obtained together with a good coupling with light. This opens the way towards an optical biomolecular sensor system in which a modification/adhesion on the free nanostructured surface can be easily detected by a simple, near normal reflectance measurement performed from the substrate side, despite the relatively large gold thickness. The simple configuration allows for a

F. Marabelli (✉) • L. Fornasari
Physics Department, University of Pavia, 27100 Pavia, Italy
e-mail: franco.marabelli@unipv.it; lucia.fornasari@unipv.it

A. Valsesia
Plasmore s.r.l., via Deledda 4, 21020 Ranco (Varese), Italy

Institute for Health and Consumer Protection, European Commission, Joint Research Center, via Enrico Fermi 2749, Ispra 21027, Italy
e-mail: andrea.valsesia@gmail.com

S. Giudicatti
Physics Department, University of Pavia, 27100 Pavia, Italy

Institute for Integrative Nanosciences, IFW Dresden, Helmholtzstraße 20, Dresden 01069, Germany
e-mail: s.giudicatti@ifw-dresden.de

P. Pellacani
Plasmore s.r.l., via Deledda 4, 21020 Ranco (Varese), Italy

Department of Applied Physics, Universidad Autònoma de Madrid, Cantoblanco, Madrid 28049, Spain
e-mail: Paola.pellacani@plasmore.com

A. Frangolho
Plasmore s.r.l., via Deledda 4, 21020 Ranco (Varese), Italy
e-mail: Ana.frangolho@plasmore.com

D. Comoretto (ed.), *Organic and Hybrid Photonic Crystals*,
DOI 10.1007/978-3-319-16580-6_16

surface plasmon resonance (SPR) imaging configuration and enables the real-time multiplexed detection of several analytes. The sensing performance of the surfaces (sensitivity to refractive index change and to the adhesion of molecular monolayers) has been tested using standard spectroscopic techniques. The electromagnetic field's spatial distribution within the nanostructures and its intensity enhancement have been numerically calculated by finite difference time domain (FDTD) simulations. The results, including the calculated reflectance spectra, are in good agreement with the experimental data.

Keywords Nanostructured surfaces • Colloidal lithography • Nanocavities • Surface plasmon resonance • Optical response • Sensing • Imaging • Multiplexing

1 Introduction

Real-time sensing of biomolecules or of interesting organic molecules in environment or food control is a very attractive target and many efforts have been made in the last decades to develop a reliable, fast, and cheap detection method.

Several examples of sensing are reported in some chapters of this book too. The advantages of early detection of biological agents are evident in medical diagnostics or analysis of contaminants, either chemical or bacteriological, in water, food, or beverages [1–4]. Beside low cost and high sensitivity, also portability, user-friendliness (no- or limited-sample preparation required), and multiplexing (simultaneous multiparametric analysis capability) are important requirements [5–8]. In this context, optically based systems and the use of organic materials play a relevant role. Nowadays, the progress in optical technology allows for simple and convenient detection instruments. Meanwhile, in addition to the advantages of flexibility and reduced costs, an organic platform is naturally compatible with biosystems due to chemical affinity and bonding mechanisms [9–11].

The insertion of metals, usually noble metals like gold or silver, into a hybrid organic and metallic nanostructured system allows the exploitation of surface plasmonic resonances (SPR) [12–16]. A more extended presentation of plasmonic principles and structures can be found in the chapter by Brückner et al. and in the chapter by Romanov. Plasmonic excitations are electromagnetic (e.m.) modes supported by the collective oscillations of the free charges of a metal [17]. At the interface between the metal and a dielectric medium the e.m. field is evanescent in the normal direction to the interface surface and is depending on the dielectric permittivity of the medium [18]. Then, the modes, even when propagating along the surface, are strongly confined at the interface, yielding to field enhancement effects. Such an enhancement is one of the major advantages in using plasmonics for detection purposes. As a matter of fact, several detection systems based on SPR are already commercial [14]. An extended surface can support propagating plasmonic waves, the so-called surface plasmon polaritons (SPP), characterized by a smaller, and even much smaller, wavelength than the one of a conventional e.m. wave having the same frequency [18]. Moreover, SPP are typically longitudinal waves, i.e. the e.m. field is oscillating along the propagation direction. Then, it is quite difficult to directly couple an optical signal to SPP and some matching mechanism has to be used. The largest part of SPR applications are working with the so-called Kretschmann configuration [19]: a very thin (less than 50 nm) metallic film is deposited over a prism and the Attenuated Total Reflection configuration is used to obtain an angle dependent measure of the plasmonic excitation at a fixed wavelength. On the other hand, metallic nanostructures can support localized plasmonic modes which are more easily coupled to optical waves and exhibit huge field enhancement, but whose intrinsic spectral broadening makes them less sensitive to the dielectric medium permittivity due to either dissipation effects or a size dispersion difficult to be controlled [20–25]. Then, few examples of Localized SPR can be found among industrial applications, apart from the use of metallic nanoparticles to enhance a Raman signal (Surface Enhanced Raman Spectroscopy) [21, 26–28].

Nevertheless, a periodic pattern in a hybrid metal-dielectric system can be exploited to either overcome the matching problem between optics and SPP or support the excitation of mixed, nearly localized plasmonic modes to be used for SPR detection [29–33]. Some instruments exploiting the properties of a grating of holes perforating the metallic film to couple SPP modes are working in an Optical Transmission configuration ([34]; Xie et al. 2010; [35–37]). Extraordinary transmittance can take place, i.e. an enhancement of the transmittance signal not scaled with the geometrical free aperture of the holes [38–42]. The disadvantage of such a configuration is the need of passing through the fluidic cell, which can introduce interference effects and reduce the sensitivity [43].

In the following we will present the main features of a system based on a two-dimensional array of hybrid polymer and metal nanostructures used for the detection of biomolecules in a simple reflectance configuration with imaging capabilities. The system has been patented for commercial application under the trademark Imaging Nanoplasmonics™. The presentation will be developed as follows: the preparation and the morphology of the nanopatterned surface of the sensor chip will be described. Then the optical response and its physical origin will be illustrated. A further paragraph will be devoted to the detection mechanism and the sensitivity performance. Finally the detection of an antibiotic molecule in a solution will be used to show the multiparametric possibilities of the instrument.

2 Sample Preparation and Structures

The systems we investigated are prepared by a nanofabrication process based on plasma-enhanced chemical vapor deposition and colloidal lithography [44]. A ppAA film—90–150 nm thick—was deposited on a glass substrate by PE-CVD using acrylic acid (purity > 99 %; Sigma–Aldrich) as liquid precursor. The capacitive coupled plasma source operated at a pressure of 7.5 Pa with a constant monomer flow. The power for plasma generation was provided by a pulsed RF power supply operating at 13.56 MHz and 50 W. Spin coating or, more recently, Langmuir–Blodgett technique was then used to cover the film with a monolayer of polystyrene (PS) nanospheres arranged in a 2D hexagonal close-packed lattice [45–47]. PS beads (Sigma–Aldrich) have usually a diameter of 500 nm with a nominal size dispersion of 10 %.

In order to transfer the geometry of the mask to the underneath layer, the sample was exposed to oxygen plasma etching in a magnetic pole enhanced-ICP source with inductive power $P = 150$ W and DC bias $|V| = 100$ V; the etching time was calculated in order to completely remove the uncovered ppAA areas, while nanoparticles diameter was reduced (in this way, no continuous ppAA film remains on the substrate, but an ordered array of ppAA pillars). Figure 1 shows a scanning electron microscopy (SEM) image of the pillar structure. The residual colloidal mask is still visible on top of each pillar. A combined control of the ppAA layer thickness and etching time allows tuning of the pillar size and shape [47]. We analyzed several series of samples prepared with different deposition parameters. Etching time mainly controls the pillar size; nevertheless, depending on the ratio between layer thickness and pillar size, etching also affects the pillar shape. As a matter of fact, two different types of pillars were obtained: the ones having a more cylindrical shape with a relatively flat top surface and the ones characterized by a more conical shape and a tapered top. An example is shown in panels (a) and (b) of Fig. 2 (after sonication and mask removal).

Afterward, a gold layer is deposited on the device by physical vapor deposition. Gold thickness is a further parameter which can be tuned to be slightly larger or smaller than the pillar height. In any case, a peculiar small (about 10 nm height) ring of gold is formed around the pillar's top.

Finally, the residual colloidal mask was removed using lift-off by an ultrasonic bath in ultra-pure water.

Fig. 1 SEM image of plasma-polymerized acrylic acid (*ppAA*) pillars as obtained after plasma etching through a colloidal mask. The residual spheres of the colloidal mask are still visible on top of each pillar

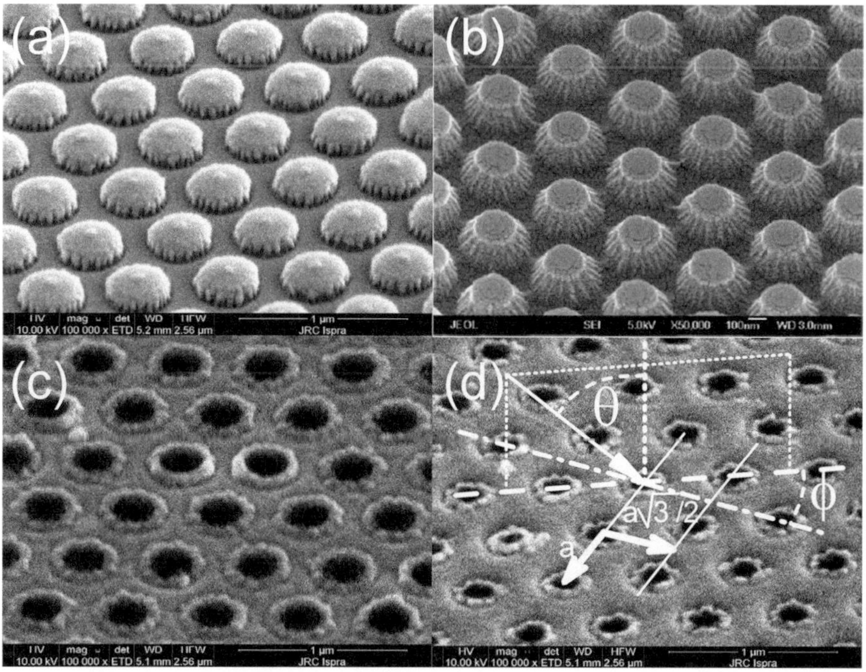

Fig. 2 SEM images of chip surfaces after mask residual removal: (**a**), (**b**) before gold deposition (with cylindrical- (**a**) or truncated conical- (**b**) pillar shape); (**c**), (**d**) after gold deposition. In panel (**d**) a sketch of the angle notation identifying the plane- and the angle-of incidence is reported. The two representative crystallographic distances are shown too

The obtained nanostructured surfaces were characterized by atomic force microscopy and (SEM) before (in order to study the pillars configuration), and after (to characterize the final sample surface) the gold deposition. The resulting structure (Fig. 2) is a gold film perforated by polymeric pillars arranged in a 2D hexagonal array. The lattice constant is determined by the initial diameter of the PS beads (usually 500 nm). The apertures on the gold surface where the pillar's top is emerging have a diameter ranging between 150 and 200 nm.

3 Plasmonic and Optical Features

The peculiar morphology described above is determining the optical response of the whole structure. As a matter of fact, the details of the pillar's shape, their aspect ratio, and the thickness of the gold layer are responsible of a series of well-defined, but relatively secondary, features affecting the transmittance and reflectance spectra. However, the general implant of the plasmonic response can be understood by considering the SPP running on the gold interfaces (either with the glass substrate or with the external medium at the exposed surface) and the localized plasmonic resonances around each hole (pillar) in the gold surface.

Optical Reflectance and Transmittance spectra have been measured over the spectral range 8000–25000 cm^{-1} (wavenumbers) or 400–1000 nm (wavelengths). Measurements have been performed with a Fourier Transform spectrometer coupled with a home made reflectometer allowing variable angle of incidence with angular resolution of less than 1° and a spot size of about 150 μm. Then, a study of the mode dispersion as a function of the incidence angle has been possible [44, 48, 49].

The energy dispersion of SPP modes of a gold interface with a dielectric medium can be described by the relationship between the wavevector k_{SPP} and the frequency ω

$$k_{SPP}(\omega) = \frac{\omega}{c}\sqrt{\frac{\varepsilon_m(\omega) \times \varepsilon_d}{\varepsilon_m(\omega) + \varepsilon_d}}, \tag{1}$$

where $\varepsilon_m(\omega)$ and ε_d are the dielectric function of the metal (gold) and of the dielectric medium (in our case: glass at the interface with the substrate, air—or water when a flow cell is used—at the exposed surface).

The coupling with the incident radiation having frequency ω becomes possible for TM polarization when the following condition is satisfied:

$$\vec{k}_{SPP}(\omega) = \vec{k}_0(\omega, \phi) \times \sin(\theta) + \vec{G}_{nm}, \tag{2}$$

where θ and ϕ are the angles defining the incidence direction on the surface—θ is the azimuthal angle from the normal to the surface and ϕ is the radial angle between the plane of incidence and a symmetry axis lying above the surface (a sketch is shown in Fig. 2d). In our case of a hexagonal lattice of period a, the symmetry line has been chosen as the one identifying the family of the closest pillar lines separated by a distance of $a\sqrt{3}/2$. $K_0\sin(\theta)$ is the wavevector of the incident beam projected to the surface within the plane of incidence, whereas G_{nm} is the wavevector of the reciprocal lattice identified by the indexes n and m.

Then, the effect of the periodic array, and of the wavevectors G_{nm}, is the folding of the dispersion relation within the first Brillouin zone of the lattice, opening the possibility of exciting SPP modes with a beam impinging at a small angle of incidence.

The resulting dispersion lines for the smallest wavevectors G_{nm} are reported in Fig. 3a, b, together with the intensity maps of the measured transmittance spectra for different orientations of the surface with respect to the plane of incidence, corresponding to the values of $\phi = 0°$ and $\phi = 30°$ and for different polarization.

Transmittance and reflectance spectra, measured on both sides of the sample, for the orientation $\phi = 0°$ and TM polarization are shown in Fig. 4. It is worth to notice that, despite some differences in the relative intensity, all the spectra show corresponding features at the same energies. All the optical spectra, independently

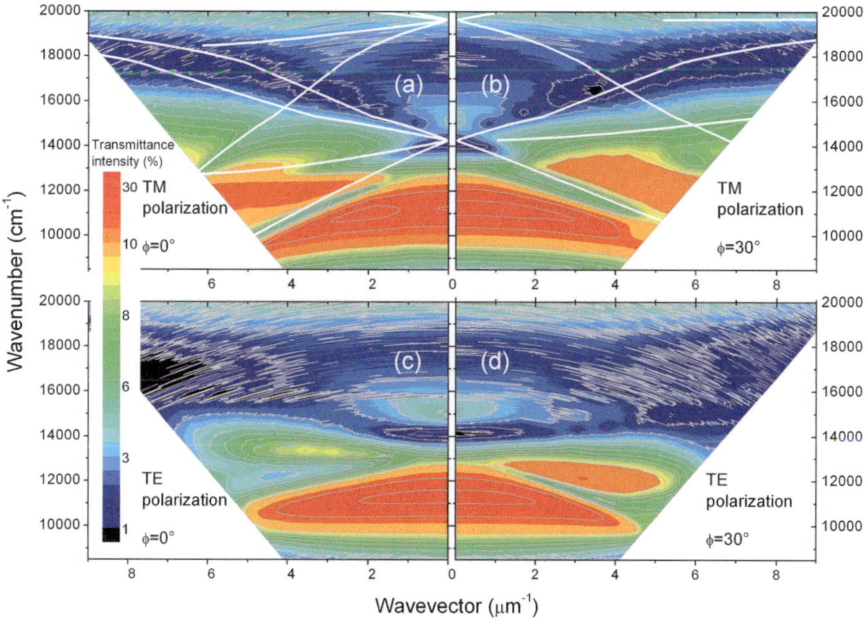

Fig. 3 Maps of the optical transmittance as a function of wavenumber and wavevector projection $k_0 \sin\theta$ for different polarization and different orientation of the plane of incidence with respect to the symmetry of the pillar lattice. Panel (**a**) TM and $\theta = 0°$; (**b**) TM and $\theta = 30°$; (**c**) TE and $\theta = 0°$; (**d**) TE and $\theta = 30°$. Theoretical dispersion curves of the surface plasmon polaritons propagating at the gold/glass and gold/air interfaces are shown as *white lines* in panels (a) and (b) corresponding to TM polarization

of orientation and polarization, are dominated by a large structure (a maximum in transmittance and a minimum in reflectance) around 11,000 cm^{-1}. Beside this structure and at higher photon energies, some other features can be seen exhibiting a progressive evolution when the incidence angle is increased. This is particularly evident for TM polarization. As shown in Fig. 3a, b, the dispersion lines calculated for SPP of a gold interface with glass ($\varepsilon_d = 2.25$) or air ($\varepsilon_d = 1$) (the white lines in the panels corresponding to TM polarization) can be easily used as a guide for eyes to follow the dispersion of the measured transmittance structures. Then, it is likely to ascribe the spectral features associated to such a dispersion to SPP modes of both gold interfaces.

On the other hand, the main structure at about 11,000 cm^{-1} (roughly corresponding to a wavelength of 900 nm) remains visible in TE polarization too. A localized electromagnetic mode can be calculated for a dielectric sphere of ppAA embedded in gold using the Mie model [20]. The first dipole mode ($L = 1$, degenerate) is obtained at the observed energy when using a sphere diameter of 100 nm, not far from the pillar radius. The next mode in energy, with $L = 2$, occurs at about 15,500 cm^{-1} (650 nm), where a second feature is observed at the lowest incidence angles. The removal of the degeneracy due to the distorted cylindrical shape of the

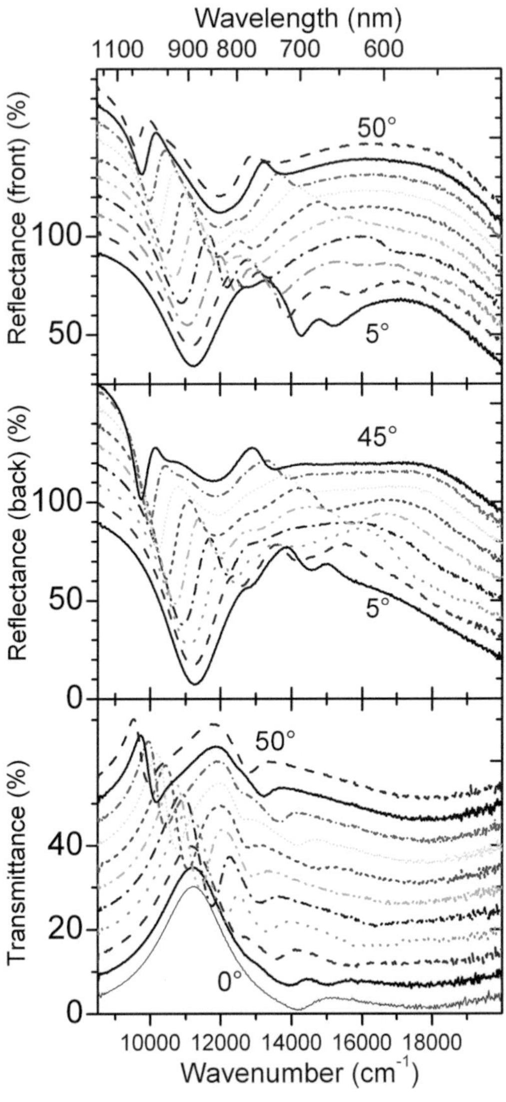

Fig. 4 Spectra of the optical transmittance and of the reflectance from the front- and the back side of a sample for the orientation $\phi = 0°$ and TM polarization as a function of the incidence angle. Curves measured at intervals of 5° in the angle of incidence have been vertically translated

pillars, as well as the interaction with the propagating SPP modes with the formation of mixed modes and anticrossing features in the dispersion, could then explain the features observed in the maps of Fig. 3.

Support to this picture is provided by finite difference time domain (FDTD) numerical simulations which, beside being able of reproducing the experimental spectra with a good richness of details, allow the calculation of the Electric field distribution [48–50]. A sketch of the typical field distribution for the mode at 900 nm is reported in Fig. 5. One can notice that the largest field intensity is

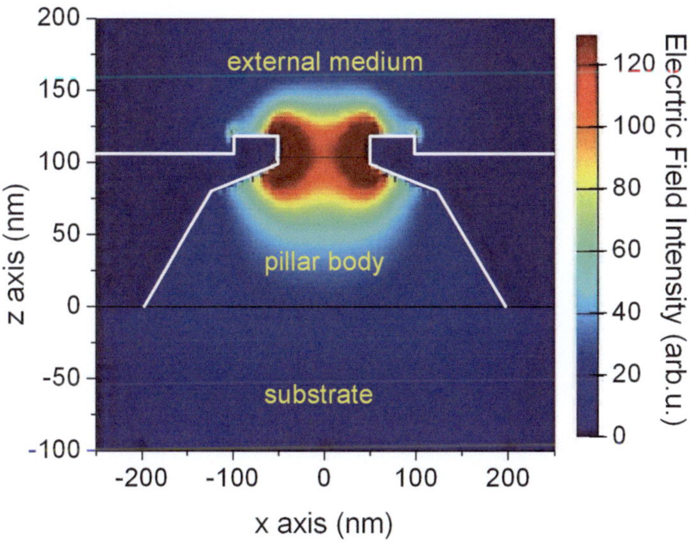

Fig. 5 Electric field intensity distribution on a vertical section of a pillar at the wavelength corresponding to the localized plasmon resonance at 900 nm as obtained by FDTD calculation. The profile of the gold film is sketched with a *white line*

concentrated within the area occupied by the pillar, but close to the surface, where the gold is surrounding the top of the pillar. This particular position at the interface between the polymer and the external medium makes the plasmonic resonance particularly sensitive to changes in the medium refractive index or to the adhesion of molecules at the interface.

4 Detection Mechanism

A large change, mainly a spectral red shift of the whole optical spectra is observed when the surface is put into contact with water. Figure 6 reports the dispersion maps of the reflectance measured from the side of the substrate when the upper surface is in contact with air (left) or with water (right). As expected, the dispersion of SPP modes is deeply changed when passing from air to water. Whereas the modes corresponding to the gold glass interface must keep the same wavelength dispersion, the ones related to the gold/air interface disappear and new features can be noticed at smaller energies corresponding to modes of the gold/water interface. Moreover, a relatively large red shift is affecting the large low energy structure which has been associated with the localized mode. The effect is clearly shown in Fig. 7 for the near normal incidence reflectance. When water is made flow over the chip surface the reflectance spectrum exhibits a clear red shift of the main minimum. The shift implies an increase of the relative reflectance intensity between

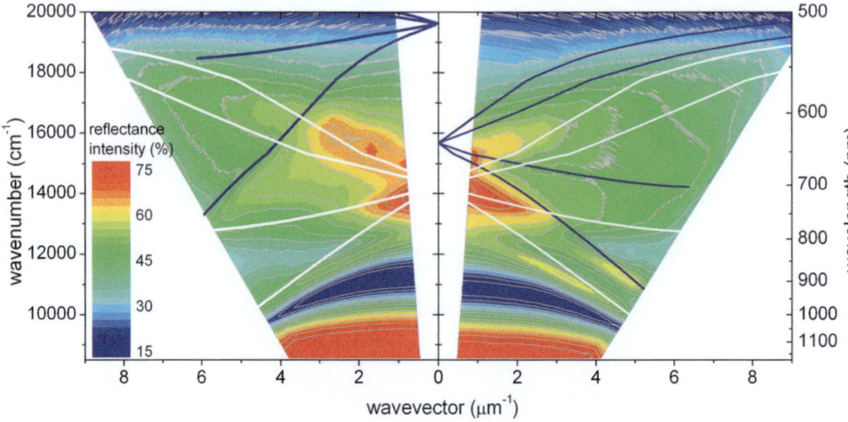

Fig. 6 Dispersion maps of the reflectance measured from the side of the substrate when the upper surface is in contact with air (*left*) or with water (*right*). The dispersion curves of the SPP modes at the gold/glass (*white lines* in *both left* and *right panel*), gold/air (*dark lines* in the *left panel*), and gold/water (*dark lines* in the *right panel*) are also shown for comparison

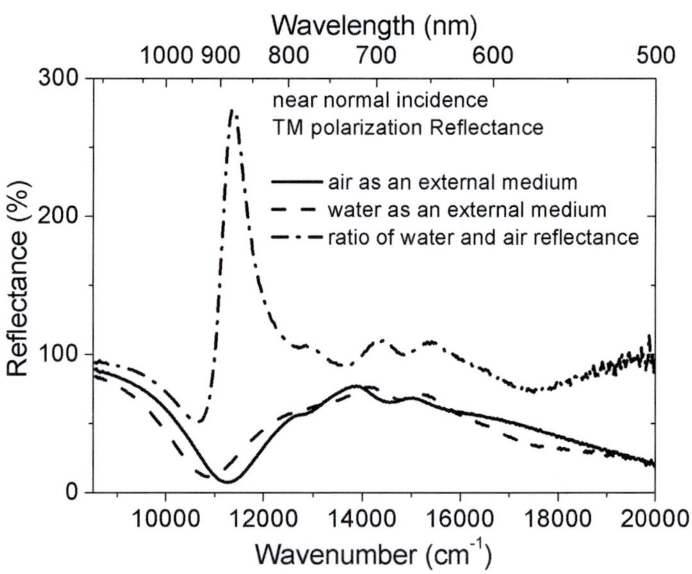

Fig. 7 Near normal reflectance measured from the back side of a chip surface when exposed to air (*solid line*) or water (*dashed line*). The *dash-dotted line* is the ratio of the water/air reflectance

11,000 and 12,000 cm^{-1} when passing from air to water. A decrease is instead observed between 10,000 and 11,000 cm^{-1}. This is well described by the ratio of the reflectance spectra with and without water, as shown in Fig. 7. One can choose of considering either the intensity increase corresponding to a spectral shift of the minimum or the intensity decrease at lower frequencies due to the same shift or,

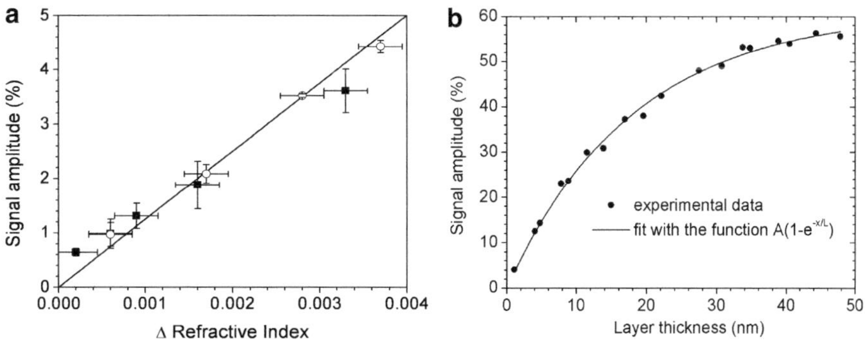

Fig. 8 Panel (**a**): Chip response to salt (*black squares*) and ethanol (*white circles*) solutions with different concentrations. The plot displays the amplitude changes in reflectance referred to the reflectance of distilled water as a function of the refractive index difference. Panel (**b**): Chip response to the progressive deposition of polyelectrolyte monolayers as a function of the total thickness of the deposited film. The resulting saturation curves is described by the function $(1-\exp(-x/L))$ with the parameter L of about 30 nm

even better, the combination of both effects. The use of both signal increase and signal decrease at two different wavelengths, allowing the application of a differential criterion (the difference of intensity at the two wavelengths) makes the measure quite stable with respect to spurious and unwanted intensity fluctuations affecting homogeneously the whole spectrum. The intensity change has been shown to be proportional to the refractive index variation occurred in the external medium at the chip surface [51]. Several tests have been done by making flow pure water on the surface and, then, a diluted solution of known refractive index, containing salt, alcohol, or glycerol. Some results are presented in the left panel of Fig. 8. The measured signal scales linearly with the refractive index change. Then, it has been possible to detect refractive index variations down to 10^{-6} refractive index units (RIU) [51].

The good sensitivity has been confirmed also with respect to the deposition of molecular layers above the surface. Using polyelectrolyte solutions of poly (diallyldimethylammonium chloride) (PDDA, Sigma Aldrich, St. Louis, Missouri) and of poly(sodium 4-styrenesulfonate) (PSS, Sigma Aldrich), diluted at 2 % in water, single molecular layers were formed for each deposition step. By alternating positively (PDDA) and negatively (PSS) charged molecules, a series of layers has been progressively build up to form a film on the surface. The signal evolution has been registered as a function of the film thickness. The resulting data are reported in the right panel of Fig. 8. One can notice that the signal intensity tends to saturate for film thicknesses above a few tens of nanometers. As a matter of fact, when the film thickness increases, the interface at which changes are occurring is progressively moved away from the gold surface and the top of the pillar where the electromagnetic field of the plasmonic mode has its maximum. Then, the saturation of the signal in the right panel of Fig. 8 is following the decay law of the field above the gold surface. From the fitting of the data with the formula $S(x) = A(1-e^{-x/L})$, one can estimate the penetration depth of the field $L \approx 30$ nm.

5 Instrumental Configuration and Imaging

The relatively wide spectral region over which an increase (decrease) of the reflectance signal can be detected allows for the use of a cheap quasi-monochromatic source like a LED to select the interesting spectral interval and monitor the reflectance intensity without using a monochromator or a filter. Actually, it is easy to find LEDs centered at 850 or 870 nm (or at 920 nm). This fact considerably simplifies the optical configuration. Using a LED source to properly illuminate the chip from the back side and collecting the reflected light on an imaging capable detector, a CMOS or a CCD camera, one can map the optical response of different points over the chip surface and enable the system to perform a simultaneous multichannel analysis (multiplexing). Since a typical biorecognition method exploits the selective binding reaction of a couple antigen-antibody (ag-ab), and the largest part of the target molecules are, or can be identified with, an antigen or an antibody, it becomes possible to pre-functionalize different areas of the chip surface with a series of selected bioreceptors, and obtain a multiplexing detection analysis for the chosen series of markers [52].

An example of this way of working is shown in Fig. 9. Using a micro spotter (a non-contact dispensing Scienion S3 sci-FLEXARRAYER spotter) an array of

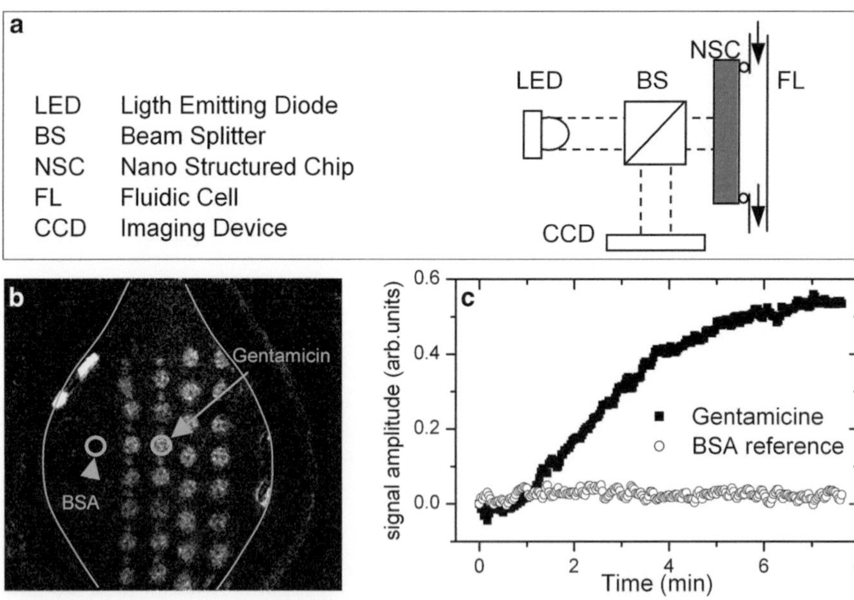

Fig. 9 (**a**) Sketch of the measuring device in the imaging configuration. (**b**) An example of plasmonic image of a chip surface functionalized with gentamicin (bright spots) and BSA (as a blocking agent) molecules after injection of antibody-gentamicin. The profile of the fluidic cell is put into evidence by a *white line*. (**c**) Evolution of the signal integrated over one of the bright spots as a function of time during the injection of the antibody-gentamicin. The equivalent signal of a BSA functionalized region is shown for comparison

four columns, each with 8 circular spots (having a radius of about 100 nm) of gentamicin, an antibiotic used in veterinary science, has been deposited on the chip surface. The remaining, non-spotted surface has been blocked with bovine serum albumine (BSA). BSA molecules should saturate all the non-specific binding sites for proteins on the surface. Just at the beginning, an image of the surface is taken as a reference; then, a solution containing antibody-gentamicin (ab) is made to flow above the surface. A typical concentration of 1 mg/ml is used. During the ab injection the differences occurring at the surface image are continuously registered pixel by pixel. An example is shown in Fig. 9b. The spots functionalized with gentamicin become progressively brighter with respect to the other surface regions. The intensity change (normalized to the intensity registered at the same pixel in the image taken before the beginning of the flowing process) integrated over a disk of the same area in correspondence of a gentamicin spot and of a BSA blocked region is evolving in time according to the example illustrated in Fig. 9c. The plot is representing the progressive adhesion of ab-gentamicin above the functionalized area of the spot. Starting from the moment when the ab solution begins to flow above the surface, the curve is well described by the kinetic law $S(t) = K(1 - e^{-t/T})$, where K is directly proportional to the molecular concentration of the flowing solution and T is the average binding time that a molecule in the solution needs to interact and adhere to the surface. Usually and for practical use T is in the order of a few minutes. Mixing a liquid sample with the antibody solution in a competitive assay (based on a competition for the antibody binding sites between the free antigen existent in the the antigen immobilized on the surface) gentamicin concentrations down to 20 ppb can be detected.

It is worth to notice that, despite the similar mathematical expression, there is no relationship between the field decay described before and this kinetically driven diffusion process. It is also important that the biorecognition event is highly specific and efficient in order to obtain a fast and reliable detection of the desired analytes and/or markers [9–11]. This is opening a huge and very active research field concerning this kind of biochemical interactions which is far beyond the scope of the present manuscript. However, the presented technique offers a unique possibility of exploring these problems and evaluating affinity or interference effects among similar classes of proteins.

6 Conclusions

The use of the plasmonic response of nanostructured hybrid polymer/metal systems appears to be very favorable for detection purposes and, in particular, for biosensor applications. The use of a periodic array offers the possibility of simplifying the optical configuration and implement imaging (and multiplexing) capabilities. New developments can be envisaged by the exploitation of field enhancement effects in SERS or fluorescence applications.

Acknowledgements The research described above has been developed with the contribution of several students and researchers. The authors are indebted to Dr Francois Rossi and Dr Pascal Colpo of the Joint Research Center of the European Commission in Ispra for their support and collaboration. Dr. Gerardo Marchesini offered his competences in biotechnological problems and as a manager of Plasmore s.r.l. Silvio Vendrame, Rita Therishod, and Alessio Carmine Scipione gave their contribution during the preparation of their diploma thesis. Francesco Floris is now taking the heritage of the past work to explore new effects and develop new applications of the plasmonic systems in his PhD thesis work. All of them deserve the author acknowledgements.

References

1. A. Qureshi, Y. Gurbuz, J.H. Niazi, Sens. Actuators B **171–172**, 62–76 (2012)
2. A.P.F. Turner, Chem. Soc. Rev. **42**(8), 3184–3196 (2013)
3. K.R. Rogers, Anal. Chim. Acta **568**(1–2), 222–231 (2006)
4. B. Van Dorst et al., Biosen. Bioelectron. **26**(4), 1178–1194 (2010)
5. Y.B. Shin et al., Sens. Actuators B Chem. **150**(1), 1–6 (2010)
6. X. Fan et al., Anal. Chim. Acta **620**(1–2), 8–26 (2008)
7. E.A. Smith, R.M. Corn, Appl. Spectrosc. **57**(11), 320A–332A (2003)
8. E. Ouellet, L. Lund, E.T. Lagally, Methods Mol. Biol. **949**, 473–490 (2013)
9. A. Einhauer, A. Jungbauer, J. Chromatogr. A **921**(1), 25–30 (2001)
10. F. Bretagnol et al., Plasma Processes Polym. **3**, 443–455 (2006)
11. A. Valsesia et al., Anal. Chem. **80**, 7336–40 (2008). doi:10.1021/ac801021z
12. J.N. Anker et al., Nat. Mater. **7**(6), 442–453 (2008)
13. X.D. Hoa, A.G. Kirk, M. Tabrizian, Biosen. Bioelectron. **23**(2), 151–160 (2007)
14. J. Homola, Chem. Rev. **108**(2), 462–493 (2008)
15. H. Šípová, J. Homola, Anal. Chim. Acta **773**, 9–23 (2013)
16. C. Situ et al., Trends Anal. Chem. **29**(11), 1305–1315 (2010)
17. H. Rather, *Surface Plasmons on Smooth and Rough Surfaces and on Gratings*. Springer Tracts in Modern Physics, vol. 111 (Springer, Berlin/Heidelberg, 1988)
18. W.L. Barnes, J. Opt. A Pure Appl. **8**(4), S87–S93 (2006)
19. E. Kretschmann, H. Raether, Z. Naturforsch. **23A**, 2135–2136 (1968)
20. A. Boardman, *Electromagnetic Surface Modes* (Wiley, New York, 1982)
21. S.A. Maier, *Plasmonics: Fundamentals and Applications* (Springer Science+Business Media LLC, New York, 2007)
22. J. Parsons et al., Phys Rev B **79**, 073412 (2009). doi:10.1103/PhysRevB.79.073412
23. M. Piliarik et al., Opt. Express **20**(1), 672–680 (2012)
24. K.A. Willets, R.P. Van Duyne, Annu. Rev. Phys. Chem. **58**, 267–297 (2007)
25. T.W. Teperik et al., Opt. Express **14**, 11964–11971 (2006)
26. G. McNay et al., Appl. Spectrosc. **65**(8), 825–837 (2011)
27. K. Kneipp, M. Moskovits, H. Kneipp, *Surface-Enhanced Raman Scattering Physics and Applications* (Springer, Berlin, 2006)
28. J.D. Caldwell et al., ACS Nano **5**(5), 4046–4055 (2011)
29. M. Najiminaini et al., Appl. Phys. Lett. **100**, 043105 (2012). doi:10.1063/1.3679173
30. J.P. Monteiro et al., Sens. Actuators B Chem. **178**, 366–370 (2013)
31. M.J. Kofke et al., Appl. Phys. Lett. **94**, 023104 (2009). doi:10.1063/1.3067835
32. A. Lesuffleur et al., Appl. Phys. Lett. **90**, 243110 (2007). doi:10.1063/1.2747668
33. W.A. Murray, S. Astilean, W.L. Barnes, Phys Rev B **69**, 165407 (2004). doi:10.1103/PhysRevB.69.165407C
34. A.G. Brolo et al., Langmuir **20**(12), 4813–4815 (2004)
35. S. Scarano et al., Biosen. Bioelectron. **25**(5), 957–966 (2010)

36. A. De Leebeeck et al., Anal. Chem. **79**(11), 4094–4100 (2007)
37. H. Im et al., Anal. Chem. **84**(4), 1941–1947 (2012)
38. T.W. Ebbesen et al., Nature **391**, 667–669 (1998). doi:10.1038/35570
39. F.J. Garcia-Vidal et al., Rev. Mod. Phys. **82**, 729–787 (2010)
40. K.J. Klein Koerkamp et al., Phys. Rev. Lett. **92**, 183901 (2004). doi:10.1103/PhysRevLett.92.183901
41. K.L. Van der Molen et al., Phys Rev B **72**, 045421 (2005). doi:10.1103/PhysRevB.72.045421
42. J. Martinez-Perdiguero et al., Procedia Eng. **47**, 805–808 (2012)
43. N.C. Lindquist et al., Ann. Phys. **524**(11), 687–696 (2012)
44. S. Giudicatti et al., Phys. Status Solidi A **207**(4), 935–942 (2010)
45. C. Jian et al., Appl. Surf. Sci. **270**, 6 (2013)
46. A. Valsesia et al., J. Phys. D Appl. Phys. **40**(8), 2341 (2007)
47. C.M. Hsu et al., Appl. Phys. Lett. **93**, 133109 (2008)
48. S. Giudicatti et al., J. Opt. Soc. Am. B **29**(7), 1641–1647 (2012)
49. S. Giudicatti, F. Marabelli, P. Pellacani, Plasmonics **8**(2), 975–981 (2013)
50. A. Taflove, S.C. Hagness, *Computational Electrodynamics—The FDTD Method*, 2nd edn. (Artech House, Norwood, 2000). A commercial software, Lumerical FDTD (www.lumerical.com), was used
51. B. Bottazzi et al., J. Biomed. Opt. **19**, 017006 (2014)
52. T. Campbell, G. Kim, Biomaterials **28**(15), 2380–2392 (2007)
53. Xie et al., Solid State Commun. **150**, 2162–2167 (2010). doi:10.1016/j.ssc.2010.09.004

Colloidal Photonic Crystals for Active Laser Applications

Seiichi Furumi

Abstract Interest in photonic crystals (PCs) is increasing from both the scientific and technological viewpoints of photonics research because they have intriguing prospects for manipulating the flow of light at will. So far, there have been many reports on a wide variety of methodologies with regard to the fabrication of 3D-PC structures by top-down processing such as lithographic techniques. In this context, colloidal crystals (CCs)—highly ordered 3D architectures of colloidal particles of polymers, silica, and so forth—have received tremendous interest as one of the alternative and facile fabrication techniques of 3D-PCs. The monodispersed microparticles have an intrinsic capability to assemble well-ordered structures on the substrate by bottom-up processing. When a particle diameter in the CCs corresponds to several hundred nanometers of the light wavelength, the photonic bandgaps (PBGs) can be visualized as Bragg reflection colors.

This chapter describes an overview of recent developments in fabrication and uses of CC structures of organic and polymer materials for active laser applications. When light-emitting materials are combined in the CCs, the stimulated laser action at PBG band edge(s) or within the PBG wavelength can be generated by optical excitation. Moreover, the optically excited laser action can be controlled by external stimuli due to the self-organization of organic and polymer materials in the CC structures. This chapter highlights not only the research backgrounds of CC structures as PCs, but also the experimental results of their active laser applications. We believe that a wide variety of CC structures will have leading roles in the next generation of photonic devices of organic and polymer materials.

Keywords Colloidal crystals • Photonic band-gap • Emission • Laser • Polymers • Flexible photonic crystals

S. Furumi (✉)
Faculty of Science, Department of Applied Chemistry, Tokyo University of Science (TUS), 1-3 Kagurazaka, Shinjuku-ku, Tokyo 162-8601, Japan

Applied Photonic Materials Group, National Institute for Materials Science (NIMS), 1-2-1 Sengen, Tsukuba, Ibaraki 305-0047, Japan
e-mail: furumi@rs.tus.ac.jp

D. Comoretto (ed.), *Organic and Hybrid Photonic Crystals*,
DOI 10.1007/978-3-319-16580-6_17

1 Introduction

Light–matter interactions at ground state and electronically excited state are of paramount importance not only in natural phenomena such as photosynthesis, but also in optoelectronic devices such as organic light-emitting diodes, organic light-emitting transistors, and molecular photovoltaic cells [1]. Therefore, the photophysical processes in organic condensed matters have attracted long-standing interest from standpoints of both fundamental science and technological development (see 1st, 11th, 14th, and 15th chapters). Recently, the stimulated emission from organic and polymer solid states is a significant research topic for exploitation in new laser applications [2–5]. Such organic and polymer lasers have intrinsic unique capabilities such as on-demand tuning of laser peak wavelength, mechanical flexibility of laser-cavity structures, and simple fabrication processing at low-cost. Numerous efforts have been devoted to the research development of light-emitting organic materials with highly optical gains as well as laser-cavities with high optical quality in order to efficiently generate the stimulated laser emission, aiming at the fabrication of low-threshold organic solid-state lasers. Over four decades ago, Kogelnik and Shank have demonstrated the optically excited laser emission from a light-emitting gel film containing a Rhodamine 6G dye with a periodic interference structure inscribed by a holographic technique [6]. This kind of laser is generally referred to as the distributed feedback (DFB) laser (see fourth chapter) [7]. Following the first demonstration of organic DFB lasers, there have been heretofore many reports on various laser-cavity structures, such as the distributed Bragg reflector (DBR) [8], micro-disks [9], and micro-ring structures [10], as represented in Fig. 1. In order to reduce the threshold excitation energy for laser action, emitted photons from light-emitting materials should be confined within micro-architectures. Photonic crystals (PCs) can serve as tailored dielectric environments in which light propagation is modified by the periodic spatial micro-structures of different dielectric materials, thereby leading to significant increase in the light–matter interactions [11, 12]. Typically, there are forbidden regions for photons appeared in the dispersion spectrum. Such spectral regions are known as photonic bandgaps (PBG). So far, many reports have made on a wide variety of fabrication methodologies of 3D-PCs through the top-down processing (see ninth chapter).

Among various 3D-PC structures, colloidal microparticles with a diameter of several hundred nanometers are most excellent materials for the easy fabrication of 3D-PCs. This is because they have the intrinsic ability to accumulate the 3D ordered micro-architectures on a flat surface by the bottom-up processing of colloidal microparticles [13–19]. Such 3D ordered micro-architectures are the so-called colloidal crystals (CCs). As shown in Fig. 2a, the CCs can be recognized as 3D-PCs from a standpoint of the PC research field (see 2nd, 5th, and 19th chapters). On the basis of the filling ratio of colloidal microparticles, such CC structures can be classified into two types: closed-packed CCs and non-closed-packed CCs, as shown in Fig. 2b. One of the most notable properties of 3D-PCs is that they have an

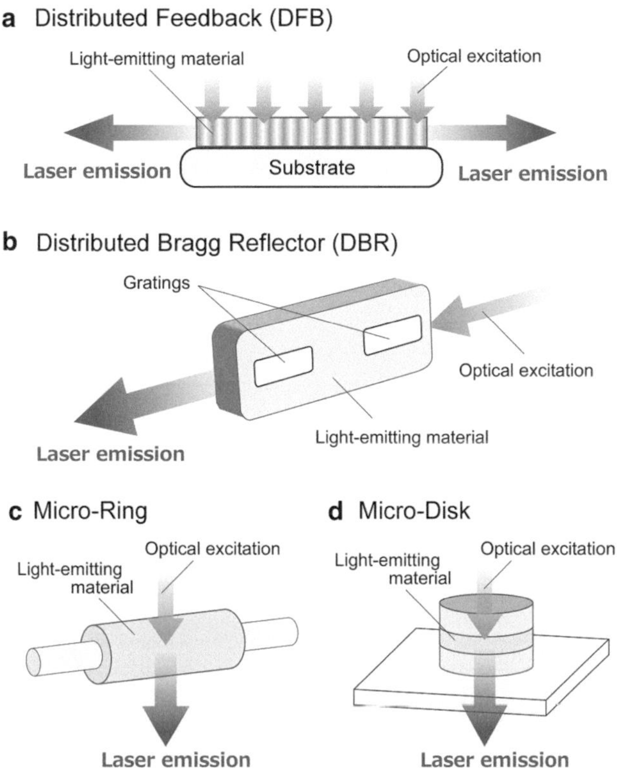

Fig. 1 Illustrative representation of various laser-cavity structures embedded with light-emitting organic and polymer materials: (**a**) the distributed feedback, (**b**) distributed Bragg reflector, (**c**) micro-ring, and (**d**) micro-disk structures

outstanding potential to perfectly forbid light propagation within a selected wavelength range of PBG and in any direction by means of the complete PBG (or full PBG) [20]. From a theoretical viewpoint, the CCs have potentiality to exhibit the complete PBG by adjusting high contrast of optical refractive indices between the microparticles and backgrounds. Research interest in self-assembled CCs is currently intensified with respect to both fundamental science and applied technologies. This is because no specific facilities are required for fabrication of CCs as 3D-PCs due to the bottom-up processing. Since the first example of artificial CC structures by Lopez and colleagues [21], intriguing attempts have established a wide variety of fabrication procedures of uniform CC structures in straightforward ways [13–19]. Thus, the CCs have been the focus of considerable effort owing to high-throughput fabrication techniques of large-size 3D-PCs at low-cost toward the practical applications of photonic devices.

When the colloidal microparticle has a diameter equal to several hundred nanometers, uniform CCs allow the PBG to be visualized as reflection colors.

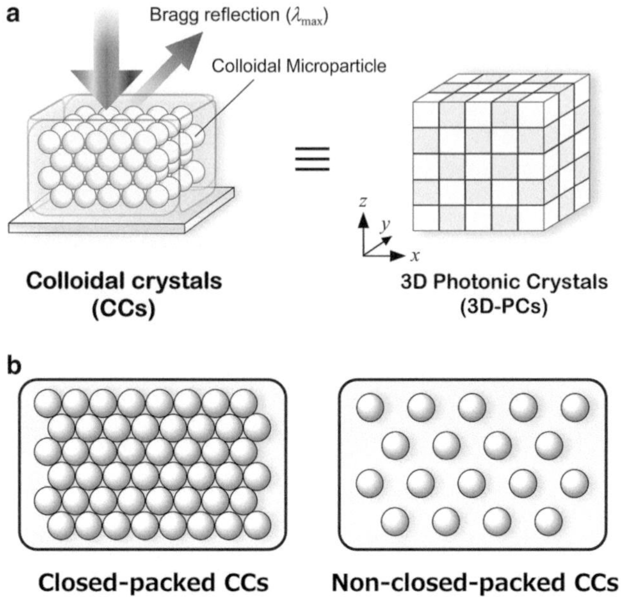

Fig. 2 (**a**) Illustrative representation of 3D micro-architectures of colloidal microparticles, which are so-called as colloidal crystals (*CCs*). Such CC structures can be regarded as 3D photonic crystals (*3D-PCs*). (**b**) Illustrative representation of two types of CCs: closed-packed CCs (*left*) and non-closed packed CCs (*right*)

This is Bragg reflection. The current research trend in CC devices is placed on the development of stimuli-responsive soft CC structures with functional organic and polymer materials that shift the reflection peak wavelength according to the changes in external stimuli such as temperature, electric field, and so on. Such reflection spectral changes can be practically applied to optical reflection sensors and display devices. Interestingly, Asher and colleagues have successfully developed colloidal crystal polymer hydrogels detecting quantitatively environmentally toxic and clinical substances [22]. Prolonged challenges for research and development of tunable CCs should realize a diversification of new devices for next-generation optoelectronics.

This chapter deals with the recent advancements in research on CCs for active lasers by organic and polymer materials. The following sections introduce the chemical synthesis of monodispersed colloidal microparticles, fabrication, and optical properties of CCs, and soft and active laser applications by utilizing CCs combined with functional organic and polymer materials. The main topics of this chapter include that one is the laser emission from the flexible cavity structures fabricated by all-polymer CCs and another is the widely tunable laser emission from a dye-doped CC film permanently stabilized by a polymer gel with an ionic liquid (IL). The final section mentions a conclusion of this article and discusses the future prospects of soft and active CC laser devices.

2 Chemical Synthesis of Microparticles and Fabrication of CCs

In order to fabricate the uniform CC structures, monodispersed colloidal microparticles as the building blocks of CCs are key components (see fifth chapter). It is plausible that the diameter and distribution of the synthesized colloidal microparticles are critical parameters in the fabrication of highly optical quality CC structures. Previously, the research progress in colloidal chemistry has enabled to synthesize monodispersed colloidal microparticles with truly spherical shape and diameters of several micrometers or less. From a material viewpoint, colloidal microparticles can be divided into inorganic and polymer microparticles.

Among inorganic microparticles, uniform colloidal silica microparticles are one of the most frequently investigated examples due to relatively simple synthetic procedures. Typically, the silica microparticles can be obtained by precipitation reactions involving the two steps of nucleation and growth. Stöber and co-workers have succeeded in for the first time a facile synthesis procedure for through sol-gel chemistry of tetraethyl orthosilicate (TEOS) [23]. The microparticle diameter can be controlled over a wide range from several tens of nanometers to several micrometers by changing the concentration of TEOS. Furthermore, there have been many reports on synthesis procedures for other functional inorganic microparticles, such as magnetite, titania, and zinc sulfide [13–19]. Inorganic particles are appropriate for stable CCs due to the chemical resistance and thermal durability.

In contrast, polymer microparticles have several advantages over inorganic microparticles as the following points. For example, monodispersed microparticles of polystyrene (PS) and poly(methyl methacrylate) (PMMA) can readily be synthesized by emulsion polymerization. In addition, they can be functionalized by mixing functional organic materials such as fluorescence dyes and magnetic nanomaterials in the monomer. Furthermore, the physical properties can be controlled by adopting the appropriate monomers. Addition of crosslinkable divinylbenzene monomer during the emulsion polymerization of styrene monomer enables the enhancement of thermal stability due to increase in the glass transition temperature of chemically crosslinked poly(styrene-*co*-divinylbenzene).

Figure 3 shows the experimental results of monodispersed PS particles with a diameter of several hundred nanometers by emulsion polymerization in the presence of sodium dodecyl sulfate (SDS) as a surfactant due to the micelle aggregation [24]. We have successfully synthesized monodispersed PS microparticles with diameters of ca. 200 nm by emulsion polymerization of styrene monomer with SDS [25, 26]. Figure 3a shows scanning electron microscopy (SEM) images of discrete single PS microparticles synthesized by emulsion polymerization with various SDS concentrations. As shown in images, all of the PS microparticles were almost spherical with a diameter of ca. 200 nm. Emulsion polymerization of styrene monomer in a range of SDS concentration from 0.67 to 1.5 g/l produced the PS microparticles with diameters from 195 to 260 nm [abbreviations: PS-(A)–PS-(F)]. Figure 3b is the changes in the PS microparticle diameter

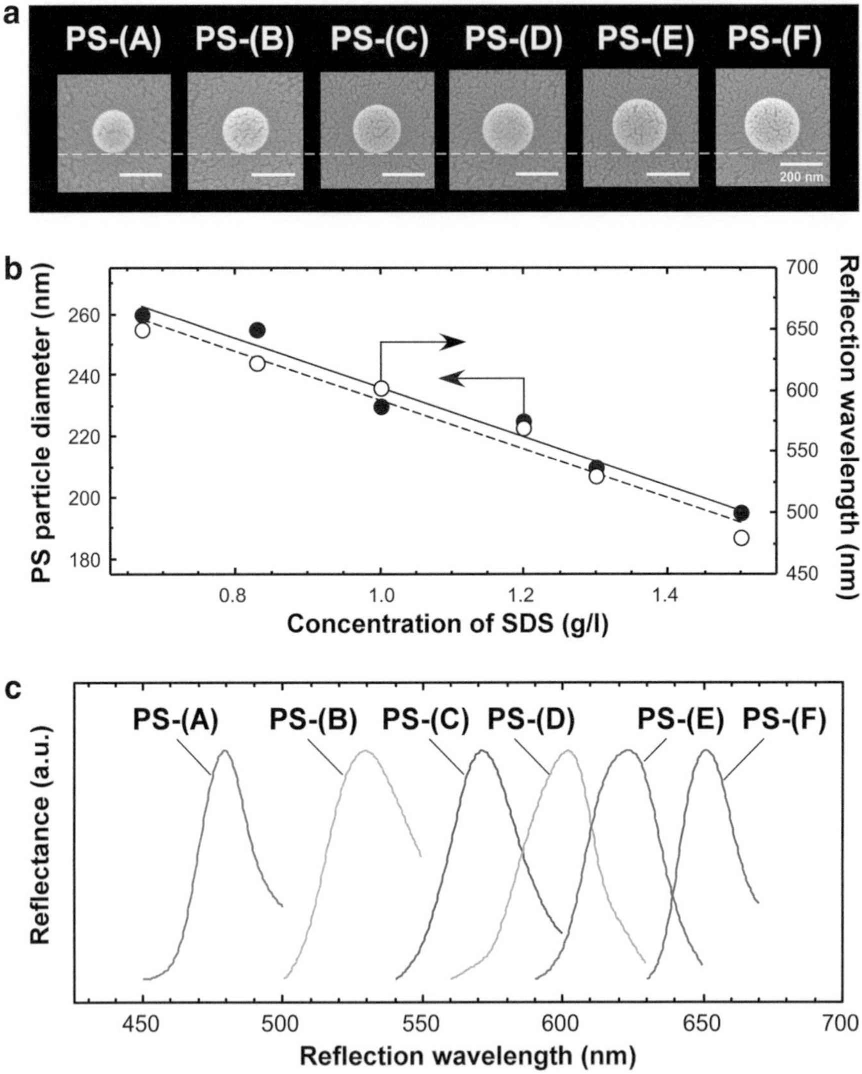

Fig. 3 (**a**) Scanning electron microscopy (*SEM*) images of discrete single polystyrene (*PS*) microparticles. These PS microparticles were synthesized by emulsion polymerization with sodium dodecyl sulfate (*SDS*) in the concentration range from 0.67 to 1.5 g/l, which resulted in PS-(A)–PS-(F) microparticles with diameters ranging from 195 to 260 nm. All scale bars correspond to 200 nm. (**b**) Changes in diameter of synthesized PS microparticles as a function of the SDS concentration of emulsion polymerization (*blue circles*). When the CC films were fabricated by the synthesized PS microparticles of PS-(A)–PS-(F), the reflection peak wavelengths observed from the CC films were also dependent on the SDS concentration (*red circles*). (**c**) Reflection spectra of the CC films prepared with PS-(A)–PS-(F) microparticles. Reproduced with permission from [25]. Copyright 2013, Royal Society of Chemistry

as a function of concentration of SDS in emulsion polymerization, suggesting that the microparticle diameter is strongly depended on the stoichiometric concentration of SDS. We found that the microparticle diameter is inversely proportional to the SDS concentration. Subsequently, we fabricated uniform CC films by using the synthesized PS-(A)–PS-(F) microparticles. Figure 3c shows reflection spectra of the CC films embedded in a poly(dimethyl siloxane) (PDMS) matrix. The CC films of PS-(A)–PS-(F) exhibited characteristic reflection bands, originating from the PBGs, in a wide visible wavelength range from 475 to 650 nm. Therefore, the reflection observed from CC films could be visualized as blue, green, and red colors. In addition, the reflection peak wavelengths are dependent on the diameters of PS microparticles because the maximum reflection wavelengths of CC films have a linear correlation with the microparticle diameter. Accordingly, the reflection peak wavelengths varied inversely with the SDS concentration (red circles, Fig. 3b). In this way, uniform CC films can be easily prepared with PBG reflection wavelengths in an entire visible region from 400 to 800 nm—from violet to red reflection colors—by using PS microparticles with diameters from 150 to 300 nm.

The following sections highlight research topics of soft and active CC lasers by utilizing functionalities of organic and polymer materials.

3 Laser Emission from CCs with Light-Emitting Planar Defects

Previously, we found a new potential use for high-quality polymer CCs to generate low-threshold laser emission by optical excitation [27]. Figure 4 illustrates the conceptual structure of laser-cavity by using polymer CC films. We designed and fabricated a CC laser-cavity with a light-emitting planar defect between a pair of polymer CC films. As materials in the light-emitting planar defect, we used a Rhodamine 640 dye (Rh) as a light-emitting material and a relatively low-molecular weight poly(ethylene glycol) diacrylate (PEG-DA) as a polymer matrix. A uniform CC film of PS microparticles with a diameter of ca. 200 nm was stabilized by an elastic PDMS matrix. After we injected the fluid mixture of Rh, PEG-DA and photopolymerization initiator between a pair of the CC films, a durable CC laser-cavity with the light-emitting planar defect was prepared by photopolymerization of PEG-DA. In fact, the intermediate light-emitting planar defect of Rh/PEG-DA was localized between a pair of CC films by the SEM observation from a cross-sectional direction (right picture, Fig. 4).

Subsequently, we evaluated the laser emission characteristics from this CC laser-cavity by optical excitation with second-harmonic generation (SHG) light at 532 nm from a pulsed neodymium:yttrium aluminium garnet (Nd:YAG) laser beam. Figure 5a shows the changes in emission spectra before and after the optically excited laser emission. At a relatively low excitation energy of ca. 70 nJ/pulse, we observed partial modification in a broad fluorescence spectrum

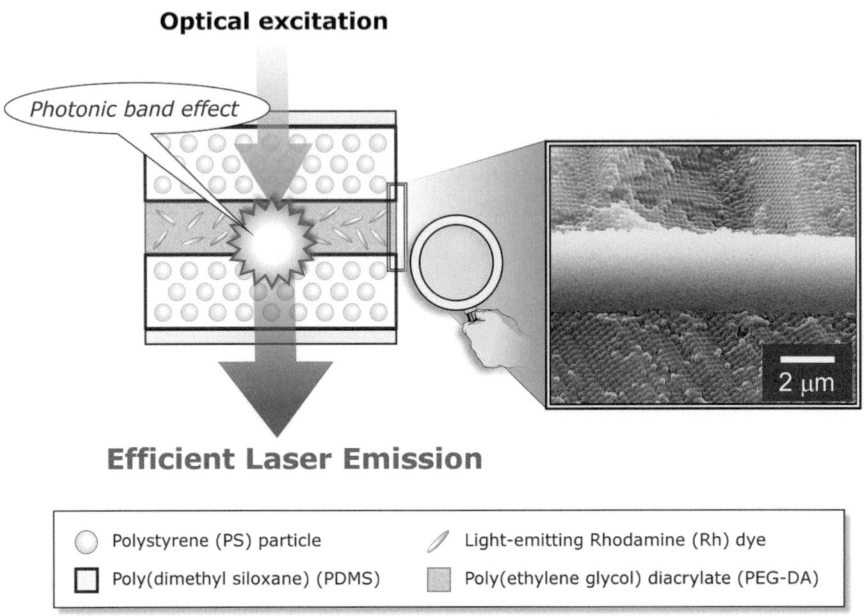

Fig. 4 Conceptual representation of a flexible CC laser-cavity structure, which consists of a light-emitting planar defect embedded between polymer CC films. By optical excitation of the CC laser-cavity, a single and very narrow laser emission peak appeared at the low-threshold excitation power. The right-hand picture shows a cross-sectional SEM image of the magnified light-emitting planar defect between the CC films. The scale bar corresponds to 2 μm. Reproduced with permission from [27]. Copyright 2007, Wiley-VCH

of Rh/PEG-DA (blue curve, Fig. 5a). Such anomalous emission stems from strong localization of the emitted photons within the reflection band of CC films. When the optical excitation energy exceeded ca. 210 nJ/pulse, the broad emission spectrum was suddenly changed to be a single laser emission peak at approximately 611 nm (red curve, Fig. 5a). At this stage, the laser emission peak showed an emission spectral linewidth as narrow as 0.17 nm. The high-resolution emission spectrum is shown in the inset of Fig. 5a. Furthermore, by using an original optical microscopy system, we also observed microscopic emission images before and after the optically excited laser emission. As can be seen in Fig. 5b, we found drastic color changes from pale to glaring orange in the emission images.

In this study, the threshold excitation energy to generate the laser emission was ca. 130 nJ/pulse. The pulse duration and the focused diameter of the excitation beam were set at 3 ns and 200 μm, respectively. Therefore, the lasing-threshold excitation peak power was estimated to be ca. 140 kW/cm^2. In order to compare quantitatively this threshold excitation power with those of the previous reports, the laser characteristics of various CC laser-cavities fabricated by the other groups [28–32] and our group [27, 33] are compiled in Table 1. As evident from this list, this threshold power of 140 kW/cm^2 is two orders of magnitude lower than that of

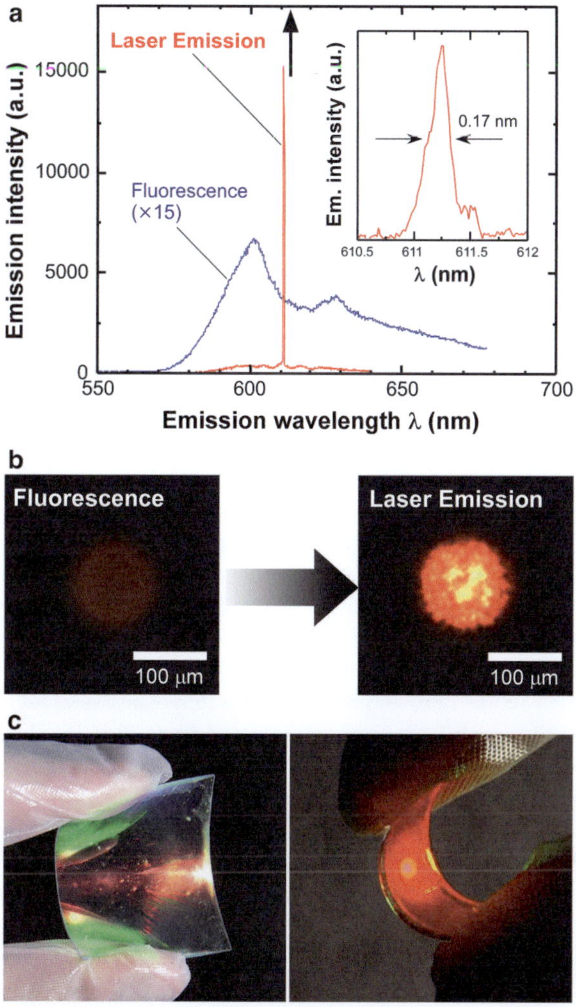

Fig. 5 (**a**) Changes in emission spectra from the CC laser-cavity by optical excitation with a pulsed 532 nm light with an energy of 70 nJ/pulse (*blue curve*) and 210 nJ/pulse (*red curve*). The intensity of the emission spectrum by optical excitation with 70 nJ/pulse is magnified 15 times to confirm the spectrum shape (*blue curve*). The *black arrow* denotes the theoretical defect-mode wavelength estimated by the SWA technique. The *inset* of this figure represents the high-resolution emission spectrum around 611 nm. (**b**) Changes in microscopic emission color images before (*left*) and after laser emission (*right*). The scale bar corresponds to 100 μm. (**c**) Photographs of a flexible all-polymer CC film on a PET sheet (*left*) and laser emission from a lithe CC laser-cavity by optical excitation (*right*). As evident from the *right* picture, we successfully demonstrated the optically excited laser emission when the laser-cavity was bent under mechanical stress. Reproduced with permission from [27]. Copyright 2007, Wiley-VCH

Table 1 Comparison of laser emission characteristics of colloidal crystal (CC) laser-cavities fabricated by the other researchers [28–32] and our group [27, 33]

Ref. No.	Researchers	Colloidal crystals (particle/background materials)	Threshold powers to be laser emission	λ ($\Delta\lambda$)	*Q* value
[28]	Vardeny et al.	Silica/Air	11 GW/cm^2	688 nm (0.2 nm)	3,400
[29]	Foulger et al.	PS/PEG	45 MW/cm^2	613 nm (2 nm)	310
[30]	Duan et al.	Silica/Air	24 MW/cm^2	589 nm (1.7 nm)	350
[31]	Yamada et al.	Silica/PAm gel	25 MW/cm^2	590 nm (0.6 nm)	4,000
[32]	Yang et al.	Silica/ETPTA	600 kW/cm^2	606 nm (3.5 nm)	170
[27]	Furumi et al.	PS/PDMS	140 kW/cm^2	611 nm (0.17 nm)	3,600
[33]	Furumi et al.	PS/PAm gel	4.5 MW/cm^2	653 nm (0.06 nm)	10,800

Note: λ laser emission wavelength, $\Delta\lambda$ spectral linewidth at laser emission wavelength, *PS* polystyrene, *PEG* poly(ethylene glycol), *PAm gel* poly(acrylamide)gel, *ETPTA* ethoxylated trimethylolpropane triacrylate, *PDMS* poly(dimethylsiloxane)

the previously investigated CC systems [28–32]. This experimental fact implies that the emitting photons from the planar defect is effectively confined in the orthogonal direction of the CC laser-cavity due to the PBG effect of CC films, leading to the laser emission by low-threshold optical excitation.

The optical mechanism to generate the laser emission from this CC laser-cavity can be rationalized in terms of the defect-mode localized in the PBG range of CC film. Such defect-mode laser emission is definitely consistent with several precedents for 1D-PCs with light-emitting planar defects [34–36]. In this CC laser-cavity, the theoretical defect-mode wavelength can be calculated on the basis of the scalar-wave approximation (SWA) technique [37]. The theoretical defect-mode wavelength was found to be 613 nm (black arrow, Fig. 5a). This theoretical defect-mode wavelength is in fairly good agreement with the laser emission wavelength experimentally observed from this CC laser-cavity structure.

Taking account of the fact that the polymer materials exhibit mechanical flexibility (see sixth and seventh chapters), we prepared a lithe CC laser-cavity structure by using all-polymer materials. A CC film of PS microparticles stabilized by a PDMS elastomer was formed on a poly(ethylene terephthalate) (PET) sheet (left picture, Fig. 5c). Therefore, we successfully demonstrated the optically excited laser emission from the all-polymer CC laser-cavity even at the bent shape (right picture, Fig. 5c). From technological viewpoints, it would be greatly advantageous to be able to manufacture the all-polymer laser devices with such intrinsic features as flexibility, easy processability, ultra-light weight, low-cost, and so forth [38–40].

As mentioned above, the potential use of CCs to generate the laser emission by low-threshold optical excitation has become apparent. As an extension of our findings, we have recently developed the dynamic photoswitching of micropatterned laser emission from a CC laser-cavity by the photochromic reaction [41]. As shown in Fig. 6a, this CC laser-cavity structure comprises a photochromic light-emitting planar defect of both photochromic 1,2-bis[2-methylbenzo[*b*]thiophen-3-yl]-3,3,4,4,5,5-hexafluoro-1-cyclopentene (BMTH) and light-emitting

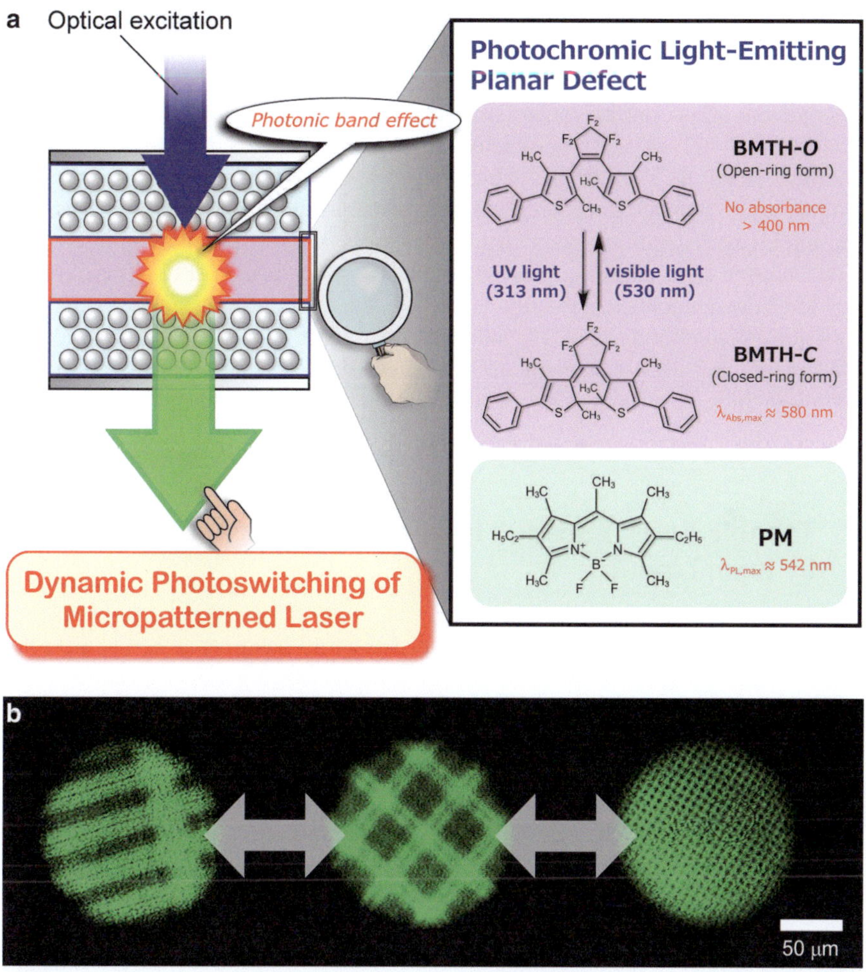

Fig. 6 (**a**) Conceptual representation of reversible photoswitching of a micropatterned laser emission. The CC laser-cavity is composed of a photochromic light-emitting planar defect between a pair of CC films. The right-hand figure shows chemical structures of the photochromic light-emitting planar defect between a pair of CC films. We adopted a pyrromethene complex (*PM*) as a light-emitting material, and a photochromic diarylethene derivative (*BMTH*) was used as a photochromic material. (**b**) Micropatterned laser emission images from a CC laser-cavity that was beforehand irradiated with 313 nm light through various photomasks. The scale bars correspond to 50 μm. Reproduced with permission from [41]. Copyright 2012, Royal Society of Chemistry

1,3,5,7,8-pentamethyl-2,6,-diethylpyrromethene-difluoroborate complex (PM) complex (see 18th chapter). Appropriate addition of BMTH with PM leads the reversible photoswitching of a broad fluorescence emission in intensity of PM by the photochromic reaction of BMTH between open-ring and closed ring form [42].

Interestingly, when the CC laser-cavity was alternately irradiated with 313 and 530 nm light, a single laser emission peak could dynamically switched by the photochromic reaction of BMTH. As the most important performance, we could observe the high-contrast micropatterned laser emission images with a high-resolution of only a few micrometers. Figure 6b shows the various images of micropatterned laser emission, which were reversibly erased and rewritten by alternating photoirradiation with 313 and 530 nm light through photomasks. We can envisage that unique photoswitchable lasers are realized by combining the photochromic light-emitting materials in the 1D-, 2D-, and 3D-PCs. These results provide new guidelines not only to realize the on-demand switching of laser emission by combining photochromic and light-emitting materials in a variety of PC systems, but also to manufacture the small-sized and high-density all-optical photoswitching devices in next-generation optoelectronics.

4 Tunable Laser Emission from CC Gel Films Stabilized by Ionic Liquid

Soft CC films with polymer hydrogels are known to have capability to tune the reflection peak wavelength by external stimuli such as temperature, pH, electronic field, and so on [13–19]. Typically, the void space between the 3D-ordered colloidal microparticles in the CC film is filled with stimuli-responsive polymer hydrogels. According to this research concept, Asher and colleagues have developed for the first time a soft CC film with a poly(acrylamide) hydrogel matrix with tunability of reflection wavelength by applying mechanical stress [43]. In this system, the mechanical tuning range of reflection wavelength was from 538 to 573 nm. Later, Yamanaka and colleagues have succeeded in a wide tuning of reflection peak from 460 to 810 nm by using charged CC films (see 19th chapter) [44]. These reports have motivated us to demonstrate the tunable laser emission from the soft CC polymer gel films over a wide wavelength range by applying mechanical stress.

For this purpose, we prepared a high-quality CC film of PS microparticles stabilized by a polymer hydrogel. Previously, Sawada and colleagues have serendipitously discovered the facile formation of single-crystal-like CC structures with a long-range orientational order by strong shear flow of colloidal suspension [45]. In this study, a non-close-packed uniform CC polymer hydrogel film of PS microparticles with a diameter of ca. 120 nm was also prepared by shearing a colloidal suspension in a flat capillary cell with a thickness of 100 μm. By adding polymer hydrogel precursors in the colloidal suspension, we fabricated a uniform CC film stabilized by a thermo-insensitive polymer hydrogel of poly(*N*-methylolacrylamide-*co*-N,N′-methylenebisacrylamide) [1], Fig. 7a. As can be seen in Fig. 7b, this CC film stabilized by a polymer hydrogel 1 exhibited very softness and flexibility in an excess of water. Confocal laser scanning microscopy (CLSM) observation enabled clear visualization of the uniform CC structure at the

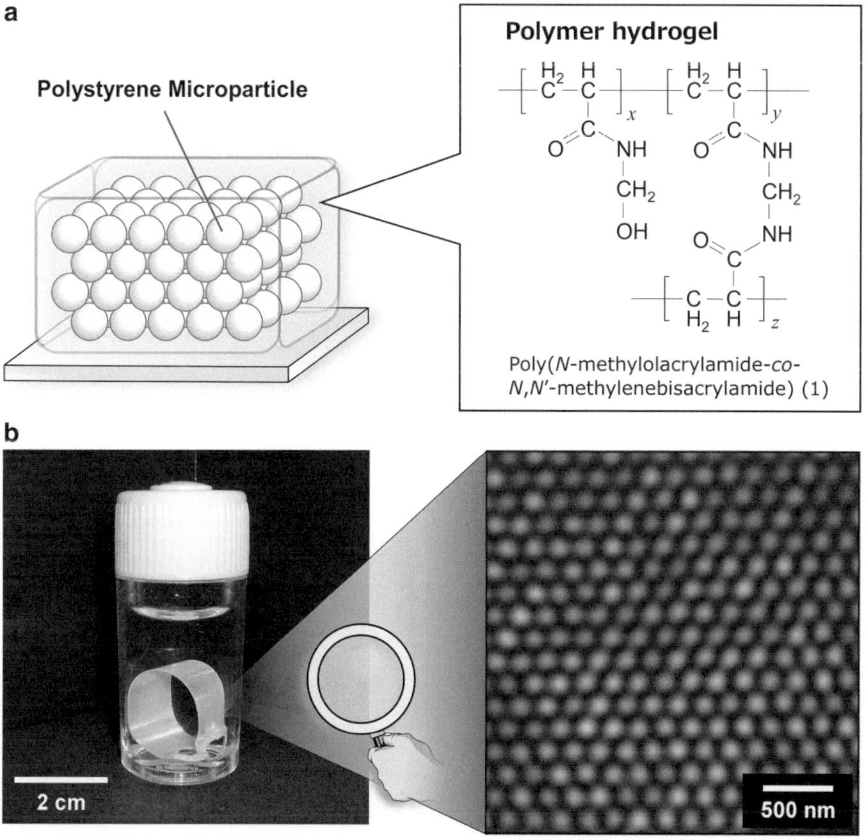

Fig. 7 (**a**) Illustrative representation of a CC hydrogel film, in which the 3D ordered structure of PS microparticles was stabilized by a polymer hydrogel matrix of poly(*N*-methylolacrylamide-*co*-*N*,*N*′-methylenebisacrylamide) (1). (**b**) Photograph of a soft CC hydrogel film prepared by shearing a colloidal suspension between a pair of glass substrates with a gap of 100 μm. (**b**) High-magnification confocal laser scanning microscopy (*CLSM*) image of the CC hydrogel film. Reproduced with permission from [25]. Copyright 2013, Royal Society of Chemistry

microscopic scale. Figure 7b shows a CLSM image of CC hydrogel film, suggesting that the PS microparticles adopt the well-ordered and non-close-packed structure in a polymer hydrogel matrix.

Generally, this type of CC film stabilized by polymer hydrogels is very vulnerable under dry atmospheric conditions. It is easily predictable that the CC polymer hydrogel film is stored in the ambient atmosphere, whereupon Bragg reflection peak completely disappears due to the natural evaporation of water. Such phenomenon is a serious drawback for CC gel films in order to fabricate the photonic devices.

To overcome this technical issue, we have successfully obtained a permanently stable CC polymer gel film for tunable laser applications by adopting a non-volatile ionic liquid (IL) [33]. Figure 8a represents this underlying research concept.

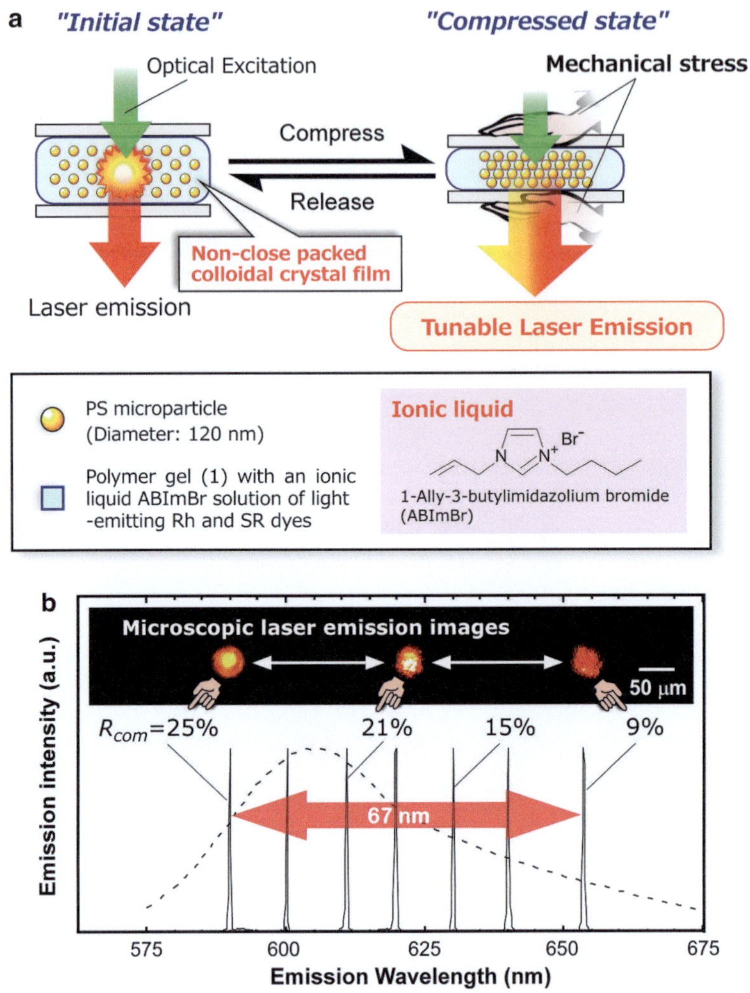

Fig. 8 (**a**) Conceptual representation of a widely tunable laser emission from of a CC gel film. The cavity structure comprises non-close-packed CC gel film of PS microparticles stabilized by a solution of light-emitting Rhodamine 640 (*Rh*) and Sulforhodamine B (*SR*) dyes in an ionic liquid (*ABImBr*). (**b**) Changes in laser emission spectra from a CC gel film with an Rh + SR/ABImBr solution as a function of ratio of mechanical compression. The CC gel film was stepwise compressed at from 9 to 25 % in the film thickness direction. The *dashed curve* denotes a broad fluorescence spectrum of the Rh + SR/ABImBr solution by excitation with a continuous 532 nm light. The *insets* show the microscopic color images of laser emission at 588 nm (*left*), 610 nm (*middle*), and 655 nm (*right*) tuned by mechanical stress. The scale bar correspond to 50 μm. Reproduced with permission from [33]. Copyright 2011, Wiley-VCH

The CC laser-cavity structure and the preparation procedure are very simple. A non-close-packed CC film stabilized by a polymer hydrogel of 1 was immersed in an IL solution of Rhodamine 640 (Rh) and Sulforhodamine B (SR) dyes. As an IL, we used 1-allyl-3-butylimidazolium bromide (ABImBr) with good miscibility with water. Therefore, ABImBr leads to the swelling state of polymer gel of 1 [46]. The CC gel film with Rh and SR could be permanently stabilized by ABImBr even under dried conditions. At initial state, this CC-L gel film showed a reflection band at 690 nm. By applying mechanical stress in the film thickness direction, the reflection band was continuously blue-shifted to 560 nm due to the geometric reduction of distance between PS microparticles. For instance, the reflection band was shifted from 690 to 645 nm when the CC gel film was compressed at ca. 10 % in the film thickness. It is anticipated that the optically excited laser emission is efficiently generated because this reflection band completely overlaps a broad fluorescence spectrum of Rh and SR (dashed curve, Fig. 8b).

We found the single laser emission peak from the CC gel film with Rh + SR/ABImBr by optical excitation. When the excitation energy exceeded a threshold value, the broad fluorescence spectrum was suddenly changed to be the single laser emission. The emission intensity became increasingly intensified, accompanied by an abrupt spectral narrowing from 45 to 0.06 nm [33]. In order to compare quantitatively this laser emission spectral linewidth with those in the precedents, the quality factor (Q) was evaluated from the high-resolution laser emission spectrum. Generally, the Q value can be estimated as the following equation [47].

$$Q = \lambda/\Delta\lambda \tag{1}$$

where λ and $\Delta\lambda$ denote the laser emission wavelength and spectral linewidth, respectively. According to Eq. (1), the Q value of laser emission observed from this CC gel film was estimated to be 1.1×10^4. To our knowledge, this Q value is the highest of various CC structures reported to date, as listed in Table 1 [28–32]. Such a single and narrow lasing peak was achieved by fabricating a high-quality CC gel film.

This single laser emission peak appeared at 653 nm, corresponding to around the longer reflection band wavelength edge of CC gel film. Such laser emission takes place due to resonant enhancement of the density of states (DOS) in the emitting photons near the reflection band edge of the PBG. Taking into account the mechanical tunability of the reflection wavelength of this CC gel film, tuning of the laser emission peak was then attempted by mechanically compressing the CC gel film. Figure 8b shows the changes in laser emission spectra for the CC gel film with Rh + SR/ABImBr as a function of the compression ratio of film thickness. When the compression ratio increased stepwise from 9 to 25 %, the single laser emission peak was continuously blue-shifted from 655 to 588 nm due to the changes in PBG reflection wavelengths. Such changes in laser emission peak were analyzed using the least squares method. It was turned out that this mechanical shift of lasing wavelength obeys the following equation:

$$\lambda_{\text{laser}} = 688 - 3.85R_{\text{com}} \tag{2}$$

where λ_{laser} and R_{com} are the laser emission wavelength and the compression ratio of film thickness, respectively. The correlation coefficient was calculated to be a high value of 0.99. In this way, the laser emission peak can be tuned by mechanical stress according to Eq. (2). In this CC gel film with Rh + SR/ABImBr, the tuning range of laser emission peak was 67 nm. Although there have been hitherto many previous reports on the mechanical tuning of laser emission peak for various cavity structures fabricated by elastic polymers [48–51], this laser emission tuning range presented here is comparatively wide.

Furthermore, we arranged a original optical microscopic system in order to observe color changes of laser emission at the microscopic scale as a function of the compression ratio [52]. As compressing the CC gel film, the laser emission color showed changes from dark-red, reddish-orange, and to orange, corresponding to the laser peak wavelengths at 655, 610, and 588 nm, respectively. It is worth noting that when the compression stress was released, the laser emission peak and color could be reversibly revived to its initial-state due to the intrinsic elastic feature of polymer gel of 1 in this CC gel film.

5 Conclusions and Outlook

This chapter has described the recent progress in active CC lasers combined with functional organic and polymer materials. Colloidal microparticles have a unique capability to spontaneously assemble themselves from a fluid colloidal suspension, resulting in the formation of highly ordered CC structures as 3D-PCs. Lately, the research on CCs has garnered a great deal of attention from both fundamental and technological perspectives. To date, tremendous efforts have established various methodologies to prepare large-scale and highly optical quality CC films at low-cost due to the facile fabrication processes. Concerning the photonic device applications, a wide variety of reflection sensors and displays have also been reported by combining CC films with functional materials that leads to the visualization of various external stimuli as changes in the reflection colors.

As the expansion of photonic applications with CCs, this chapter highlights an outline of new potential for CCs in active organic and polymer lasers that are responsive to external stimuli. We introduce novel utility of polymer CCs as flexible and widely tunable lasers by low-threshold optical excitation. When a CC laser-cavity consists of a light-emitting planar defect between a pair of polymer CC films, optical excitation at a low-threshold power brings about a single and narrow laser emission peak in the PBG wavelength range due to the defect-mode. Moreover, this CC laser-cavity can be fabricated by all-polymer materials. It is possible to generate the optically excited laser action, even when the all-polymer CC-L cavity is bent under mechanical stress. Moreover, the reversible photoswitching of a micropatterned laser emission can be realized using the CC films combined with photochromic and light-emitting organic materials. As another topic of active organic lasers, we have successfully demonstrated the widely tunable laser emission from non-close-packed CC gel films stabilized by non-volatile IL. When a soft CC film is permanently stabilized by a polymer gel with a non-volatile IL solution of light-emitting materials, optical excitation brings about a single and narrow laser emission peak near the longer PBG wavelength due to the band-edge effect. Moreover, application of mechanical stress to the CC film enables the continuous and reversible tuning of laser emission peak across a wide wavelength range due to the geometric reduction of the microparticle distance in CC film. We believe that the research and development of CCs will providc promising breakthroughs for miniaturized and switchable devices combined with functional self-assembling organic materials toward the future generation of optoelectronic devices.

Acknowledgements The author (S. F.) expresses sincere thanks to Prof. D. Comoretto (Università degli Studi di Genova) for planning this book project. S. F. also acknowledges Drs. T. Sawada, H. T. Miyazaki, H. Fudouzi and Y. Sakka and Ms. T. Terui (NIMS) and Prof. T. Kanai (Yokohama National University) for their helpful advices and technical supports. This work was supported in part by the Sekisui Chemical Grant Program for Research, the Strategic Information and Communications R&D Promotion Programme (SCOPE) project from the Ministry of Internal Affairs and Communications (MIC), the Grant-in-Aid for Scientific Research (B) and Young Scientist (A) from the Japan Society for the Promotion of Science (JSPS) and the Precursory Research for Embryonic Science and Technology (PRESTO) Research Program from the Japan Science and Technology Agency (JST).

References

1. N.J. Turro, V. Ramamurthy, J.C. Scaiano, *Modern Molecular Photochemistry of Organic Molecules* (University Science, California, 2010)
2. F. Hide, M.A. Díaz-García, B.J. Schwartz et al., Science **273**, 1833 (1996)
3. N. Tessler, G.J. Denton, R.H. Friend, Nature **382**, 695 (1996)
4. V.G. Kozlov, V. Bulovic, P.E. Burrows et al., Nature **389**, 362 (1997)
5. A. Dodabalapur, E.A. Chandross, M. Berggren et al., Science **277**, 1787 (1997)
6. H. Kogelnik, C.V. Shank, Appl. Phys. Lett. **18**, 152 (1971)
7. C.V. Schank, J.E. Bjorkholm, H. Kogelnik, Appl. Phys. Lett. **18**, 395 (1971)
8. I.P. Kaminov, H.P. Weber, E.A. Chandross, Appl. Phys. Lett. **18**, 497 (1971)
9. S.V. Frolov, M. Shkunov, Z.V. Vardeny et al., Phys. Rev. B **56**, R4363 (1997)
10. M. Kuwata-Gonokami, R.H. Jordan, A. Dodabalapur et al., Opt. Lett. **20**, 2093 (1995)
11. E. Yablonovitch, Phys. Rev. Lett. **58**, 2059 (1987)
12. S. John, Phys. Rev. Lett. **58**, 2486 (1987)
13. Y. Xia, B. Gates, Y. Yin, Y. Lu, Adv. Mater. **12**, 693 (2000)
14. D.J. Norris, E.G. Arlinghaus, L. Meng et al., Adv. Mater. **16**, 1393 (2004)
15. P.V. Braun, S.A. Rinne, F. García-Santamaría, Adv. Mater. **18**, 2665 (2006)
16. Q. Yan, L. Wang, X.S. Zhao, Adv. Funct. Mater. **17**, 3695 (2007)
17. S. Furumi, H. Fudouzi, T. Sawada, Laser Photon. Rev. **4**, 205 (2010)
18. C.I. Aguirre, E. Reguera, A. Stein, Adv. Funct. Mater. **20**, 2565 (2010)
19. J.F. Galisteo-López, M. Ibisate, R. Sapienza et al., Adv. Mater. **23**, 30 (2011)
20. A. Blannco, E. Chomski, S. Grabtchak et al., Nature **405**, 437 (2000)
21. H. Miguez, C. López, F. Meseguer et al., Appl. Phys. Lett. **71**, 1148 (1997)
22. J.H. Holtz, S.A. Asher, Nature **389**, 829 (1997)
23. W. Stöber, A. Fink, E. Bohn, J. Colloid Interface Sci. **26**, 62 (1968)
24. M.E. Woods, J.S. Dodge, I.M. Krieger, J. Paint Technol. **40**, 541 (1968)
25. S. Furumi, J. Mater. Chem. C **1**, 6003 (2013)
26. S. Furumi, Polym. J. **45**, 579 (2013)
27. S. Furumi, H. Fudouzi, H.T. Miyazaki et al., Adv. Mater. **19**, 2067 (2007)

28. M.N. Shkunov, Z.V. Vardeny, M.C. DeLong et al., Adv. Funct. Mater. **12**, 21 (2002)
29. J.R. Lawrence, Y. Ying, P. Jiang et al., Adv. Mater. **18**, 300 (2006)
30. F. Jin, C.-F. Li, X.-Z. Dong et al., Appl. Phys. Lett. **89**, 241101 (2006)
31. H. Yamada, T. Nakamura, Y. Yamada et al., Adv. Mater. **21**, 4134 (2009)
32. S.-H. Kim, S.-H. Kim, W.C. Jeong et al., Chem. Mater. **21**, 4993 (2009)
33. S. Furumi, T. Kanai, T. Sawada, Adv. Mater. **23**, 3815 (2011)
34. R. Ozaki, T. Matsui, M. Ozaki et al., Appl. Phys. Lett. **82**, 3593 (2003)
35. M.H. Song, B. Park, K.-C. Shin et al., Adv. Mater. **16**, 779 (2004)
36. J. Yoon, W. Lee, J.-M. Caruge et al., Appl. Phys. Lett. **88**, 091102 (2006)
37. K.W.-K. Shung, Y.C. Tsai, Phys. Rev. B **48**, 11265 (1993)
38. F.D. Stasio, L. Berti, M. Burger et al., Phys. Chem. Chem. Phys. **11**, 11515 (2009)
39. F. Scotognella, A. Monguzzi, F. Meinardi et al., Phys. Chem. Chem. Phys. **12**, 337 (2010)
40. G. Canazza, F. Scotognella, G. Lanzani et al., Laser Phys. Lett. **11**, 035804 (2014)
41. S. Furumi, H. Fudouzi, T. Sawada, J. Mater. Chem. **22**, 21519 (2012)
42. M. Irie, Chem. Rev. **100**, 1685 (2000)
43. S.A. Asher, J. Holtz, L. Liu et al., J. Am. Chem. Soc. **116**, 4997 (1994)
44. Y. Iwayama, J. Yamanaka, Y. Takiguchi et al., Langmuir **19**, 977 (2003)
45. T. Sawada, Y. Suzuki, A. Toyotama et al., Jpn. J. Appl. Phys. **40**, L1226 (2001)
46. T. Kanai, S. Yamamoto, T. Sawada, Macromolecules **44**, 5865 (2011)
47. J.D. Joannopoulos, P.R. Villeneuve, S. Fan, Nature **386**, 143 (1997)
48. H. Finkelmann, S.T. Kim, A. Muñoz et al., Adv. Mater. **13**, 1069 (2001)
49. M.R. Weinberger, G. Langer, A. Pogantsch et al., Adv. Mater. **16**, 130 (2004)
50. B. Wenger, N. Tétreault, M.E. Welland et al., Appl. Phys. Lett. **97**, 193303 (2010)
51. P. Görrn, M. Lehnhardt, W. Kowalsky et al., Adv. Mater. **23**, 869 (2011)
52. S. Furumi, N. Tamaoki, Adv. Mater. **22**, 886 (2010)

Stimuli-Responsive Self-Organized Liquid Crystalline Nanostructures: From 1D to 3D Photonic Crystals

Ling Wang and Quan Li

Abstract Bottom-up fabrication of photonic nanostructures from self-organizing soft materials with tailored optical properties is currently a burgeoning area of research. The liquid crystals capable of forming self-organized, tunable one-dimensional (1D) or three-dimensional (3D) photonic nanostructures represent such an elegant system because of their unique ability to control the propagation of light by various external stimuli. This chapter is confined to survey the recent progress in the stimuli-responsive self-organized liquid crystalline nanostructures for 1D and 3D photonic applications. For the 1D self-organized liquid crystalline photonic crystals, i.e. cholesteric liquid crystals, they show Bragg reflection of circularly polarized light. Their tunable properties under different stimuli, such as heat, light, electric field, and chemical environment, are mainly introduced. For the 3D self-organized liquid crystalline photonic crystals, we focus on their tunable cubic nanostructures derived from blue phase liquid crystals and tunable spherically symmetric nanostructures based on liquid crystal microdroplets.

Keywords Soft photonic crystal • Self-organized nanostructure • Liquid crystal • 1D and 3D • Stimuli-response

1 Introduction

Liquid crystals (LCs) are known as a class of soft materials with an intermediate state between the solid and liquid phases, which exhibit unique thermal, mechanical, optical, and electrical properties due to the self-organization of mesogenic molecules with the characteristics of both regularity of crystalline solid and the fluidity of isotropic liquid [1]. Nowadays, LCs have become ubiquitous in our daily life, ranging from miniature mobile telephones to high-definition flat-panel displays. Such technologies are based on the reorientation of LC material in response to applied electric field, which results in a change in its observed optical properties.

L. Wang • Q. Li (✉)
Liquid Crystal Institute and Chemical Physics Interdisciplinary Program,
Kent State University, Kent, OH 44242, USA
e-mail: qli1@kent.edu

D. Comoretto (ed.), *Organic and Hybrid Photonic Crystals*,
DOI 10.1007/978-3-319-16580-6_18

The majority of present commercial displays are based on the LCs exhibiting a nematic phase. Introducing chirality into the nematic LCs can lead to novel self-organized helical superstructures, i.e. cholesteric LC or chiral nematic phase. Fundamentally, the key feature that these cholesteric LC materials possess is the existence of a photonic bandgap (PBG) for visible light, even near-infrared light. The photonic band structures have attracted considerable attention in recent years because of their ability to control the propagation of light under various external stimuli, such as light, heat, electric field, magnetic field, and pressure [2, 3]. This PBG is analogous to the separation of electronic bands in semiconductors, and can exist in one-dimensional (1D) cholesteric LCs, or three-dimensional (3D) periodic superstructures, such as blue phases and LC droplets, depending on the number of directions in which the periodicity exists.

1.1 Nematic and Cholesteric Liquid Crystals

The least-ordered LC phase is the nematic, which has only long-range orientational order. In this case, the long axes of the molecules point in the same direction (on average), which is defined by a unit vector known as the director, ***n*** (Fig. 1). The optical axis lies along the same direction as the director, and hence it is optically uniaxial [4]. Owing to the microstructure of the phase, many of the macroscopic properties are anisotropic. For example, nematic LCs exhibit a birefringence, Δn, because the refractive index of the extraordinary ray, n_e (often denoted $n_{\|}$ as it is along the director), is different to that of the ordinary ray, n_o (often denoted $n_{\perp}$ because it is orthogonal to the director), thus $\Delta n = n_e - n_o$.

A variant of the nematic phase is the cholesteric phase, i.e. chiral nematic phase, which spontaneously forms a macroscopic helical structure either when the LC

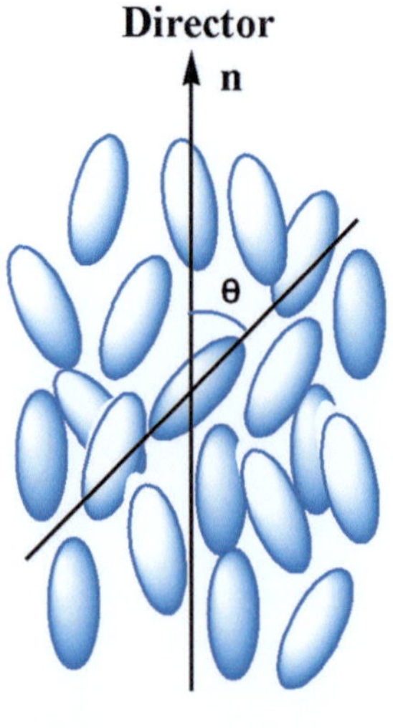

Fig. 1 Schematic drawing of molecular order present in a nematic liquid crystalline phase

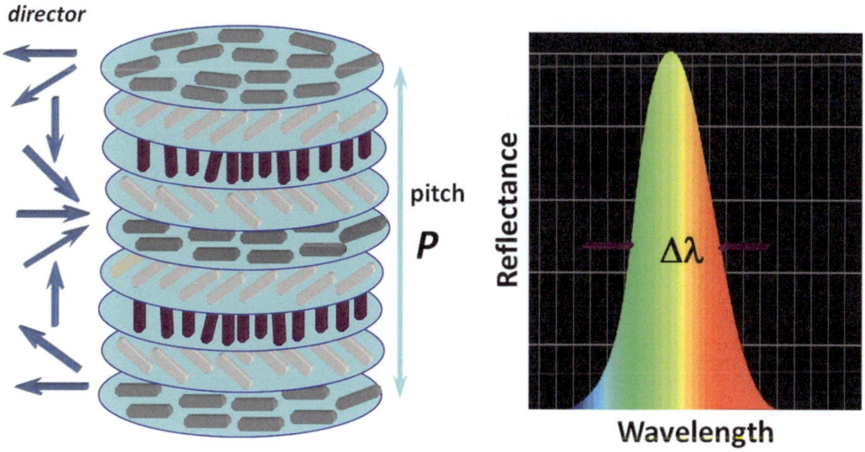

Fig. 2 Illustration of molecular arrangement of cholestric LC and its reflection property

molecules are inherently chiral or when chirality is externally introduced. The presence of chirality causes the director profile to assume a twisted configuration throughout the medium. As shown in Fig. 2, the spatial orientation of the molecules in cholesteric LCs changes in a helical fashion along a specific direction called helical axis. This helical orientation of the director is non-superimposable on its mirror image, thus making the phase chiral. The helical superstructure of cholesteric LCs is characterized by helical pitch and handedness. The pitch (P) is the distance over which the director rotates a full 360°. Handedness describes the direction in which the molecular orientation rotates along the helical axis and by convention it is expressed as sign (−) and (+), which represent left handedness and right handedness, respectively.

The most important optical property of cholesteric LCs is their selective reflection of light according to the Bragg law. The selective reflection wavelength λ is defined by $\lambda = n \times P$, where P is the pitch length of the helical structure and n is the average refractive index of the LCs. The reflection bandwidth $\Delta\lambda$ is given by $\Delta\lambda = \lambda_{\max} - \lambda_{\min} = (n_e - n_o)\ P = \Delta nP$. Within the bandwidth, right-circularly polarized light is reflected by a right-handed helix, while left-circularly polarized light is transmitted; outside the bandwidth both polarization states are transmitted. As is known, cholesteric LCs can be fabricated by two main methods. The first and simplest one is based on chiral mesogens, which can exhibit the cholesteric phase over a certain temperature range. The cholesteric phase produced by this method consists of pure material and thus may exhibit advantages such as good uniformity, enhanced stability, and fast response time to stimuli. However, it suffers from above-ambient-temperature processing conditions and the physical properties sometimes are not suitable for device applications. The second, most often adopted, strategy is to dope chiral molecules into achiral nematic LC, and the resulting mixture can self-organize into a helical superstructure and form a cholesteric phase

[5]. The ability of a chiral dopant to twist an achiral nematic LC phase is defined as helical twisting power (HTP, β) and is quantified as: $\beta = 1/(Pc)$, where c is the chiral dopant concentration. The major advantage of this method is that the pitch of cholesteric phase can be tuned by adjusting the doping concentration of the guest molecule, especially for reflection color tuning that requires the pitch length within the range of hundreds of nanometers. Moreover, a small amount of chiral dopant generates much lower synthetic cost, and the dopant can be mixed with different LC hosts to generate cholesteric LC mixtures with different properties.

1.2 Blue Phase Liquid Crystals

Liquid crystalline blue phases (BPs) are among the most interesting self-organized structures in the field of LCs, which are known to self-organize into periodic cubic lattices. The name "blue phase" is due to the blue appearance of the phase when it was first observed in cholesteryl benzoate by F. Reinitzer. However BPs can in fact have other colors. This phase exists for only a few degree Celsius before the cholesteric material clears to the isotropic liquid. Similar to cholesteric LCs, BPs can be observed either in pure chiral mesogens or can be induced from achiral LCs by adding chiral dopants [6]. The cubic structures are characterized by the double twist cylinders (DTC) which are stabilized by disclinations. Normally when chirality in LC system is high enough, DTC is more stable than a simple twist structure. However, DTCs cannot be densely packed to fill the entire space and disclination lines appear at the interstices among the cylinders to form a stabilized 3D cubic structure (Fig. 3). There are three types of BPs reported, namely BP III, BP II, and BP I, which are usually observed during cooling from the isotropic to the cholesteric phase in the order of decreasing temperature. BPI and BP II have body-centered cubic and simple cubic structures, respectively, while BP III is amorphous

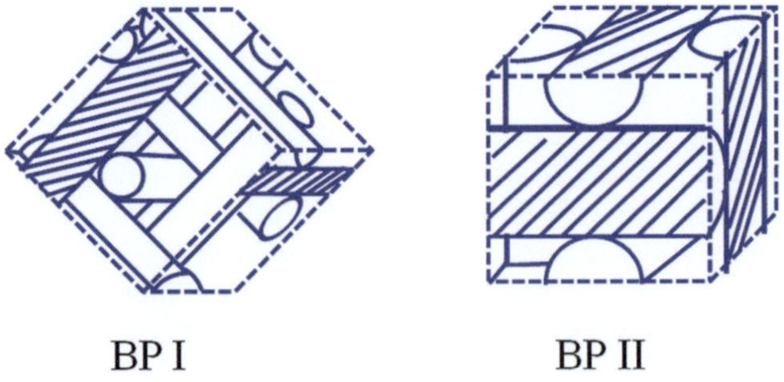

Fig. 3 Schematic representation of the lattices of BPI and BPII. Reproduced with permission from [3]. Copyright 2014 Springer, Heidelberg

in nature [7–9]. BP I and BP II are periodic in three dimensions and the periodicity is typically on the order of the wavelength of visible light. These nanostructures enable BPs to reflect light in visible region and have been recognized as promising soft self-assembled 3D photonic crystals in recent years.

1.3 Self-Organized Liquid Crystal Microdroplets

Micro- and nano-scale particles have recently attracted considerable attention in several research areas, such as optical trapping and manipulations [10, 11], microfluidics, and photonics [12, 13]. Confining LC in micrometer-sized systems allows a wide range of thermodynamically stable supramolecular configurations exhibiting various defect topologies. The particular structure depends on the anchoring of the director at the drop surface, the elastic properties of the LC, the size of the drop, and the presence of external fields [14–16]. Among all these factors, the molecular anchoring at the surface of the drop, which determines the boundary conditions, is the first of all the aspects that should be considered for classifying the drop type, as it fixes the topological constraints on the LCs. For LCs confined to spherical droplets, there are several possibilities (Fig. 4). For nematics with perpendicular boundary conditions, the texture may be radial with a hedgehog point defect in the center or axial with a ring defect lying on the equator. For nematics with parallel boundary conditions, the texture may be bipolar (with two surface defects at the poles) with a randomly oriented axis. If, however, the droplet is cholesteric, the texture is the Frank-Pryce texture, in which the twist axes are everywhere radial [17, 18]. Both nematic and cholesteric LC droplets have great potential for photonic applications such as optical microcavities or microresonators and the omnidirectional 3D lasing.

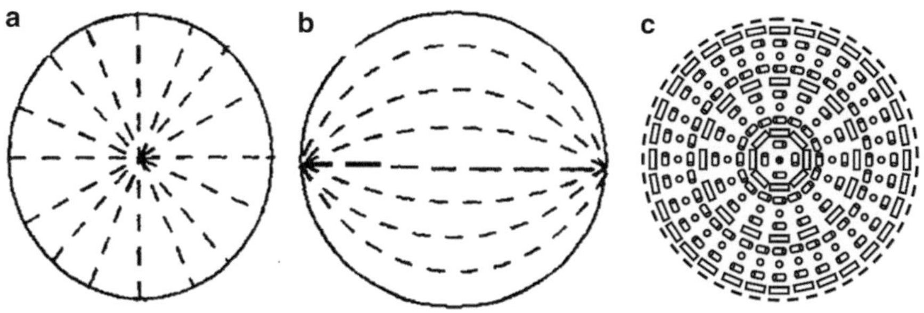

Fig. 4 Radial (**a**) and bipolar (**b**) orientation of the nematic director and (**c**) the molecular-packing model of cholesteric LC within a liquid-crystal droplet. Reproduced with permission from [18]. Copyright 1997 American Physical Society

2 Stimuli-Responsive Self-Organized One-Dimensional Liquid Crystalline Superstructures

2.1 Thermoresponsive Cholesteric Liquid Crystals

As is known, LC phases can be triggered by external environment such as temperature variation or the influence of both temperature and solvent, according to which LCs can be broadly classified as thermotropic or lyotropic LCs. Nearly all LC devices are based on thermotropic LCs. Given the thermotropic nature of LCs, it is not surprising that thermally induced effects in LCs have been examined for many years. Thermally induced optical effects in cholesteric LCs can take a variety of forms, often related to the underlying thermotropic behavior of the system. The reflection of these materials can be thermally switched as the temperature sensitive mixture transitions from cholesteric LC phase to isotropic phase [19]. A topic of considerable interest has been thermally induced tuning, especially in mixtures which present Smectic (Sm) LC phase to cholesteric LC phase transition [20]. The slight differences in the orientation of the smectic phases have a profound effect on the range of thermal tuning observed in a Sm to cholesteric LC phase transition [21]. The manifestation of thermally induced color changes has been widely employed commercially for a variety of "sensors" including mood rings, disposable thermometers, novelty films and clothing and aerodynamic flow sensors.

Recently, hydrogen-bonded cholesteric LCs have also attracted a large amount of interest and been found to be one promising candidate for thermoresponsive 1D soft photonic crystal [22]. Yang and co-workers made cholesteric LC mixtures composed of a cholesteryl ester mixture and the hydrogen-bonded chiral dopant complex. These cholesteric LC mixtures have a selective red reflection band that shifts to the blue after the sample has been heated to 60 °C. This blue reflection color is a result of an increase in the helical twisting power of the cholesteryl ester mixture and the disruption of hydrogen-bonded complex [23]. Wei and co-workers also made thermally responsive reflective color films based on chiral hydrogen-bonded complexes [24]. Single-layer films were constructed containing both right- and left-handed cholesteric LC ("double-handed") structures by utilizing an imprinted right-handed helical polymer template. This template was made by photopolymerizing a nonreactive cholesteric LC mixture in which the reactive mesogen had been dispersed. After this film had been refilled with a cholesteric LC mixture composed of a hydrogen-bonded complex of the chiral pyridine derivative and 4-hexyloxybenzoic acid forming a left-handed helix, a double-handed helical structure was obtained. The pitch of this supramolecular mixture was temperature sensitive. Interestingly, when this mixture was heated to 165 °C, the reflection intensity approached 100 % meaning that both left and right-handed circularly polarized light is reflected within the same reflection band. When the temperature was increased, two reflection bands were observed, one of which corresponds to the pitch of the polymer template and the other one to the temperature-sensitive hydrogen-bonded complex. These results show that it is possible to construct temperature-tunable, hyper-reflective LC films with simultaneous multicolor reflection.

2.2 *Electroresponsive Cholesteric Liquid Crystals*

The dielectric nature of liquid crystalline materials enables the alignment of molecules in response to the applied electric fields. Due to the inherent optical anisotropy, electrically induced macroscopic reorganizations of LCs dramatically change the resulting optical properties, which are widely used in the display industry. Improvements in the refractive index, birefringence, temperature stability, required driving voltage as well as the manufacturing base have driven the commercial success of LC displays. The electrically tunable reflectivity of cholesteric LC materials makes them strongly applicable in tunable color filters, tunable LC lasers, reflective displays, etc. Herein, the current advances in the electrically tunable reflectors, switchable reflectors, and tunable optical diodes based on cholesteric LCs will be mainly reviewed.

2.2.1 Electrically Tunable Reflection Color

Several methods have been explored to enable electric-field induced shift (tuning) of the reflection color in cholesteric LCs. Color changes in cholesteric LC reflectors have been observed by using interdigitated electrodes, polymer stabilization, and recently, negative dielectric anisotropy LCs. One of the primary interests in electrically tunable cholesteric LC reflectors is the development of full color, i.e. red, green, and blue, and addressable cholesteric LC reflectors to decrease the complexity of stacked red-green-blue cholesteric LC pixels for full color displays.

Li et al. [25] firstly realized the tuning of cholesteric LC reflection color by using interdigitated electrodes which generated transverse electric fields (parallel to the substrate). These electrically tunable cholesteric LC reflectors have demonstrated tuning range on the order of 100 nm albeit with high voltages and complex electrode geometries. Crawford et al. [26] explored these in-plane switching systems using polymer stabilization to improve switching performance. A number of researchers have also used polymer stabilization to enable tuning of the reflection color [27]. Coles et al. [28] demonstrated directly addressable wavelength tuning of the PBG from a hybrid structure consisting of an achiral nematic LC and a periodic polymer template. The wavelength tuning was found to be fast (about 43 ms) and broadband (greater than 100 nm). In their experiment, a polymer template was prepared by photopolymerizing a monomer that was dispersed within a cholesteric pre-mixture, which initially defined the periodicity or pitch of the structure, but was then subsequently removed. After that, an achiral nematic LC with a higher birefringence and a positive dielectric anisotropy was injected into the polymer-template-containing cell. A detailed schematic of the preparation and subsequent mechanism for tuning is depicted in Fig. 5. The switching mechanism for this hybrid system is based upon a simple dipole reorientation of the nematic LC. Under the influence of an electric field applied in the viewing direction, the achiral LC

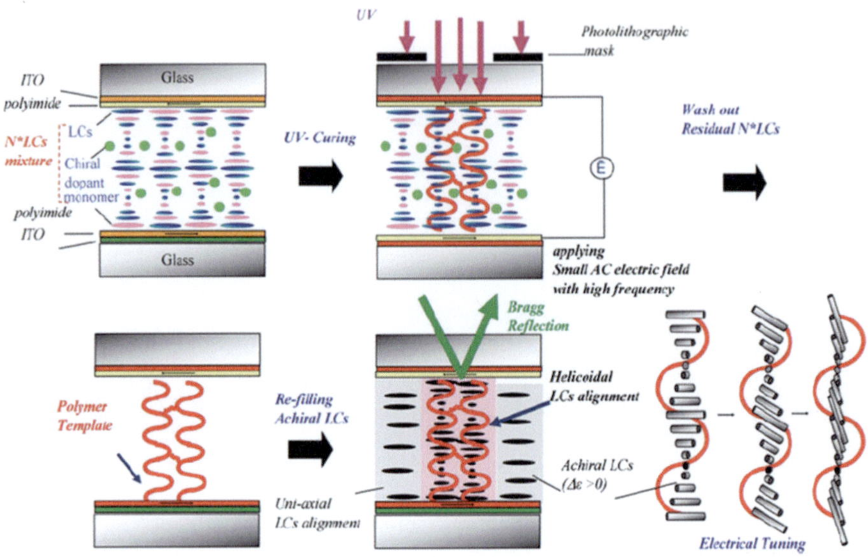

Fig. 5 Illustration outlining the procedure for preparing the photonic band structure from an achiral nematic LC and a polymer network. Reproduced with permission from [27]. Copyright 2010 John Wiley & Sons

within the polymer matrix reoriented to align with the field direction. The rigid polymer network, on the other hand, remains unperturbed.

Furthermore, several reports demonstrated electric-field tuning in systems using the negative dielectric nematic LCs as host material for cholesteric LCs. Due to the negative dielectric anisotropy, a planar aligned cholesteric LC helix based on these materials will have an aligned dipole moment along the direction of the direct current (DC) field. Thus, application of a low to moderate strength DC field does not reorient the molecules and therefore does not destroy the pitch. Observation of slight changes in the "as processed" selective reflection color upon application of a field has been attributed to defect annealing giving rise to a slightly tighter pitch which results in a small blue shift in the color [29, 30]. More recently, Bunning et al. [31] presented electrically tunable color changes in a polymer stabilized cholesteric negative LCs where modest DC field strengths can induce significant color tuning. The cells are prepared by filling a planar aligned cell with a mixture containing negative nematic LC, chiral dopant, photo-initiator, and chiral monomer. The mixture is then photopolymerized by exposing the sample to UV light. Figure 6a shows the reflection spectra at various electric field strengths taken from the positive electrode of the cell. A plot of the reflection wavelength as a function of the electric field strength is shown in Fig. 6b. Above the threshold voltage (>1 V/μm) the color undergoes a red shift that is linearly dependent with increasing field strengths. The reflection wavelength tuning occurs over hundreds of nanometers while maintaining good contrast at a relatively low value of electric field. Evidence indicates that the effect is caused by the movement of locally trapped charges on the polymer surface in response to a DC electric field.

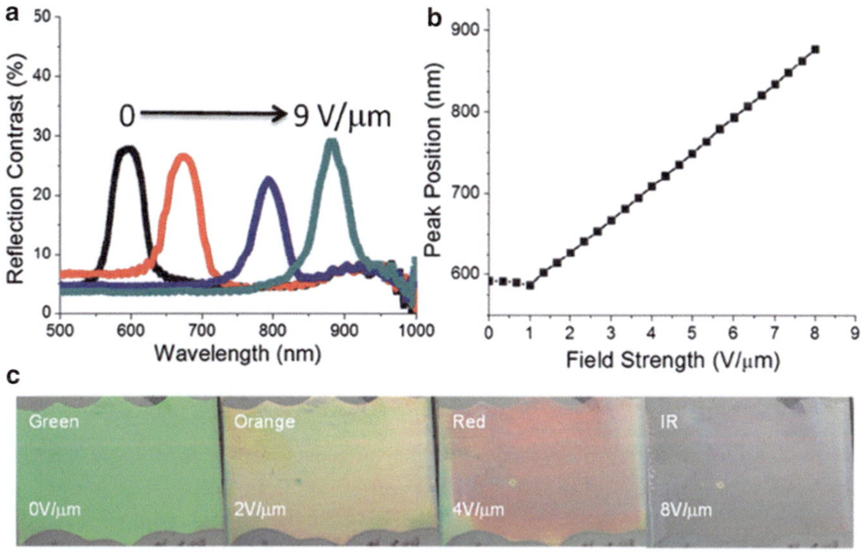

Fig. 6 Large range color tuning observed upon application of DC fields. (**a**) The reflection spectra at DC electric fields strengths ranging from 0 to 9 V/μm. (**b**) A plot of the central reflection peak position (nm) versus the applied DC field strength (V/μm). (**c**) Photographic images of color change of the cell as DC field strength is increased from 0 to 8 V/μm. Reproduced with permission from [31]. Copyright 2013 John Wiley & Sons

2.2.2 Electrically Switchable Reflectors

Application of an alternating current (AC) field across conductive glass substrates typically reorients positive nematic LCs from a planar to a homeotropic orientation. In the case of cholesteric LCs, reorientation from the planar to the homeotropic orientation destroys the periodicity of the phase along the optical axis and serves to switch the reflection off. Unfortunately, upon removal of the field the cholesteric LC transitions into a long-lived, metastable focal conic state. A number of methods have been explored to avoid this scattering state and allow for direct binary on/off switching.

The first and most common method involves polymerization of liquid crystal monomers to stabilize the cholesteric LC helix. Polymer stabilized cholesteric LC (PSCLC) is typically undertaken using liquid crystalline monomers in concentrations of 5–10 wt%, which polymerize to effectively template the cholesteric LC helix. Like conventional cholesteric LCs, application of an AC field reorients the PSCLC from planar (reflective) to homeotropic (clear) state. However, upon removal of the field, the template polymer network acts as an elastic restoring force that returns the PSCLC to its original state. Interestingly, polymer stabilization method can also be used to enable control over the bandwidth of the resulting reflector [32, 33], and the cholesteric LC film thus prepared can be easily switched between the transparent state and full-color reflective state as shown in Fig. 7.

Fig. 7 Electrically switchable mirror based on a polymer stabilized cholesteric LC with a pitch gradient. (**a**) In reflection state; (**b**) In transparent state; (**c**) In half-reflection/transmission state. Courtesy: Kent Optronics

The second common approach to develop binary on/off switchable cholesteric LC reflectors is the employment of dual frequency LCs. This approach was first demonstrated in 1984 by Gerber who reported cholesteric LC reflection switching based on adjustment in the frequency of the driving AC voltage [34]. In this special class of LC materials, the host LC transitions from positive dielectric anisotropy to negative dielectric anisotropy above a crossover frequency. The direction of the helix realigns when the frequency surpasses a threshold value that results in switching from reflective to clear. The advantage of dual frequency LCs is rapidly switchable that do not have the scatter or haze which are observed in PSCLCs.

Recently, Yang et al. [35] reported an interesting cholesteric LC/chiral ionic liquid (CIL) composite that can be switched electrically between a transparent, a scattering, and a reflective state in the visible region. All these states can be memorized for some time after removing the applied electric field. Upon the application of a DC electric field followed by an AC electric field, the bandwidth of the reflection can be controlled accurately and reversibly by adjusting the intensity of DC electric field. One of the potential applications of this composite is E-paper. This kind of E-paper exhibits a brilliant color by reflecting visible light around it without the need of backlight and is promising in realizing the low power consumption. Furthermore, by adjusting the content of the chiral dopants, switching of near-infrared (NIR) light between the transmission and the reflection states will be achieved, which can be used as smart switchable reflective windows for the dynamic control of solar light to block unwanted NIR light when it is necessary.

2.2.3 Electrically Tunable Optical Diodes

A diode is an essential element in most electronic circuits, permitting a current in the forward direction and blocking it in the backward direction. We can call a device an optical diode if it transmits light in one direction and blocks it in the reverse direction (non-reciprocal transmission). A variety of optical diodes have been proposed and demonstrated. One well-known device is an optical isolator based on linear polarizers and a Faraday rotator, which provides a high extinction ratio with a low insertion loss. Other examples include devices based on photonic crystals: a 1D photonic crystal structure with a spatial gradient in the linear refractive index together with a nonlinear optical medium [36] and a 2D photonic crystal waveguide with an asymmetric array of nonlinear defect rods [37]. Mechanisms using the competing effects of self-absorption and stimulated emission in a fluorescent dye with a concentration gradient, magnetic effects on helical periodic medium and absorbing multilayer systems have been theoretically found to show optical diode effects [38–40]. Despite such extensive studies, the concept of tunability in optical diode operation has not been exploited. As one of the important forms of soft photonic crystal, cholesteric LCs with periodic helical structures possess not only good tunability, but also selective reflection for circularly polarized light (CPL). The selective reflection in the optical properties of cholesteric LCs has recently been used to design an optical device that permits the propagation of a band of wavelengths in one direction and blocks it in the other, acting as an optical diode analogous to an electric diode. This kind of optical diode using cholesteric LCs is unique and attractive, as it can be simply fabricated, shows good non-reciprocity and can be electrically tunable [41–43].

To obtain the novel optical diode effect, Takezoe et al. [41] designed and fabricated a new optical heterojunction anisotropic structure (OHAS) consisting of an anisotropic layer sandwiched between two polymer cholesteric LC (PCLC) layers with different periodicity of helix (hetero-PBG) as shown in Fig. 8a. A half-wave phase retarder layer made of planar nematic LC is sandwiched between two PBG cholesteric LC layers with the same sense of helix handedness but different pitches ($p_1 < p_2$). The dispersion relations for forward and backward propagation of light are readily obtained by introducing an anisotropic dielectric tensor that

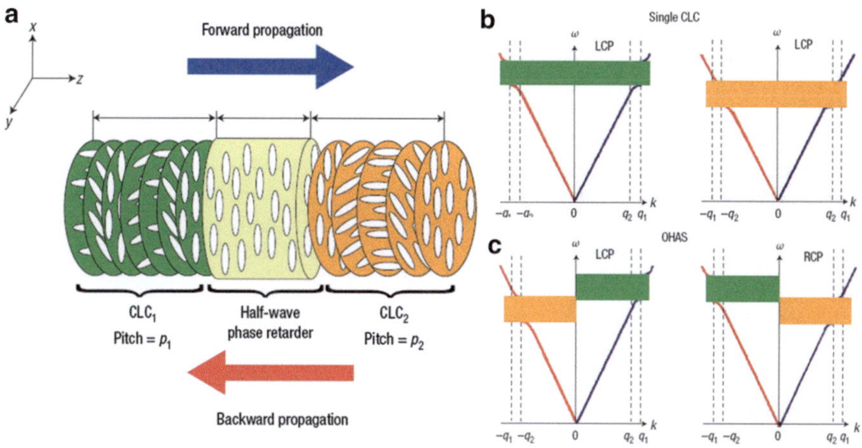

Fig. 8 Structure and principle of the optical diode: (**a**) Hetero-PBG structure with an anisotropic layer sandwiched between two L-helical cholesteric LC (CLC) layers with different pitches. (**b**) Two dispersion curves of LCP light for CLC_1 (*green*) and CLC_2 (*orange*). The bandgaps are the same for forward and backward propagations: there is no optical diode effect. (**c**) Dispersion curves for a hetero-PBG structure with an anisotropic layer. Spectral positions of PBG depend on the direction of light propagation, showing optical diode performance or non-reciprocal transmission. Reproduced with permission from [41]. Copyright 2005 Nature Publishing Group

depends on propagation direction as well as on position, as shown in Figs. 8b and 8c. In Fig. 8b, two dispersion curves for left circular polarization (LCP) light in cholesteric LC 1 (CLC_1) and 2 (CLC_2), which have a left helix, are shown (left and right panels, respectively). Green and orange stripes are the bandgaps due to CLC_1 and CLC_2, respectively. Blue and red lines are dispersions for forward and backward light propagation, respectively. Figure 8c shows the dispersion curves of the OHAS structure for LCP light (left) and circular polarization (RCP) light (right). Note that the handedness of CPL refers to the light coming into the OHAS, not the outgoing light. They found that PBG positions depend on the propagation direction, showing optical diode behavior (non-reciprocal transmission). For two opposite circular polarizations, the PBGs for forward and backward light propagations are interchanged. For both polarizations, the transmittance is either zero or 100 % when the anisotropic layer behaves as a half-wave phase retarder, indicating perfect optical diode performance. With decreasing Δn, the transmitted intensity changes for forward-propagating light only, resulting in no optical effect for $\Delta n = 0.00$. Because of the use of nematic LC as an anisotropic layer, Δn can easily be varied. With its advantages of easy fabrication and electro-tunability, the new OHAS thus prepared has wide potential application in photonic devices that make use of propagation-direction dependent operation.

2.3 Photoresponsive Cholesteric Liquid Crystals

Light irradiation is a much more precise and effective stimulus to tune the self-organized superstructures of cholesteric LCs when photosensitive molecules are infiltrated into the system. This approach can also leverage the benefits of light-directed effects—remote, temporal, and spatial control. The photochemistry at the root of these effects can occur on a picosecond timescale, allowing the potential of rapid optical responses [44]. As will be discussed, a number of approaches exist to generate photo-induced optical effects in cholesteric LCs including reflection color modulation, photo-induced phase transition, and photo-induced handedness inversion.

2.3.1 Optically Tunable Reflection Color

The tuning of reflection wavelength in cholesteric LCs holds great promise for applications in devices such as tunable color filters and reflectors, tunable LC lasers, and reflective displays. Such system can be achieved by doping a photoresponsive chiral material into a LC host to form a self-organized, optically tunable helical superstructure, i.e. a photoresponsive cholesteric LC phase. The resulting macroscopic helical superstructure can reflect light selectively according to Bragg's law and can be tuned by light of appropriate wavelength. In 1971, Sackmann [45] firstly reported a light-induced reversible color change in non-halide based cholesteryl mixture doped with azobenzene. The selective reflection of this cholesteric LC mixture blue shifted from 610 to 560 nm upon irradiation with 420 nm light. Although the tuning range is quite narrow, this seminal work provides a possible pathway to the reflection tuning in photoresponsive cholesteric LCs. After that, various material systems have been reported for this purpose, which include: chiral mesogens doped with achiral azobenzene molecules [46, 47]; pure photoresponsive chiral mesogens [48]; nematic LC doped with photoresponsive chiral dopants [49–54]. Recently, the studies are mainly focused on the design and synthesis of powerful chiral dopants which exhibit high HTPs as well as large HTP variations upon light irradiation. With the continuous efforts especially in the past decade, many chiral azobenzene dopants have been developed and used for phototuning the reflection colors of cholesteric LCs with achievements such as large tuning range, very high HTPs, with low doping concentrations.

Tamaoki et al. [50] reported a series of planar chiral dopants which were employed in commercially available nematic LC to achieve phototunable reflection colors. This compound was designed based on an azobenzenophane compound having conformational restriction on the free rotation of naphthalene moiety to impose an element of planar chirality. Due to the good solubility, moderately high HTPs, and large changes in HTPs during photoisomerization of this molecule, a fast photon mode reversible color control in induced cholesteric LCs was achieved. Li et al. [51–54] have developed a series of binaphthyl-derived axially chiral azobenzene compounds. These chiral switches have superior solubility in LC host and exhibit high HTP values and considerable differences in HTP among their various states, and they were all found capable of phototuning the reflection color across the full visible spectrum [53].

The above azobenzene derivatives can transform from *trans* form to *cis* form upon UV light irradiation and the reverse process can occur either thermally or photochemically by visible light irradiation (Fig. 9). The slow but unavoidable thermal relaxation gives rise to the problem of lack of stability and controllability. Recently, a major breakthrough in designing high HTP diarylethene dopants with thermal stability was achieved by Li et al. [55, 56] with the development of diarylethene derivatives as shown in Fig. 10. The open and closed isomers of

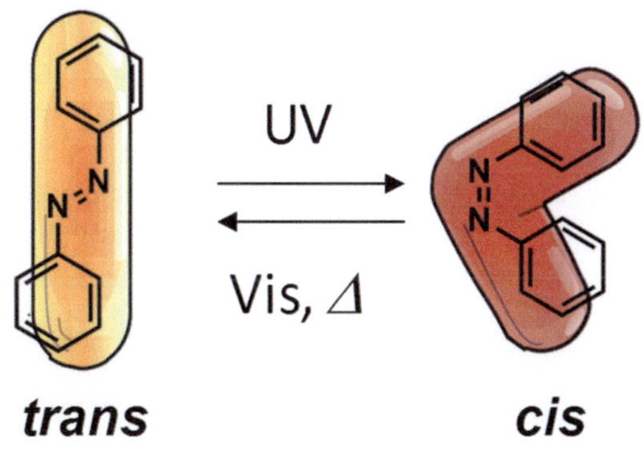

Fig. 9 Schematic description on the photoisomerization of the azobenzene molecule

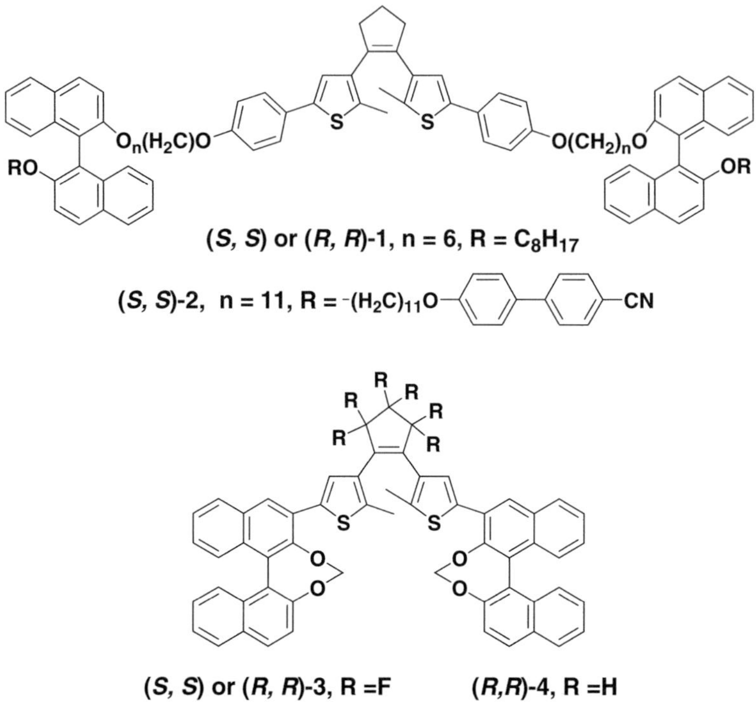

Fig. 10 Chemical structures of chiral diarylethene derivatives

compound **1** exhibited high HTPs of 92 and 77 μm^{-1} in 5CB, respectively. The relatively small change in HTP only brought out a reflection tuning range of ~50 nm. The compound (*S*, *S*)-**3** was found to possess remarkable changes in HTPs during photoisomerization in addition to the very high HTPs of 104 μm^{-1} at the initial state. A doping concentration of 7.7 wt% enabled the phototuning of reflection colors to achieve three primary red, green, and blue colors. Although the phototuning of reflection color has been documented with azobenzene and overcrowded alkene dopants, this is the first example of a diarylethene dopant for this purpose with the advantage of thermal stability at both states. As shown in Fig. 11, three primary red, green, and blue colors can be observed simultaneously in a single thin film based on different UV irradiation time facilitated by masking at different areas: red, no irradiation; green, irradiated for 10 s; blue, irradiated for 25 s (Fig. 11a–d). After driving the background color to blue by UV irradiation, the red and green reflection colors can be recorded through visible light irradiation for different time (Fig. 11e–h). Moreover, the optically addressed images can be erased by light irradiation when desired, and the cell is rewritable for many times due to the excellent fatigue resistance.

The use of high-energy UV light might result in material decomposition and poor penetration through the substrate, etc. Utilizing NIR light would be much more

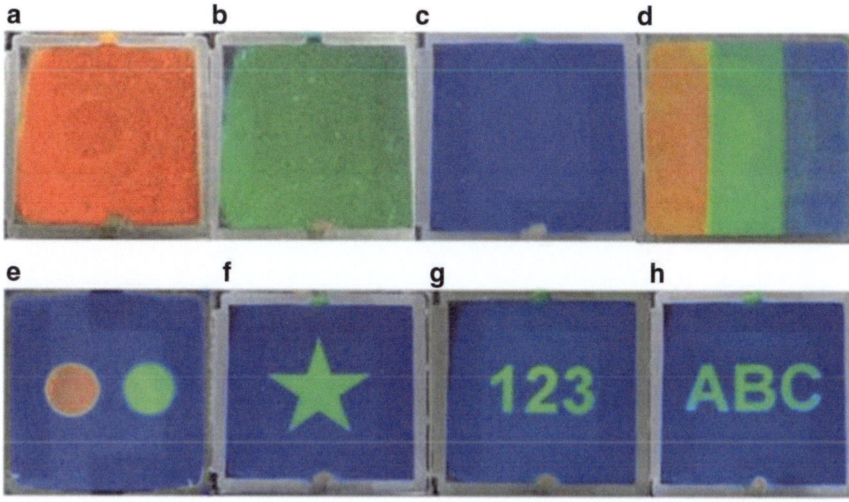

Fig. 11 Real cell images of an 8 μm thick planar cell (2.1 cm × 2.5 cm) filled with 7.7 wt% (*S*, *S*)-3 in E7. Reproduced with permission from [55]. Reflection color change: (**a**) red; (**b**) green; (**c**) blue; (**d**) red-green-blue colors in one cell; and (**e–h**) optically addressed images on blue background. Copyright 2012 American Chemical Society

desirable than either UV or visible light in the fields such as life science, materials science, and aerospace due to its superior penetration and invisibility for remote activation of materials at a specific time and location with relatively low interference and high precision. Extensive efforts have been made to design chiral azobenzene switches that can be triggered with longer wavelengths of light, i.e. visible or even NIR light. Recent research has pointed out that the photoisomerization can also be attained by visible light [57], and reversible visible light tuning reflection color of cholesteric LCs has also been achieved [58]. Interestingly, Li et al. [59] introduced the reversible NIR-directed reflection in a self-organized helical superstructure loaded with upconversion nanoparticles (UCNPs) composed of β-$NaGdF_4$: 70 % Yb^{+3}, 1 % Tm^{+3} @ β-NaGdF4 core-shell nanostructures. In their experiment, the novel self-organized helical superstructure was fabricated by doping a light-driven chiral azobenzene switch and UCNPs into a LC host composed of commercially available components. The resulting mixture was capillary-filled into a 10 μm thick planar aligned cell with antiparallel rubbing direction. The shift in the photonic band gaps upon NIR light irradiation was recorded using a spectrometer incorporated into the polarizing optical microscope. 3 wt% chiral switch and 1.5 wt% UCNPs doped in commercially available 80.5 wt% nematic LC E7 and 15 wt% S811 were found to exhibit reversibly tunable NIR light-directed reflections through red, green, and blue wavelength only by varying the power density of the excitation NIR light (Fig. 12a, b). The center reflection wavelength was around 435 nm at the initial state (Fig. 12c). Upon irradiation with 980 nm NIR laser at high power density, the reflection wavelength was tuned to 540 nm in 60 s and further reached a photo-stationary state in 120 s with a center reflection

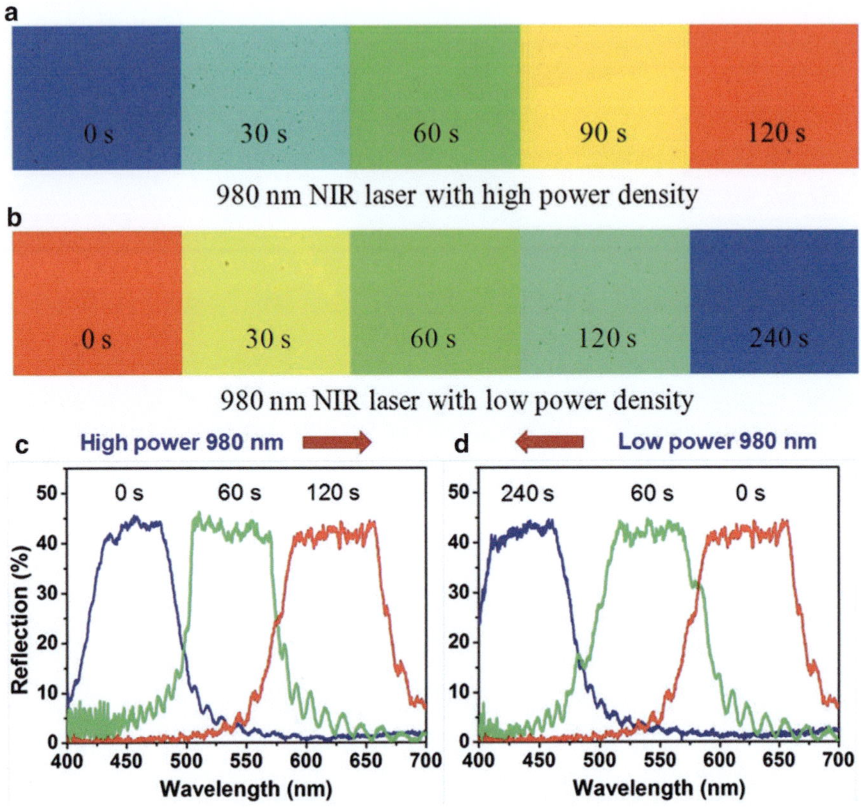

Fig. 12 Reflection colors and corresponding reflection spectra of the cholesteric LCs with 3 wt% chiral molecular switch 4 and 1.5 wt% UCNPs in a 10 μm thick planar cell at room temperature, taken from a polarized reflective mode microscope: (**a**) and (**c**) upon irradiation with 980 nm NIR laser at high power density (2 W/mm^2) and followed by (**b**) and (**d**) irradiation with 980 nm NIR laser at low power density (0.15 W/mm^2). Reproduced with permission from Ref [59]. Copyright 2014 American Chemical Society

wavelength at 625 nm. The reverse process across red, green, and blue reflection colors occurred within approximately 4 min upon irradiation with 980 nm NIR laser at low power density (Fig. 12d). The reversible tuning of reflection across red, green, and blue reflection colors was repeated many times without noticeable degradation.

2.3.2 Photo-Induced Phase Transition

LC phase transitions are known to occur when the external thermodynamic parameters are varied. For example, phase transitions in thermotropic LCs occur by the changes in temperature, while the lyotropic LCs show phase transition behavior as a

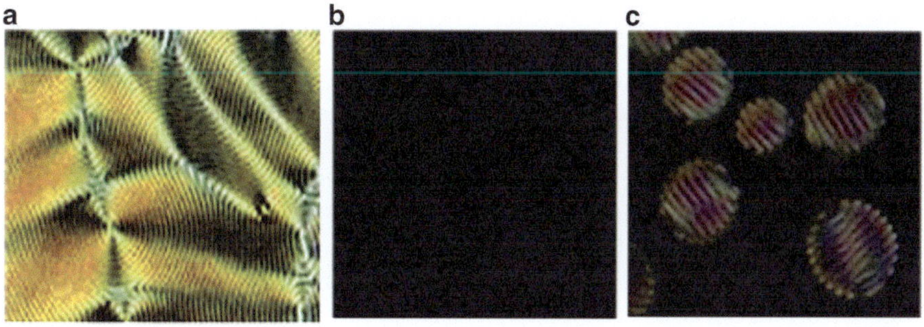

Fig. 13 Crossed polarized optical texture micrograph of the mixture of 25 % the cholesteric mesogen in 5CB on cooling at 38.9 °C (**a**, before UV irradiation; **b**, after UV irradiation for 10 s; **c**, 20 s after removal of UV light at the isotropic phase). Reprinted with permission [62]. Copyright 2005 American Chemical Society

function of both temperature and the concentration of amphiphilic molecules in the solvent. In recent years, light has been recognized as another fascinating stimulus that can induce isothermal LC phase transitions [60], where light-induced structural changes of photoresponsive molecules were used to affect the order parameters of LC phases. To date, several kinds of light-induced phase transition behavior related to the cholesteric phase have been reported, such as cholesteric (N*)–isotropic (I), and smectic (Sm)–cholesteric (N*) phase transitions.

Chanishvili et al. [61] investigated the light-induced effect in cholesteric mixtures composed of azobenzene or azoxybenzene nematic host and chiral dopants. The transition from cholesteric to isotropic state was firstly observed in the photoisomerization of host molecules upon UV irradiation. Li et al. [62] also reported the photo-induced isotropic (PHI) state in the systems containing azobenzene-derived chiral mesogens. A cholesteric polygonal fingerprint texture was observed at 38.9 °C, as shown in Fig. 13a. UV irradiation with 365 nm was generated by a UV gun with the intensity of 10 mW/cm^2. Within 10 s under irradiation, this sample transforms to the isotropic phase as evidenced by texture change as shown in Fig. 13b. It is shown by this experiment that the conversion from the *trans* to the *cis* configuration of the dopant leads to destabilization of the liquid crystalline phase of the mixture. Removal of UV light immediately leads to the reverse process of chiral nematic domain formation from the isotropic phase appearing as droplet nucleation followed by coalescence (Fig. 13c). The reversion to the original polygonal fingerprint texture in Fig. 13a was reached within approximately 2 h at room temperature in the dark. In addition, PHI states have also been demonstrated with laser irradiation [63] or by using multi-component systems containing azobenzene LCs [64] and chiral dithienylcyclopentene dopant [65].

The transition from the cholesteric to the nematic phase can be caused by irradiation with CPL or linearly polarized light (LPL) with the racemization of chiral switches or motors. Both overcrowded alkene and chiral diarylethene switches have been proven to be the good candidates for amplification of chirality via a chiral molecular switch from a nematic to cholesteric phase transition using a CPL [66]. In addition, N*–N transition can also be achieved by the HTP variation of the photoresponsive chiral dopants in induced cholesteric LCs. In these systems, the HTPs of the chiral dopants have to exhibit dramatic variation between its two states. At a certain doping concentration, the isomer with higher HTP can efficiently induce a cholesteric phase, while the isomer with lower HTP only generates an apparent nematic phase. The light-induced isomerization between these two forms can reversibly switch the LC phase between the cholesteric and the nematic. A number of diarylethene derivatives have also been reported as this kind of chiral dopant [67, 68].

Another type of phase transition related to cholesteric phase is starting from the smectic phase. This type of transition is mainly observed in chiral mesogenic compounds. Joly et al. [69] conducted a systematic research on the phase transition behavior of a series of chiral azobenzene LCs. A photo illumination–temperature phase diagram was presented to interpret the phase transition sequence including the SmC*–N*–I transition. Tamaoki and co-workers reported some chiral azobenzenes and diphenylbutadiene based LC dimers exhibiting light-induced phase transition behavior starting from smectic phases [70, 71]. For example, the dimers **5** (Fig. 14), which were synthesized by connecting a cholesteryl group with 1,4-diphenylbutadiene unit, were found to experience isothermal phase transition from SmA to N* and further to the isotropic state upon UV light irradiation at 366 nm [70]. The phase transitions starting from the smectic phase were also observed in other doped systems [72, 73]. The smectic LC host 8CB, doped with chiral molecules and azobenzene compounds, showed the phase transition from smectic to cholesteric due to the photo-induced isomerization of azobenzene molecules, and the prolonged irradiation drove the phase transition further to the isotropic phase. Matsui et al. [72] also reported a chiral smectic C (SmC*)–cholesteric (N*) phase sequence in azo-dye doped ferroelectric LCs.

2.3.3 Photo-Directed Handedness Inversion

The helical superstructures of cholesteric LCs, as mentioned earlier, are characterized by both pitch length and handedness. The control over the reflection wavelength by tuning the pitch length of cholesteric LCs has been intensively

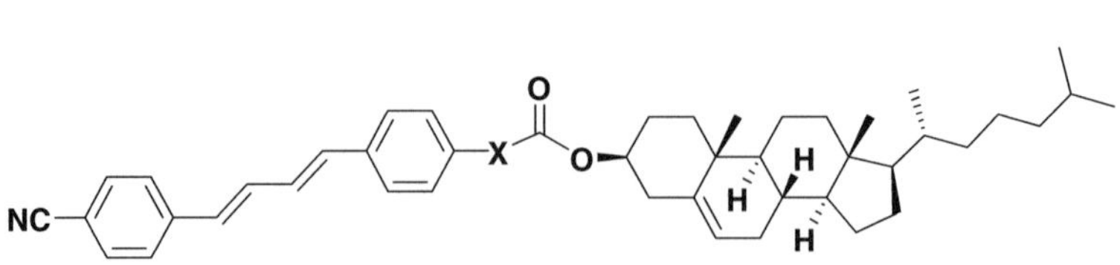

5 X = O, O$(CH_2)_8$O, O$(CH_2)_{11}$O or O$(CH_2)_{12}$O

Fig. 14 Chemical structure of photoresponsive diphenylbutadiene based LC dimer **5**

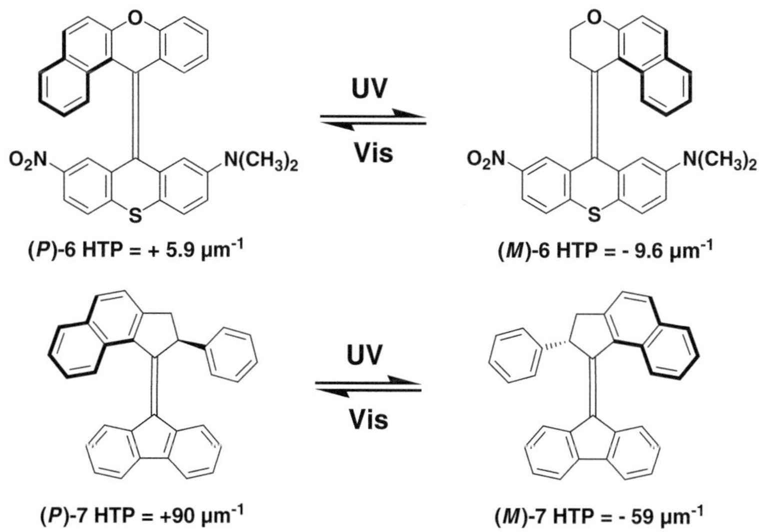

Fig. 15 Photoisomerization of overcrowded alkene dopants **6** and 7

investigated. Since cholesteric LCs only reflect CPL with the same handedness as the cholesteric helix, achieving the handedness inversion in response to external stimuli would hold great potential in the development of stimuli-responsive materials for future technological applications, especially the areas in which CPL is involved. Light-driven induction of helicity and inversion of handedness in self-organized helical superstructures have the unique property of selective reflection of CPL and the advantages of remote, spatial, and temporal controllability. Many strategies have been successfully demonstrated for photo-controlled handedness switching of the photoresponsive cholesteric LCs. Herein we will mainly focus on the most efficient systems which use single photoresponsive chiral dopant.

Chiral overcrowded alkenes are known as a family of photoresponsive molecules which are more likely to induce handedness inversion of the cholesteric helix when used as chiral dopants. A variety of these types of compounds have been synthesized and characterized by Feringa et al. [74–76]. Taking the compound **6** as an example (Fig. 15, top) [74], its initial HTP at (*P*)-*cis* form in nematic M15 is +5.9 μm^{-1}, that is, it induces a right-handed helix. Irradiation with UV light led to photoisomerization from the *cis* form to the *trans* form with a photo-stationary state (PSS) exhibiting an overall HTP of −9.6 μm^{-1}. However, the drawback of this compound is the low HTPs at both states. When the fluorene group was introduced into this type of molecule, compound **7** was found not only to induce helix inversion but also to exhibit high HTPs at both the states (Fig. 15, bottom).

Li et al. [77] reported a fast photon mode reversible handedness inversion of a self-organized helical superstructure, i.e. cholesteric LC phase, using chiral cyclic azobenzene compound. Recently, they further reported a novel design strategy based on the chiral conflict and equilibrium shifting between multiple chiral

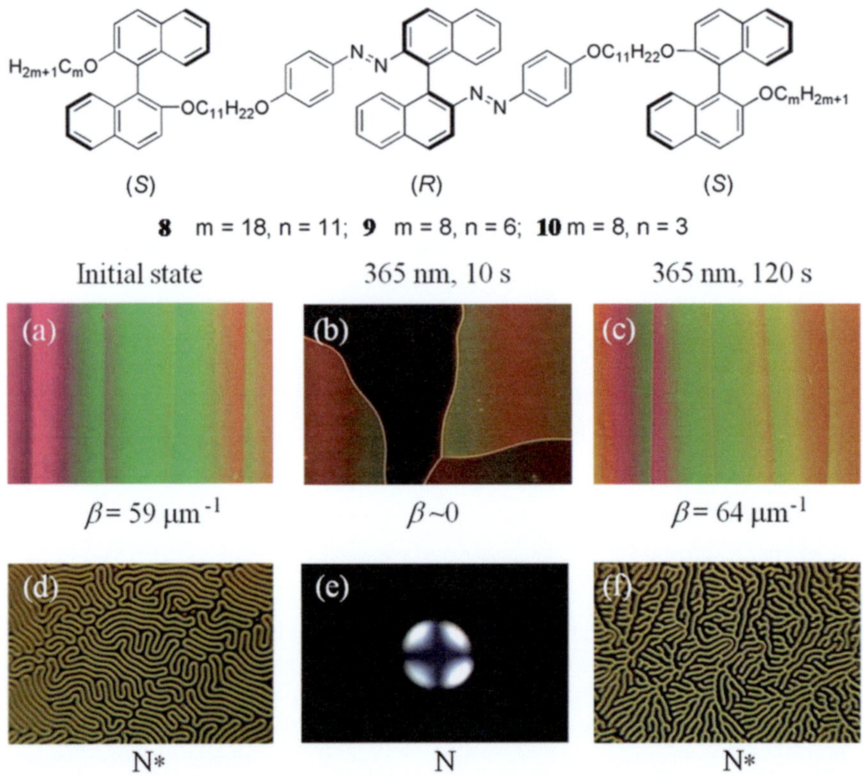

Fig. 16 Handedness inversion of 0.27 mol% (*S*, *R*, *S*)-**8** in 5CB observed in wedge cell (**a**–**c**) and homeotropic cell (**d**–**f**) from right-handed cholesteric LC (**a**, **d**) to left-handed cholesteric LC (**c**, **f**) through transient nematic phase (**b**, **e**). Reproduced with permission from [78]. Copyright 2013 John Wiley & Sons

moieties in a single molecule (Fig. 16) [78]. A series of azobenzene derivatives (*S, R, S*)-**8-10** bearing axially chiral binaphthyl units of opposite chiral configurations were synthesized. They were found to confer handedness inversion in different LC hosts upon light irradiation with moderate to high HTPs, where the chirality of each chiral moiety does not need to be reversed. Since the *R* and *S* configuration of unbridged binaphthyls will normally induce right-handed and left-handed cholesteric LCs, the design strategy is to incorporate multiple chiral binaphthyls with opposite chirality into a single molecule functionalized with photoresponsive groups, and hopefully the helix inversion can occur based on the chiral conflict and equilibrium shift between these chiral moieties.

More recently, two binaphthyl-derived diarylethene dopants were demonstrated exhibiting the handedness inversion behavior with thermal stability when employed as the chiral dopant of cholesteric LCs (Fig. 17) [79, 80]. The open form of (*S, S*)-**11** induced a right-handed cholesteric LC in the commercially available nematic host 5CB, while the handedness could be switched to left-handed upon UV irradiation

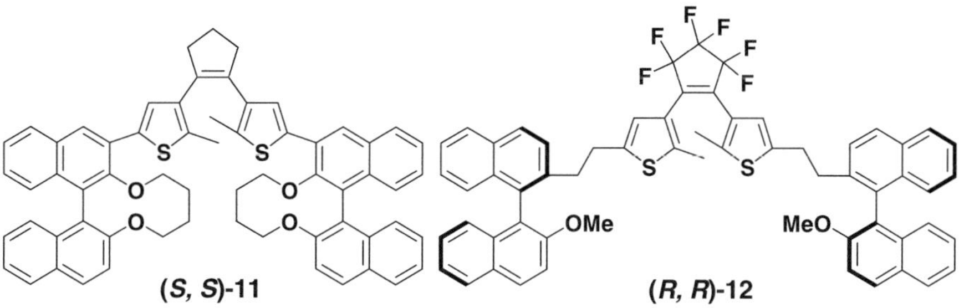

Fig. 17 Chemical structures of binaphthyl-derived diarylethene dopants **11** and **12**

and the HTP changes from the initial +54 to −75 μm^{-1} at PSS. For compound (*R*, *R*)-**12**, handedness inversion was observed in another commercially available nematic host PCHs instead of 5CB. Its initial HTP at the open form in nematic PCHs is −6.9 μm^{-1}, that is, it induces a left-handed helix. Irradiation with UV light led to photoisomerization to the close form with a PSS exhibiting an overall HTP of +7.5 μm^{-1}. It is believed that the reversible handedness inversion in these systems originated from the conformation change of binaphthyl moieties as well as the stereospecific intermolecular interactions between the binaphthyl moieties and the surrounding nematic LC molecules.

2.4 Chemically Responsive Cholesteric Liquid Crystals

In recent years, cholesteric LCs that reflect circular polarized light at a given wavelength as a result of the self-organizing molecular helices have also proven to be attractive to construct chemical sensors that do not require batteries. The pitch of the helix in these cholesteric LCs determines the wavelength of reflection and can be modified in response to the environment, resulting in a color change. These appealing smart materials will find wide applications in the built environment, the packaging industry, and health care. Herein, we would like to introduce several interesting chemical sensors based on cholesteric LCs for detecting chemical atmosphere, such as gas, humidity, pH, and so forth.

Han et al. [81] reported an approach using responsive chiral compounds as dopants for cholesteric LCs to obtain monitors for gaseous carbon dioxide and oxygen. The presence of the analyte induces visible color changes in cholesteric LC thin films. As shown in Fig. 18, the design of carbon dioxide monitoring or sensor is based on the reversible reaction of an analyte with a chiral dopant containing diamine groups to form a carbamate, while the oxygen monitoring is based on the irreversible oxidation of a binaphthalene dithiol derivative to the corresponding disulfide by molecular oxygen. In either case, the chiral dopant was designed to undergo significant conformational change upon reaction with the analyte, resulting

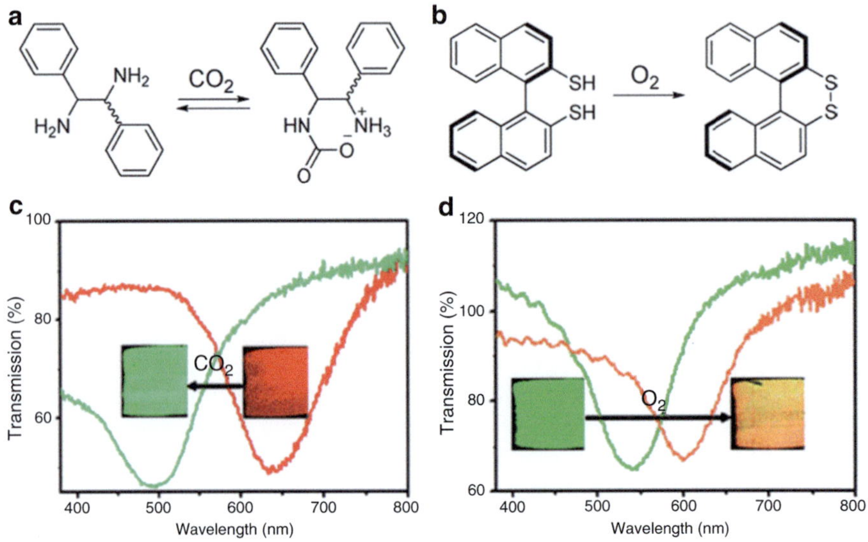

Fig. 18 Molecular structures of (**a**) and (**c**) CO_2– and (**b**) and (**d**) O_2-responsive chiral dopants and the reactions with the corresponding gas. Reproduced with permission from Ref [81]. Copyright 2010 American Chemical Society

in a change in HTP with concomitant change in pitch of the cholesteric LC phase upon exposure to the gas. In their experiment, monitoring of CO_2 was achieved by doping a commercial cholesteric liquid crystalline mixture (E7) with 1.6 mol% of the 1:1 complex of an optically pure diamine with a TADDOL derivative. Upon exposure to carbon dioxide, the reflection band of a thin film of the mixture shifted from 637 to 495 nm as a consequence of dissociation of the complex after carbamate formation of the diamine. An O_2 monitor was obtained by doping E7 with a chiral binaphthyl dithiol derivative and a nonresponsive codopant. The reflection band of the oxygen monitor film changed from 542 to 600 nm, due to the conformational change accompanying oxidation of the dithiol to disulfide. These monitoring mechanisms hold promise for application in smart packaging, where carbon dioxide and oxygen are of special interest because of their roles in food preservation.

Saha et al. [82] prepared an irreversible humidity sensor based on bis-(binaphthylenedioxy)silane as a reactive chiral dopant in E7. As shown in Fig. 19, the chiral compound is very sensitive to hydrolysis. Upon exposure to humid air, the silane compound hydrolyzes, yielding two binaphthol compounds. The silane compound and the hydrolyzed product have a similar helical twisting power, $-34.9\ \mu m^{-1}$ versus $-31.7\ \mu m^{-1}$, respectively. Nevertheless, a significant change of the reflection band was observed when a film containing 10.1 % silane dopant in E7 was exposed to humid air. The reflection of the initial blue film red-shifted to orange-red within a few minutes and the film turned colorless within an hour. The binaphthyl products are poorly soluble in the E7 matrix, resulting in

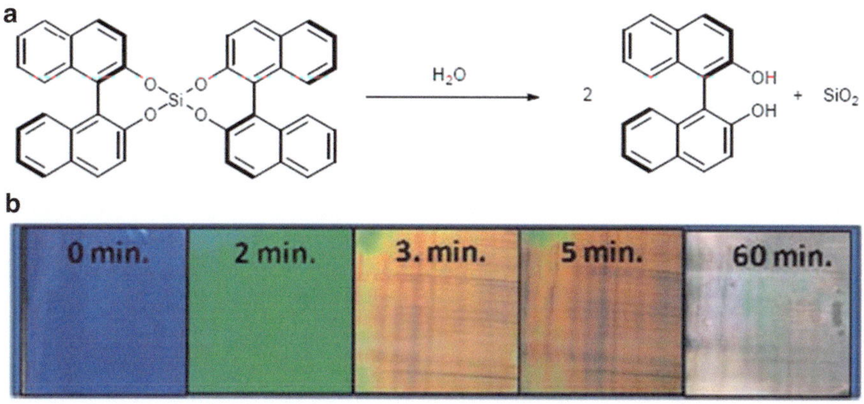

Fig. 19 (**a**) The reaction of bis-[binaphthylen-(2.2′)-dioxy]-silan with water to form (*R*)-(+)-1,1′-bi(2-naphthol) and SiO_2; (**b**) photographs indicating color changes upon exposure to air (relative humidity of 75 %) of a cholesteric LC film of E7 containing 10.1 % of bis-[binaphthylen-(2.2′)-dioxy]-silane. Reproduced with permission from Ref [82]. Copyright 2012 Royal Society of Chemistry

precipitation. The rate of the reflection-shift was found to be sensitive to the relative humidity. However, regardless of the relative humidity, a colorless film was obtained after long exposure even to a low relative humidity, indicating good time-analyte integrating properties. Fast-responsive humidity sensors have been recently made by Broer et al. [83], in which the reflection color changes between green and yellow depending on the relative humidity. To prepare the humidity-responsive cholesteric LC films, a chiral liquid crystalline mixture was used, which contained both chemically (polymerizable acrylate groups) and physically (carboxylic acid groups, which form H-bonded dimers) crosslinkable groups. Thin layers were deposited by dissolving the cholesteric LC mixture in THF (1:1) and bar-coating onto a polyimide-coated rubbed glass substrate. After photopolymerization at a temperature ($T_{curing} = 53$ °C) within the chiral nematic phase, planar cholesteric LC films were fabricated with thicknesses of 5 – 7 μm. The cholesteric LC film on glass was transformed to the polymer salt by treatment with KOH (0.05 M) for 10 min. The reflection color changes from green to red and finally to colorless when the KOH solution penetrates into the film. Subsequently, the film was dried in air, yielding again a green-reflecting film at ambient conditions. The optical sensors, thus prepared, can act as battery-free humidity sensors that respond to relative humidities ranging from 3 to 83 % in a fast and reversible manner. The change in reflection band is a result of the change in helix pitch in the film due to absorption and desorption of water, resulting in swelling/deswelling of the film material.

More recently, chemically activated hydrazone-based switches have been used for preparing the pH-responsive cholesteric LCs [84]. Interestingly, cholesteric LC films have been also applied for the discrimination between methanol and ethanol by the cholesteric LC films based on the diversity in molecular affinity of ethanol and methanol with the hydrogen-bridged cholesteric LC polymer networks [85]. The cholesteric LC networks are molecular-helix-based, 1D bandgap materials with a discrete reflection band in the visible part of the spectrum that depends on the pitch of the molecular helix. The changes in positions of the reflection bands of the cholesteric LC network accurately discriminate between the alcohol types and provide information on their ratio in case they are blended. Furthermore, Stroganov et al. [86] have fabricated a cation sensitive polymer stabilized cholesteric LC composite using crown ether moieties in the cholesteric LC polymer network. A printable hydrogen-bonded cholesteric LC polymer film has also been recently reported as a sensor for detection of gaseous trimethylamine [87]. However, all the applications mentioned above are still far from being usable in daily life. Often the selectivity, sensitivity, and signal changes do not meet the requirements set for the specific applications. Most sensors suffer from cross-selectivity and/or need co-analytes to work properly. Nevertheless the prospects are wide open for further explorations.

3 Stimuli-Responsive Self-Organized Three-Dimensional Liquid Crystalline Superstructures

3.1 Liquid Crystalline Cubic Superstructures

3.1.1 Modulation of 3D Cubic Nanostructures

BPs that self-organize into a periodic cubic lattice have recently been recognized as soft artificial 3D photonic crystals [88–92]. Compared with the photonic crystals fabricated by expensive and scale-prohibitive semiconductor manufacturing processes, the self-organization inherent in these “soft” materials is one that has the promise of overcoming inherent disadvantages without any complicated fabrication process. Additionally, being “soft” enables the materials to respond to various stimuli, resulting in dynamic photonic properties that have applications in the growing area of photonic integrated structures.

Given the thermotropic nature of BP LCs, it is not surprising that almost all the 3D cubic structures are thermo-responsive, but the shift of PBG wavelength is very narrow, usually less than 50 nm. Recently, Choi et al. [93] reported a wide (more than 100 nm) and reversible shift of PBG in a BP system as a function of temperature. Figure 20a shows typical POM images at three different temperatures during the cooling process. The observed platelet colors changed with temperature, and specifically the platelet color gradually changed from a short wavelength to a long wavelength with decreasing temperature. Reflection spectra collected at different temperatures for the cubic BP during the cooling process are shown in Fig. 20b. This reflection band is associated with the (200) reflection of cubic lattice vectors. As shown in Fig. 20b, a well-defined PBG is observed for the cubic BP, and

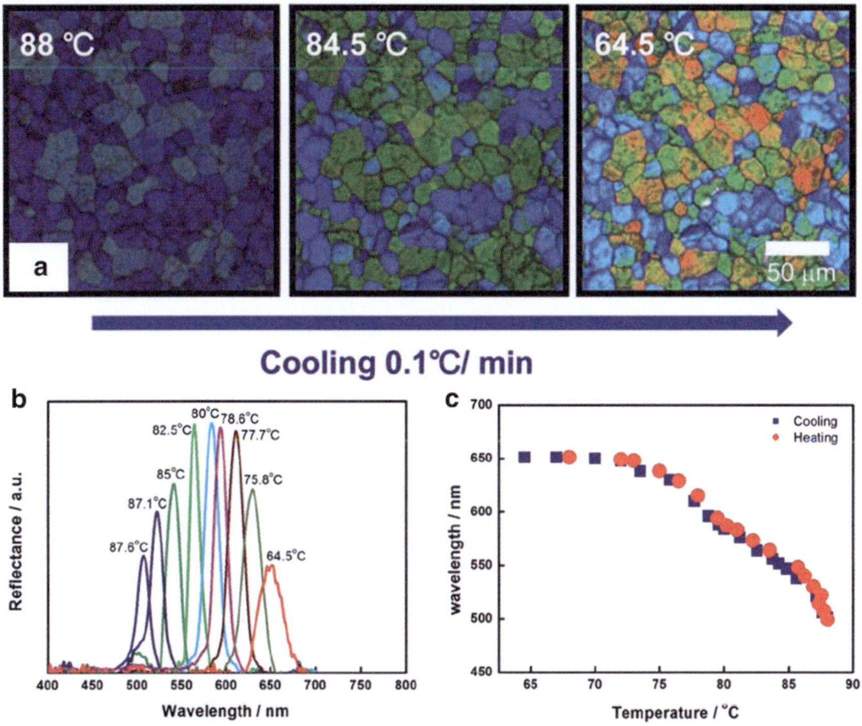

Fig. 20 (**a**) Typical POM images at three different temperatures during the cooling process. (**b**) Reflection spectra collected at different temperatures for the cubic BP during the cooling process. (**c**) Central wavelength of the PBG associated with the (200) reflection of cubic lattice vectors during cooling and heating processes between 88 and 63 °C. Reproduced with permission from [93]. Copyright 2013 John Wiley & Sons

the existence of the PBG persists with changes in temperature. The central wavelength of the PBG associated with the (200) reflection of cubic lattice vectors during cooling and heating processes between 88 and 63 °C is plotted in Fig. 20c. The change in wavelength with temperature during the heating process was in good agreement with changes during cooling. A shift in wavelength of about 150 nm was observed in a temperature range of about 25 °C.

BPs show three distinct transformations under the electric field, that is, Kerr effect, electrostriction and field-induced phase transitions [94]. The electrostriction is also known as the electric-field induced shift in Bragg reflective wavelength. For most of the cases, the tuning PBG wavelength is very narrow, around 50 nm. Lu et al. [95] reported a full visible color switching in reflected wavelength in the polymer stabilized BPs (PSBP). A color change in response to the applied voltage of a PSBP cell can be easily observed in a single-pixel cell in which the electrode area is switched to reflect a different color while the surrounding area with no electrodes remains the original blue color as shown in Fig. 21. The incorporation of a small amount of

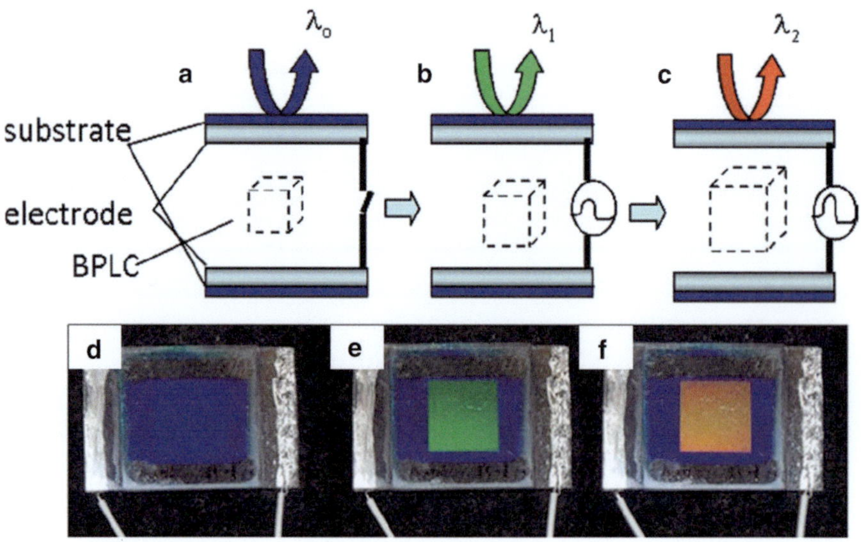

Fig. 21 Schematic illustration of electrically switched color of a PSBP cell in which the blue reflected wavelength (**a**) is switched to reflect a green color (**b**) and red (**c**) with an increase in applied voltage. The incorporation of polymer network maximizes the structural integrity and prevents structural deformation before reaching the critical field. Corresponding photographs of a one-pixel PSBP cell with 10 μm cell gap is operated in the reflective mode with appearance of color by Bragg reflection under (**d**) 0 V (*blue*), (**e**) 33 V (*green*), and (**f**) 40 V (*red*), respectively. Reproduced with permission from [95]. Copyright 2010 Optical Society of America

polymer network stabilizes against the occurrence of lattice deformation and broadens the range of electrically switched color in the visible spectrum.

Interestingly, Coles et al. [96] recently reported a stretchable BP LC gel whose optical properties can be manipulated by an applied strain. This kind of a self-assembled 3D photonic crystal was fabricated by the in situ photopolymerization of a mixture that included liquid crystalline monomer and di-acrylate reactive mesogens. As shown in Fig. 22, a color change is readily apparent in the gel films upon stretching. The observed mechano-chromic behavior is consistent with the imposed lateral stretch, inducing a reduction of the layer thickness and hence a reduction of the effective photonic crystal-lattice periodicity along the viewing direction due to the approximate volume conservation of the material. They also found that, unlike its undistorted counterpart, the mechanically deformed blue phase exhibits a Pockels electro-optic effect, which sets out new theoretical challenges and possibilities for low-voltage electro-optic devices.

Compared with thermal-, electrical-, and mechano-responsive BPs, optically tunable self-organized 3D cubic nanostructured BPs are receiving increasing attention especially in recent years [97]. The optical reflection wavelength tuning is usually quite narrow. Very recently, a breakthrough of full visible range reflection phototuning in BPs was achieved by Li et al. [98] using an axially chiral

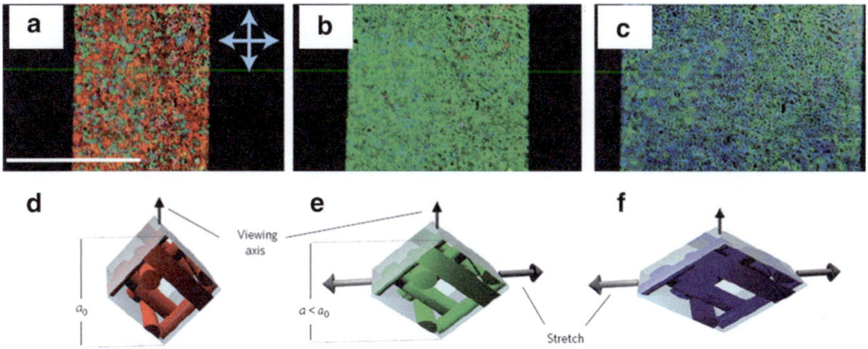

Fig. 22 Polarizing optical microscope images of a 20 μm-thick layer placed across a variable-width aperture: (**a**) The unstretched state exhibits a characteristic blue-phase texture. *Pale blue arrows* denote the orientation of the crossed polarizers (scale bar, 1 mm); (**b**) and (**c**), as the aperture is widened the sample is stretched and a color change induced. (**d**), (**e**), and (**f**), Schematic diagrams of a typical deformation of the blue-phase-I unit cell, viewed along the [011] direction and stretched along an axis perpendicular to it. Reproduced with permission from [96]. Copyright 2014 Nature Publishing Group

azobenzene dopant. The initial phase of the doped BP was the BP II with (100) lattice confirmed by the Kossel diagram as shown in Fig. 23c (0s) with a blue reflection color as shown in Fig. 23b (0s). Upon 408 nm light irradiation, the photoisomerization of the azobenzene dopant enlarged the size of the BP lattice and shifted the corresponding reflection toward longer wavelength. The rapid reverse process can occur photochemically with 532 nm light irradiation. Figure 23b shows BP images with various durations of pumping observed under a crossed polarizing optical microscope in the reflection mode. Upon 408 nm light irradiation, the reflection color of the BP thin film changed from blue (BP II) to red (BP I) within 15 s. The BP thin films were also indexed using a Kossel diagram as shown in Fig. 23c. Both blue and green images in Fig. 23b corresponded to (100) lattices of BP II, while both yellow-green and red images represented (110) lattices of BP I. Figure 23c also exhibited the variation of the Kossel diagram with different time upon 408 nm irradiation. As the irradiation time increased from 0 to 15 s, the circle patterns of the (100) lattice of the BPII enlarged, switching to diamond-shaped pattern corresponding to (110) lattice of BPI, and gradually shrinking. Increasing the dopant concentration shifted the initial BP reflection to the UV region, which was tuned across the entire visible region with broad range. Upon 408 nm light irradiation, the reflection bands of BPII red-shifted from the UV to 520 nm. Once the BPII transformed into the BPI, the corresponding reflection bands discontinuously jumped to 560 nm, which was further red-shifted from 560 to 710 nm within BPI state. Red, green, and blue reflections were successfully achieved in this self-organized 3D cubic nanostructured single thin film.

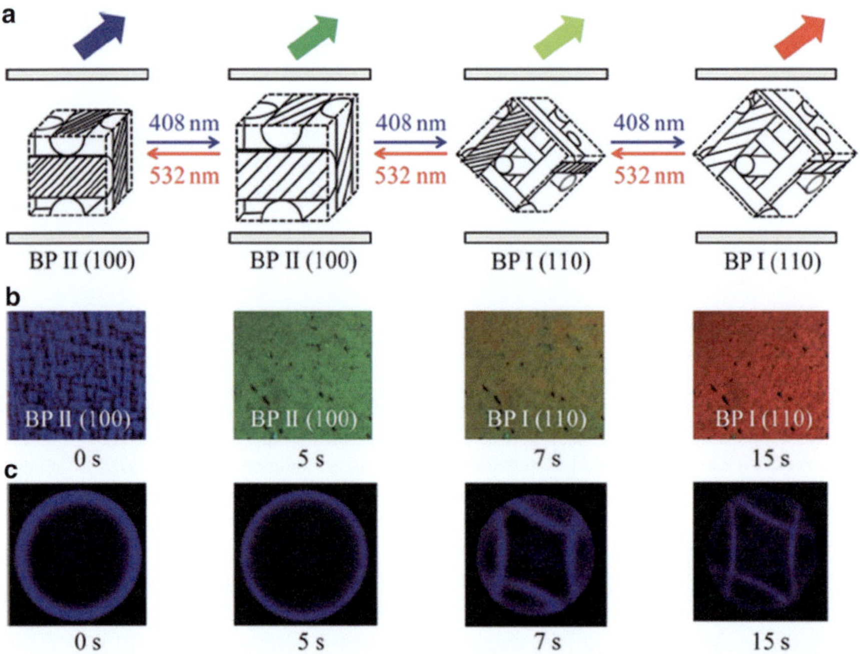

Fig. 23 (**a**): Schematic illustration of the phototuning of the BP. (**b**): Reflection color images BPs upon 408 nm light irradiation at 40 °C. (**c**): The corresponding Kossel diagrams with (**b**). Reproduced with permission from [98]. Copyright 2013 John Wiley & Sons

3.1.2 Template for Fabricating 3D Photonic Crystals

Since the introduction of the concept of 3D photonic crystals and PBG materials, scientists have pursued realization of photonic circuits in analogy to semiconductor electronics. This poses a major technological challenge for nanofabrication. The past decade has witnessed intensive research efforts related to the design and fabrication of photonic crystals by self-assembly approaches. One of the major limitations for the approaches to building 3D photonic materials is to scale the materials up from the nanoscale to the device dimensions. Exploiting the interplay between structural and optical properties, colloidal lattices or polymer networks embedded in BPs are promising candidates for such materials. Ravnik et al. [99] have presented extensive computer modeling to demonstrate that the BPs can serve as an excellent 3D template for colloidal spheres.

Recently, Coles et al. [100] have reported a promising approach for fabrication of materials with nanoscale features by the transfer of liquid crystalline BP cubic structure to polymer networks. The fabrication process is shown in Fig. 24. Essentially, a cast of the BP was formed using a polymer network. In stage A, BP I self-assembled on cooling from the isotropic phase, showing a characteristic platelet texture. The cell was then illuminated with ultraviolet light to polymerize the

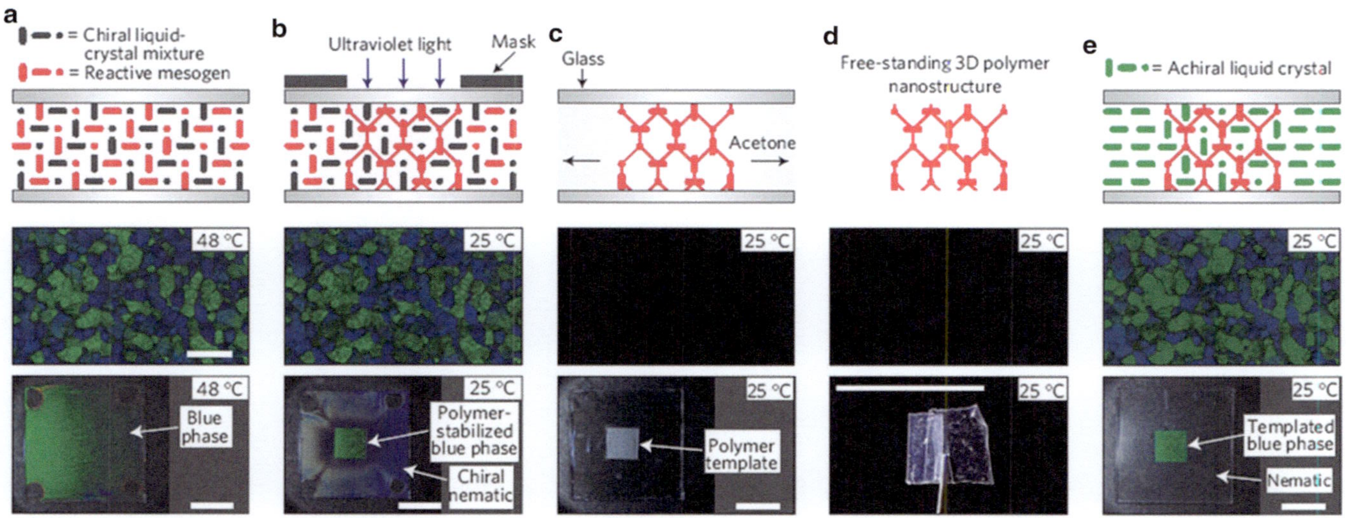

Fig. 24 Formation of the 3D nanostructured polymer. (*Top*) Schematic diagram of the procedure. (*Middle*) The corresponding transmission optical polarizing microscopy images (scale bar, 100 μm) and (*Down*) photographs of cell (scale bars, 5 mm) at each step. Reproduced with permission from [100]. Copyright 2012 Nature Publishing Group

reactive mesogens (Fig. 24, stage B). To provide a direct comparison between polymerized and unpolymerized regions, a mask was used such that only the center of the cell was illuminated. In stage C, the cell was placed in acetone, typically for 16–22 h to remove, by diffusion, the LCs and remaining unpolymerized free reactive-mesogen/photo-initiator mixture. When the cell was removed from the acetone, any acetone remaining in the cell was left to evaporate at room temperature. The cell could then be opened and the polymer was removed using a razor blade. A solid, free-standing, polymer structure was thus obtained (Fig. 24, stage D). To confirm conclusively that the polymer retained the 3D structure of the BP, an unopened cell was refilled with the nematic LC (5CB), as shown in Fig. 24, stage E. In the non-ultraviolet-exposed regions of the cell, where no polymer remained, a nematic phase was observed, as expected (Fig. 24, stage E). In the ultraviolet-exposed regions, which contained the crosslinked polymer, a BP-like structure was observed, such template BPs have unprecedented thermal stability in the range −125 to 125 °C. Furthermore, BP-templated laser was successfully observed, which further demonstrated the 3D structure of the template and the use of such a template in photonic devices. Achieving laser emission from BP structures is non-trivial. One major benefit is that laser emission in three orthogonal directions can be obtained simultaneously by virtue of the three-dimensional PBG. This was first demonstrated using BP II, which exhibits a double-twist structure [101].

3.2 *Liquid Crystalline Spherically Symmetric Superstructures*

3.2.1 Nematic Liquid Crystal Droplets

Polymer dispersions of nematic LC droplets have been studied extensively in the past, which are well known as polymer dispersed LCs (PDLCs) [102]. In the case of PDLCs, the droplets are usually made by the phase separation method. This means that initially the LC is dissolved in a polymer and the droplets are formed by polymerization-, thermally- or solvent-induced phase separation. In this case the droplets are very small, usually less than one micrometer, very dense and sometimes also of irregular shapes. Recently, Muševič et al. [103] reported the individual nematic LC droplets embedded in a polymer matrix which can be used as the tunable, low-loss whispering-gallery-mode (WGM) microresonators.

The LC microresonators are spherical LC microdroplets, suspended in a supporting polymer matrix. They were prepared by mechanically mixing a small amount (~5 %) of fluorescently labeled nematic LC E12 (a commercial mixture of cyanobiphenyl nematic LCs) and polydimethylsiloxane (PDMS). A micrograph of a typical microdroplet of E12 in PDMS is shown in Figs. 25a and 26b. A uniaxial cross in Fig. 25b clearly indicates that the internal structure of the director field, describing molecular orientation inside a spherical cavity, is radial, as shown schematically in Fig. 25c. This means that nematic LC molecules are aligned perpendicularly to the surface of the microcavity, which creates splay elastic

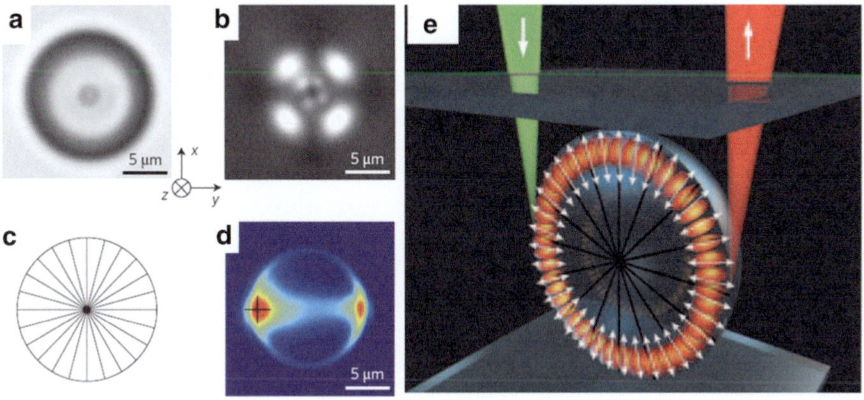

Fig. 25 Light in liquid-crystal microdroplets. (**a**) Microdroplet of nematic LC E12 in PDMS. (**b**) The same droplet under crossed polarizers. (**c**) Structure of the director field inside the microdroplet, deduced from (**b**). The lines (that is, the director field) present the direction of orientation of LC molecules. (**d**) Detected light intensity under illumination by a strongly focused beam of the Art laser tweezers, illuminating the submicrometer sized spot near the rim of the droplet, indicated by the *black cross*. (**e**) Schematic view of the excitation of WGMs in a LC droplet by a focused Art laser (*green light* illuminating the spot on the droplet), and detection of the radiated WGM light (*red light*). Reproduced with permission from [103]. Copyright 2009 Nature Publishing Group

distortion of the nematic LC; and a point topological defect, called the radial hedgehog defect, appears in the center of the droplet. When the dye-doped E12 microdroplet of ~10 mm diameter was illuminated with a tightly focused 514 nm laser beam near its edge, a bright spot was observed on the other side of the droplet as can be seen in Fig. 25d. A stripe of light in between these two points, which is in fact a light ring along the circumference of the microcavity, is also clearly visible. The above observations confirm that WGMs were observed in a LC droplet as shown in Fig. 25e.

After showing that LC droplets can support WGMs, the tunable properties of the droplets were further investigated. Electric field can change the nematic director orientation inside the droplets, which has an effect on the average refractive index for the light traveling inside the droplet. By applying moderate field of up to 2.6 V/μm the WGMs were tuned almost 20 nm for the modes near 600 nm [103, 104]. Furthermore, the WGMs in a LC droplet can also be tuned by temperature and mechanical deformation [105], and they can also be used as versatile chemical sensor [106].

3.2.2 Cholesteric Liquid Crystal Droplets

Spherical Bragg-reflector microcavities, based on an alternating series of low- and high refractive-index material concentric shells, are particularly appealing because of their perfect rotational symmetry in three dimensions, where the photonic band gap is expected to be independent of the direction of light propagation and the light is strongly confined in all directions [107, 108]. However, unfortunately, it is quite difficult to manufacture any 3D solid-state microcavity [109]. There have been several attempts to produce 3D Bragg-onion resonators in the solid-state structures by chemical synthesis [110] or by combining etching and chemical vapor deposition [108].

Instead of building a solid-state 3D microlaser, Muševič et al. [111] created a "soft" 3D Bragg-onion microlaser by dispersing a microdroplet of dye-doped cholesteric LCs into an isotropic carrier fluid. A typical cholesteric droplet is shown in Fig. 26a, b, the alternating light-dark concentric shells were clearly observed which are due to the radial modulation of the index of refraction. The radial period corresponds to one half of the cholesteric pitch which can be

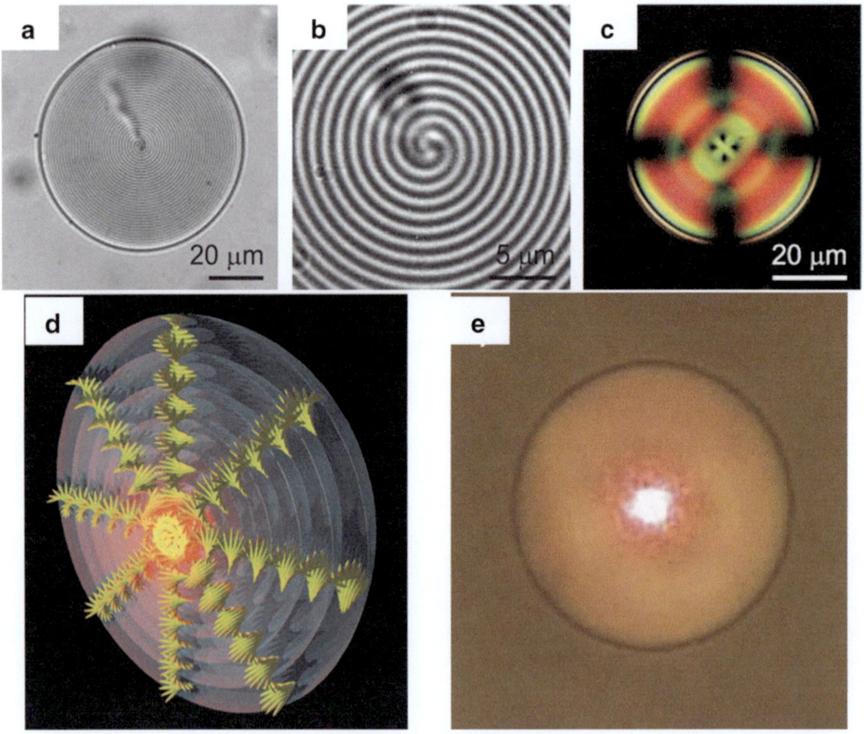

Fig. 26 Cholesteric LC droplets micrographs. (**a**) A cholesteric LC droplet with the pitch of $p = 2.2$ µm in glycerol. The perfect structure with nearly concentric shells is clearly visible. From the center to the surface there is a defect line, that is out of focus. (**b**) A magnification of the central part of a droplet when viewing along the defect line. (**c**) A short pitch cholesteric LC droplet viewed in crossed polarizer. (**d**) The schematic view of the arrangement of cholesteric LC molecules in a cholesteric microdroplet with parallel anchoring of the LC molecules at the surface. (**e**) Omnidirectional (3D) lasing in a cholesteric droplet illuminated by laser pulses ($\lambda = 532$ nm) and a weak white background illumination. Reproduced with permission from [111]. Copyright 2010 Optical Society of America

adjusted by changing the concentration of chiral dopant. When irradiating a single cholesteric LC microdroplet by a pulsed laser, a bright red laser spot suddenly appeared in the center of the droplet, as shown in Fig. 26e. The radiating light became very strong by further increasing the pump-light intensity. Interestingly, the laser light was emitted from the center of the cholesteric LC microdroplet in all directions, thus can act as a coherent, point-like, and omnidirectional source of light. Moreover, the lasing wavelength depends solely on the natural helical period of the cholesteric LCs and can be easily tuned by various stimuli. The cholesteric LC onion microlasers could also be combined with optical fiber waveguides to collect radiating light into the waveguides. The proposed procedure of making a cholesteric LC onion microlaser by mechanical mixing is simple and straightforward and can produce millions of microlasers for soft-matter photonic devices.

If there is a monodisperse core-shell type emulsion with a cholesteric LC shell, it enables us to exploit the full potential of the omnidirectional laser resonator because of the following characteristics: (1) the mono-dispersity prevents the widening of width of the lasing wavelength induced by the poly-dispersity of the droplet size; (2) various dyes can be dissolved in the core part of the core-shell type emulsion for the distributed Bragg reflection (DBR) type laser action; (3) output direction of the laser from the core-shell emulsion with a thin cholesteric LC shell is tunable by the positional control of irradiated area. Uchida et al. [112] reported the fabrication of the cholesteric LC shells as a water-oil-water (W/O/W) double emulsion using a microfluidic device. The oil phase of the W/O/W emulsion is made up of a cholesteric LC phase. The photonic structure in the cholesteric LC shell is confirmed by the spatial mapping of fluorescence emission spectra of a dye dissolved in the aqueous core region with a two-dimensional imaging spectrometer. By changing the combination of dyes and cholesteric LCs, three types of laser modes (DFB, DBR, and WGM) using cholesteric LC shells are also presented. Since the center of the core-shell structure is made not of the cholesteric LC phase but of the isotropic aqueous phase, cholesteric LC helical axis would remain perpendicular to the shell surface even if the shell is deformed. The PBG could be fine-tuned by the external field, such as the magnetic and electric fields. Furthermore, a variety of dyes, either hydrophilic or hydrophobic, are utilizable as the laser dye for the cholesteric LC shell resonators.

More recently, Li et al. [113] reported the photoresponsive monodisperse cholesteric liquid crystalline micro-shells fabricated through the capillary-based microfluidic method from the photoresponsive cholesteric LCs (Fig. 27). The photoresponsive cholesteric LCs have been formulated by doping a novel axially chiral azobenzene switch into the commercially available LC host E7. In the presence of DCM dye, these micro-shells exhibited photonic band edge laser emission in all directions due to the radial disposition of the cholesteric LC helices within the shells. The laser emission wavelength self-tunes from ~590 to ~640 nm when pumped due to the simultaneous photoisomerization of the chiral molecular switch. After stored in darkness for 12 h, the emitting

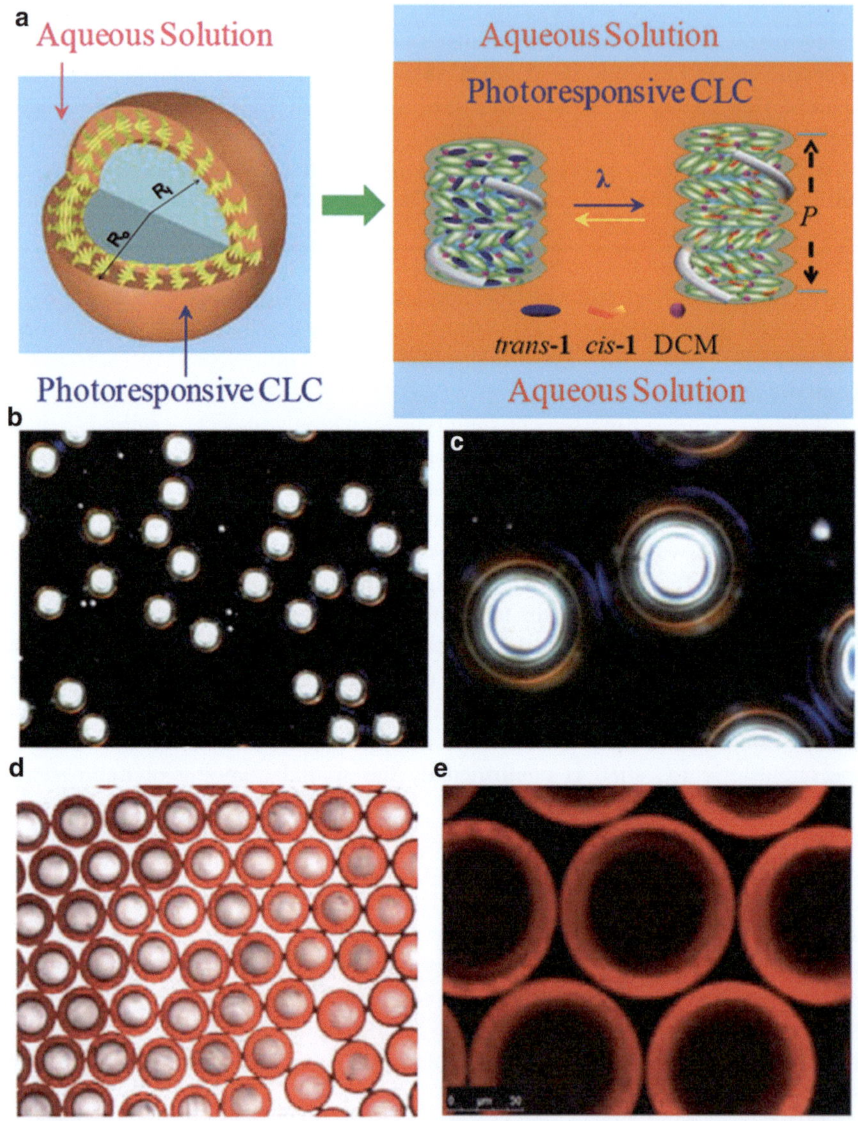

Fig. 27 (**a**) Schematic structure of photoresponsive monodisperse cholesteric LC micro-shell as a water/oil/water double emulsion (*left*) and schematic mechanism of phototunable lasing enabled by light-driven chiral molecular switch (*right*). Cross-polarized (**a**, **b**) and confocal (**c**, **d**) images of the cholesteric LC shells with $R_o = 58$ μm and $R_i = 40$ μm taken under microscope with different magnification. Reproduced with permission from [113]. Copyright 2014 John Wiley & Sons

wavelength fully recovered to its initial state. These new organic soft and dynamic photonic band gap structures may be useful for biological imaging and sensing in addition to 3D PGB applications. This study has opened up some interesting possibilities. For example, appropriate combination of materials by judicious choice may lead to phototunablc lasing across full visible range, thus providing omnidirectional red, green, and blue lasers from the same micro-shell.

4 Conclusion and Outlook

Liquid crystalline photonic materials are intriguing due to their unique self-organized superstructures and the abilities of being tuned by various external stimuli. This chapter has demonstrated the recent progress of stimuli-responsive self-organized liquid crystalline superstructures for 1D and 3D photonic applications. Cholesteric LCs can be considered as 1D photonic crystal, which can respond to various stimuli such as heat, light, electric field, and chemical environment. 3D cubic superstructures of blue phase LCs and spherically symmetric superstructures of LC microdroplets are fascinating for tunable "soft" 3D photonic crystals, which have the promise of overcoming inherent disadvantages of the traditional expensive and scale-prohibitive solid-state counterparts without any complicated fabrication process. Importantly, LC-based photonic crystals open a new area of research called *soft matter photonics*. All living creatures have extraordinary complexity and are made mostly of soft matter, where the discrete building blocks are self-organized by multiple and complex interactions. In this context, artificial photonic circuits made up of a single self-assembled soft matter component seem to be highly promising. So serious attempts should be directed to realize them in a facile manner. Additionally, bottom-up approach in this case would enable cheap, large-scale production of such photonic structures without the need of complex fabrication procedures such as lithography that is limited to planar geometries. Such soft structures would also have a number of useful properties such as tunability, real-time reconfigurability, and self-healing. Actually there is still much work to be done in fabricating such kind of self-organized soft photonic structures, and it would be very exciting if the whole integrated photonic structure in the future could be made from soft materials by using different LC materials such as smectic, ferroelectric, discotic, blue phase and others, plus different shapes of LC droplets and by applying various external stimuli, an innumerable combinations are conceivable.

Acknowledgements The preparation of this chapter benefited from the support to Quan Li by US Air Force Office of Scientific Research (AFOSR), US Department of Energy (DOE), US Army Research Office (ARO), US Department of Defense Multidisciplinary University Research Initiative (DoD MURI), US National Aeronautics and Space Administration (NASA) and US National Science Foundation (NSF), and Ohio Third Frontier. We thank all the Li's lab current and former members and his collaborators, whose names can be found in the references, for their significant contributions in this project.

References

1. Q. Li (ed.), *Liquid Crystals Beyond Displays: Chemistry, Physics, and Applications* (Wiley, New Jersey, 2012)
2. Q. Li (ed.), *Intelligent Stimuli-Responsive Materials: From Well-Defined Nanostructures to Applications* (Wiley, New Jersey, 2013)
3. Q. Li, *Nanoscience with Liquid Crystals: From Self-Organized Nanostructures to Applications* (Springer, Heidelberg, 2014)
4. P. Oswald, P. Pieranski, *Nematic and Cholesteric Liquid Crystals* (Taylor and Francis, Boca Raton, 2005)
5. S. Pieraccini, S. Masiero, A. Ferrarini, G.P. Spada, Chem. Soc. Rev. **40**, 258–271 (2011)
6. H. Kikuchi, in *Liquid Crystalline Blue Phases*. Liquid Crystalline Functional Assemblies and Their Supramolecular Structures (Springer, Berlin, 2007), pp. 99–117
7. L. Wang, W. He, X. Xiao et al., Small **8**, 2189–2193 (2012)
8. L. Wang, W. He, X. Xiao et al., J. Mater. Chem. **22**, 2383–2386 (2012)
9. B. Li, W. He, L. Wang et al., Soft Matter **9**, 1172–1177 (2013)
10. K. Dholakia, T. Cizmar, Nat. Photon. **5**, 335–342 (2011)
11. M. Padgett, R. Bowman, Nat. Photon. **5**, 343–348 (2011)
12. G.M. Whitesides, Nature **442**, 368–373 (2006)
13. D. Psaltis, S.R. Quake, C. Yang, Nature **442**, 381–386 (2006)
14. P.S. Drzaic, *Liquid Crystal Dispersions* (World Scientific, Singapore, 1995)
15. F.C. Mackintosh, T.C. Lubensky, Phys. Rev. Lett. **67**, 1169–1172 (1991)
16. J.H. Erdmann, S. Zumer, J.W. Doane, Phys. Rev. Lett. **64**, 1907–1910 (1990)
17. P.G. de Gennes, J. Prost, *The Physics of Liquid Crystals* (Oxford University Press, Oxford, 1995)
18. F. Xu, P.P. Crooker, Phys. Rev. E **56**, 6853–6860 (1997)
19. R.A. Soref, J. Appl. Phys. **41**, 3022–3026 (1970)
20. R.S. Pindak, C.-C. Huang, J.T. Ho, Phys. Rev. Lett. **32**, 43–46 (1974)
21. F. Zhang, D.K. Yang, Liq. Cryst. **29**, 1497–1501 (2002)
22. D.J. Broer, C.M. Bastiaansen, M.G. Debije et al., Angew. Chem. Int. Ed. **51**, 7102–7109 (2012)
23. W. Hu, H. Cao, L. Song et al., J. Phys. Chem. B **113**, 13882–13885 (2009)
24. J. Guo, H. Wu, F. Chen et al., J. Mater. Chem. **20**, 4094–4102 (2010)
25. Z. Li, P. Desai, R.B. Akins, G. Ventouris et al., Proc. SPIE-Int. Soc. Opt. Eng. **4658**, 7–13 (2002)
26. M.J. Escuti, C.C. Bowley, G.P. Crawford et al., Appl. Phys. Lett. **75**, 3264–3266 (1999)
27. H. Yu, B.Y. Tang, J. Li et al., Opt. Express **13**, 7243–7249 (2005)
28. S.S. Choi, S.M. Morris, W.T.S. Huck et al., Adv. Mater. **22**, 53–56 (2010)
29. T.H. Lin, H.C. Jau, C.H. Chen et al., Appl. Phys. Lett. **88**, 061122 (2006)
30. V. Natarajan, J.M. Wofford, V.P. Tondiglia et al., J. Appl. Phys. **103**, 093107 (2008)
31. M.E. McConney, V.P. Tondiglia, L.V. Natarajan et al., Adv. Opt. Mater. **1**, 417–421 (2013)
32. D.J. Broer, J. Lub, G.N. Mol, Nature **378**, 467–469 (1995)
33. R.A.M. Hikmet, H. Kemperman, Nature **392**, 476–479 (1998)
34. P.R. Gerber, Appl. Phys. Lett. **44**, 932–934 (1984)
35. W. Hu, H. Zhao, L. Song et al., Adv. Mater. **22**, 468–472 (2010)
36. M. Scalora, J.P. Dowling, C.M. Bowden et al., J. Appl. Phys. **76**, 2023–2026 (1994)
37. S.F. Mingaleev, Y.S. Kivshar, J. Opt. Soc. Am. B **19**, 2241–2249 (2002)
38. S. Mujumdar, H. Ramachandran, Opt. Lett. **26**, 929–931 (2001)
39. A.H. Gevorgyan, Tech. Phys. **47**, 1008–1013 (2002)
40. A.H. Gevorgyan, Tech. Phys. Lett. **29**, 60–68 (2003)
41. J. Hwang, M.H. Song, B. Park et al., Nat. Mater. **4**, 383–387 (2005)
42. M.H. Song, B. Park, S. Nishimura et al., Adv. Funct. Mater. **16**, 1793–1798 (2006)
43. M.H. Song, B. Park, K.C. Shin et al., Adv. Mater. **16**, 779–783 (2005)

44. T.J. White, M.E. McConney, T.J. Bunning, J. Mater. Chem. **20**, 9832–9847 (2010)
45. E. Sackmann, J. Am. Chem. Soc. **93**, 7088–7090 (1971)
46. U.A. Hrozhyk, S.V. Serak, N.V. Tabiryan et al., Adv. Funct. Mater. **17**, 1735–1742 (2007)
47. H. Akiyama, V.A. Mallia, N. Tamaoki, Adv. Funct. Mater. **16**, 477–484 (2006)
48. M. Mathews, R.S. Zola, D.-K. Yang et al., J. Mater. Chem. **21**, 2098–2103 (2011)
49. T. Yoshioka, T. Ogata, T. Nonaka et al., Adv. Mater. **17**, 1226–1229 (2005)
50. M. Mathews, N. Tamaoki, J. Am. Chem. Soc. **130**, 11409–11416 (2008)
51. Q. Li, L. Green, N. Venkataraman et al., J. Am. Chem. Soc. **129**, 12908–12909 (2007)
52. X. Chen, L. Wang, Y. Chen et al., Chem. Commun. **50**, 691–694 (2013)
53. J. Ma, Y. Li, T. White et al., Chem. Commun. **46**, 3463–3465 (2010)
54. Q. Li, Y. Li, J. Ma et al., Adv. Mater. **23**, 5069–5073 (2011)
55. Y. Li, A. Urbas, Q. Li, J. Am. Chem. Soc. **134**, 9573–9576 (2012)
56. Y. Li, M. Wang, A. Urbas, Q. Li, J. Mater. Chem. C **1**, 3917–3923 (2013)
57. A.A. Beharry, O. Sadovski, G.A. Woolley, J. Am. Chem. Soc. **133**, 19684–19687 (2011)
58. Y. Wang, A. Urbas, Q. Li, J. Am. Chem. Soc. **134**, 3342–3345 (2012)
59. L. Wang, H. Dong, Y. Li et al., J. Am. Chem. Soc. **136**, 4480–4483 (2014)
60. T. Ikeda, J. Mater. Chem. **13**, 2037–2057 (2003)
61. A. Chanishvili, G. Chilaya, G. Petriashvili et al., Mol. Cryst. Liq. Cryst. **409**, 209–218 (2004)
62. Q. Li, L. Li, J. Kim et al., Chem. Mater. **17**, 6018–6021 (2005)
63. M.-H. Wu, C.-C. Chu, M.-C. Cheng et al., Mol. Cryst. Liq. Cryst. **557**, 176–189 (2012)
64. U.A. Hrozhyk, S.V. Serak, N.V. Tabiryan et al., Adv. Mater. **19**, 3244–3247 (2007)
65. K. Rameshbabu, A. Urbas, Q. Li, J. Phys. Chem. B **115**, 3409–3415 (2011)
66. N.P.M. Huck, W.F. Jager, B. deLange, B.L. Feringa., Science **273**, 1686–1688 (1996)
67. C. Denekamp, B.L. Feringa, Adv. Mater. **10**, 1080–1082 (1998)
68. T. Yamaguchi, T. Inagawa, H. Nakazumi et al., Chem. Mater. **12**, 869–871 (2000)
69. G. Joly, A. Anakkar, Liq. Cryst. **26**, 1251–1255 (1999)
70. S. Abraham, V.A. Mallia, K.V. Ratheesh et al., J. Am. Chem. Soc. **128**, 7692–7698 (2006)
71. V.A. Mallia, N. Tamaoki, Chem. Soc. Rev. **33**, 76 (2004)
72. T. Matsui, K. Nakayama, M. Ozaki, K. Yoshino, Appl. Phys. Lett. **76**, 1228–1230 (2000)
73. J. Guan, M. Zhang, W. Gao et al., ChemPhysChem **13**, 1425–1428 (2012)
74. B.L. Feringa, N.P.M. Huck, H.A. van Doren, J. Am. Chem. Soc. **117**, 9929–9930 (1995)
75. R.A. van Delden, N. Koumura, N. Harada, B.L. Feringa, Proc. Natl. Acad. Sci. U. S. A. **99**, 4945–4949 (2002)
76. J. Vicario, M. Walko, A. Meetsma, B.L. Feringa, J. Am. Chem. Soc. **128**, 5127–5135 (2006)
77. M. Mathews, R.S. Zola, S. Hurley et al., J. Am. Chem. Soc. **132**, 18361–18366 (2010)
78. Y. Li, M. Wang, T.J. White et al., Angew. Chem. Int. Ed. **52**, 8925–8929 (2013)
79. H. Hayasaka, T. Miyashita, M. Nakayama et al., J. Am. Chem. Soc. **134**, 3758–3765 (2012)
80. Y. Li, X. Xue, M. Wang et al., Angew. Chem. Int. Ed. **52**, 13703–13707 (2013)
81. Y. Han, K. Pacheco, C.W. Bastiaansen et al., J. Am. Chem. Soc. **132**, 2961–2967 (2010)
82. A. Saha, Y. Tanaka, Y. Han et al., Chem. Commun. **48**, 4579–4581 (2012)
83. N. Herzer, H. Guneysu, D.J.D. Davies et al., J. Am. Chem. Soc. **134**, 7608–7611 (2012)
84. X. Su, S. Voskian, R.P. Hughes, I. Aprahamian, Angew. Chem. Int. Ed. **52**, 10934–10939 (2013)
85. C.-K. Chang, C.M.W. Bastiaansen, D.J. Broer et al., Adv. Funct. Mater. **22**, 2855–2859 (2012)
86. V. Stroganov, A. Ryabchun, A. Bobrovsky et al., Macromol. Rapid Commun. **33**, 1875–1881 (2012)
87. J.E. Stumpel, C. Wouters, N. Herzer et al., Adv. Opt. Mater. **2**, 459–464 (2014)
88. L. Wang, W. He, M. Wang et al., Liq. Cryst. **40**, 354–367 (2013)
89. L. Wang, W. He, X. Xiao et al., J. Mater. Chem. **22**, 19629–19633 (2012)
90. A. Mazzulla, G. Petriashvili, M.A. Matranga et al., Soft Matter **8**, 4883–4885 (2012)
91. L. Wang, W. He, Q. Wang et al., J. Mater. Chem. C **1**, 6526–6531 (2013)
92. L. Wang, L. Yu, X. Xiao et al., Liq. Cryst. **39**, 629–638 (2012)

93. S.T. Hur, B.R. Lee, M.J. Gim et al., Adv. Mater. **21**, 3002–3006 (2013)
94. D.C. Wright, N.D. Mermin, Rev. Mod. Phys. **61**, 385–432 (1989)
95. S.Y. Lu, L.C. Chien., Opt. Lett. **35**, 562–564 (2010)
96. F. Castles, S.M. Morris, J.M.C. Hung et al., Stretchable liquid-crystal blue-phase gels. Nat. Mater. **13**, 817–821 (2014)
97. X. Chen, L. Wang, C. Li et al., Chem. Commun. **49**, 10097–10099 (2013)
98. T.-H. Lin, Y. Li, C.-T. Wang, H.-C. Jau et al., Adv. Mater. **25**, 5050–5054 (2013)
99. M. Ravnik, G.P. Alexander, J.M. Yeomans et al., Proc. Natl. Acad. Sci. U. S. A. **108**, 5188–5192 (2011)
100. F. Castles, F. V. Day, S. M. Morris et al., Nat. Mater. **11**, 599–603 (2012)
101. W. Cao, A. Munoz, P. Palffy-Muhoray et al., Nat. Mater. **1**, 111–113 (2002)
102. J.W. Doane, N.A. Vaz, B.G. Wu et al., Appl. Phys. Lett. **48**, 269–271 (1986)
103. M. Humar, M. Ravnik, S. Pajk et al., Nat. Photonics **3**, 595–600 (2009)
104. I. Musevic, M. Skarabo, M. Humar, J. Phys. Condens. Matter **23**, 284112 (2011)
105. I. Musevic, M. Humar, Proceedings of SPIE **7955**, 795509 (2011)
106. M. Humar, I. Musevic, Opt. Express **19**, 19836–19844 (2011)
107. K.G. Sullivan, D.G. Hall, Phys. Rev. A **50**, 2701–2707 (1994)
108. Y. Xu, W. Liang, A. Yariv et al., Opt. Lett. **29**, 424–426 (2004)
109. A. Tandaechanurat, S. Ishida, K. Aoki et al., Appl. Phys. Lett. **94**, 171115 (2009)
110. I. Gourevich, L.M. Field, Z. Wei et al., Macromolecules **39**, 1449–1454 (2006)
111. M. Humar, I. Musevic, Opt. Express **18**, 26995–27003 (2010)
112. Y. Uchida, Y. Takanishi, J. Yamamoto, Adv. Mater. **25**, 3234–3237 (2013)
113. L. Chen, Y. Li, J. Fan et al., Adv. Opt. Mater. **2**, 845–848 (2014)

Gel-Immobilized Colloidal Photonic Crystals with Tunable Properties

Toshimitsu Kanai

Abstract Colloidal crystals, three-dimensional periodic arrays of monodisperse colloidal particles, have attracted considerable attention owing to their novel optical applications as photonic crystals. In particular, the possibility to tune the optical properties of colloidal crystals immobilized in polymer gels by adjusting the gel size through an external stimulus is a very attractive feature. Therefore, these crystals are expected to have applications such as tunable photonic crystals, and biological and chemical sensors for monitoring changes through the optical stop-band wavelength or reflection color. This chapter provides an overview of our recent work on the preparation and tunable properties of gel-immobilized colloidal crystals with high optical quality. We developed an air-pulse-drive system for preparing a large single-crystal-like colloidal crystal film. Polycrystalline charge-stabilized colloidal crystals could be converted into single-crystal-like ones through a flow-induced shear effect by running the suspension in a flat capillary cell. These crystals could be subsequently immobilized in a hydrogel network by a photopolymerization technique that preserves the high crystalline quality. Tuning of their properties by an external stimulus, such as solvent exchange, temperature change, and mechanical stress, is described. The optical stop-band wavelength of the crystals can be tuned over a wide wavelength region while preserving high spectral quality.

Keywords Colloidal crystals • Photonic crystals • Polymer gels • Tunable properties • Flexible photonic crystals

1 Introduction

Monodisperse colloidal particles spontaneously form three-dimensional periodic arrays in a liquid medium. This ordered state is called colloidal crystals, by analogy with atomic crystals [1–8]. Colloidal crystals function as photonic crystals because of the spatial periodicity of the refractive index between the particles and

T. Kanai (✉)
Faculty of Engineering, Yokohama National University, 79-5 Tokiwadai, Hodogaya, Yokohama, Kanagawa 240-8501, Japan
e-mail: tkanai@ynu.ac.jp

D. Comoretto (ed.), *Organic and Hybrid Photonic Crystals*,
DOI 10.1007/978-3-319-16580-6_19

surroundings [9–13]. Thus far, photonic crystals have been predominantly fabricated using a lithographic technique [14, 15], and various innovative optical applications such as photonic chips [16], low-threshold lasers [17], and superprisms [18] have been reported. While the crystal structure of self-assembled colloidal crystals cannot be designed at will as well as those fabricated through a lithographic technique, self-assembly is one of the most promising approaches for fabricating photonic crystals because large crystals can be prepared at a low cost. In addition, three-dimensional photonic crystals with optical stop bands in the visible light region can be prepared readily by using submicrometer colloids. Since the optical stop band of colloidal crystals is observed as the Bragg reflection, the crystals composed of submicrometer colloids exhibit strong reflection color. Therefore, these are expected to have applications as not only photonic crystals that function in the visible region, but also color pigments that are not degraded by UV light.

The Bragg diffraction wavelength is expressed as [19, 20]

$$m\lambda = 2n_{\mathrm{c}}d_{\mathrm{hkl}}\sin\theta \tag{1}$$

where m is an integer, λ is the wavelength of light, n_{c} is the refractive index of the crystal, d_{hkl} is the interplanar spacing of the hkl lattice planes, and θ is the Bragg diffraction angle. As shown in Eq. (1), the Bragg wavelength of the colloidal crystal at a specific angle of incident light is determined from the value of the refractive index and lattice constant of the crystal. Since it is difficult to change the refractive index of materials significantly, adjusting the lattice constant, which is determined by the particle size and particle volume fraction, is essential for setting the desired optical stop-band wavelength. Therefore, it is meaningful to classify colloidal crystals into two categories based on packing: tightly or loosely packed.

The former comprises opal-type colloidal crystals [1–4], in which monodisperse particles are aligned with closest packing in a dry state. By evaporating the liquid medium of the colloidal suspension on a substrate, the particles spontaneously assemble to form the closest-packed structure. The inverse structure, which is called the inverse opal [21–23], can be obtained by infiltration of a different material with a high refractive index into the void space in the opal structure. Since a high contrast in refractivity between the particles and surroundings is preferable in applications as photonic band-gap materials, this type of colloidal crystal has been extensively studied. The latter comprises charge-stabilized colloidal crystals, in which charged particles form crystalline structures, but with a low packing density in the liquid medium owing to repulsive electrostatic interactions among the particles [5–7]. A great advantage of the charge-stabilized system is that the lattice constant of crystals can be significantly altered by changing the particle volume fraction. Thus, the Bragg wavelength of the crystals can be tuned over a wide range. Furthermore, a technique to immobilize colloidal crystals in a polymer gel has been recently developed. This imparts stability, as well as unique tunable properties, to the crystal structure (Fig. 1) [24, 25]. In general, polymer gels undergo a significant change in size in response to external stimulus, for example, a change in temperature and pH, and mechanical stress. Thus, the lattice constant or

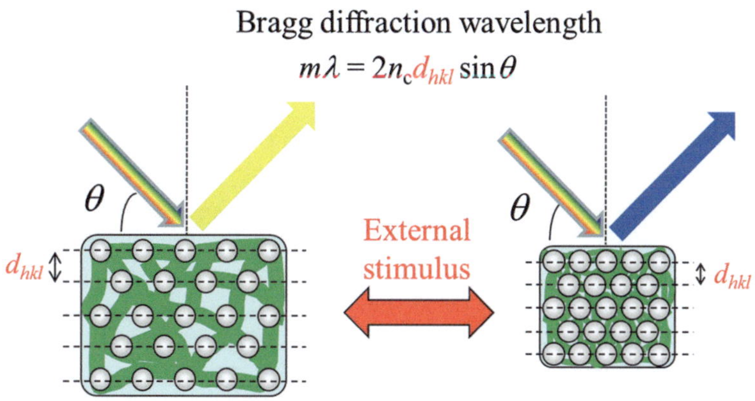

Fig. 1 Schematic image of tunable colloidal photonic crystals immobilized in polymer gels

Bragg wavelength of the colloidal crystals embedded in the gel can be altered on demand. This is very useful for applications such as tunable photonic crystals, and chemical and biological sensors for monitoring changes in a specific environment through the Bragg wavelength or diffraction color [26–28].

Despite the usefulness of gel-immobilized colloidal photonic crystals, the development of their practical applications is less advanced because of the difficulties in preparing large single-crystalline colloidal crystals immobilized in a gel network. In this chapter, our fabrication method is introduced, which involves the preparation of a large single-crystal-like colloidal crystal film through a flow-induced shear effect and its subsequent immobilization by photopolymerization without deterioration of the crystalline quality. In addition, tuning of the properties of the gel-immobilized colloidal crystal films by an external stimulus, such as solvent exchange, temperature change, and mechanical stress, is described.

2 Preparation of Gel-Immobilized Colloidal Crystal Films with High Optical Quality

2.1 Air-Pulse-Drive Preparation of Single-Crystal-Like Colloidal Crystals

Sawada et al. [29] were the first to demonstrate that a centimeter-order single-domain colloidal crystal could be prepared by running the colloidal suspension strongly into a flat capillary cell. This single crystallization method is based on the shear flow effect: [30, 31] the strong flow of the suspension into the cell generates shear stress in the suspension, which induces the formation of a microstructure of colloidal crystals with a long-range orientational order. Although this method provides surprisingly uniform and transparent colloidal crystals in the cell, it requires an abrupt stoppage of the strong flow because a slow tapering deteriorates the quality of the crystalline texture. Owing to the difficulty in controlling such a peculiar flow condition, a large single-domain crystal could not be prepared in a quantitative and highly reproducible fashion. To overcome this problem, we developed an air-pulse-drive system shown in Fig. 2 [32, 33]. This system is primarily composed of a flat capillary cell made of fused quartz (internal dimensions: 0.1 mm thick, 9 mm wide, 70 mm long), digital pulse regulator (ML-606GX, Musashi Engineering Inc.), and air compressor. An aqueous suspension of charged polystyrene or silica particles (particle diameter: ~200 nm; particle volume fraction: ~10 %) in polycrystalline state is loaded into a plastic syringe, which is connected to the digital pulse regulator attached to the air compressor. The syringe is connected to one side of the flat capillary cell, and then, the suspension is injected into the cell by a short air pulse generated by an electronically controlled valve action in the regulator. As shown in the inset in Fig. 2, the time profile of the air pulse pressure (ΔP, difference from atmospheric pressure) is mostly a plateau followed by a steep decline. This sharp decrease in pressure can suppress the slow tapering of the flow. By adjusting the pulse height in terms of pressure, the strength of the pulsed flow of the suspension is controlled quantitatively. Thus, a quantitative study of flow effects on the crystalline quality as a function of pulse

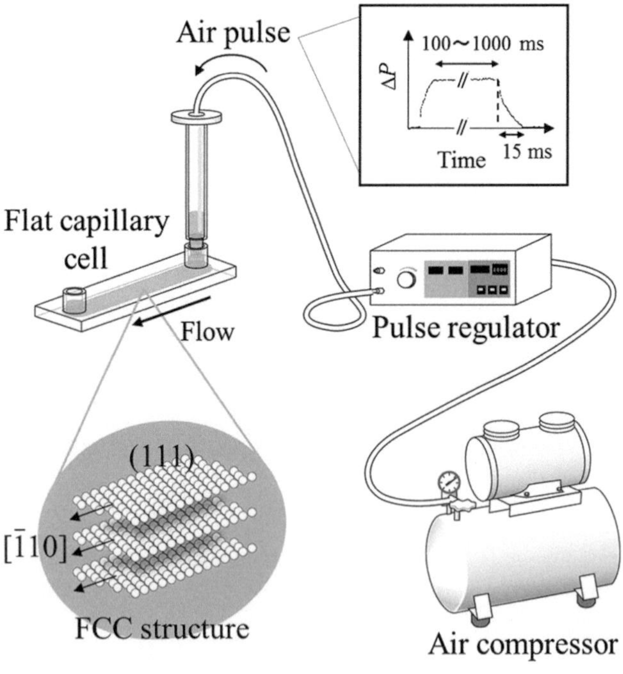

Fig. 2 Schematic diagram of the air-pulse-drive system for the preparation of single-crystal-like colloidal crystals (Reproduced from [32] with permission from Wiley-VCH Verlag GmbH & Co. KGaA)

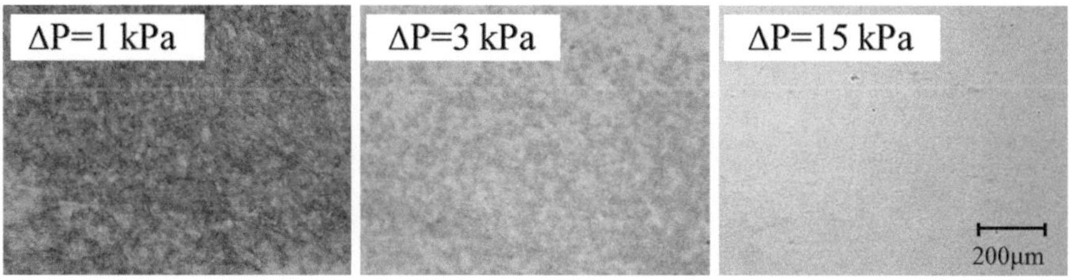

Fig. 3 Images of colloidal crystals at indicated ΔP values from transmission optical microscopy (Reproduced from [32] with permission from Wiley-VCH Verlag GmbH & Co. KGaA)

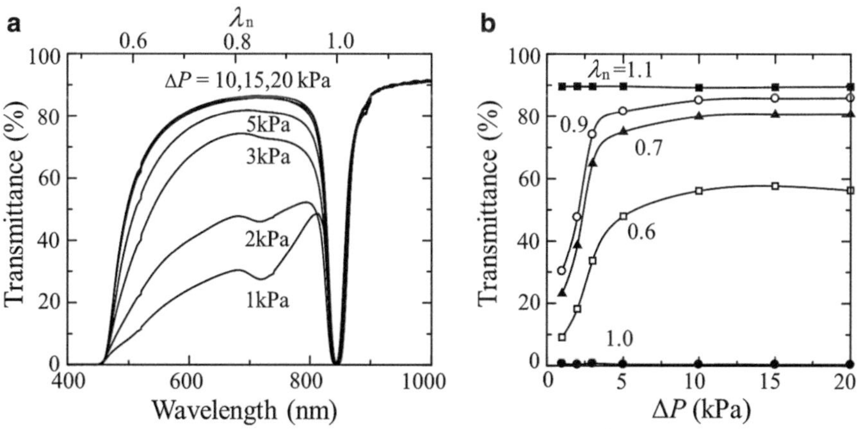

Fig. 4 (**a**) Transmission spectra of colloidal crystals processed at various ΔP values. λ_n is the wavelength normalized to the dip wavelength. (**b**) ΔP-dependent changes in transmittance at representative wavelengths derived from (**a**) (Reproduced from [32] with permission from Wiley-VCH Verlag GmbH & Co. KGaA)

pressure can be carried out and large single-domain crystals can be prepared with high reproducibility.

While reflection measurements have been predominantly used to characterize colloidal crystals, transmission measurements were carried out in this case because this method is more sensitive to disorder owing to the use of light transmitted through the crystals. Figure 3 shows images from transmission optical microscopy of colloidal crystals formed at representative ΔP values. An irregular polycrystalline texture was observed at low ΔP (1 kPa). Increasing ΔP, however, enhanced uniformity in the texture, and an almost uniform texture was formed at $\Delta P = 15$ kPa.

Figure 4a shows the transmission spectra of colloidal crystals processed at various ΔP values for incident light normal to the cell surface. A deep dip was observed at 845 nm, which was attributed to the Bragg diffraction from the (111) lattice planes of the face-centered cubic (FCC) structure parallel to the cell surface. Upon increasing ΔP, only the transmittance in the wavelength region shorter than

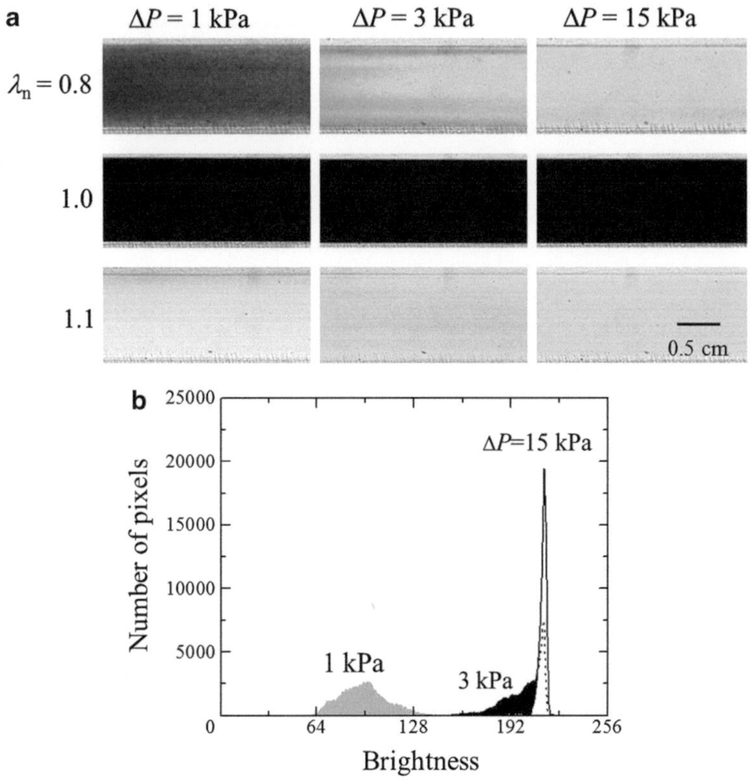

Fig. 5 (**a**) Single-wavelength images for representative wavelengths taken under transmitted illumination of colloidal crystals processed at indicated ΔP values. (**b**) ΔP-dependent changes in brightness distribution of single-wavelength images for $\lambda_n = 0.8$ in (**a**). A part of the distribution for 3 kPa overlaps with that for 15 kPa, which is indicated by the *dashed line* (Reproduced from [32] with permission from Wiley-VCH Verlag GmbH & Co. KGaA)

the dip increased and then reached a saturation point corresponding to a high value of about 80 % at $\Delta P = 10$ kPa. This convergent behavior can be clearly seen in Fig. 4b, where the transmittance at several representative wavelengths is plotted as a function of ΔP.

The crystals processed above $\Delta P = 10$ kPa exhibited excellent spectral qualities in terms of transmittance as well as spatial uniformity. Figure 5a shows the spatial distribution of the transmittance with brightness at representative wavelengths above, at, and below the dip ($\lambda_n = 1.1$, 1.0, and 0.8, respectively). The images were taken at different ΔP values for the same sample with an in-plane image resolution of 100×25 μm. In the image, the absence of uniformity in the optical properties is detected from the contrast in brightness [34]. For wavelengths above and at the dip, the transmittance of the entire sample area was very uniform regardless of ΔP. In contrast, for the wavelength below the dip ($\lambda_n = 0.8$), a non-uniform texture with low brightness was observed at low ΔP; however, both

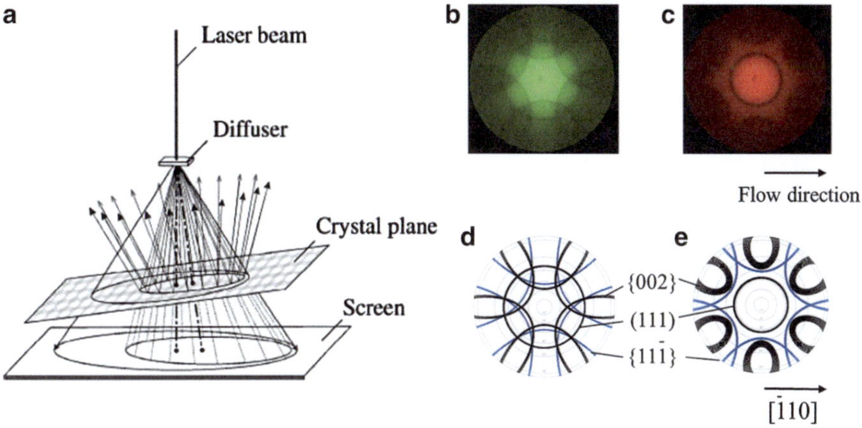

Fig. 6 (**a**) Schematic diagram of the formation of the Kossel pattern in the transmission geometry. The experimentally obtained Kossel patterns for the colloidal crystal at (**b**) 543.5 nm and (**c**) 632.8 nm. The simulated Kossel patterns at (**d**) 543.5 nm and (**e**) 632.8 nm (Reproduced from [35] with permission from The Japan Society of Applied Physics)

uniformity and brightness increased with increasing ΔP. This was also demonstrated quantitatively in Fig. 5b where the number of pixels at a particular brightness (brightness histogram) at different ΔP values for $\lambda_n = 0.8$ was compared, with a sharper peak representing higher uniformity. These data indicate that the crystals processed above $\Delta P = 10$ kPa reached a highly uniform state from a spectroscopic standpoint.

An analysis of the Kossel lines in the transmission geometry was also performed to examine the crystal structure, particularly over its entire thickness (Fig. 6a) [35]. Figure 6b, c shows the transmitted Kossel images for the colloidal crystal obtained at wavelengths of 543.5 and 632.8 nm, respectively. The crystal exhibited distinct hexagonally symmetric patterns, which were also observed in different places in the uniform texture. Assuming that the crystal forms an FCC structure with (111) twin-stacking faults, and the (111) planes lie parallel to the cell surface and the $[\bar{1}10]$ direction is parallel to the flow direction, Kossel lines for 111, 002, and $11\bar{1}$at 543.5 and 632.8 nm were numerically simulated as shown in Fig. 6d, e, respectively. The simulated patterns were in good agreement with the experimental ones, indicating that almost the entire cell was filled with a single-domain crystal of fixed crystallographic orientation determined by the cell geometry.

Considering these data, the observed single crystallization can be explained as a consequence of the transition of polycrystalline domains toward a highly oriented structure with increasing ΔP owing to the shear flow effect. For the FCC structure, the longest Bragg wavelength occurred when the incident light was normal to the (111) planes, which corresponds to the dip ($\lambda_n = 1.0$) in Figs. 4 and 5. Since the sample prepared at low ΔP included polycrystalline domains with random orientations, various Bragg reflections occurred at wavelengths below the dip. Thus, the non-uniform texture with low transmittance was observed at low ΔP only below the dip. Upon increasing ΔP, the polycrystalline domains of the crystals were aligned by the shear stress generated in the cell, resulting in an increase in the transmittance and spatial uniformity. The saturation behavior of the transmittance and uniformity, and the symmetric Kossel patterns observed above $\Delta P = 10$ kPa suggest that the polycrystalline domains were oriented highly enough to be regarded as a single-crystal-like structure, with a fixed orientation determined by the cell geometry.

2.2 Immobilization of Colloidal Crystals in a Hydrogel Network While Preserving High Optical Quality

The next challenging process is the immobilization of the single-crystal-like colloidal crystals, formed by the flow-induced method, in the hydrogel network without deterioration of the crystalline quality. Gelation reagents (*N*-methylolacrylamide (NMAM) as monomer, *N*,*N'*-methylenebisacrylamide as crosslinker, and camphorquinone as polymerization photoinitiator) were added to the colloidal suspension before injecting it into the flat capillary cell [36]. The suspension was then bubbled with Ar gas to remove dissolved oxygen, which deactivated the photoinitiator. After running the suspension in the cell, the formed single-domain crystal was uniformly irradiated with high-brightness blue light-emitting diode (LED) arrays to polymerize the dissolved gelation reagents.

Figure 7a shows the transmission spectra of the colloidal crystals before and after gelation. By properly adjusting the gelation conditions, including the oxygen removal procedure, and the irradiation intensity and time, an excellent spectral profile, i.e., a deep dip due to the stop-band and high transmittance at the pass-band wavelength, could be preserved after gelation. The crystal also preserved high spectral uniformity over a large area after gelation, which was confirmed from the single-wavelength images shown in Fig. 7b. The obtained gel-immobilized colloidal crystals are sufficiently solid to be handled with tweezers and can be removed from the fabrication cell as a self-standing film (Fig. 7c) [37].

3 Tuning the Properties of Gel-Immobilized Colloidal Crystals by External Stimulus

3.1 Tuning by Solvent Exchange

In general, the change in size or degree of swelling of the gel depends on its affinity to the solvent. For example, the poly(*N*-methylolacrylamide) (PNMAM) gel swells well in water but shrinks considerably in ethanol. When a water-ethanol mixture is used as the swelling solvent for the gel, the degree of swelling can be tuned by adjusting the ethanol concentration. Therefore, the lattice constant or Bragg

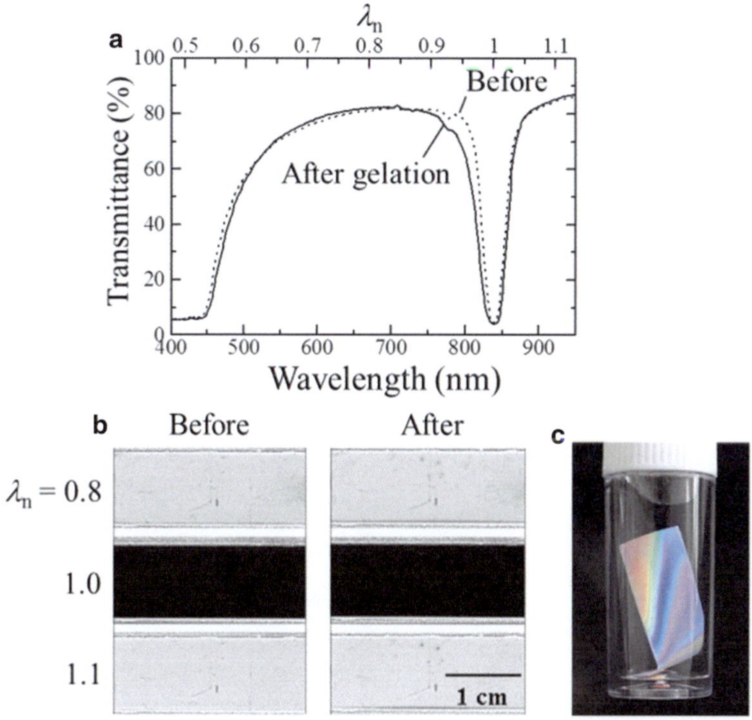

Fig. 7 (**a**) Transmission spectra of the colloidal crystals before and after gelation. (**b**) Single-wavelength images taken under transmitted illumination before and after gelation. The incident light was normal to the cell face. (**c**) Photograph of the gel-immobilized colloidal crystal film soaked in water (Reproduced from [33] with permission from The Ceramic Society of Japan)

wavelength of the colloidal crystals immobilized in the gel can also be tuned through the change in gel size [36, 38]. Figure 8a shows the photographs of a circular fragment cut from the PNMAM-immobilized colloidal crystal film soaked in water-ethanol solution at various ethanol concentrations. As the ethanol concentration increased, the film shrank gradually until the concentration was 60 %. Upon further increase in the concentration from 60 to 80 %, the size or degree of swelling of the gel film decreased significantly, resulting in a change in the Bragg reflection color (Fig. 8a, b). The change in the Bragg wavelength (Fig. 8c, d) is in good accordance with the change in the degree of swelling (Fig. 8b), which confirms that the change in the Bragg reflection color was caused by the change in gel size.

3.2 *Tuning by Temperature Change*

When colloidal crystals are immobilized in a thermosensitive polymer gel, the Bragg wavelength can be tuned by changing the temperature. For example, a

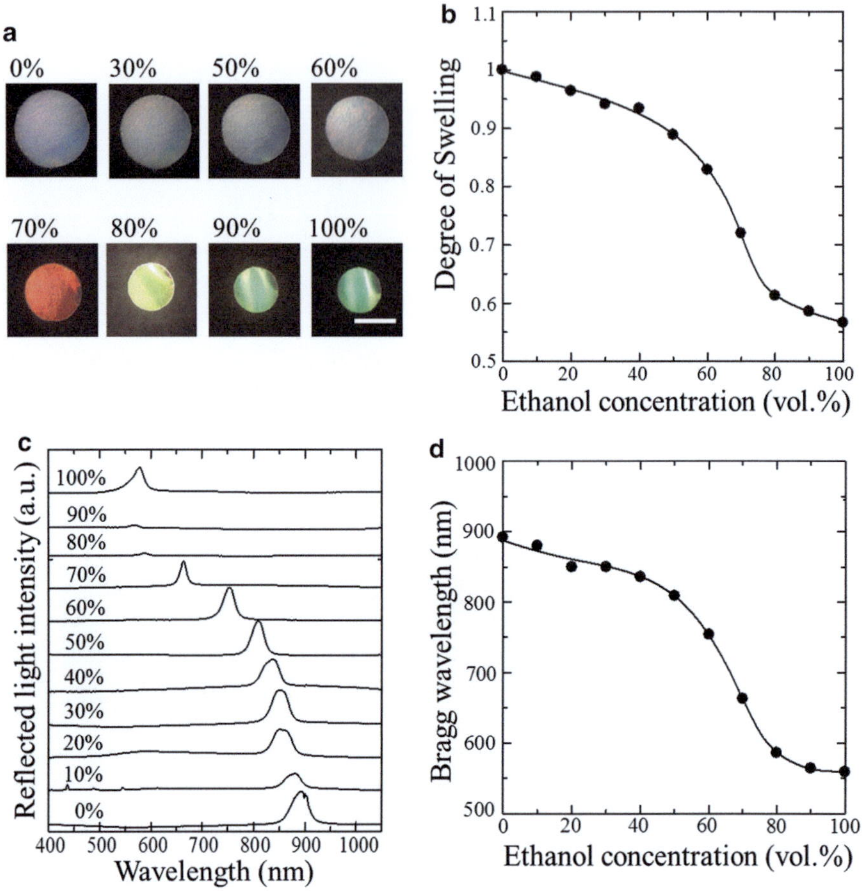

Fig. 8 (**a**) Photographs of the gel-immobilized colloidal crystal film at various ethanol concentrations. The length of the scale bar is 1 mm. (**b**) Degree of swelling as a function of ethanol concentration. (**c**) Reflection spectra at various ethanol concentrations. (**d**) Bragg reflection wavelength as a function of ethanol concentration. The curves in (**b**) and (**d**) serve as guides (Reproduced from [38] with permission from The Society of Chemical Engineers, Japan)

representative thermosensitive gel, poly(*N*-isopropylacrylamide) (PNIPAM), exhibits volume phase transition at around 32 °C [39]. The PNIPAM-immobilized colloidal crystal films were prepared by combining the flow-induced method with the photopolymerization technique described in Sect. 2, but using *N*-isopropylacrylamide (NIPAM) instead of NMAM. The changes in the degree of swelling and Bragg wavelength during heating were then examined [40]. Figure 9a, b shows the plots of the degree of swelling and Bragg wavelength, respectively, of colloidal crystals immobilized in the PNIPAM and PNMAM gel films as a function of temperature. During heating from 10 to 60 °C, the PNIPAM gel film began to shrink significantly at around 32 °C, resulting in a large shift in the Bragg wavelength. In contrast, the PNMAM gel film did not exhibit such thermosensitivity.

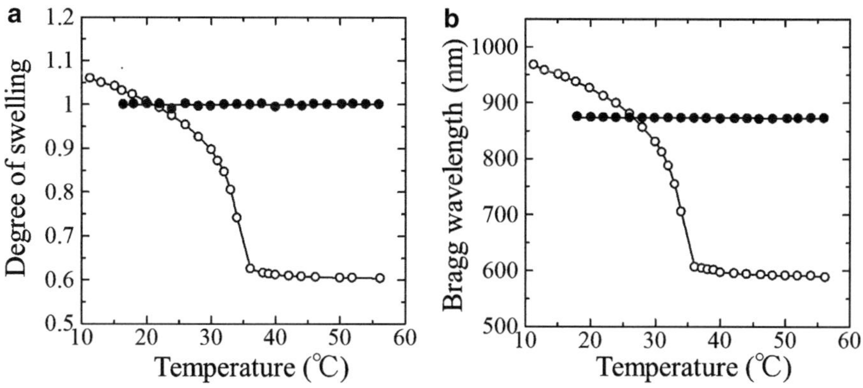

Fig. 9 Plots of the (**a**) degree of swelling and (**b**) Bragg wavelength of the gel-immobilized colloidal crystal film as a function of temperature (*black circle*: PNMAM; *white circle*: PNIPAM). The curves in (**a**) and (**b**) serve as guides (Reproduced from [40] with permission from The Royal Society of Chemistry)

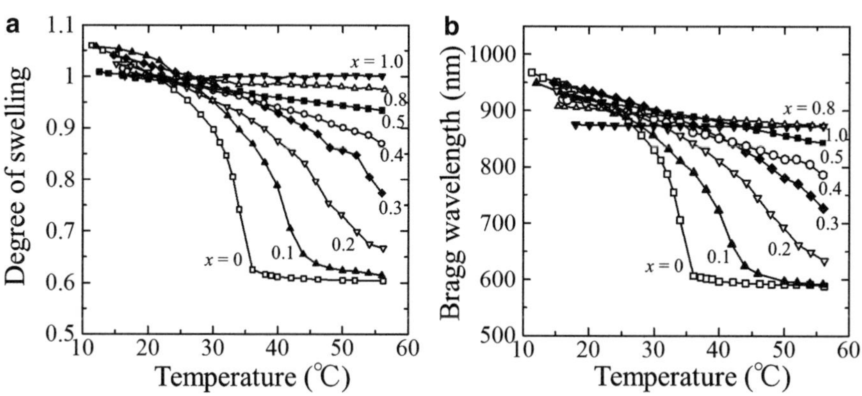

Fig. 10 Plots of the (**a**) degree of swelling and (**b**) Bragg wavelength of gel-immobilized colloidal crystal film at different x values as a function of temperature (Reproduced from [40] with permission from The Royal Society of Chemistry)

Thermosensitivity can be controlled by immobilizing the colloidal crystals in a copolymer hydrogel composed of the thermosensitive NIPAM and non-thermosensitive NMAM and varying the mixing ratio of the two gels. Figure 10a, b shows the plots of the degree of swelling and Bragg wavelength, respectively, of the colloidal crystals immobilized in the copolymer hydrogel film at different mole fractions of NMAM (x) as a function of temperature. With increasing x, the degree of swelling and Bragg wavelength changed gradually with temperature, and the transition temperature increased; furthermore, these parameters intriguingly exhibited a linear dependence on temperature above $x = 0.4$ (Fig. 11 and Table 1). In particular, the crystal at $x = 0.5$ exhibited a linear

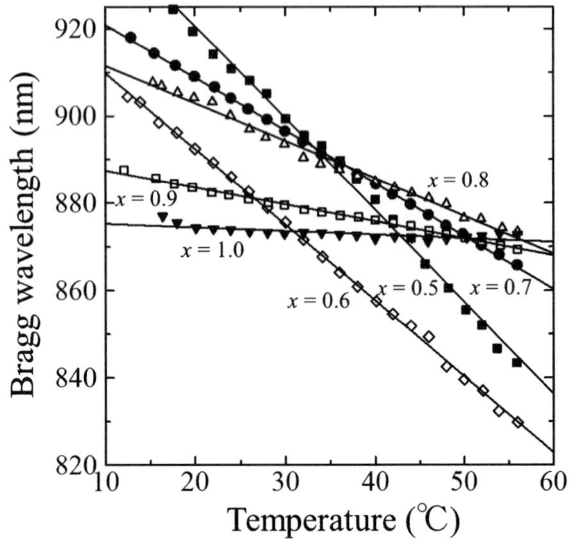

Fig. 11 Plots of the Bragg wavelength of gel-immobilized colloidal crystal film at $x \geq 0.5$ as a function of temperature. The *solid line* is the best-fit line obtained using the least-squares method (Reproduced from [40] with permission from The Royal Society of Chemistry)

Table 1 Slope of the fitted line and correlation coefficient determined using the least-squares method (Reproduced from [40] with permission from The Royal Society of Chemistry)

x	Slope (nm/°C)	Correlation coefficient
0.5	−2.1	0.998
0.6	−1.7	0.999
0.7	−1.2	0.999
0.8	−0.86	0.995
0.9	−0.39	0.998
1	−0.083	0.777

thermosensitivity of −2.1 nm/°C, which is ten times that of conventional tunable photonic crystals. Although it is challenging to gain a theoretical understanding of the linear behavior of these materials, achieving linear thermosensitivity is potentially useful in practical applications as tunable photonic crystals and temperature sensors.

3.3 *Tuning with Ionic Liquids*

The tunability of gel-immobilized colloidal crystals is based on the change in size of the gel matrix swollen in the solvent. In other words, if the swelling solvent evaporates, such tuning is not achieved. Therefore, in general, the gel-immobilized colloidal crystals are used in a sealed container. We found that ionic liquids were very useful solvents for gel-immobilized colloidal crystals to overcome the

Table 2 Degree of swelling of gel-immobilized colloidal crystal films soaked in various ionic liquids (Reproduced from [41] with permission from American Chemical Society)

Ionic liquid		Degree of swelling
1-Butyl-3-methylimidazolium acetate	Hydrophilic	1.06
1,3-Diallylimidazolium bromide	Hydrophilic	1.04
1-Hexyl-3-methylimidazolium chloride	Hydrophilic	1.01
1-Ethyl-3-methylimidazolium acetate	Hydrophilic	0.99
1-Ethyl-3-methylimidazolium dicyanamide	Hydrophilic	0.95
1-Methyl-3-octylimidazolium chloride	Hydrophilic	0.92
1-Butyl-3-methylimidazolium thiocyanate	Hydrophilic	0.82
1-Ethyl-3-methylimidazolium bis(trifluoromethylsulfonyl)imide	Hydrophobic	<0.6
1,3-Diallylimidazolium bis(trifluoromethanesulfonyl)imide	Hydrophobic	<0.6
Trihexyltetradecylphosphonium chloride	Hydrophobic	<0.6
Trihexyltetradecylphosphonium dicyanamide	Hydrophobic	<0.6

evaporation problem [41, 42]. Ionic liquids are salts in the liquid state at room temperature [43–45]. These have received increased attention as new solvents because of their attractive features such as non-volatility and high thermal stability. In addition, their characteristics can be adjusted by changing the cation–anion combination. Thus, if ionic liquids are used as solvents for gel-immobilized colloidal crystals, the possibility of designing colloidal photonic crystals with non-volatile solvent is increased.

The gel was found to swell in hydrophilic ionic liquids, with the swelling volume varying with the type of ionic liquid as shown in Table 2. Interestingly, the gel swells to a greater extent in some hydrophilic ionic liquids than in water. For example, when water was replaced with 1,3-diallylimidazolium bromide, the gel-immobilized colloidal crystal film swelled to 1.04 times its initial size (Fig. 12). The peak in the reflection spectrum shifted to a longer wavelength, from 807 nm to 956 nm, without degradation of spectral quality.

The tuning of the Bragg wavelength for conventional gel-immobilized colloidal crystals swollen in aqueous solutions is restricted to the wavelength region shorter than this wavelength because crystals achieve the widest lattice spacing in this solvent. On the other hand, tuning to a wider wavelength range can be achieved using an ionic liquid as solvent because the Bragg wavelength in 1,3-diallylimidazolium bromide (hydrophilic), for example, would be longer than in water. In fact, the swelling volume of the gel and Bragg wavelength could be varied over a wide range by mixing a hydrophobic ionic liquid with a hydrophilic one. As shown in Fig. 13, upon decreasing the volume fraction (x_{Br}) of 1,3-diallylimidazolium bromide (hydrophilic) in the mixture with 1,3-diallylimidazolium bis(trifluoromethanesulfonyl)imide (hydrophobic), the

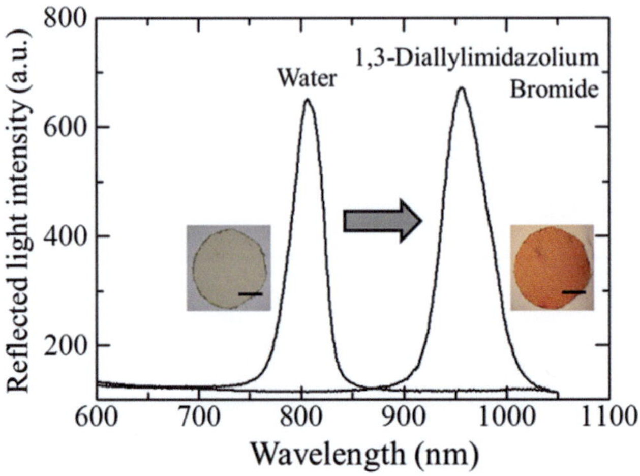

Fig. 12 Reflection spectra and microscope images of gel-immobilized colloidal crystal film before and after replacement of water in the gel with 1,3-diallylimidazolium bromide. The lengths of the scale bars are 1 mm (Reproduced from [41] with permission from American Chemical Society)

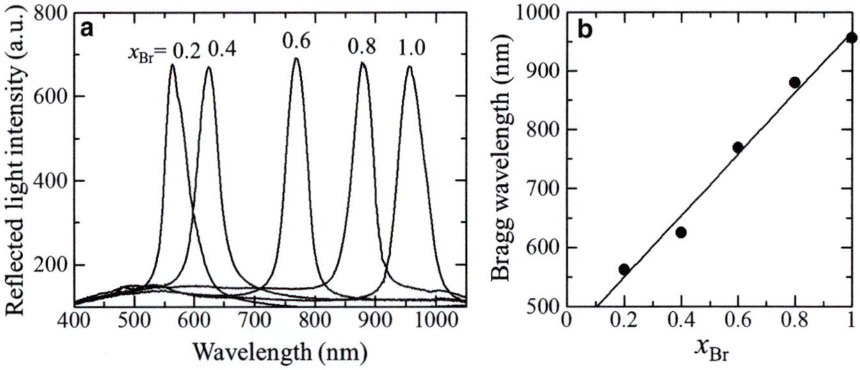

Fig. 13 (**a**) Reflection spectra of gel-immobilized colloidal crystal film immersed in a mixture of 1,3-diallylimidazolium bromide and 1,3-diallylimidazolium bis(trifluoromethanesulfonyl)imide at various ratios. x_{Br} is the volume fraction of 1,3-diallylimidazolium bromide measured prior to mixing. (**b**) Bragg wavelength derived from (**a**) as a function of x_{Br} (Reproduced from [41] with permission from American Chemical Society)

Bragg reflection peak shifted to shorter wavelengths over a range of about 400 nm while preserving spectral quality. Furthermore, the swelling size and Bragg wavelength surprisingly showed a linear dependence on the mixing ratio. In general, however, the swelling-shrinking phenomenon that depends on the change in the mixing ratio of the solvents is known to be a type of phase transition, and the gel size or Bragg wavelength is known to be a strongly

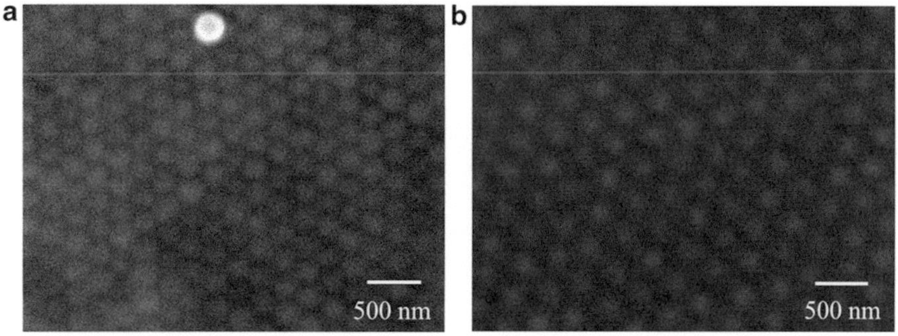

Fig. 14 Top-view SEM images of gel-immobilized colloidal crystal film with (**a**) $x_{Br} = 0.2$ and (**b**) $x_{Br} = 0.8$ (Reproduced from [41] with permission from American Chemical Society)

non-linear function of the mixing ratio, as already demonstrated for the water-ethanol system in Sect. 3.1. Although a theoretical understanding of this linear dependence poses a serious challenge for future investigation, it is quite advantageous for adjusting the optical stop-band of the crystals for practical applications.

Since ionic liquids do not evaporate even in high vacuum, the particle arrays in the swollen gel can be observed directly using routine scanning electron microscopy (SEM). Figure 14 shows the SEM images of the gel-immobilized colloidal crystals swollen in different mixing ratios of the hydrophilic and hydrophobic ionic liquids. For $x_{Br} = 0.2$, relatively dense arrays of colloids were observed, while for $x_{Br} = 0.8$, loose arrays of particles were seen. From the SEM images, the distance between the nearest-neighbor particles for $x_{Br} = 0.2$ and $x_{Br} = 0.8$ was estimated to be 247 and 335 nm, respectively. These values are in good agreement with those determined using the Bragg wavelength. These data directly verify the presence of distance-controlled particle arrays in the swollen gel.

3.4 *Tuning by Mechanical Stress*

The Bragg wavelength of gel-immobilized colloidal crystals can also be manipulated by applying mechanical stress. For example, uniaxially compressing the crystals between two glass substrates can vary the Bragg wavelength easily. We fabricated a mechanical compression apparatus to compress the crystals uniformly (Fig. 15). The film sample was placed between two parallel quartz substrates, each fixed on a hole in the steel plate, and uniformly compressed using three micrometers. Using the micrometers and springs, the distance between parallel substrates could be adjusted reversibly with an accuracy of ± 0.2 μm. Ionic liquids were used as swelling solvents for the gel-immobilized colloidal crystal films to prevent

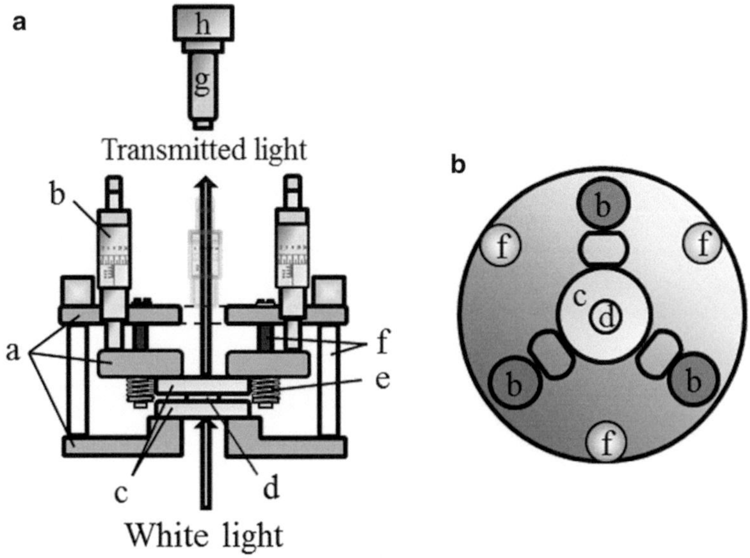

Fig. 15 Schematic diagram of (**a**) side and (**b**) top views of the compression apparatus (*a*, stainless steel plate; *b*, micrometer; *c*, quartz substrate; *d*, film sample; *e*, spring; *f*, support rod; *g*, spectrometer; *h*, CCD camera)

instability of the crystals owing to the evaporation of solvent from the open space between the substrates.

Figure 16a shows the change in the transmission spectrum of the gel-immobilized colloidal crystal film with $x_{\mathrm{Br}} = 1.0$ by compression in increments of 2 μm. As the film is compressed, i.e., the distance between substrates (d) decreased, the dip gradually became shallower and shifted to a lower wavelength because of the reduction of the interplanar spacing of (111) planes parallel to the film surface. Conservation of the spectral profile, such as the dip shape and high transmittance at the pass-band wavelength, even after a large shift of more than 100 nm indicates that the film had excellent tunable properties. The film also exhibited high spectral uniformity under compression, which was visually shown in the single-wavelength images (Fig. 16c) and spatial distribution images of the Bragg wavelength (Fig. 16d) in the entire sample area. After reduction of the distance between substrates by compression, the uniform textures were maintained at the pass-band wavelengths while a slight decrease in uniformity was observed at the stop-band wavelength. The uniformity of the Bragg wavelength also increased slightly with decreasing distance between substrates as shown in Fig. 16d; however, the standard deviation was 8 nm at most, suggesting that high spectral uniformity was maintained under compression.

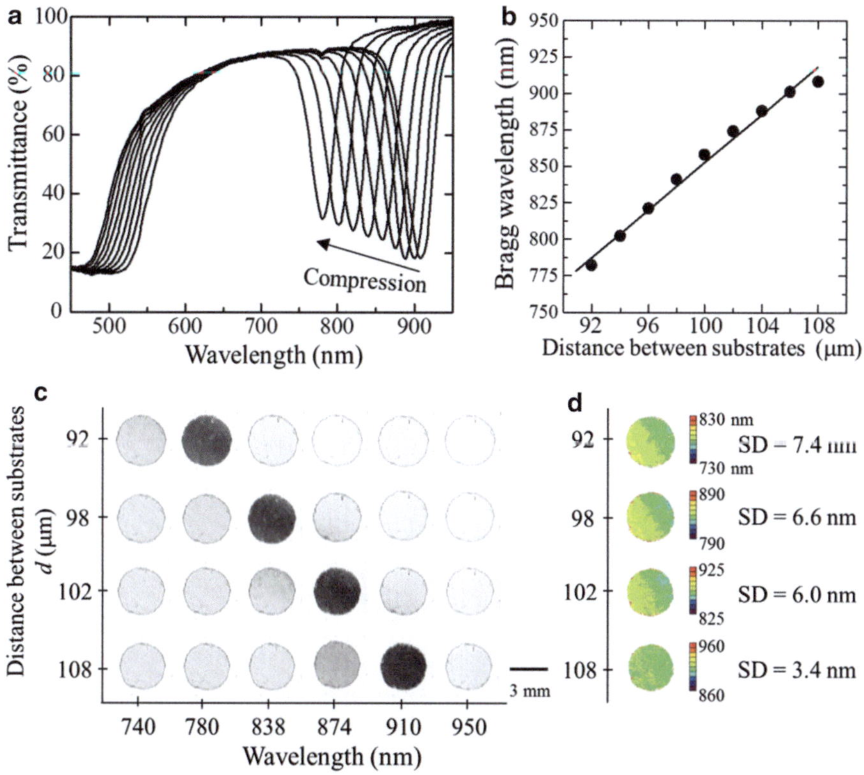

Fig. 16 (**a**) Average transmission spectra in the entire sample area for the gel-immobilized colloidal crystal film with $x_{Br} = 1.0$ under mechanical compression in increments of 2 μm. The dark level of the transmittance increased owing to the characteristics of the CCD spectrometer. (**b**) Plot of the Bragg diffraction wavelength derived from (**a**) as a function of the distance between substrates. (**c**) Single-wavelength images for representative wavelengths and (**d**) spatial distribution images of the Bragg wavelength for the film under compression

Similarly excellent spectral tuning was possible for films with other x_{Br} values as shown in Fig. 17. Bragg wavelengths were lower for films with lower x_{Br} before compression and further shifted to lower values upon compression while preserving high spectral quality (Fig. 17a, b). Standard deviations of the Bragg wavelengths for all films under compression were less than 8 nm (Fig. 17c). These data indicate that these materials are potentially useful as tunable photonic crystals and stress sensors.

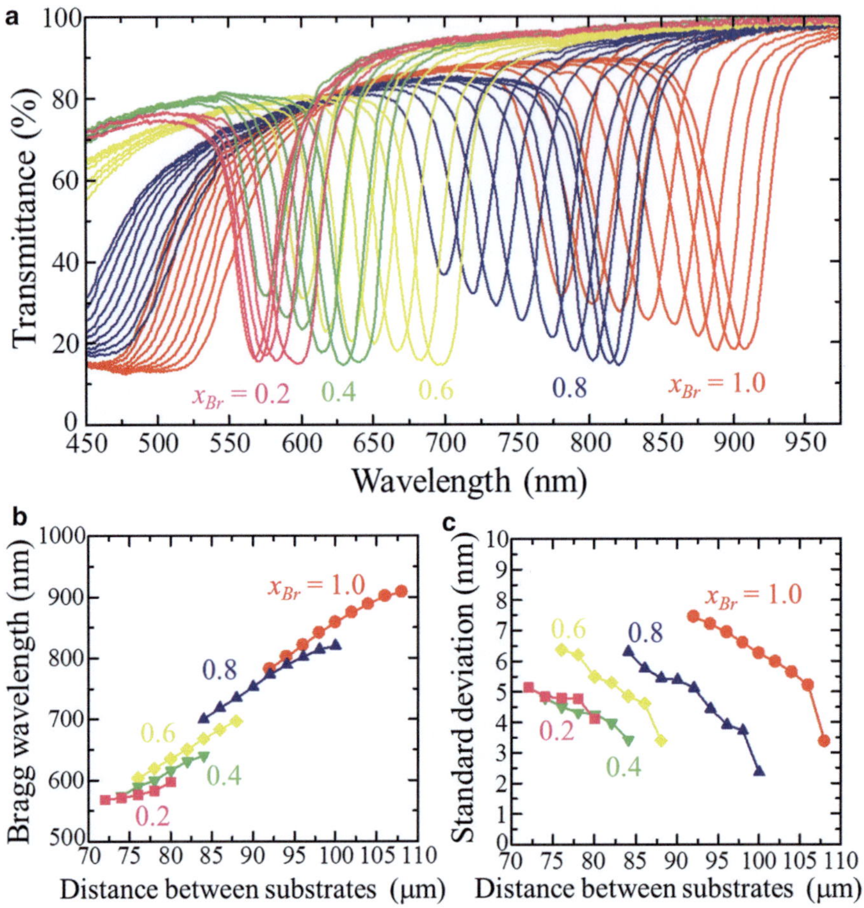

Fig. 17 (**a**) Average transmission spectra in the entire sample area for the gel-immobilized colloidal crystal film with different x_{Br} values under mechanical compression in increments of 2 μm. The dark level of the transmittance increased owing to the characteristics of the CCD spectrometer. (**b**) Plots of the Bragg diffraction wavelength derived from (**a**) and (**c**) corresponding standard deviations for the film with different x_{Br} values as a function of the distance between substrates

4 Conclusions

In this chapter, our recent work on the preparation and tunable properties of the gel-immobilized colloidal crystal film with high optical quality was introduced. The optical stop-band wavelength could be tuned over a wide wavelength region by the external stimulus, such as solvent exchange, temperature change, and mechanical stress, while preserving high spectral quality. The film is potentially useful as tunable photonic crystals and sensors for monitoring the changes in the environments through the Bragg wavelength or reflection color.

Acknowledgments Most of the work was carried out at the National Institute for Materials Science (NIMS) hosted by Dr. Tsutomu Sawada. The author gratefully acknowledges support from JSPS KAKENHI (Grant Nos. 22686063 and 25289237), Mazda Foundation, Ogasawara Science and Technology Foundation, and Iketani Science and Technology Foundation.

References

1. K.P. Velikov, C.G. Christova, R.P.A. Dullens et al., Science **296**, 106 (2002)
2. G.A. Ozin, S.M. Yang, Adv. Funct. Mater. **11**, 95 (2011)
3. F. Garcia-Santamaria, H.T. Miyazaki, A. Urquia et al., Adv. Mater. **14**, 1144 (2002)
4. P. Ni, P. Dong, B. Cheng et al., Adv. Mater. **13**, 437 (2001)
5. P. Pieranski, Contemp. Phys. **24**, 25 (1983)
6. A.K. Arora, B.V.R. Tata (eds.), *Ordering and Phase Transitions in Charged Colloids* (VCH, New York, 1996)
7. A.P. Gast, W.B. Russel, Phys. Today **51**, 24 (1998)
8. S. Furumi, H. Fudouzi, T. Sawada, Laser Photon. Rev. **4**, 205 (2010)
9. E. Yablonovitch, Phys. Rev. Lett. **58**, 2059 (1987)
10. S. John, Phys. Rev. Lett. **58**, 2486 (1987)
11. K. Ohtaka, Phys. Rev. B **19**, 5057 (1979)
12. J.D. Joannopoulos, R.D. Meade, J.N. Winn, *Photonic Crystals* (Princeton University Press, Princeton, 1995)
13. K. Sakoda, *Optical Properties of Photonic Crystals* (Springer, Berlin, 2001)
14. Y. Akahane, T. Asano, B. Song et al., Nature **425**, 944 (2003)
15. T. Baba, Nat. Photon **2**, 465 (2008)
16. S. Noda, K. Tomoda, N. Yamamoto et al., Science **289**, 604 (2000)
17. O. Painter, R.K. Lee, A. Scherer et al., Science **284**, 1819 (1999)
18. H. Kosaka, T. Kawashima, A. Tomita et al., Phys. Rev. B **58**, 10096 (1998)
19. P.A. Hiltner, I.M. Krieger, J. Phys. Chem. **73**, 2386 (1969)
20. T. Kanai, T. Sawada, K. Kitamura, Langmuir **19**, 1984 (2003)
21. J.E.G.J. Wijnhoven, W.L. Vos, Science **281**, 802 (1998)
22. M. Muller, R. Zentel, T. Maka et al., Adv. Mater. **12**, 1499 (2000)
23. H. Miguez, E. Chomski, F. Garcia-Santamaria et al., Adv. Mater. **13**, 1634 (2001)
24. E.A. Kamenetzky, L.G. Magliocco, H.P. Panzer, Science **263**, 207 (1994)
25. J. Yamanaka, M. Murai, Y. Iwayama et al., J. Am. Chem. Soc. **126**, 7156 (2004)
26. J. Holtz, S.A. Asher, Nature **389**, 829 (1997)
27. H. Fudouzi, Y. Xia, Adv. Mater. **15**, 892 (2003)
28. Y. Iwayama, J. Yamanaka, Y. Takiguchi et al., Langmuir **19**, 977 (2003)
29. T. Sawada, Y. Suzuki, A. Toyotama et al., Jpn. J. Appl. Phys. **40**, L1226 (2001)
30. B.J. Ackerson, Nature **281**, 57–60 (1979)
31. B.J. Ackerson, J.B. Hayter, N.A. Clark et al., J. Chem. Phys. **84**, 2344 (1986)
32. T. Kanai, T. Sawada, A. Toyotama et al., Adv. Funct. Mater. **15**, 25 (2005)
33. T. Kanai, J. Ceram. Soc. Jpn. **120**, 87 (2012)
34. T. Kanai, T. Sawada, K. Kitamura, Chem. Lett. **34**, 904 (2005)
35. T. Kanai, T. Sawada, I. Maki et al., Jpn. J. Appl. Phys. **42**, L655 (2003)
36. A. Toyotama, T. Kanai, T. Sawada et al., Langmuir **21**, 10268 (2005)
37. T. Kanai, T. Sawada, J. Yamanaka, J. Ceram. Soc. Jpn. **118**, 370 (2010)
38. H. Yano, H. Sugiyama, T. Sawada, et al., Kagaku kogaku ronbunshu **41**, 43 (2015)
39. B.R. Saunders, B. Vincent, Adv. Colloid Interface Sci. **80**, 1 (1999)

40. H. Sugiyama, T. Sawada, H. Yano et al., J. Mater. Chem. C **1**, 6103 (2013)
41. T. Kanai, S. Yamamoto, T. Sawada, Macromolecules **44**, 5865 (2011)
42. S. Yamamoto, T. Sawada, T. Kanai, Chem. Lett. **41**, 495 (2012)
43. T. Welton, Chem. Rev. **99**, 2071 (1999)
44. R. Sheldon, Chem. Commun. **23**, 2399 (2001)
45. R.F. de Souza, J.C. Padilha, R.S. Goncalves et al., Electrochem. Commun. **5**, 728 (2003)

Tailoring Optical Spectra of Colloidal Photonic Crystals by Designed Surface Modes

Sergei G. Romanov

Abstract Functionalization of photonic crystals can be sought by designing the strongly localized surface excitations. In this chapter we address engineering of surface states created with the help of metal mirrors attached at crystal boundaries. We used colloidal crystals as three-dimensional photonic crystal templates and explored the respective modifications of optical transmission and reflectance spectra.

If the mirror is flat, changes of optical spectra are mostly quantitative and associated with recycling of light in a crystal volume. In order to prepare the optical defect state at the crystal surface the accomplishing of 2π phase shift along the light round trip at the photonic bandgap wavelength is required. The additional dielectric spacer inserted between the Bragg and the metal mirrors is the straightforward solution of this task. This defect provides tunneling of photons through the photonic bandgap and ensures up to 50 times increase of transmission. Tuning of the defect transmission band can be achieved by changing the spacer and the metal mirror. Oppositely, if the metal film adopts the periodical corrugation of the opal surface, the separate spacer is no longer needed due to automatic fulfillment of the resonance conditions. Such metal-terminated crystals normally possess the optical Tamm surface states at their stop-bands.

Another sort of surface states is the surface plasmon polaritons that are localized at interfaces of the metal film. Their dispersions satisfy the phase matching conditions between light wavevectors and vectors of the reciprocal crystal lattice. Tunneling of surface plasmons through corrugated metal films deposited on the surface of colloidal crystal provides the bands of enhanced transmission with their summit called the extraordinary light transmission. These surface modes can be tuned by changing the parameters of photonic crystals. Independence of surface plasmon polaritons on the diffraction resonances of photonic crystal templates allows to consider such hybrid structures as plasmonic-photonic hetero-crystals.

S.G. Romanov (✉)
Institute of Optics, Information and Photonics, University of Erlangen-Nuremberg, Haber-str. 9a, 91058 Erlangen, Germany

Ioffe Physical Technical Institute, 194021, Polithekhnicheskaya ul., 26, St. Petersburg, Russia
e-mail: sergei.romanov@physik.uni-erlangen.de

D. Comoretto (ed.), *Organic and Hybrid Photonic Crystals*,
DOI 10.1007/978-3-319-16580-6_20

Finally, the plasmons localized at metal caps on colloidal spheres can be resolved in spectra of hybrid crystals as non-dispersive bands that modify the appearance of diffraction and plasmon resonances of hybrid crystals.

Overall, magnitudes of surface mode contributions in optical spectra of hybrids are comparable to that of diffraction resonances of three-dimensional photonic crystals. Hence, surface modes establish the useful toolbox for the purposive tailoring of photonic crystal properties.

Keywords Photonic crystal • Surface states • Tamm state • Surface plasmon polaritons • Localized plasmons • Transmission and reflectance spectra

1 Introduction

Photonic crystals (PhCs) can be understood as lattices of photonic atoms. If the light wave impinges on such lattice, it experiences the scattering since the symmetry of the incident waves is in mismatch with the symmetry of the lattice. As the result of the scattering, one fraction of the incident light is rejected by PhC but another is allowed to propagate without further scattering. In the solid state physics such propagating waves are known as Bloch waves and in optics they are called eigenmodes. In order to ensure the lossless propagation, the electromagnetic field of eigenmodes acquires the complex structure compatible to all fine details of PhC topology. In turn, the total collection of eigenmodes constitutes the photonic bandgap (PBG) diagram of PhC [1]. Insight into PhC theory is given in first chapter of this book.

Diffraction of eigenmodes in PhCs provides strong resonances in transmission/ reflectance spectra, which lend themselves for investigation and exploitation. The common way to modify the optical response of PhC is to change eigenmodes by amending the structure and/or the composition of the lattice. Such parameters as the geometry, connectivity and volume fraction of photonic atoms, the lattice symmetry, and the refractive index are targeted in the first instance.

Alternative way to design functional PhCs is embedding artificial defects. In lucky case this defect adds its eigenmode in the photonic bandgap (PBG), where it becomes a localized state. This approach mimics resonant optical cavities, which are frequently applied to engineer efficient light emitters and detectors [2]. In this respect, the point (0-dimensional or 0D) defect in a 3D lattice is the most advantageous configuration since it provides the omnidirectional confinement of the light in the defect cavity. However, the technological complexity of defect formation is dramatically escalated with increasing the PhC dimensionality [3]. That is why the planar defects in 1D Bragg reflector providing only the unidirectional confinement remain the most widely used architectures in commercial optical devices since late 70s of the last century [4]. Planar defects were also considered in 3D PhCs [5]. Nonetheless, fabrication of even the simplest defects in 3D PhCs is a serious challenge in terms of the mutual alignment of lattices in a PhC sandwich and the surface roughness. Would it be possible to avoid such problems but preserve the reach physics of 3D PhCs? In this chapter we address these issues.

In this chapter we will consider the 3D PhCs based on opal-like colloidal crystals. Opals are used as templates for preparing arrays of nanostructures since early 1980s and as photonic crystals since mid-1990s [6]. Nowadays, they are often used for the proof-of-principle investigations. Opals represent the PhCs with 3D photonic bandgap (PBG) structure. In terms of the PhC topology each sphere in the opal crystal nicely represents the photonic atom. In turn, the Mie scattering resonances of such spheres resemble the electronic levels of common atoms. Joining photonic atoms in a crystal results in the hybridization of Mie excitations. In other words, Mie resonances are transformed in photonic bands, which constitute the PBG structure in a similar manner as electronic orbitals of atoms are transformed in the electron energy bands in atomic solids.

On the one hand, owing to self-assembly, spheres in opals are crystallized in the intrinsically stable lattice with the unique face-centered cubic symmetry. In general, this fixed lattice topology cannot be changed without the aid of external templates [7]. On the other hand, the structural sameness leads to the uniformity of their optical properties that is preserved for the wide range of refractive indices and sizes of spheres. Poor susceptibility to modification is one of major disadvantages of opals compared to PhCs fabricated by nanolithography. Design of functional defects is one way to partially overcome this drawback.

The defects can be built in the opal volume either due course of crystal assembling [8, 9] or by means of post-crystallization processing, e.g., using electron beam lithography or two-photon polymerization [10, 11]. Unfortunately, the intrinsic lattice irregularities mask the contribution of artificial 0D and 1D defects to the optical response of opal-based PhCs. Oppositely, the optical signature of planar 2D defects can be strong enough to justify their incorporation in thin opal films. The figure of merit in this sense is achieving the comparable strength of defect-related and diffraction resonances.

The design of integrated photonic circuits assumes assembling PhCs of different functionalities on one substrate. It seems impractical to fabricate for each particular function the special crystal of specific symmetry and/or material and/or the embedded defect. Ideally, it would be of advantage to fabricate all crystals on chip using the same technology and then to apply post-crystallization processing for shaping their functionalities. The obvious method that ensures the preservation of the PhC core is to make instrumental the interfaces of PhC. Provisionally, different services can be attained through manipulations with the PhC surface [12].

Let us take into account that the surface of 3D PhC is the planar defect, where the abrupt change of the lattice periodicity takes place. The electromagnetic energy flow impinging on PhC excites the crystal eigenmodes at the surface. Under certain conditions the specific resonances of a planar defect (the PhC surface) can be also excited, so that they also accumulate a part of the incident flux. The latter excitations can be considered as providing the gate action, which applies further limitations on the light coupled to PhC. Moreover, this gate function can differ significantly from simple filtering due to hybridization of volume and surface excitations. Thus, by gaining a control on the surface resonances one can amend PhC functionality.

In general, the surface mode is the electromagnetic excitation, which is localized at both sides of the interface between materials of different optical properties. The reason for specific excitations is the fact that such boundary breaks the translational invariance. For example, the surface waves can propagate along the interface between materials with positive and negative permittivities. In this case the field penetration in a medium with a negative permittivity (e.g., metal) is suppressed due to its purely imaginary field wavenumber that is the familiar mirror action. To achieve the wave decay in a medium with a positive permittivity, the total internal reflection is needed (like in the waveguide it occurs at grazing incidence of light on the boundary), so that the tangential component of the wave vector exceeds in modulus of the wave vector of the light cone. Hence, the surface wave is the evanescent one, i.e., it is bounded to the interface [13].

Let us to replace the medium of the negative permittivity with the PhC. The eigenmodes of PhC are the Bloch modes that are invariant with respect to translations of the underlying lattice. At bandgap frequencies they do not carry the energy and their magnitudes exponentially decay from the boundary in depth of PhC. Oppositely, the eigenmodes of the homogeneous medium with the positive permittivity are plane waves. On the free space side of the boundary the PhC eigenmodes should be transformed either in plane waves capable to traverse the free space or in surface waves, if the condition of the total internal reflectance is fulfilled. In this configuration the requirement of opposite signs of dielectric permittivities of materials at the interface is lifted up and PhC takes a role of medium with the negative permittivity.

Finally, the requirement of the total internal reflection (grazing light incidence) in the open space can be also relaxed, if PhC interfaces a material of a negative permittivity, e.g., the metal. Thus, the metal-terminated PhCs is potentially preferable configuration for maintaining the surface states in the PBG spectral range. In contrast to evanescent states, these waves can be directly excited by the incoming at any angle plane wave in the case of remaining transparency of sandwiched materials. Such surface states are called the Tamm surface modes by analogy to Tamm electron states at the surface of a semiconductor [14]. In the optical experiment the Tamm state reveals itself as the band of high transmission inside the bandgap transmission minimum. This band is associated with light tunneling that makes the transmission through Tamm state higher than that of the uncoupled system consisting of spatially separated metal film and PhC at the bandgap frequency. To be realized, this mechanism requires at least the 1D PhC.

With the increase of the PhC dimensionality, i.e., in slab 2D and 3D PhCs terminated by the corrugated metal mirror, the surface modes exist independently of the PBG. But in this case we deal with another sort of surface waves, which are truly evanescent. These modes are the conventional surface plasmon polaritons (SPPs) located at the metal-dielectric interface. The periodical lattice on the PhC surface provides the diffraction grating as the coupling element. The latter allows the incident plane wave to excite the evanescent surface modes. The optical signature of SPPs is the unique correlation between the SPP dispersions and the symmetry of PhC surface lattice.

Such metal-terminated PhCs are called the hybrid metal-dielectric PhCs. In hybrid PhCs both the surface and volume modes are omnipresent, moreover, their interaction with each other gives rise to complex hybridization of different excitations. The special case of hybrid architectures is the structures assembled from 3D PhCs and thin metal films [15]. These photonic materials are semitransparent that leads to superimposing of different resonances and to a multitude of attainable optical phenomena. In what follows we will review the optical properties of 3D metal film-terminated colloidal PhCs and describe the contributions of different surface states.

2 Experimental Techniques

Thin film colloidal crystals for our studies have been prepared using the self-assembly of spherical hydrophilic beads (SiO_2, PMMA, polystyrene) on hydrophilic glass substrates [16]. Detailed overview of growth and properties of opals is given in 2nd, 5th, 17th, and 19th chapters of this book. The improvement of opal crystallinity compared to the standard vertical drawing technique was achieved by applying acoustic noise agitation to the vessel containing the suspension of spheres [17].

Opal can be considered as the 3D grating in a form of the lattice of tightly packed monodisperse dielectric spheres (Figs. 1 and 2a). Resonances in optical spectra of opal appear due to light diffraction in the volume of this grating. Each set of crystal planes can be considered as the separate Bragg mirror and it becomes involved in diffraction, when the wavelength of the incident light becomes comparable to the interplane distance. All these resonances are seen in transmission spectra, but only few of them can be observed in reflectance spectra, because the corresponding diffraction orders cannot escape the opal due to total internal reflection. The pattern of diffraction resonances rapidly changes with changing the incidence angle and the wavelength of incoming light following the Bragg law. Only a small fraction of these resonances can be associated with the directional PBGs (also called the stop-bands), the spectral interval of which is free of optical modes [18]. Other resonances do not fulfill the latter requirement, because the opal grating merely splits the incident beam on diffraction orders propagating along different directions.

In order to prepare metal-dielectric hybrids we used two configurations of metal films. To interface the opal film with the flat metal film, the opals have been crystallized on metal-coated substrates. Typically, there was 20/100 nm thick gold or silver layers deposited on glass substrates by means of magnetron spattering. Also a SiO_2 spacer can be added on top of a metal film and used as an external cavity.

20/100 nm thick metal films of another topology have been prepared by depositing metal on top surface of opal films. By virtue of preparation such film replicates the topology of the underlying opal surface. Since opals are typically crystallized

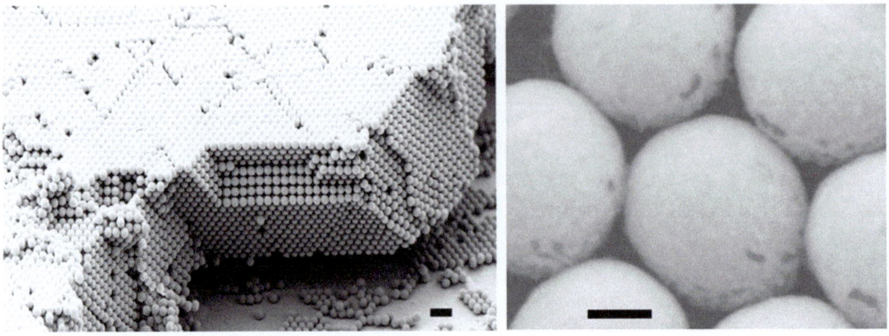

Fig. 1 *Left panel*—Top-coated with 50 nm Ag film the opal film (430 nm PMMA beads) on a glass substrate. Scale bar—1,000 nm. *Right panel*—30 nm thick Ag caps on 625 nm SiO_2 spheres—the metal semishells. Metal bridges are resolved. Scale bar—200 nm. No additional metal coating was applied to the sample before taking the SEM image

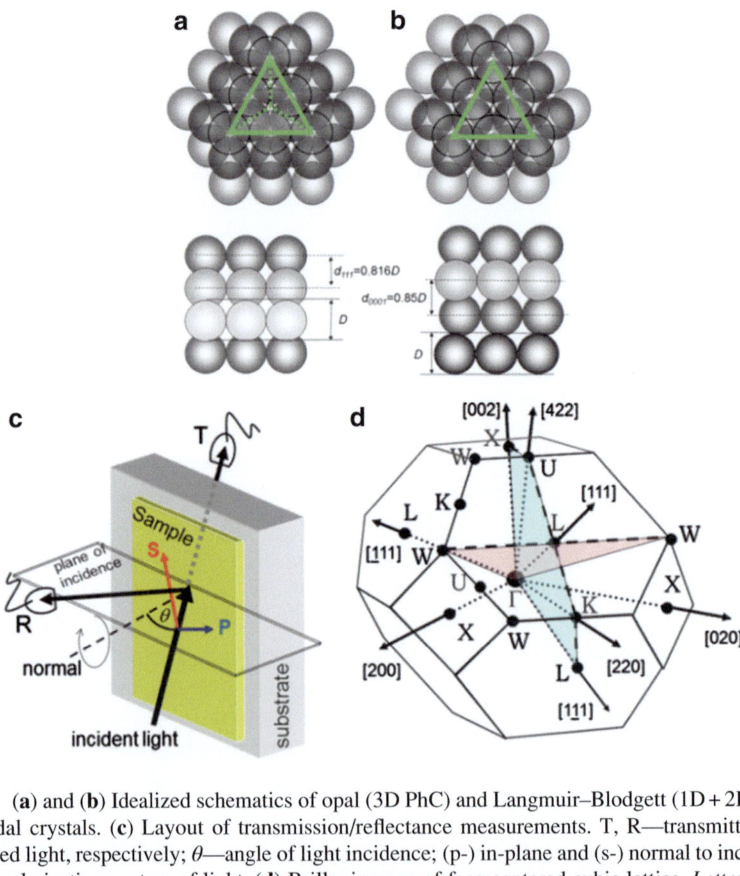

Fig. 2 (**a**) and (**b**) Idealized schematics of opal (3D PhC) and Langmuir–Blodgett (1D + 2D PhC) colloidal crystals. (**c**) Layout of transmission/reflectance measurements. T, R—transmitted and reflected light, respectively; θ—angle of light incidence; (p-) in-plane and (s-) normal to incidence plane polarization vectors of light. (**d**) Brillouin zone of face-centered cubic lattice. *Letters* label the symmetry points. *Shaded cross-sections* correspond to scanned directions of light incidence. *Arrows* are labeled by Miller indices of opal planes in the real space lattice

with the (111) plane of the fcc lattice facing the surface, the metal film acquires the form of the hexagonal lattice of electrically connected metal semispheres (Fig. 1, right). Thus, the metal component of metal-coated opal is the quasi-planar 2D diffraction grating with spacing equal to the sphere diameter. Such periodically corrugated or perforated metal films are known as plasmonic crystals, which eigenmodes are surface plasmon polaritons [19]. Correspondingly, a metal-coated opal can be considered as a complex photonic architecture consisted of seriesly connected 3D photonic and 2D planar plasmonic crystals.

In order to have the reference samples lacking 3D diffraction, semi-ordered ensembles of colloidal particles have been prepared using the Langmuir–Blodgett (LB) technique. LB method assumes formation of monolayers of monodisperse hydrophobic spheres at the water–air interface and subsequent monolayer-by-monolayer transferring them to a substrate. In LB colloidal ensembles the successive monolayers are arranged not in registry with each other [20] (Fig. 2b). Such arrays are conveniently described as PhCs with (1 + 2)-dimensionality, because their optical properties resemble those of the 1D Bragg mirror for transversal and those of 2D slab PhC for the tangential light propagations [21, 22]. Metal films were deposited on LB films similarly to preparation of opal-based hybrids.

In order to perform the optical characterization, the samples were illuminated by a collimated beam of ~1 mm in diameter from a tungsten lamp. Angle-resolved transmission/reflectance spectra were acquired using the CCD spectrometers at different angles of light incidence, θ, namely, in the range $-80°/80°$ with respect to the film normal. Linear polarizer and analyzer were used to obtain spectra in s- or p-polarizations that are characterized by the orientation of the electrical field in the incident light either perpendicular or parallel to the plane of incidence, respectively (Fig. 2c). In this review we will discuss the optical spectra obtained by the interrogation of opal crystals in the ΓLKLUX cross-section of the Brillouin zone (Fig. 2d). In what follows the diffraction resonances will be labeled according to the Miller indices of the fcc lattice planes.

The dispersions of diffraction resonances for different families of opal crystal planes were approximated using the Bragg–Snell law $\lambda_{\mathrm{hkl}} = 2d_{\mathrm{hkl}}n_{\mathrm{eff}}\sqrt{1 - \sin^2 r_{\mathrm{hkl}}}$, where D is the sphere diameter, $d_{\mathrm{hkl}} = \sqrt{2}D/\sqrt{(h^2 + k^2 + l^2)}$ is the interplane distance for (hkl) planes, and r_{hkl} is the incidence angle of light with respect to the different crystal planes. In the case of (111) planes oriented along the substrate, r_{111} angle relates to the external incidence angle via Snell law $n_{\mathrm{eff}} \sin(r_{111}) = n_{\mathrm{air}} \sin(\theta)$. The effective index of the opal refraction n_{eff} was obtained by fitting experimentally obtained central wavelengths of the (111) resonance to the Bragg–Snell law. In order to allow the comparison of samples assembled from spheres of different diameters, some spectra are plotted against the reduced frequency D/λ.

3 Opals on a Flat Metal-Coated Substrate

So far colloidal PhCs on metal-coated substrates have been studied mostly in association with monolayers of spheres [23–26]. The grating-excited SPP modes, the hybridization of dielectric and plasmonic guided modes, and the Mie resonances of spheres have been detected in reflectance and transmission spectra. The strongly enhanced light confinement in monolayers and the related improvement of quality factors of quasi-guided modes are the main conclusions of these studies.

Being applied to 3D colloidal PhCs, the same approach improves the light confinement leading to higher magnitude of Fabry–Perot oscillations, greater attenuation at diffraction resonances [27], and the multitude of optical resonances [28]. Such samples will be further abbreviated as, e.g., opal-metal, where metal assumes silver or gold films. One can see that transmission/reflectance spectra of the opal-Au sample are dramatically different from those of the same opal assembled on a glass substrate (Fig. 3). This is in the first instance the result of the enhanced reflectance (reduced leakage of light to the substrate) promoting bouncing of light between opal film boundaries (recycling of transmitted light). If the metal film thickness is below 100 nm, the whole structure lends itself for transmission measurements. The transmission spectra are shaped according to the transparency of Au film that drops at $\lambda \geq 508\,\text{nm}$ due to interband transitions in the gold film [29]. However, in spite of expectations, no features pointing to the presence of surface modes were observed in the spectra of this architecture so far.

One can analyze the simplified 1D PhC (multilayer or Bragg mirror) model to estimate the resonance conditions in hybrid structures. The Bragg mirror is the well-known architecture possessing some PhC properties (Fig. 4). Reader can refer to 4th, 7th, 8th, 11th, 13th, and 14th chapters of this book to learn more about principles and applications of Bragg mirrors. The incident light experiences constructive interference outside the Bragg mirror at the resonance wavelength and the destructive one—inside it, i.e., its intensity exponentially decreases from surface in depth [30]. In order to produce the localized mode in the multilayer stop-band, the planar defect should generate the constructive interference at the same wavelength, i.e., to form the Fabry–Perot resonator. The resonance (the standing wave) in the defect requires the 2π-proportional phase shift for the round trip of light inside the multilayer including the resonator spacer.

In the first approximation, the opal structure along its [111] axis, which is normal to the substrate, can be represented as the stack of ¼-wavelength-thick ($t_i = {}^{1}\!/\!_{4}n_i \times \lambda_0, \lambda_0$ – resonance wavelength) layers of refractive indices n_1 and n_2. This multilayer interfaces the glass substrate and the air via the layers of the lower n_1 index as it happens in opal films (Fig. 4a). In the bare opal, the distractive interference takes place at stop-bands (Fig. 4c). Replacing the glass substrate with the metal mirror does not change the phase shift and, thus, creates no additional state in the stop-band (Fig. 4b, c). This phase shift argument explains the observations.

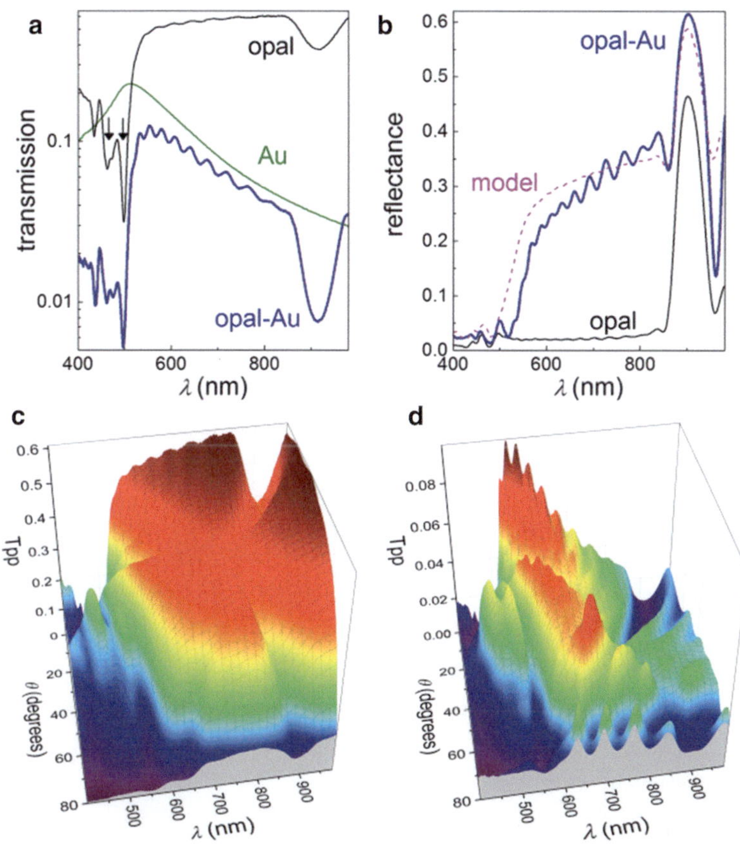

Fig. 3 (**a**) and (**b**) Transmission and reflectance spectra of opal-Au (opal on Au) hybrid (430 nm PMMA spheres, 50 nm Au film) in comparison to spectra of the opal on a glass substrate and to transmission of 50 nm Au film. All spectra are obtained at normal light incidence. The model reflectance spectrum has been calculated as $R_{\text{opal-Au}} = R_{\text{opal}} + 0.7T_{\text{opal}}$. (**c**) and (**d**) Comparison of angle-resolved p-polarized transmission spectra of opal and opal-Au hybrid. In both hybrids the opal films are of 9 (111) planes in thickness

The common approach to introducing planar defects in opals exploits the idea of heterostructuring [6, 31–33]. It assumes crystallization of the opal film, preparation of the defect layer and subsequent resuming of the opal growth to produce the sandwich-like structures [34]. Unfortunately, this approach cannot guarantee the lateral alignment of sequentially crystallized 3D lattices. The corresponding mismatch of the electromagnetic field distributions can be the reason for the reported poor resolution of the defect band in optical spectra.

In order to obtain the required resonance conditions one can, e.g., insert the spacer of the high index dielectric in between the Bragg and the metal mirrors (Fig. 4d, e). This spacer performs as the Fabry–Perot cavity, the thickness, and the refractive index of which should ensure the 2π-multiple phase shift for the light

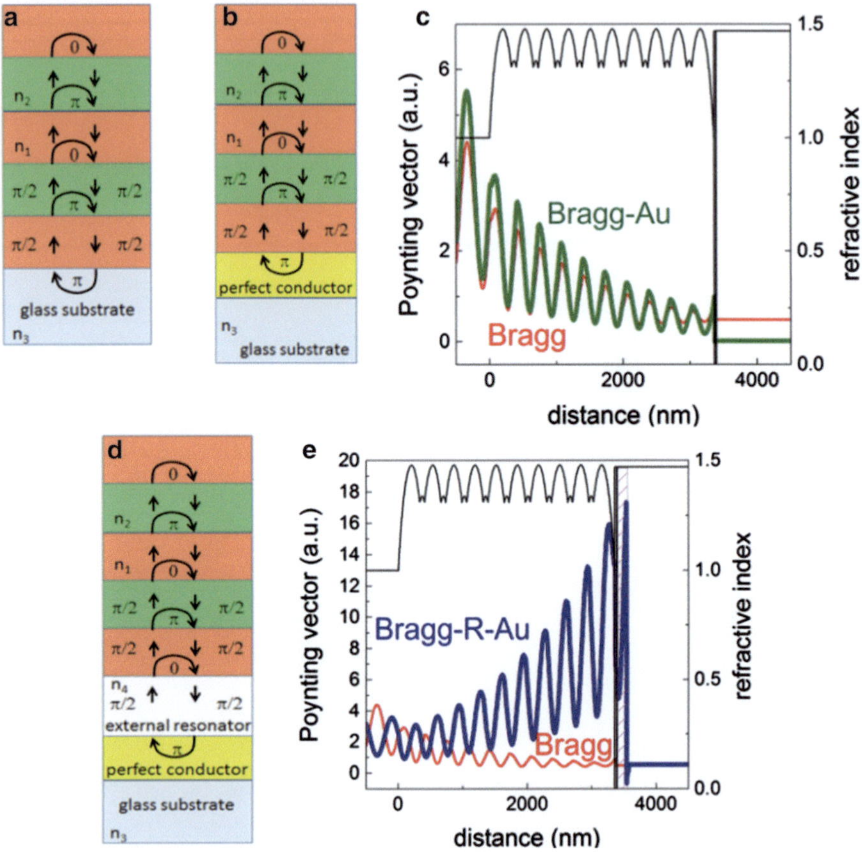

Fig. 4 (**a**) and (**b**) ¼-wavelength Bragg mirrors on bare and on metal-coated glass substrates. $n_1 < n_2 < n_3$—refractive indices. *Arrows* show the directions of the round trip and associated phase shifts. (**c**) Decay of the energy flux in depth of the Bragg mirror on bare and Au-coated glass substrates at the Bragg resonance (in the stop-band). Poynting vector is calculated at incidence angle of 10^{-5} degree. The rounded refractive index profile used for simulations is adjusted to that in opal film along [111] axis [53]. (**d**) and (**e**) Round trip and Poynting vector in the cross-section of Bragg mirror-resonator-metal mirror architecture when the Bragg resonance matches the cavity resonance wavelength. $\lambda/4$-cavity is positioned at the distance of 3,400 nm from the surface of the Bragg mirror

round trip at the Bragg resonance of the multilayer. Indeed, the discussed resonance belongs to the class of surface modes, because the respective electromagnetic field decays on both sides of the cavity (Fig. 4e). But this is not the evanescent wave, because it can be excited directly by the incident light. The resonance of such asymmetric cavity effectively mimics the functionality of the defect surrounded by two similar Bragg mirrors due to mirror image of the multilayer by the metal film. The defect resonance lends its localized mode for light tunneling through the stop-band. In corresponding spectra of opal-R-Au (R—stands for the resonator) this

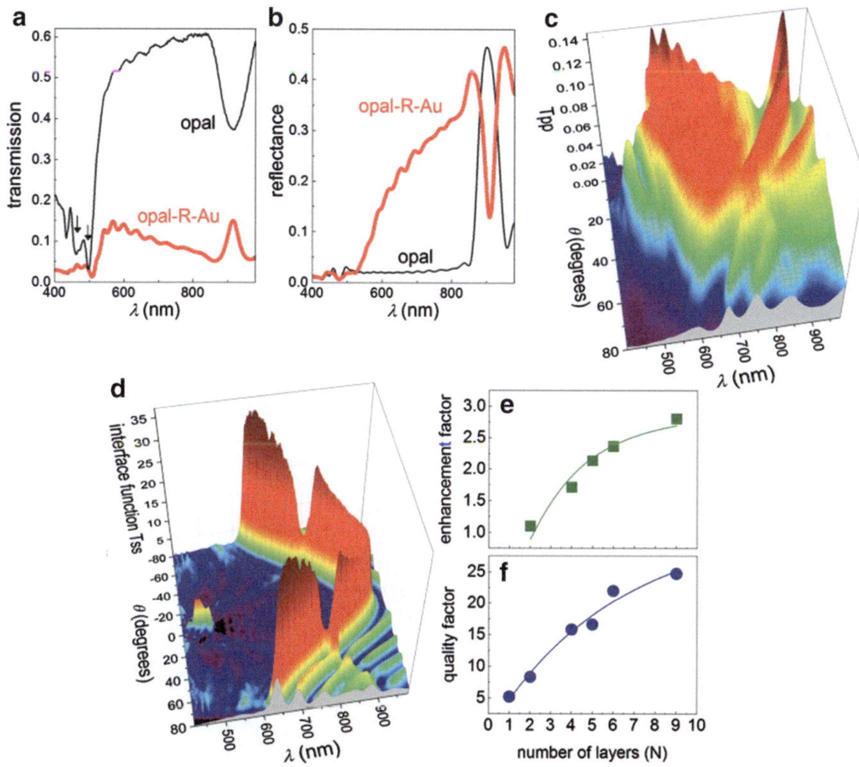

Fig. 5 (**a**) and (**b**) Transmission and reflectance spectra of opal-R-Au hybrid (9 (111) layer-thick opal film, 280 nm SiO_2 spacer, 50 nm Au film on glass substrate) in comparison to spectra of the opal on a glass substrate. All spectra are obtained at the normal light incidence $\theta = 0°$. (**c**) Angle-resolved transmission spectra in p-polarized light. (**d**) The rectified transmission enhancement (interface function) in opal-R-Au hybrid in s-polarized light. (**e**) and (**f**) Transmission enhancement factor in the center of the opal stop-band and the quality factor of the defect band resonance at $\theta = 0°$ plotted as a function of the number of (111) planes

defect band manifests itself as the transmission peak and reflectance minimum, which appear in the opal stop-band (Fig. 5a, b). The broadening of the resonance band compared to model calculations is the result of the disorder of the opal lattice.

In order to visualize the net effect of the resonant cavity, it is instructive to construct the so-called interface function by normalizing the optical spectra of the hybrid architecture on the spectra of their components [35]. Hence, by dividing the spectra of the opal-R-Au (Fig. 5c) on those of the bare opal film of the same thickness and of the flat Au film all unmodified spectral features become eliminated from the graph. The rectified transmission spectra reveal two sharp bands coinciding the (111) and (222) diffraction resonances of the opal film (Fig. 5d). This picture shows 20 and 5 times higher transmission through the respective defect modes. While these values are far below the calculated enhancement for the ideal Bragg mirror, the model does not take into account the 3D profile of the refractive index modulation and the intrinsic disorder in the lattice of polydisperse dielectric spheres.

The observed transmission enhancement in the defect mode of opal-R-Au (Fig. 5e) is about five times higher compared to the same parameter in the case of a planar defect sandwiched in between two identical opal films [9, 35, 36]. The quality factor of this resonance in opal-R-Au depends on the number of (111) planes in the opal mirror and saturates at value of ~30 for 50 (111) planes (Fig. 5f) [27].

The 3D nature of the opal mirror manifests itself at the avoided band crossing of (111) and (200) (or $(\bar{1}11)$) diffraction resonances occurring along the light propagation towards the edge of the Brillouin zone [37]. At this anticrossing region the continuity of the defect band is interrupted and the transmission enhancement drops to zero because the Bragg mirror is missed (Fig. 5d). This observation is important, in the first place, for stressing the difference between dimensionalities of the Bragg mirror and the opal PhCs. Secondly, the diffraction of the incident light at the opal surface should excite the surface plasmons at the metal-dielectric interface, which were indeed observed in spectra of a monolayer-on-metal architectures [23, 25, 26]. However, no respective surface modes obeying the specific SPP dispersion have been discriminated in opal-R-Au hybrids. This can be provisionally explained by simultaneous excitation of uncorrelated SPPs by diffracted light and by light that is randomly scattered in opal.

4 Corrugated Metal Film on Opal Surface: Tamm States

Architectures prepared by depositing thin metal (Au, Ag) films on the opal surface show a very rich set of resonances in their optical spectra. Such samples will be further abbreviated as, e.g., Au-opal (Au on opal). In this section we will discuss the spectral range around (111) Bragg resonance, i.e., the same that was considered in the Sect. 3.

If the flat metal film is attached to the opal surface, the respective Bragg mirror should be drawn with the metal film interfacing the low-index layer (Fig. 6a). This configuration is identical to the case in Fig. 4b, i.e., the destructive interference at the Bragg resonance is not perturbed and the transmission spectrum will show the PBG minimum (Fig. 6b). The interference pattern changes, if the Bragg mirror interfaces the metal film by its high-index layer (Fig. 6c). The phase shift for the round trip that involves the light reflection at the metal mirror accumulates now the multiple of 2π. Thus, the Bragg multilayer and the metal mirror define the effective Fabry–Perot resonator. The peculiarity of such resonator is the absence of a designated resonant cavity: reflectors are in direct contact with each other, but the field is enhanced at the multilayer-metal interface similarly to the case in Fig. 4e [38]. Respectively, the surface wave can be supported at any angle of light incidence as long as its wavelength falls in the stop-band of the Bragg mirror

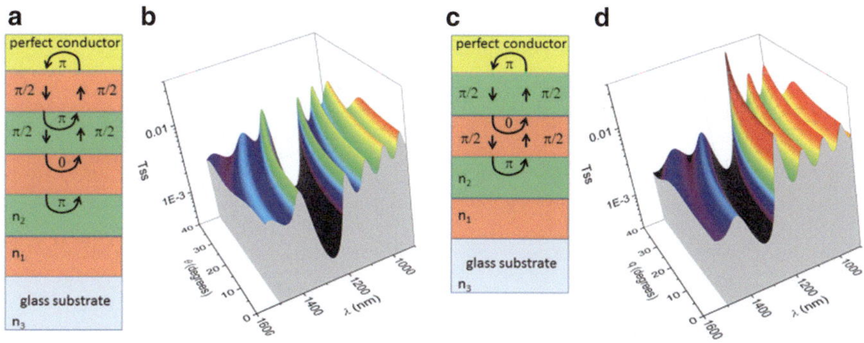

Fig. 6 (**a**) and (**b**) ¼ wavelength Bragg mirror facing the metal mirror by its low refractive index layer shows only the diffraction minimum in transmission spectrum. (**c**) and (**d**) The Bragg mirror facing the metal film by its high-index layer shows the Tamm resonant state in the stop-band. Refractive index profile is the same as in Fig. 4 (modeling courtesy A.V. Korovin)

(Fig. 6d). If so, the resonance transmission peak appears in the stop-band transmission minimum because light can tunneling through the Tamm surface state.

Experimental transmission spectra of Au-opal show the transmission increase at the long-wavelength edge of the (111) stop-band (Fig. 7a). The same applies to spectra of all studied Au- and Ag-coated samples. Modeling demonstrates that the Tamm state can be moved across the PBG spectral range, if the thickness of the metal-facing low refractive index layer is truncated (Fig. 7b, c). Let us note that, when the metal film adopts the shape of spheres, the metal mirror penetrates down to the high refractive index cross-section of the (111) layer at the top of the opal film (Fig. 1). This allows to realize the required phase shift of light that is reflected by metal mirror and obtain the respective resonator (Tamm state).

The evidence of the independence of the Tamm state on the 3D ordering of colloidal spheres comes from comparison of transmission patterns of the Au-coated LB and opal films (Fig. 7d, e). One can see the similar modification of the long-wavelength edge of the transmission minimum in Au-LB (1D + 2D PhC [39]) and Au-opal (3D PhC) spectra with respect to that of the bare template. Since the diffraction of light in the stack of monolayers remains unaffected by the metal film at its surface, the observed shift can be interpreted as the photon tunneling via the defect state localized at the long-wavelength edge of the stop-band of LB or opal PhCs. This example demonstrates that formation of the Tamm surface state requires the 1D stack of monolayers of spheres as the Bragg mirror, but these states insensitive to the lateral ordering of spheres in monolayers.

In order to give an idea about the magnitude of the Tamm state-related transmission we can rectify the Au-opal spectra by normalizing them to spectra of the bare opal and the Au coating (Fig. 8a). Theory predicts the dependence of the Tamm state transmission on the metal film thickness. First of all, the blue shift of the resonance occurs due to reduction of the thickness of the effective resonator since, e.g., 20 nm Au film is too thin to screen substantially the electromagnetic field. Second, for thicker coatings the cross-section of metal bridges between

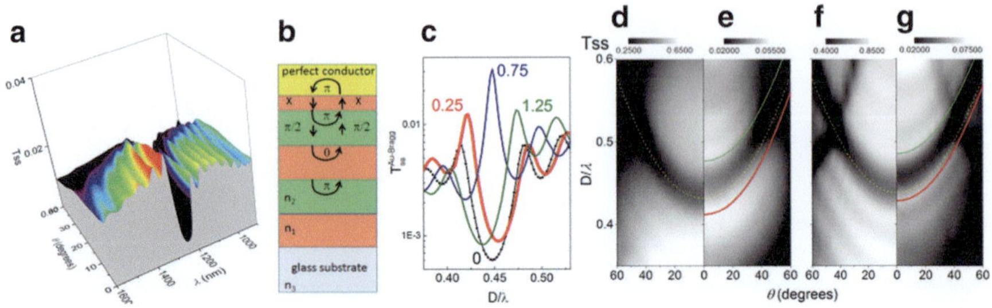

Fig. 7 (**a**) Experimental transmission spectra of Au-opal (50 nm Au on 560 nm PS spheres). (**b**) The Bragg multilayer with the truncated upper low-index layer. (**c**) The calculated wavelength of the Tamm mode in a stop-band of a multilayer as a function of the thickness of upper layer [panel (**b**)] (courtesy A.V. Korovin). (**d**) and (**e**) s-polarized light transmission spectra of bare and Au-coated LB template (50 nm Au film, 350 nm SiO_2 spheres). (**f**) and (**g**) The same for the Au-opal (50 nm Au film, 350 nm SiO_2 spheres). Dispersions of the Bragg resonance—*dashed lines*, Tamm state—*thick solid lines* and Fabry–Perot resonance—*thin solid lines*

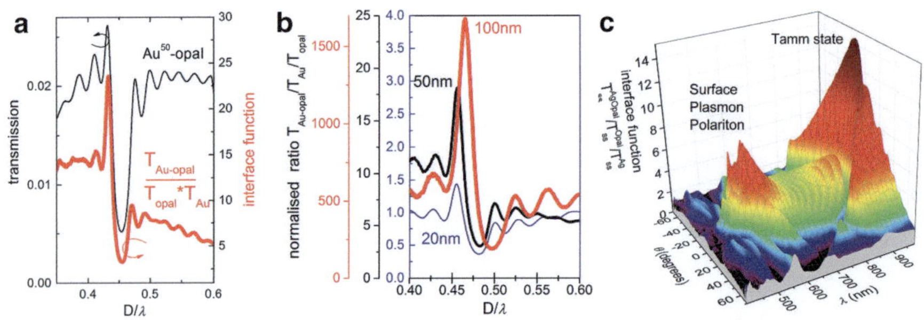

Fig. 8 (**a**) The experimental transmission spectrum of Au-opal (50 nm Au, 560 nm PS spheres) and the respective interface function at normal light incidence. (**b**) Enhancement of the Tamm mode with increasing thickness of Au layer. Coating thickness is indicated at curves. Each spectrum relates to its own vertical scale. (**c**) The rectified Tamm state-related transmission in Ag-opal (50 nm Ag, 430 nm PMMA spheres) at $\lambda \approx 890$ nm in comparison with the SPP-related extraordinary transmission band at $\lambda \approx 615$ nm at $\theta = 0^o$

semishells increases and the area of holes decreases, thus improving the reflectivity of the metal mirror. The 20 nm Au film is the poor mirror producing the weak Tamm state. Further increase of the coating thickness results in the rapid increase of the light confinement and the tunneling through the Tamm state rapidly increases (Fig. 8b). The magnitude of the transmission enhancement in these relative transmission spectra is overestimated, because normalization is made on the transmission of the continuous metal film, but the growing tendency is unambiguous. It is worth noting, the Tamm resonance transmission is almost as strong as that obtained with the help of the external cavity (Fig. 5d). Similar relative transmission spectra are obtained in the case of Ag-opal sample, which observation confirms the generality of this effect (Fig. 8c).

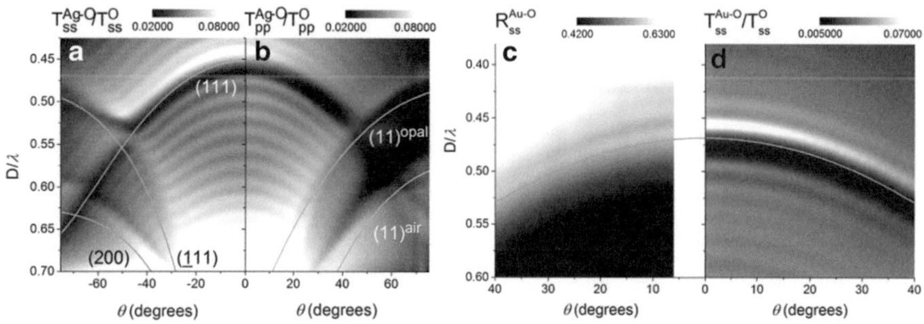

Fig. 9 (**a**) and (**b**) The relative transmission patterns of Ag-opal (50 nm Ag, 430 nm PMMA spheres) in s- and p-polarized light in ΓLKL cross-section of the Brillouin zone. Lines in panel (**a**) are the dispersions of diffraction resonances at (111), $(\bar{1}11)$ and (200) opal planes. Lines in panel (**b**) are the dispersions of (11) SPP resonances on the air (a) and opal (o) sides of Ag coating. (**c**) and (**d**) Comparison of the Tamm state appearance in reflectance from the metallized side of the sample and in relative transmission spectra of Au-opal (50 nm Au, 560 nm PS spheres), respectively

The Bragg stop-band, the Tamm band, and the Fabry–Perot resonances obey the same dispersion law that clearly points to the same origin of these phenomena (Fig. 7e, g). The Fabry–Perot oscillations are the sequence of the destructive and constructive interference resonances occurring along changing the wavelength. They correspond to light bouncing between upper and low boundaries of the PhC film, but they suffer the low quality factor due to poor reflectivity of these boundaries. Respectively, the magnitude of Fabry–Perot resonances is increased owing to metal mirror in Au-opal.

The transmission through the Tamm state depends on the light polarization and follows the polarization anisotropy of the diffraction at (111) opal planes (Bragg mirror) (Fig. 9a, b). Thus, the Tamm resonance appears more pronounced under the s-polarized illumination. It disappears at the edge of the first Brillouin zone ($\theta \approx 46^{\circ}$) due to multiple-wave Bragg diffraction for the same reason as does the resonance of the external cavity (Fig. 5d) [27]. This is the property of higher dimensionality of the opal PhC compared to that of the Bragg multilayer. In contrast, in p-polarized light the Tamm band rapidly vanishes along the incident angle increase due to vanishing (111) Bragg mirror reflectivity (Fig. 8c), but no multiple-wave Bragg diffraction takes place.

The essential difference between the Bragg and the Tamm resonances is reversing the destructive and constructive interferences in transmission and reflectance. That is why the Tamm state appears as the minimum in reflectance and as the maximum in transmission spectra (Fig. 9c, d and Fig. 5a, b). By comparing transmission and reflectance spectra of surface states in opals with the external microcavity and with the effective resonator, one can notice the weaker resolution of this resonance in the latter case. This can be the consequence of the higher dissipation of the electromagnetic energy due to corrugation of the metal mirror.

Many researches assume that the Tamm state in metal-coated Bragg multilayers is inevitably associated with oscillations of the electron density in a metal mirror (e.g., [14, 40] and recent works cited in 12th chapter of this book). They call this coupled photon-electron state as the Tamm plasmon polariton and assign to it very specific properties, like (1) the possibility of direct excitation by the incident light (since the Tamm state falls within the light cone in contrast to conventional SPPs), (2) the excitation by both s- and p-polarized light, and (3) the slow group velocity due to parabolic dispersion. As for metal-coated colloidal crystals, we did not observe so far any specific plasmonic effects in obtained transmission/reflectance spectra. Provisionally, Tamm plasmons should experience strong scattering being propagating along the metal film due to incommensurate periodicities of the profile and the plasmon wavelength.

5 Corrugated Metal Film on Opal Surface: Surface Plasmon Polaritons

Corrugated and perforated metal film on the opal surface is the plasmonic crystal (PlC) itself because it supports only periodicity-matching eigenmodes, the surface plasmon polaritons, the full set of which forms the SPP energy band structure. These SPPs are the Bloch waves in contrast to SPPs on a flat metal surface. By the spirit of preparation, the hexagonal lattice of such plasmonic crystal replicates the underlying PhC template that is in our case the hexagonal lattice of spheres of (111) opal plane. This PlC consists of metal semishells, which are sixfold interconnected by metal bridges with each other (Fig. 1). In addition to the periodic corrugation, such metal coating also includes the hexagonal lattice of triangular holes. As the result the modes of photonic and plasmonic crystals in such hybrid architecture overlap both in space and in energy.

Eigenmodes of such hybrid crystals are the SPP surface and PhC volume modes. One can also assume that coupling of photon to plasmon excitations takes place at the interface between these crystals. Calculations show that metal film intercepts more than 80 % of incoming light and transforms it in plasmonic excitations. We can distinguish three types of modes: (1) conventional eigenmodes of the all-dielectric 3D PhC; (2) SPPs of the planar PlC; (3) hybrid plasmonic-photonic modes in the transition range of interpenetrating plasmonic and photonic crystals. These resonances act together and exchange energy thus providing the complex optical response. Below we will discuss the SPP modes of metal film-terminated opals [41].

Integration of colloidal crystals and corrugated metal films has been initially driven by the idea of making self-organized materials with the extraordinary transmission (EOT) band [42]. EOT is the diffractive phenomenon allowing to transport much higher light flux through the periodically perforated and/or corrugated films compared to the flux permitted by the area of holes and the thickness of

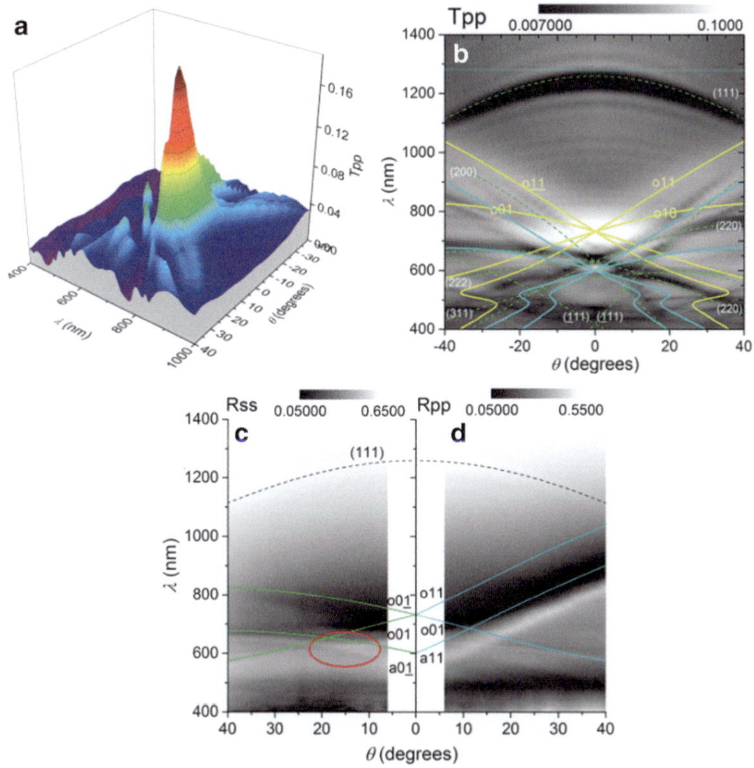

Fig. 10 (**a**) Transmission surface of Au-opal (50 nm Au film on opal assembled from 560 nm PMMA spheres, ΓLKL cross-section of the Brillouin zone) in p-polarized light. (**b**) The same transmission pattern overlaid with calculated dispersions of diffraction (*dash lines* labeled by Miller indices of fcc lattice) and SPP (*solid lines* labeled by Miller indices of hexagonal lattice, two sets of branches are shown similar to Fig. 9b) resonances. (**c**) and (**d**) Patterns of s- and p-polarized light reflected by the Au side of Au-opal. *Lines*—the dispersions of (01) (panel **c**) and (11) (panel **d**) families of SPPs at opal and air sides of the metal film, respectively. *Ellipse* is drawn around the region of avoided SPP band crossing

the film. In spite of rich variety of suggested EOT mechanisms [43], it is still unclear, which one applies to colloidal hybrid crystals.

So far, investigators dealt mostly with monolayers of metal-coated spheres rather than opals [44–46]. Eigenmodes of the monolayer are the guided modes of periodically profiled planar waveguide. They propagate in the near-field vicinity to the metal film and overlap with its SPPs so that the mode hybridization takes place [47]. In the practically interesting spectral range the light wavelength is comparable to the thickness of a monolayer, hence, all modes in metal-coated monolayers can be regarded as the surface modes.

Typical EOT peak in angle-resolved transmission spectra of metal-coated opal is shown in Fig. 10a. Away from this peak, the transparency of Au-opal sample is reduced by a factor of 30 compared to that of the bare opal film of the same thickness. The main reason of the low transmission is the high reflectance of the 50 nm thick Au film. The transmission pattern of the Au-opal hybrid shows (1) the bands of reduced transmission at λ_{hkl} related to the diffraction at planes of 3D opal lattice and (2) the bands of enhanced transmission related to SPP modes of the metal film with the dispersion $k(\omega) = \pm\sqrt{k_{\mathrm{SPP}}(\omega)^2 - \left(\frac{2\pi}{D\sqrt{3}}(2j-i)\right)^2} - \frac{2\pi i}{D}$, where $k_{\mathrm{SPP}}(\omega) = \frac{2\pi}{\lambda}\sqrt{\frac{\varepsilon_d\varepsilon_m}{\varepsilon_d+\varepsilon_m}}$, ε_m, ε_d—dielectric constants of the metal and dielectric (opal or air) at the interface [48] and (3) the enhanced by the Tamm resonance transmission at the edge of the (111) diffraction resonance (Fig. 10b).

The good agreement between the angle dispersions of diffraction resonances in the fcc lattice and experimentally obtained transmission minima of opal films is trivial nowadays. The EOT maximum at 725 nm corresponds to the node of the SPP resonances at incidence angle $\theta = 0°$ at gold film side facing the opal. At oblique light incidence the pass bands in p-polarized transmitted light follow the dispersion of $o\langle 11\rangle$ (o stands for opal) branches for SPPs at the same interface of Au film. The next bunch of SPP modes occurring at the air-facing side is blue shifted because of lower refractive index of air. Its node is centered at 605 nm. At oblique light incidence the $a\langle 11\rangle$ (a stands for air) SPPs dispersions are also associated with transmission pass bands.

In contrast to transmission spectra, the reflectance pattern of Au-opal in p-polarized light is naturally dominated by SPPs at the air-facing side of the Au film due to screening of the hybrid interior (Fig. 10d). The reflectance pattern in s-polarized light looks substantially different as a consequence of forbidden excitation of SPPs by s-polarized light (Fig. 10c). Observation of plasmons in both polarizations becomes possible due to the s-p-coupling in hexagonal lattice. Overall, the enhanced transmission in pass bands of Au-opal can be associated with surface modes of this PlC-PhC hybrid. These pass bands and the EOT peak in transmission can be interpreted in terms of the light tunneling through the metal film, which is stimulated by higher optical mode density in SPP modes.

The SPP modes are the eigenmodes of hybrid architectures that exist independently of photonic bandgaps and belong to the surface of hybrids. To clarify this statement let us compare the spectra of Ag-monolayer (Ag-ML), Ag-opal and Ag-LB, the colloidal templates of which are assembled from the same spheres and wearing the 30 nm Ag films of the similar topology (Fig. 11a, c, d) [49].

Transmission spectra of Ag-monolayer show the sequence of well-defined SPP bands, which spread over the broad angle range. The transmission magnitude at EOT peak is about 45 %. The Ag-Opal and Ag-LB samples possess the lower EOT of 20 %. The short wavelength SPP bands are suppressed due to the lower transparency of opal and LB colloidal crystals owing to (1) the increasing light scattering at lattice defects and (2) the transmission suppression by high order diffraction resonances. Moreover, the decay of the EOT peak along the angle increase appears faster in Ag-LB. The above differences can be explained by the efficient cooperation (hybridization) of unilaterally propagating SPPs and photonic modes in Ag-monolayer, which condition is not achieved in bulk colloidal crystals.

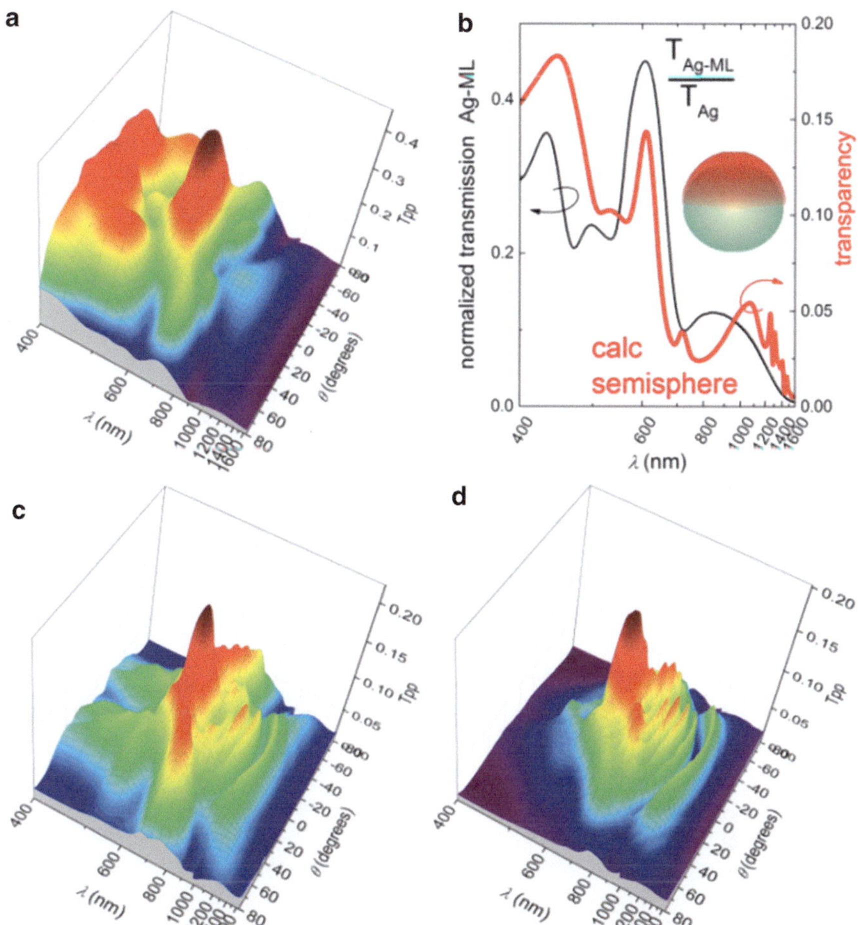

Fig. 11 (**a**) Transmission spectra of Ag-ML (30 nm Ag film on 520 nm SiO_2 spheres) in p-polarized light. (**b**) Normalized transmission of Ag-ML in comparison with finite element method calculations of the Mie resonances of a single Ag-SiO_2 semishell. (**c**) and (**d**) Transmission spectra of Ag-opal and Ag-LB in p-polarized light. The spheres and coatings are the same as for the Ag-ML of panel (**a**)

Nevertheless, the character of SPP-related transmission pass bands is the same for all these PhCs that points to the leading role of the metal film topology, i.e., the transmission spectra are mostly shaped by the surface modes. The relative strength of Tamm resonance in transmission of Ag-opal appears higher than that of the EOT SPP resonance (Fig. 8c), whereas the transmission change in absolute values looks higher for the latter resonance. This confusion can be resolved taking into account the exponentially higher reflectance of the same metal film with increasing the probe wavelength.

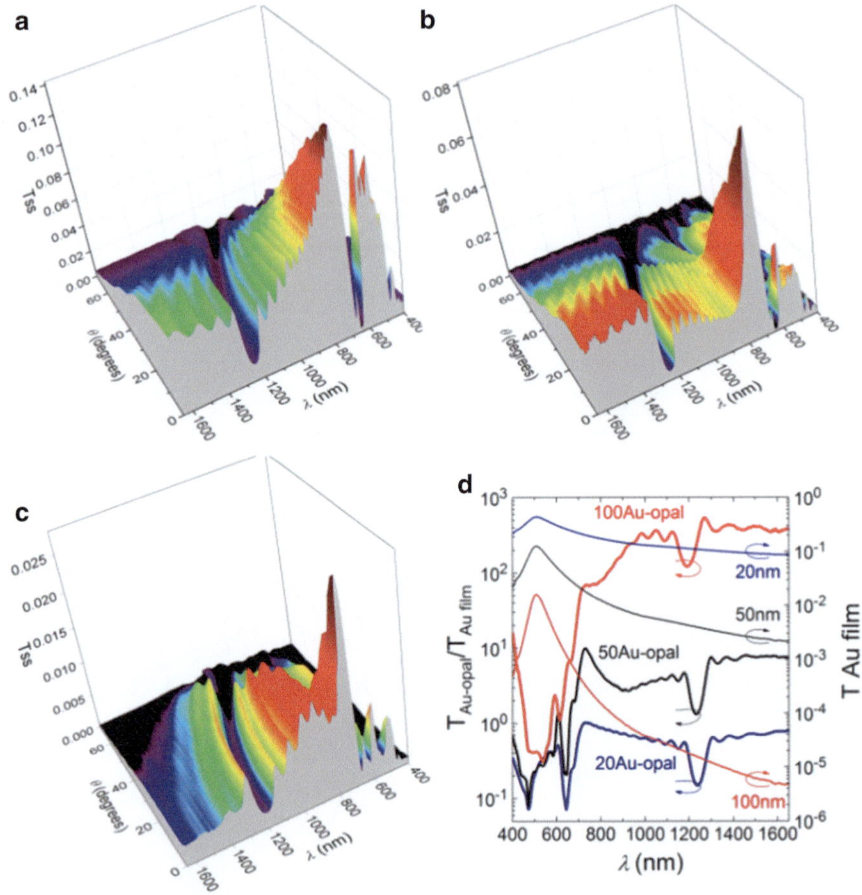

Fig. 12 (**a**), (**b**), and (**c**) Evolution of transmission spectra of Au-opal (560 nm PMMA spheres) in s-polarized light along the increase of the Au film thickness from 20 to 50 to 100 nm, respectively. (**d**) Relative transmission spectra of Au-opal with 20, 50, and 100 nm Au film thickness against transmission of flat Au films of the same thicknesses

One can also expect the strong dependence of the SPP-related transmission on the metal film thickness. On the one hand, the necessary condition for plasmon propagation is the connectivity of the metal film. If the metal film thickness is insufficient to bridge of semishells, the respective sample is EOT incapable, but the connectivity is improved for thicker films. Nevertheless, such connectivity in metal-coated colloidal spheres is very fragile (Fig. 1) and SPPs should be the short-range ones because of the deep corrugation profile. On the other hand, the SPP tunneling through the metal film is predictably reduced with the increase of the metal film thickness and the corresponding reduction of the hole openings (Fig. 12a–c). This competition assumes the optimum film thickness for maximizing the EOT magnitude.

The increase of the nominal Au film thickness from 20 to 100 nm shows the dramatic evolution of the EOT peak in the relative transmission spectra $T_{Au-opal}/T_{Au\ film}$ (Fig. 12d). This peak in normalized spectra is hardly seen for both 20 and 100 nm coatings in contrast to its clear resolution in the case of 50 nm coating. In numbers, the EOT enhancement can be estimated as 2.5; 6; 2.5 for 20, 50, and 100 nm thick coatings, respectively.

6 Localized Plasmon Resonances

Let us consider a metal semishell—the dielectric sphere with a metal cap [50]. Such semishell possesses the set of its Mie resonances. Similarly to the resonances of metal nanoshells they consist of cavity and particle plasmon resonances, but with additional asymmetric modes [51]. Increasing the ratio of the dielectric core radius to the metal shell thickness rapidly moves Mie resonances of semishells to longer wavelengths. In the array of connected semishells these localized resonances become coupled throughout the corrugated film, but they remain partly localized as well, i.e., they can be spectrally resolved.

The comparison of experimental transmission of the Ag-ML sample ($D = 560$ nm) with calculated spectrum of the single semishell shows that transmission minima of Ag-ML correspond the maxima of the scattering cross-section of the semishell (Fig. 11b). In the latter spectrum the expression $T = -\ln E$ was used to convert the calculated extinction spectrum into the transmission one. The obvious similarity of spectra of localized plasmons of a semishell and propagating plasmons in the monolayer of semishells nicely illustrates the tight-binding model used to calculate the energy spectra of solids starting from electronic levels of their constituent atoms, i.e., metal semishells represent the photonic atoms.

This minimum in the localized plasmon spectrum of a semishell is seen in transmission spectra of ML-, opal-, and LB-based samples as the transmission passband located at the long-wavelength side of the EOT peak (Fig. 11). In these spectra the antinode appears in the same range since the sphere diameters and the thicknesses of Ag coatings are same. The excitation of SPPs in this spectral range is excluded because the wavevector of the incident light is too short to provide the necessary phase matching conditions with the vectors of the reciprocal lattice of a monolayer (hexagonal facet of the Brillouin zone in Fig. 2d). Hence, the Mie resonances of semishells are responsible for the observed uneven background of transmission spectra in this wavelength range.

The transformation of transmission spectra of Au-opal samples along the metal coating thickness increase can be traced in Fig. 12. Assuming that these spectra consist of superpositions of localized and propagating resonances, the apparent change of the transmission background can be understood as the shift of the broad band of localized semishell plasmon to shorter wavelengths. This shift is fully consistent with the shift of metal nanoshell resonances occurring along the decrease of the thickness-to-radius ratio [51].

With respect to the bulk 3D PhC these localized plasmons are also the eigenmodes of the hybrid architecture. These plasmons are the non-dispersive surface modes owing to their localization in metal semishells occupying the surface of hybrid crystals. It is worth to estimate how efficiently the localized plasmons can participate in shaping the optical response of the hybrids [52]. The relative transparency of Au-opals rises dramatically with the coating thickness increase (Fig. 12d). For the 20 nm coating the minimum of the Mie resonance is positioned well outside of the measurement range and its influence is negligible. For 50 nm coating this minimum serves for the substantial transmission enhancement at the long-wavelength side of the spectrum. Its relative contribution is comparable to the EOT contribution. For the 100 nm coating the Mie scattering becomes so strong at the resonance that overrides the plasmonic EOT band and reduces dramatically the relative transmission in the respective spectral range.

7 Summary

The extreme efficiency of surface states in modifying the optical spectra of metal-coated 3D photonic crystals is pre-determined by the relatively high reflectivity of metal films. By terminating the colloidal PhCs with 20–100 nm thick metal films one can induce a multitude of optical resonances, which are absolutely alien to those of parent all-dielectric PhCs, and simultaneously enhance the contrast of intrinsic PhC resonances. The functionality of different combinations of PhCs and nanostructured metals can be considered on the basis of coupling light at the interface between materials with engineered negative and positive dielectric permittivities. In particular, the changes can be applied at will separately to both the metal and PhC counterparts of such hybrid heterojunction. The resulting functionality is provided by the interplay of Bragg resonances and Tamm surface states, cavity resonances, SPPs, and localized plasmon resonances. The native to PhCs the diffraction and the Fabry–Perot resonances experience the enhancement as well. Moreover, the Fano resonances manifesting the coupling of resonances of different origins supplement the selection of design tools.

It is amazing how dramatic is the influence of a passive layer of less than 1/10 of a lattice period thickness on PhC optical properties. This is a consequence of the strong field localization at the metal-dielectric interfaces. It is worth mentioning that coatings are not limited to metals. If, e.g., corrugated dielectric layers are applied, they can support their own guiding modes as well. In this case the hybrid architecture will be the photonic heterocrystal assembled from 3D PhC and 2D slab PhC. However, this semiconductor layer should be thick enough to accommodate the electromagnetic wave.

In terms of technology the hybrid approach has many advantages: (1) the PhC template remains intact due course of processing and preserves its optical properties, (2) the metal coating reserves the ability of further template modification by embedding PhC voids with "guest" materials for designing light emitting/

harvesting devices, non-linear light transducers, phase transformers, etc., and (3) the metal coating lend itself for post-deposition modification or patterning.

The unavoidable drawback of metal-containing PhCs is the light losses due to absorption in metal films. However, one should bear in mind that the transmission decrease demonstrated by hybrids occurs mostly due to the metal film reflectance, because the light absorption in flat metal films is less than 5 % at wavelengths longer than those of interband transitions in the electron energy band structure of metals.

We described the changes of PhC optical properties in terms of surface states that mediate the light interaction with PhCs. This description is based on the geometry of hybrid architectures and the localization of the optical field in the vicinity of metal-dielectric boundaries. In this case the new spectral features are treated as the resonances affecting the hybrid transparency by tunneling of light through surface modes. Equally well the spectrum transformation can be described using the language of hetero-crystals, where transient light flow is sequentially processed by photonic and plasmonic crystals as well as experiences coupling to different light carriers, namely, photon and plasmon modes. The latter language allows to treat light propagation in different parts of hybrid architecture as, e.g., waveguiding and diffraction, but requires stitching of propagating modes. This approach can be further extended to design very complex multiple-component architectures.

The added value of described surface states is the tight localization of the electromagnetic field. The achieved high field strength in respective modes accelerates the radiation-coupled electron transitions in light emitters, absorbers, and sensors as well as promotes the non-linear optical phenomena. This circumstance opens new prospects of purposive functionalization of hybrid photonic crystals and target-specific design of efficient photon managing devices.

Colloidal crystals were chosen in our experiments as readily accessible templates, but they do not exhaust the variety of 3D PhCs, which are susceptible to hybridization. Applying PhCs with tailored optical properties one can, on the one hand, extend the variety of engineered functionalities and, on the other hand, use the nanolithography to fabricate architectures compatible with the layout of optical processors.

Acknowledgements This work was supported by the DFG Cluster of Excellence "Engineering of Advanced Materials" in Friedrich-Alexander University of Erlangen and EU IRSES project "Phantasy."

References

1. J.D. Joannopoulos, S.G. Johnson, J.N. Winn, R.D. Meade, *Photonic Crystals: Molding the Flow of Light*, 2nd edn. (Princeton University Press, Princeton, 2008), 304 pp
2. A. Jebali, R.F. Mahrt, N. Moll, D. Erni, C. Bauer, G.-L. Bona, W. Bächtold, J. Appl. Phys. **96**, 3043 (2004)

3. P.V. Braun, S.A. Rinne, F. García-Santamaría, Adv. Mater. **18**, 2665 (2006)
4. H. Soda, K. Iga, C. Kitahara, Y. Suematsu, Jpn. J. Appl. Phys. **18**, 2329 (1979)
5. O. Toader, S. John, Phys. Rev. E **70**, 605 (2004)
6. S.G. Romanov, C.M. Sotomayor Torres, Three-dimensional lattices of nanostructures - template approach, in *Handbook of Nanostructured Materials and Technology*, vol. 4, Chap. 4, ed. by H.S. Nalwa (Academic Press, San Diego, 2000), pp. 231–323
7. A. van Blaaderen, R. Ruel, P. Wiltzius, Nature **385**, 321 (1997)
8. R. Rengarajan, P. Jiang, D.C. Larrabee, V.L. Colvin, D.M. Mittleman, Phys. Rev. B **64**, 205103 (2001)
9. F. Fleischhaker, A.C. Arsenault, J. Schmidtke, R. Zentel, G.A. Ozin, Chem. Mater. **18**, 5640 (2006)
10. P. Ferrand, M. Egen, J. Seekamp, R. Zentel, S.G. Romanov, C.M. Sotomayor Torres, Appl. Phys. Lett. **83**, 5289 (2003)
11. W. Lee, S.A. Pruzinsky, P.V. Braun, Adv. Mater. **14**, 271 (2002)
12. K. Ishizaki, S. Noda, Nature **460**, 367 (2009)
13. A.P. Vinogradov, A.V. Dorofeenko, A.M. Merzlikin, A.A. Lisyansky, Phys. Usp. **53**, 243 (2010)
14. M. Kaliteevski, I. Iorsh, S. Brand, R.A. Abram, J.M. Chamberlain, A.V. Kavokin, I.A. Shelykh, Phys. Rev. B **76**, 165415 (2007)
15. S.G. Romanov, A. Regensburger, A.V. Korovin, U. Peschel, Adv. Mater. **23**, 2515 (2011)
16. J.F. Galisteo-López, M. Ibisate, R. Sapienza, L.S. Froufe-Pérez, Á. Blanco, C. López, Adv. Mater. **23**, 30 (2011)
17. W. Khunsin, A. Amann, G. Kocher, S.G. Romanov, S. Pullteap, H.C. Seat, E.P. O'Reilly, R. Zentel, C.M. Sotomayor Torres, Adv. Funct. Mater. **22**, 1812 (2012)
18. J. Küchenmeister, C. Wolff, K. Busch, U. Peschel, S.G. Romanov, Adv. Opt. Mater. **1**, 952 (2013)
19. S.A. Maier, *Plasmonics: Fundamentals and Applications* (Springer, New York, 2007)
20. M. Bardosova, M.E. Pemble, I.M. Povey, R.H. Tredgold, Adv. Mater. **22**, 3104 (2010)
21. S.G. Romanov, M. Bardosova, M. Pemble, C.M. Sotomayor Torres, Appl. Phys. Lett. **89**, 43105 (2006)
22. S.G. Romanov, U. Peschel, M. Bardosova, S. Essig, K. Busch, Phys. Rev. B **82**, 115403 (2010)
23. L. Shi, X. Liu, H. Yina, J. Zi, Phys. Lett. A **374**, 1059 (2010)
24. X. Yu, L. Shi, D. Han, J. Zi, P.V. Braun, Adv. Funct. Mater. **20**, 1910 (2010)
25. M. Lopez-Garcıa, J.F. Galisteo-Lopez, A. Blanco, J. Sanchez-Marcos, C. Lopez, A. Garcıa-Martın, Small **6**, 1757 (2010)
26. S.G. Romanov, N. Vogel, K. Bley, K. Landfester, C.K. Weiss, S. Orlov, A.V. Korovin, G.P. Chuiko, A. Regensburger, A.S. Romanova, A. Kriesch, U. Peschel, Phys. Rev. B **86**, 195145 (2012)
27. S.G. Romanov, A. Regensburger, A.V. Korovin, A.S. Romanova, U. Peschel, Phys. Rev. B **88**, 125418 (2013)
28. V. Robbiano, M. Giordano, C. Martella, F. Di Stasio, D. Chiappe, F. Buatier de Mongeot, D. Comoretto, Adv. Opt. Mater. **1**, 389 (2013)
29. B.R. Cooper, H. Ehrenreich, H.R. Philipp, Phys. Rev. **138**, A494 (1965)
30. A. Yariv, P. Yeh, *Optical Waves in Crystals* (Wiley, New York, 1984)
31. S.G. Romanov, H.M. Yates, M.E. Pemble, R.M. De La Rue, J. Phys. Condens. Matter. **12**, 8221 (2000)
32. E. Istrate, E.H. Sargent, Rev. Mod. Phys. **78**, 455 (2006)
33. M. Egen, R. Voss, B. Griesebock, R. Zentel, S. Romanov, C. Sotomayor Torres, Chem. Mater. **15**, 3786 (2003)
34. E. Palacios-Lidón, J.E. Galisteo-López, B.H. Juárez, C. López, Adv. Mater. **16**, 341 (2004)
35. B. Ding, M. Bardosova, M.E. Pemble, A.V. Korovin, U. Peschel, S.G. Romanov, Adv. Funct. Mater. **21**, 4182 (2011)

36. K. Wostyn, Y. Zhao, G. de Schaetzen, L. Hellemans, N. Matsuda, K. Clays, A. Persoons, Langmuir **19**, 4465 (2003)
37. S.G. Romanov, T. Maka, C.M. Sotomayor Torres, M. Müller, R. Zentel, D. Cassagne, J. -Manzanares-Martinez, C. Jouanin, Phys. Rev. E **63**, 056603 (2001)
38. A.P. Vinogradov, A.V. Dorofeenko, S.G. Erokhin, M. Inoue, A.A. Lisyansky, A.M. Merzlikin, A.B. Granovsky, Phys. Rev. B **74**, 045128 (2006)
39. S.G. Romanov, Phys. Solid State **52**, 844 (2010)
40. M.E. Sasin, R.P. Seisyan, M.A. Kalitteevski, S. Brand, R.A. Abram, J.M. Chamberlain, A.Y. Egorov, A.P. Vasil'ev, V.S. Mkhrin, A.V. Kavokin, Appl. Phys. Lett. **92**, 251112 (2008)
41. B. Ding, M.E. Pemble, A.V. Korovin, U. Peschel, S.G. Romanov, Phys. Rev. B **82**, 035119 (2010)
42. T.W. Ebbesen, H.J. Lezec, H.F. Ghaemi, T. Thio, P.A. Wolff, Nature **391**, 667 (1998)
43. F.J. Garcia-Vidal, L. Martin-Moreno, T.W. Ebbesen, L. Kuipers, Rev. Mod. Phys. **82**, 729 (2010)
44. P. Zhan, Z. Wang, H. Dong, J. Sun, J. Wu, H.-T. Wang, S. Zhu, N. Ming, J. Zi, Adv. Mater. **18**, 1612 (2006)
45. L. Landström, D. Brodoceanu, K. Piglmayer, D. Bäuerle, Appl. Phys. A **84**, 373 (2006)
46. C. Farcau, S. Astilean, J. Opt. A Pure Appl. Opt. **9**, S345 (2007)
47. L. Landstrom, D. Brodoceanu, D. Bauerle, F.J. Garcia-Vidal, S.G. Rodrigo, L. Martin-Moreno, Opt. Express **17**, 761 (2009)
48. W. Khunsin, G. Kocher, S.G. Romanov, C.M. Sotomayor Torres, Adv. Funct. Mater. **18**, 2471 (2008)
49. A.S. Romanova, A.V. Korovin, S.G. Romanov, Phys. Solid State **55**, 1612 (2013)
50. R. Bukasov, T.A. Ali, P. Nordlander, J.S. Shumaker-Parry, ACS Nano **4**, 6639 (2010); A. Belardini, A. Benedetti, M. Centini, G. Leahu, F. Mura, S. Sennato, C. Sibilia, V. Robbiano, M.C. Giordano, C. Martella, D. Comoretto, F.B. de Mongeot, Adv. Opt. Mater. **2**, 208 (2014)
51. A.I. Maaroof, M.B. Cortie, N. Harris, L. Wieczorek, Small **4**, 2292 (2008)
52. V. Robbiano, M. Giordano, C. Martella, F. Di Stasio, D. Chiappe, F.B. de Mongeot, D. Comoretto, Adv. Opt. Mater. **1**, 389–396 (2013)
53. G.M. Gajiev, V.G. Golubev, D.A. Kurdyukov, A.V. Medvedev, A.B. Pevtsov, A.V. Sel'kin, V.V. Travnikov, Phys. Rev. B **72**, 205115 (2005)

Control of Photon Emission by Photonic Bandgap Engineering in Colloidal Crystals

Pieter-Jan Demeyer and Koen Clays

Abstract Photonics is generally regarded as the most promising industry to continue the technological evolution when electronic devices reach their fundamental limits on operating speeds and bandwidth. Nonetheless, more efficient manipulation of the optical processes is required to harness its full potential. Many optical devices, e.g. lasers, (organic) light emitting diodes, and (organic) photovoltaics, are based on excited state processes. To optimize the efficiency of these devices, the flow of energy should be controlled both spatially and energetically. Photonic crystals are seen as a promising class of materials that are able to fulfill both roles. In telecommunication, photonic crystal fibers have already proven their merits in providing spatial control on the optical signals. Moreover, photonic crystals have the potential to guide light on a microscopic scale and in three dimensions, which will be important in future optical devices.

In this chapter, we will detail how photonic crystals can be employed to engineer the flow of energy as well. The reduced density of optical states in the photonic bandgap region can be exploited to suppress spontaneous emission of photons from the excited state to the ground state. If the spontaneous emission is isotropic, a 3D full bandgap is required to maximize suppression. However, the realization of such bandgaps requires specific crystal structures and high refractive index contrast between the two materials constituting the photonic crystal. Usually, colloidal particles are self-assembled into a 3D photonic crystal (see fifth chapter). These so-called artificial opals are easy and cost-effective to fabricate but have only a pseudo bandgap. Thus, not all spatial directions are equally affected by the photonic crystal resulting in a strong spatial dependence of the spontaneous emission.

The photonic bandgap can be exploited to influence the excited state kinetics of various optical processes (Sect. 1). Concisely, the reduced density of optical states inside the photonic bandgap range suppresses spontaneous emission (Sect. 2) which leads to spectral redistribution and a longer lifetime of the excited state. The spectral redistribution can have a large influence on stimulated emission, as detailed in Sect. 3. On the other hand, the prolonged lifetime of the excited state affects the efficiency of energy transfers from that state. These processes are crucial in the

P.-J. Demeyer • K. Clays (✉)
Department of Chemistry, Laboratory of Molecular Photonics and Electronics, KU Leuven University, Celestijnenlaan 200D, B2425 3001, Heverlee, Flemish Brabant, Belgium
e-mail: demeyer.pj@gmail.com; koen.clays@fys.kuleuven.be

D. Comoretto (ed.), *Organic and Hybrid Photonic Crystals*,
DOI 10.1007/978-3-319-16580-6_21

working principles of lasers, light emitting diodes, and solar cells. As such, the engineering of photon emission could prove relevant to optimize the efficiency of these devices.

Keywords Colloidal photonic crystals • Spontaneous emission • Stimulated emission • Lasing • Energy transfer • Upconversion • Spectral redistribution • DOS manipulation

Photonics is generally regarded as the most promising industry to continue the technological evolution when electronic devices reach their fundamental limits on operating speeds and bandwidth. Nonetheless, more efficient manipulation of the optical processes is required to harness its full potential. Many optical devices, e.g. lasers, (organic) light emitting diodes, and (organic) photovoltaics, are based on excited state processes. To optimize the efficiency of these devices, the flow of energy should be controlled both spatially and energetically. Photonic crystals are seen as a promising class of materials that are able to fulfill both roles. In telecommunication, photonic crystal fibers have already proven their merits in providing spatial control on the optical signals [1]. Moreover, photonic crystals have the potential to guide light on a microscopic scale and in three dimensions, which will be important in future optical devices [2].

In this chapter, we will detail how photonic crystals can be employed to engineer the flow of energy as well. The reduced density of optical states in the photonic bandgap region can be exploited to suppress spontaneous emission of photons from the excited state to the ground state. If the spontaneous emission is isotropic, a 3D full bandgap is required to maximize suppression. However, the realization of such bandgaps requires specific crystal structures and high refractive index contrast between the two materials constituting the photonic crystal [3, 4]. Usually, colloidal particles are self-assembled into a 3D photonic crystal (see fifth chapter) [5]. These so-called artificial opals are easy and cost-effective to fabricate but have only a pseudo bandgap [6–8]. Thus, not all spatial directions are equally affected by the photonic crystal resulting in a strong spatial dependence of the spontaneous emission.

The photonic bandgap can be exploited to influence the excited state kinetics of various optical processes (Sect. 1). Concisely, the reduced density of optical states inside the photonic bandgap range suppresses spontaneous emission (Sect. 2) which leads to spectral redistribution and a longer lifetime of the excited state. The spectral redistribution can have a large influence on stimulated emission, as detailed in Sect. 3. On the other hand, the prolonged lifetime of the excited state affects the efficiency of energy transfers from that state. These processes are crucial in the working principles of lasers, light emitting diodes, and solar cells. As such, the engineering of photon emission could prove relevant to optimize the efficiency of these devices.

1 General Concept

Photonic crystals can be employed to tune the relative importance of various excited state processes. This general concept can be explained by considering the kinetics of the excited state. The various processes described in this chapter are schematically summarized in a Jablonski diagram, which maps the conversion of energy during the various processes using the energy states of the molecules or atoms involved (Fig. 1). While the basis of each process is explained below, the intricacies are detailed in the corresponding sections.

1.1 Excited State Processes

When the energy of an incoming photon is close to the energy difference between the ground state and an excited state, absorption can occur. After excitation, the absorbed energy needs to be released to regain the more stable ground state conformation. In the simplest case, the energy can only be converted to heat via the multiple vibrational and rotational states in the system. This process is called nonradiative decay and is always present [9].

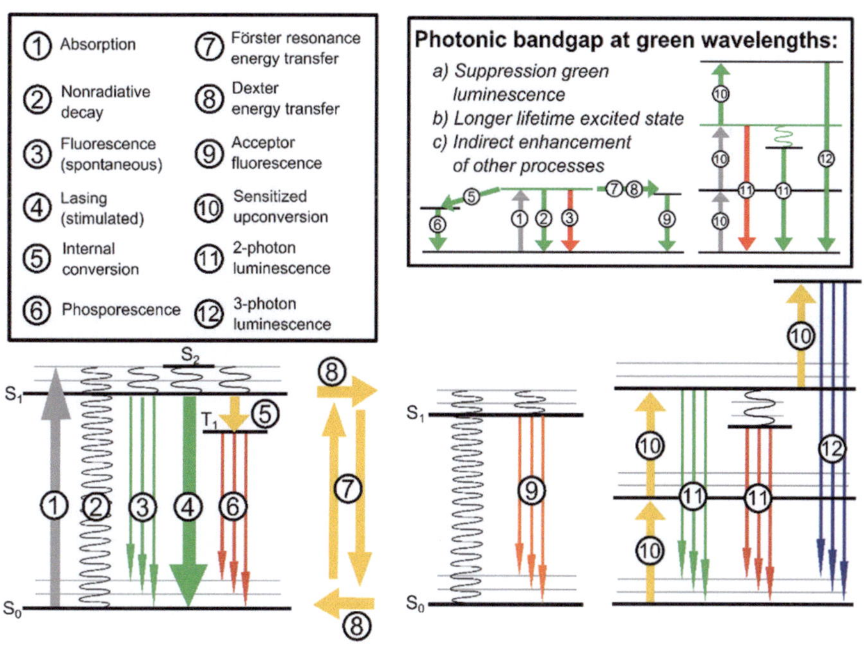

Fig. 1 Jablonski diagram of the various excited state processes discussed in this chapter. The *inset* on the *top-right* exemplifies the changes in kinetics when a photonic bandgap at green wavelengths is applied

In fluorescent molecules, there are also radiative decay pathways available. After absorption, the molecule quickly relaxes to the vibrational ground state of the electronic excited state by nonradiative decay. Then, a photon is spontaneously emitted with an energy that matches the difference between the excited state and the ground state or a vibrational state close to it. As some of the absorbed energy is converted to heat, the fluorescence peak is always spectrally red-shifted with respect to the absorption peak (Stokes shift) [10]. The amount of spontaneous emission from an excited state is directly related to the density of optical states of the environment via Fermi's golden rule (Sect. 2). As photonic crystals feature a reduced density of optical states in the frequency range of the photonic bandgap, they can suppress spontaneous emission [11]. This mechanism constitutes the primary method to control the excited state kinetics in colloidal crystals.

In contrast to spontaneous emission, stimulated emission requires the presence of a photon to induce radiative decay. The photon stimulates the emission of a second photon with exactly the same energy and phase. As the number of photons is doubled in this process, stimulated emission can establish optical amplification. Still, the stimulated emission needs to outweigh the loss mechanisms (spontaneous emission, nonradiative decay) to reach true optical gain, and hence lasing [12]. This usually requires population inversion, i.e. a higher occupancy of the excited state than that of the ground state. At least three levels are required to reach population inversion [12]. In our example (Fig. 1) the S_2-level is directly pumped via absorption. After nonradiative decay, the S_1-level is populated until its occupancy exceeds that of the ground state and population inversion is established.

When other molecules or atoms (acceptors) are in the vicinity of the excited molecule/atom (donors), the excitation energy can be transferred nonradiatively. As no energy is lost during the transfer itself, the excited state energies need to be close to each other. In general, the acceptor energy is slightly lower and the amount of spectral overlap between the donor emission and acceptor absorption is important for the efficiency of energy transfer [9]. We can distinguish between two main mechanisms for nonradiative energy transfer: Förster resonance energy transfer (FRET) and Dexter energy transfer. FRET is based on a coulombic dipole–dipole interaction between the donor and acceptor transition dipoles. The energy of the donor transition is employed to induce the acceptor excitation, without emission of a photon. Dexter energy transfer occurs through the simultaneous exchange of electrons between the donor and acceptor [9]. While the excited donor electron is transferred to the excited state of the acceptor, a ground state electron of the acceptor is brought to the ground state of the donor. The net result is a transfer of the excitation energy. After either FRET or Dexter energy transfer, the acceptor can regain its ground state conformation through fluorescence or other processes.

Energy transfers are at the basis of sensitized upconversion as well [13]. The system shown in Fig. 1 is a simplification of the emission of Er^{3+}-ions sensitized by Yb^{3+}-ions. Here, Yb^{3+} is preferentially excited, and it can transfer its energy to a nearby Er^{3+}. In this system, the Er^{3+} can be excited multiple times via the energy transfer process. It can then radiatively decay from higher energy states to the ground state. Depending on how many photons were required to reach the radiative state, we call this two- or three-photon luminescence. The name upconversion is derived from the fact that the energy of each released photon is higher than the energy of each absorbed photon. In the specific case of Er^{3+}, excitation with infrared light leads to red, green, and blue luminescence.

1.2 Decay Rates, Quantum Yields, and Lifetimes

The kinetics of excited state processes are described by decay rates, quantum yields, and lifetimes. Every process is associated with a rate constant (k_i) and their mathematical expressions strongly depend on the physical mechanisms involved. The lifetime of an excited state (τ) is determined by the rate constants of all possible decay pathways (k_j) via [10]:

$$\tau = \frac{1}{\sum_j \kappa_j} \tag{1}$$

The relative importance of each process is expressed by quantum yields (φ_i) [10]:

$$\varphi_i = \frac{k_i}{\sum_j k_j} = k_i \cdot \tau \tag{2}$$

As an example, we will now consider the hypothetical case where spontaneous emission of green luminescence is suppressed by the photonic bandgap (inset Fig. 1). The reduced local density of optical states (LDOS) inside the photonic bandgap directly decreases the rate constants of fluorescence and green two-photon luminescence (upconverted two-photon luminescence). As a consequence, the excited state lifetime is prolonged [Eq. (1)] and the quantum yield of all other processes is enhanced [Eq. (2)]. Although this provides an indirect way of controlling the excited state processes, we see that there are limitations imposed by the intrinsic kinetics of the system. When the quantum yield of fluorescence is low, suppressing it will have limited effect on the lifetime of the excited state or the quantum yields of other processes. The influence on stimulated emission is less straightforward and is more related to spectral redistribution than to an enhanced lifetime of the excited state.

In the next section we will describe the suppression of spontaneous emission in more detail. We will see that there is a strong spatial dependence of the fluorescence due to the photonic bandgap anisotropy in colloidal crystals.

2 Spontaneous Emission

The spontaneous emission from an excited state can be described using Fermi's golden rule [10]. This expression contains the LDOS, which is dependent on the frequency and location of the emitter inside the colloidal crystal and the direction of emission [14, 15]. As a consequence, the spontaneous emission of light in colloidal crystals has a modified spectral and spatial dependence. Experimentally, the detection of these phenomena occurs on the outside of the colloidal crystal. To relate the detected spontaneous emission to the LDOS simulations, the propagation through the complex photonic crystal environment should be taken into account. This includes disorder-induced diffuse light propagation and outcoupling of the emitted light.

2.1 Fermi's Golden Rule

The decay rate of spontaneous emission (k_{se}) is expressed as [16]:

$$k_{\mathrm{se}}\left(\omega,\vec{r}\right)=\frac{\pi\omega}{3\hbar\varepsilon}\left|\left\langle\Psi_1\middle|\hat{\mu}\middle|\Psi_0\right\rangle\right|^2\rho\left(\omega,\vec{r}\right) \tag{3}$$

with ω the angular frequency, ε the permittivity of the effective medium, and ρ the local density of states. The factor $\left|\left\langle\Psi_1\middle|\hat{\mu}\middle|\Psi_0\right\rangle\right|$ represents the magnitude of the transition dipole moment between the excited state (Ψ_1) and the ground state (Ψ_0). It takes into account the chemical properties of the molecular system such as orbital overlap between the two states and introduces selection rules based on symmetry and spin conservation [9]. The vibrational/rotational freedom of the molecule and the polarity of the environment also determine this molecular contribution to Fermi's golden rule. When investigating the influence of the LDOS on spontaneous emission, a reference with the exact same chemical environment is required to exclude effects related to changes in the transition dipole moments [16].

The LDOS constitutes the field contribution of Fermi's golden rule. It describes the available propagation modes for the emitted light waves. In free space, propagation modes are available in all directions but this can be drastically altered inside photonic crystal structures. A classic example of the effect of LDOS on spontaneous emission was predicted by E. Purcell. He proposed that the spontaneous emission of atoms can be strongly increased in a resonant cavity as certain locations inside the cavity have enhanced density of optical states [17].

In a 3D photonic crystal environment, periodicity dictates that all propagation modes are Bloch modes. Hence, the LDOS is given by [18]:

$$\rho\left(\omega,\vec{r}\right)=\sum\nolimits_n\int_{\mathrm{BZ}}d^3k\left|E_{n,k}(r)\right|^2\delta\left(\omega-\omega_n\left(\vec{k}\right)\right) \tag{4}$$

with k the wavevector, BZ the first Brillouin zone, and $E_{n,k}$ the Bloch modes of the photonic crystal. The band structure of an *fcc* colloidal crystal shows a region without modes along the Γ-L direction, which corresponds to the <111> direction in real space [3]. Although zero density of states is predicted, finite, and imperfect, colloidal crystals experience only a reduction of the density of states inside the photonic bandgap. Moreover, the photonic bandgap is not at the same frequency for different crystal directions. It is therefore only a pseudo bandgap and not a full bandgap.

The spatial and spectral angle-dependence of the spontaneous emission from colloidal crystals is conceptually represented in Fig. 2. For small internal angles with respect to the <111> direction (θ), the angular dependence of the photonic bandgap spectral position is given by Bragg's law for *fcc* close-packed crystals (Fig. 2, top) [8]:

$$\sqrt{\frac{8}{3}}D\cos\theta = \frac{\lambda_{\max}}{n_{\mathrm{eff}}} \tag{5}$$

In this equation, D represents the sphere diameter of the colloidal particles used during self-assembly. The wavelength of the photonic bandgap maximum ($\lambda_{\max}$) in air is divided through the refractive index of the effective medium (n_{eff}). The

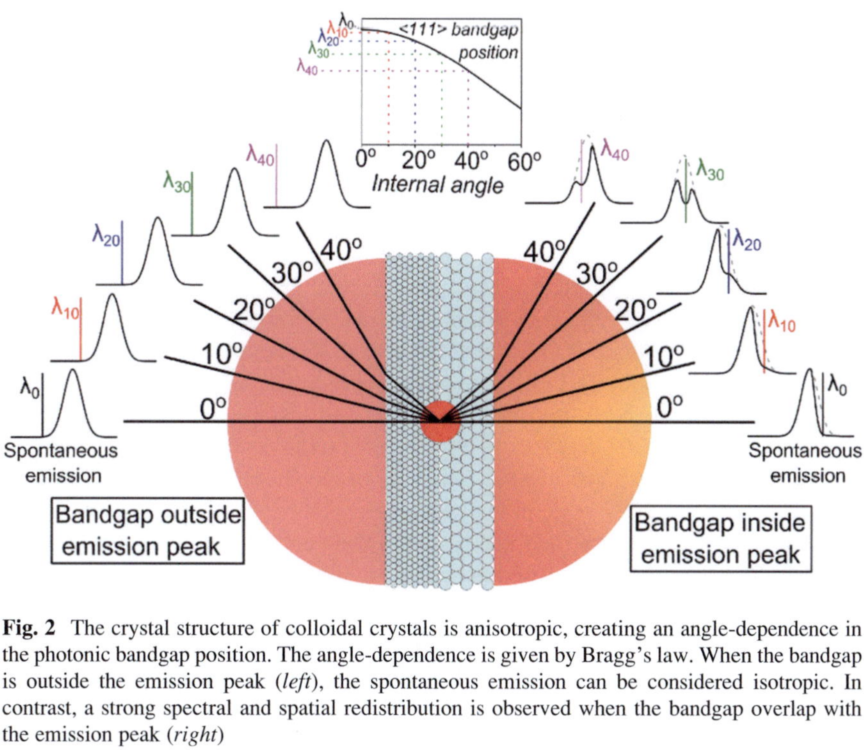

Fig. 2 The crystal structure of colloidal crystals is anisotropic, creating an angle-dependence in the photonic bandgap position. The angle-dependence is given by Bragg's law. When the bandgap is outside the emission peak (*left*), the spontaneous emission can be considered isotropic. In contrast, a strong spectral and spatial redistribution is observed when the bandgap overlap with the emission peak (*right*)

spectra become more complex at larger angles as they result from the interplay of reflections from {111} and {200} planes [19]. The angle-dependence of the photonic bandgap leads to a reduced LDOS at different frequencies for different angles. No suppression occurs and isotropic (Lambertian) emission is expected when the bandgap is outside the spectral range of emission. When the photonic bandgap positions coincide with the emission range, the spectral position of suppression will differ with the angle and pronounced shape changes in the emission spectrum can be observed (Fig. 2).

Examples of angle-dependent suppression of spontaneous emission are plentiful in literature [20–32]. Often, lifetime measurements are conducted to prove the prolongation of the excited state lifetime upon reducing the rate of radiative decay. The effect on the lifetime is usually much smaller than expected from the emission spectra as it is averaged over all directions of emission, not just the detected ones [33]. This is well exemplified in the experimental results in the chapter by V. Robbiano et al.

Under experimental conditions, angle-dependent suppression is insufficient to fully explain the obtained results. Computational analysis indicates that the LDOS is influenced not only by the crystal direction but by the position inside the photonic crystal as well (Sect. 2.2). Moreover, scattering and refraction phenomena modify the emission before detection (Sect. 2.3). These effects are usually non-negligible in real colloidal crystals and affect the observed spectral and spatial emission profiles [33, 34].

2.2 LDOS Simulation

The LDOS can be calculated using various computational approaches. Often, it is determined via eigenmode calculations. Once the eigenmodes of the photonic crystal are known, the number of states in a certain frequency range is counted [18, 35, 36]. Although straightforward, these results are only qualitatively useful since infinite photonic crystals are assumed. For finite photonic crystals, approaches based on Green's functions, transfer matrices, scattering matrices, or finite-difference time-domain (FDTD) calculations are required [33].

Computational analysis of colloidal crystals and inverse opals have discerned that the experienced LDOS is strongly dependent on the position and orientation of the emitter [14, 15, 36]. The LDOS in the bulk of the photonic crystal differs from that close to its surface. Even emitters that are outside but in close proximity to the photonic crystal experience modified LDOS. Strong enhancements of spontaneous emission have been reported when depositing emitters on the surface of colloidal crystals [37]. Directly under the surface of the colloidal crystal, the optical confinement is still weak and the LDOS resembles that of free space more closely. It has been predicted that the LDOS assumes its bulk value after approximately two lattice constants [14]. Megens et al. estimated that the contribution of the surface layers to the total emission was only 3 % in their structures [34]. Not only the proximity to the surface but also the position and relative orientation inside the unit cell is relevant when calculating the LDOS [14, 15, 36]. Although the majority of positions display a reduction of the LDOS, specific positions with enhanced LDOS can be found as well [36].

In practice, the emitters are distributed over multiple positions and the orientation of the emitter is randomized. Each molecule position is subject to a different LDOS and hence has a different life time. A weighted average of this ensemble should be taken to rationalize experimental spectra. As a consequence of this distribution, lifetime measurements of emitters inside photonic crystals feature a nonexponential decay [33].

2.3 Spatial Redistribution

While the LDOS determines the rate of spontaneous emission, the emitted light propagates through the colloidal crystal and the surrounding medium before reaching the detector. Scattering and refraction effects are therefore important to quantitatively explain experimentally determined spectra and their angle-dependence.

Self-assembled colloidal crystals inevitably contain point defects, line defects, and cracks. These imperfections give rise to background Mie-scattering (see also 15th chapter). The magnitude of this scattering background is determined by the crystal quality, the particle size, and the refractive index contrast between the two media constituting the photonic crystal [38]. Scattering events change the direction of the emitted light and can therefore wash-out the direction dependence of the LDOS by randomizing the direction of the emitted light, resulting in diffuse light [16, 34, 39–41]. The influence of the wash-out effect can be estimated by evaluating the thickness of the photonic crystal and its crystalline quality. The attenuation of the photonic bandgap results from the amount of order and is characterized by the Bragg attenuation length (L_B). In contrast, the disorder gives rise to scattering, characterized by the mean free path length of a photon (L_S). We can now distinguish between two limiting cases: photonic crystals much thinner than L_S and crystals with a thickness much larger than L_S.

In thin photonic crystals, the emitted light is not necessarily scattered before reaching the outer surface of the sample. The direction dependence of the LDOS is maintained and the amount of suppression is mainly related to the photonic bandgap strength.

In thick samples, one can assume that (multiple) scattering events occur before the emitted light can exit the photonic crystal. Still, spatial dependence is re-established by Bragg attenuation on a length scale L_B between the scattering events. As such, the position of the last scattering event before exiting the sample is crucial. If this scattering event occurs at a distance $z < L_B$, Bragg attenuation is not complete and diffuse light exits the sample. At positions $L_B < z < L_S$, the light is Bragg attenuated and the photonic bandgap influences the shape of the emission spectrum. Concisely, it is the ratio between order (L_B) and disorder (L_S) that determines the maximal suppression of spontaneous emission in thick samples. In practice, around 80 % suppression can be achieved inside the photonic bandgap range [34].

Upon exiting the photonic crystal, the emitted light has to conserve both its frequency (ω) and the wave vector component parallel to the interface ($k_{\|}$). For homogeneous media, this gives rise to Snell's law and a critical angle for internal reflection. Since the dispersion in colloidal crystals is anomalous, the out-coupling condition cannot be simply described by a critical angle and is given by [33]:

$$k_{\|} \leq \omega \varepsilon_{\text{out}} c \tag{6}$$

with ε_{out} the permittivity of the surrounding medium (e.g., air). Angles inside the colloidal crystal that do not satisfy this condition are internally reflected.

2.4 Conclusion

Spontaneous emission is strongly influenced by the LDOS of its environment, as stated by Fermi's golden rule. In the frequency range of the photonic bandgap, the spontaneous emission is suppressed due to the lower LDOS. Colloidal crystals feature an anisotropic bandgap which creates a strong spatial dependence of the spontaneous emission. A full description of the spatial dependence should take into account the position dependence of the LDOS, the influence of disorder on light propagation and the outcoupling of light at the outer surface of the photonic crystal. Despite the spatial redistribution, an overall suppression of spontaneous emission is evidenced by a prolonged excited state lifetime.

3 Stimulated Emission

Stimulated emission is at the basis of laser operation as it provides optical amplification. Like spontaneous emission, stimulated emission is suppressed when the LDOS is lowered at the transition frequency [10]. Nonetheless, photonic crystals can lower lasing thresholds through spectral redistribution. This section gives a conceptual overview of the phenomena at play when stimulated emission is introduced into photonic crystal structures. More detailed descriptions and experimental data can be found in 4th, 15th, and 17th chapters.

3.1 Lasing Requirements

Both an optical gain medium and optical feedback are required to achieve lasing. In the optical gain medium, there is competition between three main excited state decay pathways: the stimulated emission required for lasing, spontaneous emission, and nonradiative decay (Fig. 1). Stimulated emission requires the presence of a photon with similar energy as the excited state to occur. The optical feedback keeps the emitted photons in the system and thus shifts the balance from spontaneous emission toward stimulated emission [12]. The lasing threshold is reached when the optical gain (through stimulated emission) outweighs the loss mechanisms (nonradiative decay and spontaneous emission) for a certain wavelength. A narrow spontaneous emission spectrum is favorable for stimulated emission and therefore lowers the lasing threshold. As the same amount of excitation energy is distributed over fewer wavelengths, optical gain conditions are easier to reach for these wavelengths.

There are two main types of lasers [12]. In a cavity laser, the optical gain medium is contained between two reflective surfaces (mirrors or Bragg reflectors) to provide optical feedback. Alternatively, the optical gain medium can be integrated into a reflective structure (distributed feedback lasers). Colloidal crystals can be employed in the construction of both types of lasers.

3.2 *Colloidal Crystal Lasing*

Lawrence et al. demonstrated the use of colloidal crystals in a cavity-type laser [42]. A non-close-packed colloidal crystal, contained in an elastomer, constituted one of the optical feedback mirrors. In this system, the selective reflection of the pseudo bandgap determined the feedback wavelength. By applying mechanical pressure to the elastomeric photonic crystal, the photonic bandgap and hence also the lasing wavelength could be tuned.

Alternatively, 3D distributed feedback lasers can be fabricated by integrating the optical gain medium into the colloidal crystal. In this case, the stimulated emission is subject to LDOS modification. Inside the photonic bandgap stimulated emission is reduced due to a lower LDOS, much like spontaneous emission. In contrast, the LDOS is higher at the edges of the photonic bandgap, which leads to enhanced emission. Moreover, the group velocity of light is lower at the band edges [43–48]. The combination of these effects leads to spectral narrowing of the spontaneous emission spectrum and the occurrence of low-threshold lasing at the band edges of the photonic crystal when overlapped with the emission spectrum. Since the photonic bandgap is anisotropic in colloidal crystals, the lasing is highly directional [48].

A third option is to induce lasing by defect-modes inside the photonic bandgap region. The disruption of the refractive index periodicity gives rise to a pass band within the photonic bandgap [49–52]. When overlapped with the emission spectrum, the LDOS is suppressed at all wavelengths of the photonic bandgap except inside the pass band. This leads to spectral narrowing and a lower lasing threshold [53–56].

The lasing wavelength of colloidal crystal lasers can be tuned by changing the photonic bandgap position. This can be done passively by changing the size of the colloidal particles or actively, e.g., by applying mechanical stress to elastomeric colloidal crystals [42, 45, 57]. A related field of research is that of chiral liquid crystal lasers [58]. Here, the photonic bandgap is not provided by colloids but by the periodic structure of the chiral liquid crystal, which is different for left- and right-handed polarized light. As a consequence, the polarization that is affected by the photonic bandgap experiences more optical feedback and circularly polarized lasing light results.

4 Energy Transfer

The lower LDOS in the photonic bandgap was shown to reduce the rate of spontaneous emission. As a consequence, the quantum yield of other excited state processes is increased. As energy transfers are technologically relevant in, e.g., light emitting diodes and photovoltaics, this strategy seems feasible to optimize device operation. In this section various mechanisms of energy transfer processes and their susceptibility to LDOS modifications are discussed.

4.1 *Mechanisms of Energy Transfer*

When an excited molecule (donor, D) is in close proximity to a second molecule (acceptor, A) with similar excited state energy, the excitation energy can be transferred.

$$^{*}\mathrm{D} + \mathrm{A} \rightarrow \mathrm{D} + {}^{*}\mathrm{A} \tag{7}$$

The energy transfer can occur through various mechanisms (Fig. 3), the best-known being radiative energy transfer, dipole–dipole coupling (Förster), and electron-exchange (Dexter) energy transfer [9, 10].

During radiative energy transfer, the donor emits a photon that is absorbed by the acceptor. For radiative energy transfer, overlap between the emission spectrum of the donor and the absorption spectrum of the acceptor is crucial. Although the efficiency is usually low, radiative energy transfer can occur over large distances. In contrast, nonradiative transfer occurs only at small donor–acceptor separations, as it is based on electronic coupling between both molecules. When the coupling is weak, perturbation theory is valid and the rate of energy transfer (k_{et}) can be described by [9]:

$$\begin{aligned} k_{\mathrm{et}} \sim \big[& a.\langle \Psi(^{*}\mathrm{D})\Psi(\mathrm{A}) | H_{\mathrm{dd}} | \Psi(\mathrm{D})\Psi(^{*}\mathrm{A}) \rangle^{2} \\ & + b.\langle \Psi(^{*}\mathrm{D})\Psi(\mathrm{A}) | H_{\mathrm{ex}} | \Psi(\mathrm{D})\Psi(^{*}\mathrm{A}) \rangle^{2} \big] \rho_{\mathrm{DA}} \end{aligned} \tag{8}$$

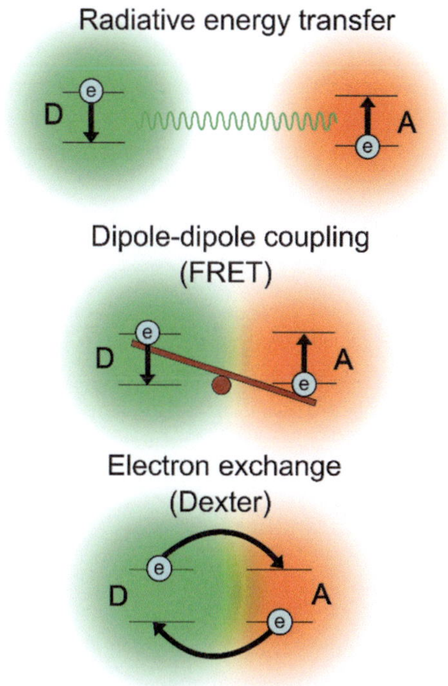

Fig. 3 Conceptual overview of the various energy transfer mechanisms between donor (D) and acceptor (A). The seesaw represents the energy transfer by transition dipole coupling

The first term represents the interaction between the wave functions before and after energy transfer through dipole–dipole coupling while the second term represents the electron-exchange interactions. The weighting factors (*a* and *b*) depend not only on the characteristics of the donor–acceptor pair but also on the distance between donor and acceptor (R_{DA}). ρ_{DA} is the density of overlapping states between donor and acceptor. The dipole–dipole coupling mechanism occurs naturally in, e.g., photosynthesis. The rate of energy transfer through dipole–dipole coupling (k_{dd}), is inversely proportional to the sixth power of the donor–acceptor separation distance [59]:

$$k_{dd} = \frac{2\pi\kappa^2}{\hbar n^4} \frac{\mu_D^2 \mu_A^2}{R_{DA}^6} J \tag{9}$$

with κ the orientation factor between the transition dipole moments of donor (μ_D) and acceptor (μ_A) and n the refractive index of the medium. The normalized spectral overlap integral (J) is related to the density of overlapping states. The rate of the electron-exchange mechanism (k_{ex}) falls exponentially with the donor–acceptor separation distance as orbital overlap is required [9].

$$k_{ex} = KJe^{-\left(\frac{2R_{DA}}{L}\right)} \tag{10}$$

K is a parameter related to the specific orbital interactions, L is the sum of the average van der Waals radii of both molecules. In most cases, one can say that dipole–dipole coupling is dominant if the separation between donor and acceptor is more than 1 nm. Further complications arise from the different requirements on spin conservation for both mechanisms. More details on these energy transfer mechanisms can be found in specific literature.

Energy transfers are at the basis of many upconversion mechanisms as well [60, 61]. Most often, lanthanide ions are integrated into a crystal structure together with a sensitizer to obtain efficient upconversion. The exact mechanisms of energy transfer between the sensitizer (e.g., Yb^{3+}) and the upconverting lanthanide ion (e.g., Er^{3+}) are not yet fully deciphered but can involve multiple donors and acceptors [60].

4.2 *Influence of the LDOS on Energy Transfers*

Radiative energy transfer is a far-field interaction that involves the propagation of a real photon with a well-specified energy. Due to the typically large separation distances between donor and acceptor, the spontaneous emission of the donor is crucial to achieve energy transfer. As noted in Sect. 2, this is influenced directly by the LDOS of the environment. Therefore, the radiative energy transfer can be suppressed when the photonic bandgap is in the region of spectral overlap between donor and acceptor. On the other hand, suppression of the donor emission outside the spectral overlap region can lead to increased radiative energy transfer through spectral redistribution of the donor emission [62, 63].

As nonradiative energy transfers are also subject to Fermi's golden rule (in the weak coupling regime), the LDOS was first thought to directly affect the rate of energy transfer as well [64]. Indeed, at the quantum-electrodynamic level, nonradiative energy transfer can be described as the exchange of virtual photons that propagate through the LDOS of the medium for a very short distance [65, 66]. Nonetheless, experimental studies have shown that the rate of dipole–dipole coupled energy transfer is independent of the LDOS [67, 68]. This discrepancy was rationalized by employing the Heisenberg uncertainty principle:

$$\Delta E \Delta t \geq \frac{\hbar}{2} \tag{11}$$

Since very short distances are involved in nonradiative energy transfer, the travelling time (Δt) of the photon is short enough to have a large spread on the energy of the photon (ΔE). Therefore, not only the LDOS at the frequency of the transition is relevant but also the average LDOS over a large range of frequencies. In most systems, the averaged LDOS closely follows the smooth energy squared dependence of a homogeneous medium and no direct effect of the LDOS on the rate of nonradiative energy transfer is expected [67]. Nonetheless, the quantum yield of energy transfer can be increased by suppressing the rate of donor spontaneous emission [see Eq. (2)] [69, 70]. This was shown experimentally by Kolaric et al. using a FRET-pair with a fixed donor–acceptor separation distance in a colloidal crystal [69].

The influence of LDOS on upconverting systems has been studied by integrating rare-earth crystals into a colloidal crystal architecture. The suppression of spontaneous emission in the photonic bandgap is often observed, [71–74] and is in some cases accompanied by increased luminescence from higher energy levels through energy transfer [75, 76]. Nonetheless, the study of upconversion is complicated by the nonlinear kinetics and multiple pathways in these systems [77, 78].

5 Conclusion

Colloidal photonic crystals feature a reduced LDOS inside their photonic bandgap. Although it is not a full bandgap, they represent an excellent research platform for the study of LDOS effects on excited state processes. The LDOS directly affects spontaneous emission inside colloidal crystals through the Fermi golden rule expression. The result is a spatial and spectral redistribution of the emitted light and an overall increase in the excited state lifetime. In colloidal crystal lasers, the spectral redistribution can be exploited to enhance stimulated emission and hence create low-threshold lasers. Finally, the prolonged excited state lifetime that results from spontaneous emission suppression can enhance the efficiency of energy transfers. This has a direct influence on the exciton diffusion length, which is relevant for photovoltaics and light emitting diodes. Moreover, energy transfers are at the basis of many upconversion processes, employed for harvesting long wavelengths in photovoltaics and the fabrication of short-wavelength lasers. By manipulating the LDOS with photonic crystals the efficiency of these devices could therefore be improved. Although these concepts have been proven experimentally, the combination with full bandgap materials would strongly increase the potential to manipulate the excited state processes (as far as the intrinsic kinetics of the system allow). This strategy is highly complementary to material design research as it does not affect the intrinsic molecular or atomic properties of the system.

References

1. P. Russell, Science **299**, 358 (2003)
2. W. Bogaerts, M. Fiers, P. Dumon, IEEE J. Sel. Top. Quantum Electron. **20**, 1 (2014)
3. J.D. Joannopoulos, S.G. Johnson, J.N. Winn et al., *Photonic Crystals: Molding the Flow of Light*, 2nd edn. (Princeton University Press, Princeton, 2008), p. 304

4. F. García-Santamaría, C. López, F. Meseguer, et al., Appl. Phys. Lett. **79** (2001)
5. O. Kruglova, P.-J. Demeyer, K. Zhong et al., Soft Matter **9**, 9072 (2013)
6. J.F. Galisteo-López, M. Ibisate, R. Sapienza et al., Adv. Mater. **23**, 30 (2011)
7. C. López, Adv. Mater. **15**, 1679 (2003)
8. L. González-Urbina, K. Baert, B. Kolaric et al., Chem. Rev. **112**, 2268 (2012)
9. N.J. Turro, V. Ramamurthy, J.C. Scaiano, *Principles of Molecular Photochemistry: An Introduction* (University Science, Sausalito, 2009)
10. J.A. Barltrop, J.D. Coyle, *Excited States in Organic Chemistry* (Wiley, Bristol, 1975)
11. E. Yablonovitch, Phys. Rev. Lett. **58**, 2059 (1987)
12. O. Svelto, D.C. Hanna, *Principles of Lasers* (Heyden, London, 1976)
13. F. Auzel, Chem. Rev. **104**, 139 (2003)
14. C. Hermann, O. Hess, J. Opt. Soc. Am. B **19**, 3013 (2002)
15. W. Vos, A. Koenderink, I. Nikolaev, Phys. Rev. A **80**, 053802 (2009)
16. A. Femius Koenderink, L. Bechger et al., Phys. Status Solidi **197**, 648 (2003)
17. E.M. Purcell, H.C. Torrey, R.V. Pound, Phys. Rev. **69**, 37 (1946)
18. K. Busch, S. John, Phys. Rev. E **58**, 3896 (1998)
19. S. Schutzmann, I. Venditti, P. Prosposito et al., Opt. Express **16**, 897 (2008)
20. P.-J. Demeyer, S. Vandendriessche, S. Van Cleuvenbergen et al., ACS Appl. Mater. Interfaces **6**, 3870 (2014)
21. A.N. Gruzintsev, G.A. Emelchenko, V.M. Masalov et al., Semiconductors **43**, 197 (2009)
22. K. Baert, B. Kolaric, W. Libaers et al., Res. Lett. Nanotechnol. **2008**, 1 (2008)
23. R.V. Nair, R. Vijaya, K. Kuroda et al., J. Appl. Phys. **102**, 123106 (2007)
24. K. Yoshino, S.B. Lee, S. Tatsuhara et al., Appl. Phys. Lett. **73**, 3506 (1998)
25. A. Blanco, C. López, R. Mayoral et al., Appl. Phys. Lett. **73**, 1781 (1998)
26. T. Yamasaki, T. Tsutsui, Appl. Phys. Lett. **72**, 1957 (1998)
27. R.A.L. Vallée, K. Baert, B. Kolaric et al., Phys. Rev. B **76**, 45113 (2007)
28. J. Li, B. Jia, G. Zhou, M. Gu, Appl. Phys. Lett. **91**, 254101 (2007)
29. K. Clays, K. Wostyn, Y. Zhao et al., J. Nonlinear Opt. Phys. Mater. **11**, 261 (2002)
30. P. Lodahl, A. Floris Van Driel, I.S. Nikolaev et al., Nature **430**, 654 (2004)
31. A. Koenderink, L. Bechger, H. Schriemer et al., Phys. Rev. Lett. **88**, 143903 (2002)
32. M. Aloshyna, S. Sivakumar, M. Venkataramanan et al., J. Phys. Chem. C **111**, 4047 (2007)
33. J. Gutmann, H. Zappe, J.C. Goldschmidt, Phys. Rev. B **88**, 205118 (2013)
34. M. Megens, J.E.G.J. Wijnhoven, A. Lagendijk et al., J. Opt. Soc. Am. B **16**, 1403 (1999)
35. R. Sprik, B.A. van Tiggelen, A. Lagendijk, Europhys. Lett. **35**, 265 (1996)
36. I.S. Nikolaev, W.L. Vos, A.F. Koenderink, J. Opt. Soc. Am. B **26**, 987 (2009)
37. Y.-Q. Zhang, J.-X. Wang, Z.-Y. Ji, D.-B. Zhu et al., J. Mater. Chem. **17**, 90 (2007)
38. G. Mie, Ann. Phys. **330**, 377 (1908)
39. H. Schriemer, H. van Driel, A. Koenderink et al., Phys. Rev. A **63**, 011801 (2000)
40. L. Bechger, P. Lodahl, W.L. Vos, J. Phys. Chem. B **109**, 9980 (2005)
41. I. Nikolaev, P. Lodahl, W. Vos, Phys. Rev. A **71**, 053813 (2005)
42. J.R. Lawrence, Y. Ying, P. Jiang et al., Adv. Mater. **18**, 300 (2006)
43. H.Y. Lin, H.K. Fu, C.L. Cheng et al., Opt. Express **16**, 16697 (2008)
44. M. Scharrer, A. Yamilov, X. Wu et al., Ultraviolet lasing in high-order bands of three-dimensional ZnO photonic crystals. Appl. Phys. Lett. **88**, 201103 (2006)
45. S. Furumi, T. Kanai, T. Sawada, Adv. Mater. **23**, 3815 (2011)
46. K. Yoshino, S. Tatsuhara, Y. Kawagishi et al., Appl. Phys. Lett. **74**, 2590 (1999)
47. S.-H. Kim, S.-H. Kim, W.C. Jeong et al., Chem. Mater. **21**, 4993 (2009)
48. M.N. Shkunov, Z.V. Vardeny, M.C. DeLong et al., Adv. Funct. Mater. **12**, 21 (2002)
49. Y. Zhao, K. Wostyn, G. de Schaetzen et al., Appl. Phys. Lett. **82**, 3764 (2003)
50. P. Massé, S. Reculusa, K. Clays et al., Chem. Phys. Lett. **422**, 251 (2006)
51. K. Wostyn, Y. Zhao, G. de Schaetzen et al., Langmuir **19**, 4465 (2003)
52. K. Zhong, P.-J. Demeyer, X. Zhou et al., J. Mater. Chem. C **2**, 8829 (2014)
53. K. Baert, K. Song, R.A.L. Vallée et al., J. Appl. Phys. **100**, 123112 (2006)

54. S. Furumi, H. Fudouzi, T. Sawada, J. Mater. Chem. **22**, 21519 (2012)
55. S. Furumi, H. Fudouzi, H.T. Miyazaki et al., Adv. Mater. **19**, 2067 (2007)
56. J.-F. Dechezelles, T. Aubert, F. Grasset et al., Phys. Chem. Chem. Phys. **12**, 11993 (2010)
57. H. Yamada, T. Nakamura, Y. Yamada et al., Adv. Mater. **21**, 4134 (2009)
58. S. Furumi, Polym. J **45**, 579 (2013)
59. D.L. Andrews, C. Curutchet, G.D. Scholes, Laser Photonics Rev. **5**, 114 (2011)
60. S. Cotton, *Lanthanide and Actinide Chemistry* (Wiley, Chichester, 2006), pp. 61–87
61. F. Auzel, Chem. Rev. **104**, 139 (2004)
62. A. Tsuchida, Y. Ohi, S. Takahashi et al., Macromol. Symp. **288**, 87 (2010)
63. K. Shibata, H. Kimura, A. Tsuchida et al., Colloid Polym. Sci. **285**, 127 (2006)
64. P. Andrew, Science **290**, 785 (2000)
65. J. Knoester, S. Mukamel, Phys. Rev. A **40**, 7065 (1989)
66. G. Juzeliūnas, D.L. Andrews, *Advances in Chemical Physics* (Wiley, New York, 2007), pp. 357–410. http://onlinelibrary.wiley.com/doi/10.1002/9780470141717.ch4/summary
67. M.J.A. de Dood, J. Knoester, A. Tip et al., Phys. Rev. B **71**, 115102 (2005)
68. C. Blum, N. Zijlstra, A. Lagendijk et al., Phys. Rev. Lett. **109**, 203601 (2012)
69. B. Kolaric, K. Baert, M. Van der Auweraer et al., Chem. Mater. **19**, 5547 (2007)
70. Z. Yang, X. Zhou, X. Huang et al., Opt. Lett. **33**, 1963 (2008)
71. Z.-X. Li, L.-L. Li, H.-P. Zhou et al., Chem. Commun. **43**, 6616 (2009)
72. Z. Yang, D. Yan, Z. Song, et al., in *2012 Symposium on Photonics and Optoelectronics* (IEEE, Shanghai, 2012), pp. 1–3
73. F. Zhang, Y. Deng, Y. Shi et al., J. Mater. Chem. **20**, 3895 (2010)
74. Z. Yang, K. Zhu, Z. Song et al., Solid State Commun. **151**, 364 (2011)
75. Z. Yang, D. Yan, K. Zhu et al., Mater. Chem. Phys. **133**, 584 (2012)
76. D. Yan, J. Zhu, H. Wu et al., J. Mater. Chem. **22**, 18558 (2012)
77. J. Suyver, A. Aebischer, S. García-Revilla et al., Phys. Rev. B **71**, 125123 (2005)
78. M. Pollnau, D.R. Gamelin, S.R. Lüthi et al., Phys. Rev. B **61**, 3337 (2000)

Index

D. Comoretto (ed.), *Organic and Hybrid Photonic Crystals*,
DOI 10.1007/978-3-319-16580-6

MIX
Papier aus verantwortungsvollen Quellen
Paper from responsible sources
FSC® C105338

If you have any concerns about our products,
you can contact us on
ProductSafety@springernature.com

In case Publisher is established outside the EU,
the EU authorized representative is:
Springer Nature Customer Service Center GmbH
Europaplatz 3, 69115 Heidelberg, Germany

Printed by Libri Plureos GmbH
in Hamburg, Germany